Fourth Edition

COLLEGE ALGEBRA
A FUNCTIONS
APPROACH

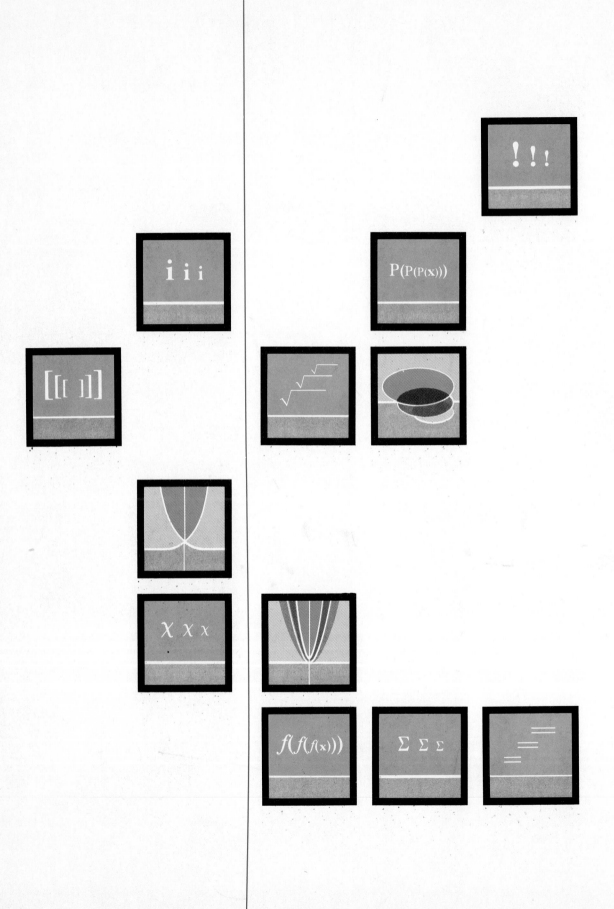

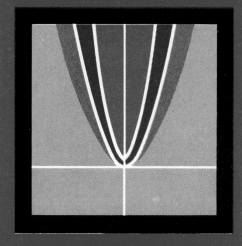

FOURTH EDITION

MERVIN L. KEEDY
Purdue University

MARVIN L. BITTINGER
Indiana University—
Purdue University at Indianapolis

COLLEGE
ALGEBRA

A FUNCTIONS APPROACH

ADDISON-WESLEY PUBLISHING COMPANY

Reading, Massachusetts ■ Menlo Park, California
Don Mills, Ontario ■ Wokingham, England ■ Amsterdam
Sydney ■ Singapore ■ Tokyo ■ Mexico City ■ Bogotá
Santiago ■ San Juan

Sponsoring Editor	●	*Susan Zorn*
Production Supervisor	●	*Susanah H. Michener*
Editorial and Production Services	●	*Quadrata, Inc.*
Text Designer	●	*Vanessa Piñeiro, Piñeiro Design Associates*
Illustrator	●	*VAP Group, Ltd.*
Art Consultant	●	*Joseph Vetere*
Manufacturing Supervisor	●	*Ann DeLacey*
Cover Illustrator	●	*Susan Schwartz*

PHOTO CREDITS   **1,** Belanger Associates   **51,** Fred Maroon/Photo Researchers, Inc.   **105,** Jeff Albertson/Stock, Boston
155, Mark Antman/Stock, Boston   **199,** W. B. Finch/Stock, Boston   **239,** source unknown
265, Frank Siteman/Stock, Boston   **303,** Oldsmobile, General Motors   **371,** source unknown
407, Ellis Herwig/Stock, Boston   **433,** Jeff Dunn/Stock, Boston.

Library of Congress Cataloging-in-Publication Data

Keedy, Mervin Laverne.
  College algebra.

  Includes index.
  1. Algebra.   I. Bittinger, Marvin L.   II. Title.
  QA154.2.K432   1986   512.9   85-26877
  ISBN 0-201-13290-7

Reprinted with corrections, December 1986

CDEFGHIJ-MU-8987

Using a functions approach, this text presents basic concepts of college-level algebra. It is appropriate for one-semester courses in college algebra, or for one-half of a full-year precalculus course in conjunction with Keedy and Bittinger's *Trigonometry: Triangles and Functions*.

WHAT'S NEW IN THE FOURTH EDITION?

In addition to suggestions for improvement made by many third-edition users, the following features have been incorporated into the text.

☐ *Augmentation of the extension exercises*. The extension and challenge exercises introduced in the third edition that required the student to have more than a simple understanding of the immediate objectives at hand have been augmented due to positive response from users and requests for more such exercises.

☐ *Condensed chapters*. Users of the third edition found Chapters 1 and 3 to be slightly too long. To remedy that problem, the material on equations, inequalities, and dimension symbols has been moved from Chapter 1 to Chapter 2. Material on inverses of functions has been moved from Chapter 3 to the beginning of Chapter 7 on Exponential and Logarithmic Functions. This shortens Chapter 3 and places the material at a more advantageous position in terms of student retention.

☐ *Chapter introductions*. To provide students with an introduction to each chapter, we have given a brief description of what is studied in the chapter, together with a photograph touching on a real-world application of mathematical concepts taken from the chapter.

☐ *Deemphasis of tables*. The advent of the calculator and computer has reduced the need for logarithmic tables, so interpolation and tables for computations have been deleted from Chapter 7 on Exponential and Logarithmic Functions.

We still feel that interpolation should be taught if time permits. No computer will ever hold all tables, and numerical approximation methods will continue to be an important part of applied mathematics. Thus we have included an appendix on interpolation. Although the tables used are logarithmic, the objective of the appendix is to learn to *interpolate*.

☐ *Summary and review sections and chapter tests.* The Chapter Reviews of the third edition have been reworked and are now called *Summary and Review* Sections. This feature has been designed to enable students to easily review each chapter's objectives. In this section, the objectives of the chapter are stated in boldface type. Following each list of objectives are exercises pertaining to them. Answers to the exercises are given at the back of the book together with section references so that students can restudy the material if necessary.

The Summary and Review section is followed by a new chapter test, the answers to which are not included in the book.

☐ *Real-world applications of probability.* We have added many real-world applications of probability theory to the final chapter of the text in order to illustrate that the applications of probability theory extend beyond games of chance.

FEATURES

Following is a list of salient features of the third edition that have been retained in this edition.

☐ *Exercise sets.* The first exercises in a set are very much like the examples in the text. These exercises are graded in difficulty, are paired, and are marked with objective references **i** , **ii** , and so on. The next exercises are considered to be extension exercises (marked ☐) and require the student to go beyond the immediate objectives. They may, for example, ask the student to synthesize either objectives in the section or those from preceding chapters with those of this particular section. For the first two types, answers to the odd-numbered exercises are given at the back of the book. The instructor can therefore easily make an assignment that is varied in terms of availability of answers. The challenge exercises are marked ■, and some of them are quite difficult. Answers to the challenge exercises are not in the text, but are given in the Instructor's Manual.

☐ *Functions and transformations.* This text thoroughly applies the concepts of function, relation, and transformation. The idea of transformation makes Chapter 3 unique and sets up the study of later material.

The concept of transformation helps to simplify topics related to quadratic functions and to graphing. In effect the notion of transformation becomes a graphing tool. It also provides a new approach to graphing inequalities with absolute value, and many of the trigonometric identities are made very simple. Instructors tell us of cases in which students are afraid that they may not be understanding the material because they grasp the ideas so readily!

☐ *Use of margins.* For student reference, the margins contain objectives for each section. They also contain developmental exercises, which have proved to be extremely effective. The text refers the student to these exercises at appropriate points in the discussion, when he or she should stop reading and do them. Then the answers can be checked at the back of the text. Thus students receive reinforcement, guidance, and practice before continuing with the text development. The first part of each exercise set is very similar to these developmental exercises. Thus the margin materials constitute a built-in study guide.

☐ *Flexibility of teaching method.* There are many ways in which to use this book. The instructor who wishes to use it in a traditional way can simply ignore the margins and have students ignore them until they do their homework. The instructor who wishes to use the lecture method primarily, but also wants to introduce some student-centered activity into the class, can easily do so by interrupting the lecture and having the students work the exercises in the margins at appropriate times. The exercises can also be used in traditional blackboard drill. On the other hand, the book is well suited for use in math labs or other systems of individualized instruction, or in any approach that is essentially self-study. The book can be used with minimal instructor guidance, so it is particularly effective for use in large classes.

ACKNOWLEDGMENTS

The authors wish to thank many people without whose committed efforts our work could not have been completed successfully. Paul Welsh (Pima Community College, East Campus), Karen Anderson, Barbara Dunn, Gloria Langer, and Jane Smith did a tremendous job checking the manuscript. Randy Becker, Judy Beecher, Ron Epperlein (Arizona State University), John Losse (Scottsdale Community College), and Sharon Walker (Arizona State University) were superb in their work on the supplements.

Many instructors provided reviewing information, and we thank them for their suggestions for improvement. They are Leonard Andrusitis (University of Lowell), Daniel Comenetz (University of Massachusetts, Boston Campus), Arleen Watkins (University of Arizona), Paul Welsh (Pima Community College, East Campus), and Edward Zeidman (Essex Community College).

M.L.K.
M.L.B.

SUPPLEMENTS

The following supplementary materials are available.

☐ *Student's Guide to Exercises.* Contains answers for all the even-numbered exercises in the exercise sets, except for the challenge exercises, and solutions to all the odd-numbered exercises, as well as answers to the chapter tests.

☐ *Instructor's Manual with Tests and Computer Programs.* Contains five alternate forms of the chapter tests and five final exams. All answers to these tests are included in this manual. Also included are the answers to the chapter tests and the challenge exercises found in the text. This manual is embellished with two types of black-line grid masters, one for rectangular graphing and one for polar graphing. These make very useful transparencies for use in class. Another option is the inclusion of computer programs, about one per chapter, which when used by the student can allow an appreciation of the power of the computer in relation to the mathematics being considered at that point in the text. The programs are in BASIC. An explanation of how "Cactusplot" (software described below) can be used with this text is also included.

☐ *Computer Software.* There are two kinds of computer software to accompany the text. For many algebra parts of the text there is an "Instructional Software for Algebra" supplement. This software is highly interactive and covers topics such as simplifying algebraic expressions, factoring, equations, inequalities, 2×2 linear systems, conic sections, radicals, complex numbers, and quadratic equations.

Another software tool that is most helpful with this text is "Cactusplot," available from:

The Cactusplot Company
1441 N. McAllister
Tempe, AZ 85281

This software can be used either by the instructor alone in class or by the student at home or in a computer lab. The software will graph *any* function found in this text, solve equations, find function values, and make tables of function values. It will graph more than one function on the screen at a given time. This is most advantageous when working with transformations.

☐ *Computerized Testbank.* Contains test items that can be randomly selected and a test constructed by a computer using the testbank. The bank is available for the Apple II series and IBM-PC computers.

☐ *Placement Test.* Helps students to be placed properly in either this book or other books by the same authors. The test is handy because it is short, although its brevity may make it appear less useful than it actually is. It was thoroughly evaluated statistically and is known to give a good screening with respect to placement into one of four categories into which the books of the series fall.

I *Arithmetic,* Fourth Edition

II *Introductory Algebra,* Fourth Edition
A Problem-Solving Approach to Introductory Algebra, Second Edition

III *Intermediate Algebra,* Fourth Edition
A Problem-Solving Approach to Intermediate Algebra, Second Edition

IV *College Algebra,* Fourth Edition
Fundamental College Algebra, Second Edition
Algebra and Trigonometry, Fourth Edition
Fundamental Algebra and Trigonometry, Second Edition
Trigonometry: Triangles and Functions, Fourth Edition

More specific placement within this book can be obtained by using a form of the final exam from the Instructor's Manual.

CONTENTS

1

BASIC CONCEPTS OF ALGEBRA   1

2

EQUATIONS, INEQUALITIES, AND PROBLEM SOLVING   51

3

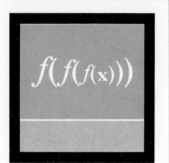

RELATIONS, FUNCTIONS, AND TRANSFORMATIONS   105

4

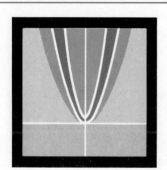

LINEAR AND QUADRATIC FUNCTIONS AND INEQUALITIES   155

5

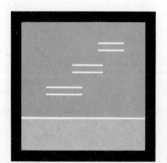

SYSTEMS OF LINEAR EQUATIONS AND INEQUALITIES   199

6

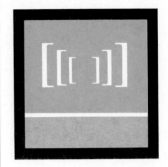

7

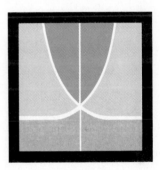

8

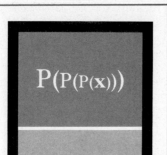

9

POLYNOMIALS AND RATIONAL FUNCTIONS   327

10

EQUATIONS OF SECOND DEGREE AND THEIR GRAPHS   371

11

SEQUENCES, SERIES, AND MATHEMATICAL INDUCTION   407

12

COMBINATORICS AND PROBABILITY   433

Fourth Edition

COLLEGE ALGEBRA
A FUNCTIONS
APPROACH

How can we use the skid marks of a car to estimate the car's speed before the brakes were applied?

BASIC CONCEPTS OF ALGEBRA

This chapter is intended to be a warmup or review. We assume that you have already studied intermediate algebra. That being the case, you should find most of this material review. We cover the properties of the real-number system and various kinds of algebraic expressions and manipulations of them—for example, how to add them, multiply them, factor them, and so on.

If your algebra background is recent, you might be able to skip this chapter, or at least most of it. To determine whether that is the case, you can work through the test at the end of the chapter. If you answer 75% to 85% of the questions correctly then it might be wise for you to go on to Chapter 2.

1

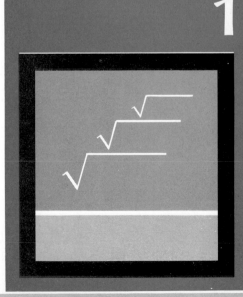

OBJECTIVES

You should be able to:

i Identify various kinds of real numbers.

ii Find the absolute value of a real number.

iii Add, subtract, multiply, and divide positive and negative real numbers. Find the additive inverse of a real number.

Consider the numbers

$$1, \frac{3}{4}, -6, 0, 19, -\frac{8}{7}.$$

1. Which are natural numbers?

2. Which are whole numbers?

3. Which are integers?

4. Which are rational numbers?

1.1

THE REAL-NUMBER SYSTEM

i REAL NUMBERS

There are various kinds of numbers. Those most used in elementary algebra are the so-called *real numbers*. Later we consider a more comprehensive system of numbers called the *complex numbers*. The real numbers are often shown in one-to-one correspondence with the points of a line, as follows.

The positive numbers are shown to the right of 0 and the negative numbers to the left. Zero itself is neither positive nor negative.

There are several subsystems of the real-number system. They are as follows.

> *Natural numbers.* Those numbers used for counting: 1, 2, 3, ...
>
> *Whole numbers.* The natural numbers and 0: that is 0, 1, 2, 3, ...
>
> *Integers.* The whole numbers and their additive inverses: 0, 1, −1, 2, −2, 3, −3, ...
>
> *Rational numbers.* The integers and all quotients of integers (excluding division by 0): for example, $\frac{4}{5}$, $\frac{-4}{7}$, $\frac{9}{1}$, 6, −4, 0, $\frac{78}{-5}$, $-\frac{2}{3}$ (can also be named $\frac{-2}{3}$, or $\frac{2}{-3}$).

DO EXERCISES 1–4 (in the margin).

Any real number that is not rational is called *irrational*. The rational numbers and the irrational numbers can be described in several ways.

> The *rational numbers* are:
>
> 1. Those numbers that can be named with fractional notation a/b, where a and b are integers and $b \neq 0$ (definition);
> 2. Those numbers for which decimal notation either ends or repeats.

EXAMPLES All of these are rational.

1. $\frac{5}{16} = 0.3125$ Ending (terminating) decimal

2. $-\frac{8}{7} = -1.142857142857\ldots = -1.\overline{142857}$ Repeating decimal. The bar indicates the repeating part.

3. $\frac{3}{11} = 0.2727\ldots = 0.\overline{27}$ Repeating decimal

> The *irrational numbers* are:
>
> 1. Those real numbers that are not rational (definition);
> 2. Those real numbers that cannot be named with fractional notation a/b, where a and b are integers and $b \neq 0$;
> 3. Those real numbers for which decimal notation does not end and does not repeat.

There are many irrational numbers. For example, $\sqrt{2}$ is irrational. We can find rational numbers a/b for which $(a/b) \cdot (a/b)$ is close to 2, but we cannot find such a number a/b for which $(a/b) \cdot (a/b)$ is *exactly* 2.

Unless a whole number is a perfect square, its square root is irrational. For example, $\sqrt{9}$ and $\sqrt{25}$ are rational, but all of the following are irrational:

$$\sqrt{3}, \qquad -\sqrt{14}, \qquad \sqrt{45}.$$

There are also many irrational numbers that cannot be obtained by taking square roots. The number π is an example.* Decimal notation for π does not end and does not repeat.

EXAMPLES   All of these are irrational.

4. $\pi = 3.1415926535 \ldots$ Numeral does not repeat.

5. $-1.10100100010000100000 1 \ldots$ Numeral does not repeat.

6. $\sqrt[3]{2} = 1.25992105 \ldots$ Numeral does not repeat.

In Example 5 there is a pattern, but it is not a pattern formed by a repeating block of digits.

DO EXERCISES 5–10.

In arithmetic we use numbers, performing calculations to obtain certain answers. In algebra, we use arithmetic symbolism, but in addition we use symbols to represent unknown numbers. We do calculations and manipulations of symbols, on the basis of the properties of numbers, which we review now. Algebra is thus an extension of arithmetic and a more powerful tool for solving problems.

ii ABSOLUTE VALUE

Before considering addition we need to recall that the absolute value of a number is its distance from 0 on a number line. We will define absolute value more precisely later in this section. The absolute value of a number a is denoted $|a|$.

EXAMPLES   Simplify.

7. $|-7|$ The distance of -7 from 0 is 7, so $|-7| = 7$.

8. $|5|$ The distance of 5 from 0 is 5, so $|5| = 5$.

9. $|0|$ The distance of 0 from 0 is 0, so $|0| = 0$.

To get the absolute value of a negative number, change its sign (make it positive). The absolute value of a nonnegative number is that number itself. Thus, $\left|\frac{2}{3}\right| = \frac{2}{3}$ and $|-9.7| = 9.7$.

DO EXERCISES 11–14.

Which of the following are rational? Which are irrational?

5. $\dfrac{-4}{5}$

6. $\dfrac{59}{37}$

7. 7.42

8. 0.47474747 . . .
(Numeral repeats)

9. 2.57340046631 . . .
(Numeral does not repeat)

10. $\sqrt{7}$

Simplify.

11. $|2|$

12. $\left|\sqrt{3}\right|$

13. $|-11.3|$

14. $\left|-\dfrac{3}{4}\right|$

---

* $\frac{22}{7}$ and 3.14 are only rational approximations to the irrational number π.

Add.

15. $-5 + (-7)$

16. $-1.2 + (-3.5)$

17. $-\dfrac{6}{5} + \dfrac{2}{5}$

18. $0.5 + (-0.7)$

19. $8 + (-3)$

20. $\dfrac{14}{3} + \left(-\dfrac{14}{3}\right)$

---

iii OPERATIONS ON THE REAL NUMBERS

Addition

Assuming that addition of nonnegative real numbers poses no problem, let us review how the definition of addition is extended to include the negative numbers.

> **1.** To add two negative numbers, add their absolute values (the sum is negative).
>
> **2.** To add a negative number and a positive number, find the difference of their absolute values. The result will have the sign of the summand with the larger absolute value. If the absolute values are the same, the sum is 0.

EXAMPLES   Add.

$$-5 + (-6) = -11,$$
$$8.6 + (-4.2) = 4.4,$$
$$-5 + 3 = -2,$$
$$-\tfrac{3}{4} + \left(-\tfrac{7}{8}\right) = -\tfrac{13}{8},$$
$$\pi + (-\pi) = 0,$$
$$-\sqrt{3} + (-4\sqrt{3}) = -5\sqrt{3},$$
$$8 + (-5) = 3,$$
$$-\tfrac{9}{5} + \tfrac{3}{5} = -\tfrac{6}{5}$$

DO EXERCISES 15–20.

Properties of Real Numbers Under Addition

In solving equations and doing other kinds of work in algebra, we manipulate algebraic symbols in various ways, such as collecting like terms. For example, instead of

$$4x + 7x$$

we might write

$$11x$$

knowing that the two expressions represent the same number no matter what x represents. In that sense the expressions $4x + 7x$ and $11x$ are equivalent. We define equivalent expressions in general.

Definition

> If two expressions represent the same number for *any* sensible replacement, the expressions are said to be *equivalent*.

By a nonsensible replacement for an expression, we mean a number which when substituted for a variable in the expression does not give us a real number. For example, the number 2 is not a sensible replacement for y in the expression

$$\frac{8 + y}{y - 2}$$

because it gives a denominator of 0, and we cannot divide by 0.

We now list for review the fundamental properties of real numbers under addition. These are properties on which algebraic manipulations are based, especially when symbols for unknown numbers are used. These properties allow us to find equivalent expressions.

Commutativity.   For any real numbers a and b,

$$a + b = b + a.$$

(The *order* in which numbers are added does not affect the sum.)

Associativity.   For any real numbers a, b, and c,

$$a + (b + c) = (a + b) + c.$$

(When *only* additions are involved, parentheses for grouping purposes may be placed as we please without affecting the sum.)

Identity.   There exists a unique real number 0 such that for any real number a,

$$a + 0 = 0 + a = a.$$

(Adding 0 to any number gives that same number.)

Inverses.   For every real number a, there exists a unique number, denoted $-a$, called the additive inverse of a, for which

$$-a + a = a + (-a) = 0.$$

Additive Inverses

Concerning additive inverses, we caution the reader about one of the most misunderstood, or confusing, ideas in elementary algebra: It is common to read an expression such as $-x$ as "negative x." This can be confusing, because $-x$ may be positive, negative, or zero, depending on the value of x. The symbol $-$, used in this way, indicates an *additive inverse*; somewhat unfortunately the same symbol may also indicate a negative number, as in -5, or it may indicate subtraction, as in $3 - x$.

CAUTION!   An initial $-$ sign, as in $-x$ or $-(x^2 + 3x - 2)$, should always be interpreted as meaning "the additive inverse of." The entire expression may be positive, negative, or zero, depending on the value of the part of the expression that follows the $-$ sign.*

EXAMPLES   Find the additive inverse.

10. $-x$, when $x = 5$ $-(5) = -5$ (negative 5)

11. $-x$, when $x = -3$ $-(-3) = 3$ The inverse of negative 3 is 3.

12. $-x$, when $x = 0$ $-(0) = 0$ The inverse of 0 is 0.

13. $-(x^2 + 4x + 2)$, when $x = 1$ $-(1^2 + 4 \cdot 1 + 2) = -7$
$-(x^2 + 4x + 2)$, when $x = -1$ $-((-1)^2 + 4(-1) + 2) = 1$

It is easily shown that $-1 \cdot x = -x$ and $-(-x) = x$ for any number x. That is, multiplying a number x by negative 1 produces the additive inverse of x, and the additive inverse of the additive inverse of a number is the number itself. We list this as a theorem.

---

*Taking the additive inverse is sometimes called "changing the sign."

Find $-x$ and $-1 \cdot x$ when:

21. $x = 6$.

22. $x = -8$.

23. $x = -3.4$.

24. Find the additive inverse $-(x^2 - 3x)$ when $x = 2$ and $x = -1$.

Multiply.

25. $4 \cdot (-6)$

26. $-\frac{7}{5} \cdot \left(-\frac{3}{5}\right)$

27. $(-2) \cdot (-3) \cdot (-4) \cdot (-6)$

Theorem 1

> For any real number x, $-1 \cdot x = -x$ and $-(-x) = x$. (Multiplying a number by -1 produces its additive inverse and the additive inverse of the additive inverse of a number is the number itself.)

DO EXERCISES 21–24.

We can use the notion of additive inverse to give a precise definition of absolute value. The absolute value of a nonnegative number is that number itself. The absolute value of a negative number is its additive inverse. We define this as follows.

Definition

> For any real number x,
>
> $$|x| = x \quad \text{if } x \geq 0,$$
>
> and
>
> $$|x| = -x \quad \text{if } x < 0.$$

This definition may be understood better if it is stated as follows. The absolute value of a number x is

a) the number x itself, if x is not negative;

b) the additive inverse of x, if x is negative.

A common source of confusion arises from interpreting $-x$ as meaning something negative, rather than "the additive inverse of x."

MULTIPLICATION

Assuming that multiplication of nonnegative real numbers poses no problem, let us review how the definition of multiplication is extended to include the negative numbers.

> **1.** To multiply two negative numbers, multiply their absolute values (the product is positive).
> **2.** To multiply a positive number and a negative number, multiply their absolute values and take the additive inverse of the result (the product is negative).

EXAMPLES Multiply.

14. $3 \cdot (-4) = -12$

15. $1.5 \cdot (-3.8) = -5.7$

16. $-5 \cdot (-4) = 20$

17. $-\frac{2}{3} \cdot \left(-\frac{4}{5}\right) = \frac{8}{15}$

18. $-3 \cdot (-2) \cdot (-4) = -24$

DO EXERCISES 25–27.

We now list the properties of the real numbers under multiplication. Again, these are properties that allow us to find equivalent expressions.

Commutativity.   For any real numbers a and b,

$$ab = ba.$$

(The *order* in which numbers are multiplied does not affect the product.)

Associativity.   For any real numbers a, b, and c,

$$a(bc) = (ab)c.$$

(When *only* multiplications are involved, parentheses for grouping purposes may be placed as we please without affecting the product.)

Identity.   There exists a unique real number 1 such that for any real number a,

$$a \cdot 1 = 1 \cdot a = a.$$

(Multiplying any number by 1 gives that same number.)

Inverses.   For each nonzero real number a, there exists a unique number, denoted $\frac{1}{a}$ or a^{-1}, called the *multiplicative inverse* or *reciprocal*, for which

$$a\left(\frac{1}{a}\right) = \frac{1}{a}(a) = 1.$$

EXAMPLES

19. The multiplicative inverse, or reciprocal, of 2 is $\frac{1}{2}$.

20. The multiplicative inverse of $-\frac{2}{3}$ is $-\frac{3}{2}$.

21. The reciprocal of 0.16 is 6.25.

There is a very special property that connects addition and multiplication, as follows.

Distributivity.   For any real numbers a, b, and c,

$$a(b + c) = ab + ac.^*$$

(This is also called the distributive law of multiplication over addition.)

Any number system having the preceding properties for addition and multiplication is called a *field*. Thus we refer to this list of properties as the *field properties*.

Many other properties important in algebraic manipulations can be proved from the field properties. We list some of these as theorems.

SUBTRACTION

Subtraction is the operation opposite to addition, as given in the following definition.

---

* The expression $ab + ac$ means $(a \cdot b) + (a \cdot c)$. By agreement, we can omit parentheses around multiplications. According to this agreement, multiplications are to be done before additions or subtractions.

Subtract.

28. $2.5 - 1.2$

29. $12 - (-5)$

30. $-\dfrac{8}{5} - \dfrac{3}{5}$

31. $-20 - (-7)$

Definition Subtraction

For any real numbers a and b, $a - b = c$ if and only if $b + c = a$ ($a - b$ is the number which when added to b gives a).

In any field, we actually subtract by adding an inverse, according to the following theorem.

Theorem 2*

For any real numbers a and b, $a - b = a + (-b)$.

Theorem 2 follows immediately from the definitions of subtraction and additive inverse. It says that to subtract a number we can add its additive inverse.

EXAMPLES Subtract.

$$8 - 5 = 8 + (-5) = 3,$$
$$3 - 7 = 3 + (-7) = -4,$$
$$8.6 - (-2.3) = 8.6 + 2.3 = 10.9,$$
$$-15 - (-5) = -15 + 5 = -10,$$
$$10 - (-4) = 10 + 4 = 14,$$
$$\tfrac{5}{9} - \tfrac{2}{9} = \tfrac{5}{9} + (-\tfrac{2}{9}) = \tfrac{1}{3}$$

DO EXERCISES 28–31.

Multiplication is distributive over subtraction in the real-number system, as the following theorem states.

Theorem 3 Distributivity

For any real numbers a, b, and c,

$$a(b - c) = ab - ac.$$

(This is the distributive law of multiplication over subtraction.)

Theorem 3 follows easily from Theorem 2 and the other distributive laws.

DIVISION

Division is the operation opposite to multiplication, as given in the following definition.

Definition Division

For any number a and any nonzero number b, $a \div b = c$ if and only if $b \cdot c = a$ ($a \div b$ is the number which when multiplied by b gives a).

---

* Theorem 2 is often used as a *definition* of subtraction. The definition of subtraction used here is more general, since it does not depend on the existence of inverses. Our definition is valid in the system of natural numbers—for example, where Theorem 2 would not even make sense since additive inverses do not exist.

In any field, we usually divide by multiplying by a reciprocal, according to the following theorem.

Theorem 4

> For any real number a and nonzero number b, $a \div b = a\left(\dfrac{1}{b}\right)$.

The definition of division parallels the one for subtraction, and Theorems 2 and 4 are also parallels.

EXAMPLES   Divide by multiplying by a reciprocal.

22. $\frac{3}{4} \div \left(-\frac{2}{3}\right) = \frac{3}{4}\left(-\frac{3}{2}\right) = -\frac{9}{8}$

23. $-\frac{6}{7} \div \left(-\frac{3}{5}\right) = -\frac{6}{7}\left(-\frac{5}{3}\right) = \frac{10}{7}$

From Theorem 4 it follows easily that the quotient of two negative numbers is positive and that the quotient of a positive number and a negative number is negative.

DO EXERCISES 32–35.

ORDER

The order of the real numbers is shown intuitively by a number line.

If a number a is pictured to the left of a number b, then a is less than b ($a < b$). In this event, if we subtract a from b, the answer will be a positive number. This idea motivates the definition of $<$.

Definition

> For any real numbers a and b, $a < b$ if and only if $b - a$ is positive.

EXAMPLES   Verify each inequality using the definition of $<$.

24. $2 < 8$ $8 - 2 = 6$ and 6 is positive.

25. $-4 < 9$ $9 - (-4) = 13$ and 13 is positive.

26. $-7 < -5$ $-5 - (-7) = 2$ and 2 is positive.

The symbol for "greater than" ($>$) is defined in terms of $<$, as follows.

Definition

> For any real numbers a and b, $a > b$ is defined to mean $b < a$.

Thus to show that $a > b$, we can show that $a - b$ is positive.

Divide.

32. $\dfrac{-20}{-5}$

33. $\dfrac{4.5}{-1.5}$

34. $-\dfrac{4}{5} \div \dfrac{3}{10}$

35. $-\dfrac{5}{6} \div \left(-\dfrac{5}{12}\right)$

THE USE OF CALCULATORS

We assume that you own a calculator and will use it while studying this book, although it is not mandatory. You will notice that certain exercises are designed for use of a calculator. They are indicated by the symbol ▦. Certain examples and discussions are similarly marked.

Keep in mind that there can be differences in answers found on the calculator because of rounding-error differences. For example, suppose you were asked to approximate $\sqrt{18}$, precise to seven decimal places. On a certain kind of calculator with an eight-digit readout the display would show

$$\sqrt{18} = 4.2426406$$

On another kind of calculator with a ten-digit readout the display would show

$$\sqrt{18} = 4.242640687$$

The answer would be found by rounding back to the seventh decimal place and would be given as

$$\sqrt{18} = 4.2426407$$

Thus there can be variance in the seventh decimal place.

EXERCISE SET 1.1

i Consider the numbers

$$-6, \quad 0, \quad 3, \quad -\tfrac{1}{2}, \quad \sqrt{3}, \quad -2, \quad -\sqrt{7}, \quad \sqrt[3]{2}, \quad \tfrac{5}{8}, \quad 14, \quad -\tfrac{9}{4}, \quad 8.53, \quad 9\tfrac{1}{2}.$$

1. Which are natural numbers?   **2.** Which are whole numbers?

3. Which are irrational numbers?   **4.** Which are rational numbers?

5. Which are integers?   **6.** Which are real numbers?

Which of the following are rational? Which are irrational?

7. $-\tfrac{6}{5}$ **8.** $-\tfrac{3}{7}$ **9.** -9.032 **10.** 3.14

11. $4.516\overline{516}$ (Numeral repeats)   **12.** $-7.323\overline{232}$ (Numeral repeats)

13. $4.303003000300003\ldots$ (Numeral does not repeat)   **14.** $6.414114111411114\ldots$ (Numeral does not repeat)

15. $\sqrt{6}$ **16.** $\sqrt{7}$ **17.** $-\sqrt{14}$ **18.** $-\sqrt{12}$

19. $\sqrt{49}$ **20.** $-\sqrt{16}$ **21.** $\sqrt[3]{5}$ **22.** $\sqrt[4]{10}$

ii Simplify.

23. $|12|$ **24.** $|-2.56|$ **25.** $|-47|$ **26.** $|0|$

iii Find $-x$ and $-1 \cdot x$, when:

27. $x = -7$.   **28.** $x = -\tfrac{10}{3}$.   **29.** $x = 57$.   **30.** $x = \tfrac{13}{14}$.

Find $-(x^2 - 5x + 3)$, when: Find $-(7 - y)$, when:

31. $x = 12$.   **32.** $x = -8$.   **33.** $y = -9$.   **34.** $y = 19$.

Compute.

35. $-3.1 + (-7.2)$ **36.** $-735 + 319$ **37.** $\tfrac{9}{2} + (-\tfrac{3}{5})$ **38.** $-6 + (-4) + (-10)$

39. $-7(-4)$ **40.** $-\tfrac{8}{3}(-\tfrac{9}{2})$ **41.** $(-8.2) \times 6$ **42.** $-6(-2)(-4)$

43. $-7(-2)(-3)(-5)$ **44.** $(-7.1)(-2.3)$ **45.** $-\tfrac{14}{3}(-\tfrac{17}{5})(-\tfrac{21}{2})$ **46.** $-\tfrac{13}{4}(-\tfrac{16}{5})(\tfrac{23}{2})$

47. $\tfrac{-20}{-4}$ **48.** $\tfrac{49}{-7}$ **49.** $\tfrac{-10}{70}$ **50.** $\tfrac{-40}{8}$

51. $\frac{2}{7} \div (-\frac{14}{3})$ **52.** $-\frac{3}{5} \div (-\frac{6}{7})$ **53.** $-\frac{10}{3} \div (-\frac{2}{15})$ **54.** $-\frac{12}{5} \div (-0.3)$

55. $11 - 15$ **56.** $-12 - 17$ **57.** $12 - (-6)$ **58.** $-13 - (-4)$

59. $15.8 - 27.4$ **60.** $-19.04 - 15.76$ **61.** $-\frac{21}{4} - (-\frac{7}{4})$ **62.** $\frac{10}{3} - (-\frac{17}{3})$

Calculate. Round to six decimal places. The symbol ▦ indicates an exercise meant to be done with a calculator.

63. ▦ a) $(1.4)^2$
$(1.41)^2$
$(1.414)^2$
$(1.4142)^2$
$(1.41421)^3$

     b) What number does the sequence of numbers 1.4, 1.41, 1.414, and so on, seem to approach as a limit?

64. ▦ a) $(2.1)^3$
$(2.15)^3$
$(2.154)^3$
$(2.1544)^3$
$(2.15443)^3$

     b) What number does the sequence of numbers 2.1, 2.15, 2.154, and so on, seem to approach as a limit?

What property is illustrated by each sentence?

65. $k + 0 = k$ **66.** $ax = xa$

67. $-1(x + y) = (-1x) + (-1y)$ **68.** $4 + (t + 6) = (4 + t) + 6$

69. $c + d = d + c$ **70.** $-67 \cdot 1 = -67$

71. $4(xy) = (4x)y$ **72.** $5(a + t) = 5a + 5t$

73. $y\left(\frac{1}{y}\right) = 1, \quad y \neq 0$ **74.** $-x + x = 0$

75. Show that subtraction is not commutative. That is, find real numbers a and b such that $a - b \neq b - a$.

76. Show that division is not commutative.

77. Show that division is not associative.

78. Show that subtraction is not associative.

79. Use the definition of absolute value to explain the meaning of the expression $|x - 3|$.

To convert from repeating decimal notation to fractional notation, consider an example such as $8.97656565\ldots$ or $8.97\overline{65}$. Let $n = 8.976565\ldots$. Then

$$10,000n = 89765.6565\ldots$$
$$\underline{100n = 897.6565\ldots}$$
$$9,900n = 88,868$$

$$n = \frac{88,868}{9900}.$$

Convert to fractional notation.

80. $0.999\overline{9}$ **81.** $3.74\overline{74}$ **82.** $18.3245\overline{245}$ **83.** $12.347\overline{652}$

84. Prove that for any real numbers $(b + c)a = ba + ca$.

85. Prove that any positive number is greater than 0.

1.2

EXPONENTIAL, SCIENTIFIC, AND ABSOLUTE-VALUE NOTATION

i INTEGERS AS EXPONENTS

When an integer greater than 1 is used as an exponent, the integer gives the number of times the base is used as a factor. For example, 5^3 means $5 \cdot 5 \cdot 5$. An exponent of 1 does not change the meaning of an expression. For example, $(-3)^1 = -3$. When 0 occurs as the exponent of a nonzero expression, we agree that the expression is equal to 1. For example, $37^0 = 1$.

OBJECTIVES

You should be able to:

i Simplify expressions with integer exponents.

ii Convert from decimal notation to scientific notation, and vice versa.

iii Simplify expressions involving absolute value, leaving as little as possible inside the absolute-value signs.

Rename with exponents.

1. $8 \cdot 8 \cdot 8 \cdot 8$

2. xxx

3. $4y \cdot 4y \cdot 4y \cdot 4y$

Rename without exponents.

4. 3^4

5. $(5x)^4$

6. $(-5)^4$

7. -5^4

8. $(3x)^0$

Simplify.

9. $(5y)^2$

10. $(-2x)^3$

11. Rename $\dfrac{1}{4^3}$ using a negative exponent.

12. Rename 10^{-4} without a negative exponent.

13. Write three other symbols for 4^{-3}.

Multiply and simplify.

14. $8^{-3} \cdot 8^7$

15. $y^7 y^{-2}$

16. $(9x^4)(-2x^7)$

17. $(-3x^{-4})(25x^{-10})$

18. $(5x^{-3}y^4)(-2x^{-9}y^{-2})$

19. $(4x^{-2}y^4)(15x^2y^{-3})$

When a minus sign occurs with exponential notation, a certain caution is in order. For example, $(-4)^2$ means that -4 is to be raised to the second power. Hence $(-4)^2 = (-4)(-4) = 16$. On the other hand, -4^2 represents the additive inverse of 4^2. Thus $-4^2 = -16$. It may help to think of $-x^2$ as $-1 \cdot x^2$, according to Theorem 1.

DO EXERCISES 1–10.

Negative integers as exponents are defined as follows.

Definition

If n is any positive integer, than a^{-n} means $1/a^n$ for $a \neq 0$. In other words, a^n and a^{-n} are reciprocals of each other.

EXAMPLES

1. $\dfrac{1}{5^2} = 5^{-2}$

2. $7^{-3} = \dfrac{1}{7^3}$

3. $5^{-4} = \dfrac{1}{5^4} = \dfrac{1}{5 \cdot 5 \cdot 5 \cdot 5} = \dfrac{1}{625}$

DO EXERCISES 11–13.

PROPERTIES OF EXPONENTS

Multiplication

Let us consider an example involving multiplication:

$$b^5 \cdot b^{-2} = (b \cdot b \cdot b \cdot b \cdot b) \cdot \frac{1}{b \cdot b} = \frac{b \cdot b}{b \cdot b} \cdot (b \cdot b \cdot b)$$

$$= 1 \cdot (b \cdot b \cdot b) = b^3.$$

Note that the result can be obtained by adding the exponents. This is true in general.

Theorem 5

For any number a and any integers m and n, $a^m \cdot a^n = a^{m+n}$.

We can use Theorem 5 to find equivalent expressions for products of exponential expressions with the same base.

EXAMPLES   Multiply and simplify.

4. $x^4 \cdot x^{-2} = x^{4+(-2)} = x^2$

5. $5^4 \cdot 5^6 = 5^{10}$

6. $c^{-3} \cdot c^{-2} = c^{-5}$

7. $a^4 \cdot a^3 = a^7$

DO EXERCISES 14–19.

Division

Let us consider an example involving division:

$$\frac{8^5}{8^3} = 8^5 \cdot \frac{1}{8^3} = 8^5 \cdot 8^{-3} = 8^{5-3} = 8^2.$$

Note that the result could also be obtained by subtracting the exponents. Here is another example:

$$\frac{7^{-2}}{7^3} = 7^{-2} \cdot 7^{-3} = 7^{-2-3} = 7^{-5}.$$

Again, the result could be obtained by subtracting the exponents. This is true in general.

Theorem 6

For any nonzero a and any integers m and n, $a^m/a^n = a^{m-n}$.

EXAMPLES Divide and simplify.

8. $\dfrac{9^{-2}}{9^5} = 9^{-2-5} = 9^{-7}$

9. $\dfrac{7^{-4}}{7^{-5}} = 7^{-4-(-5)} = 7^1 = 7$

10. $\dfrac{x^{10}}{x^8} = x^{10-8} = x^2$

11. $\dfrac{y^{-5}}{y^{-8}} = y^{-5-(-8)} = y^3$

This property can be used to show why a^0 is not defined when $a = 0$. Consider the following:

$$a^0 = a^{3-3} = \frac{a^3}{a^3}.$$

If a were 0, we would then have 0/0, which is meaningless.

DO EXERCISES 20–26.

Raising Powers to Powers

Consider this example:

$$(5^2)^4 = 5^2 \cdot 5^2 \cdot 5^2 \cdot 5^2 = 5^8.$$

The result can be obtained by multiplying the exponents. This is true in general.

Theorem 7

For any number a and any integers m and n, $(a^m)^n = a^{mn}$.

EXAMPLES Simplify.

12. $(8^{-2})^3 = 8^{-2} \cdot 8^{-2} \cdot 8^{-2} = 8^{-6}$

13. $(x^{-5})^4 = x^{-20}$

14. $(3x^2y^{-2})^3 = 3^3(x^2)^3(y^{-2})^3 = 3^3 x^6 y^{-6} = 27x^6y^{-6}$

Divide and simplify.

20. $\dfrac{4^8}{4^5}$

21. $\dfrac{5^4}{5^{-2}}$

22. $\dfrac{10^{-5}}{10^9}$

23. $\dfrac{9^{-8}}{9^{-2}}$

24. $\dfrac{y^6}{y^{-5}}$

25. $\dfrac{10y^2}{2y^3}$

26. $\dfrac{42x^7y^6}{-21y^{-3}x^{10}}$

Simplify.

27. $(3^7)^7$

28. $(8^2)^{-7}$

29. $(y^4)^{-7}$

30. $(2xy)^3$

31. $(4x^{-2}y^7)^2$

32. $(3x^4y^2)^{-3}$

33. $(10x^{-4}y^7z^{-2})^3$

Convert to scientific notation.

34. 465,000

35. 3789

Convert to scientific notation.

36. 0.000145

37. 0.00000000067

Convert to decimal notation.

38. 4.67×10^{-5}

39. 7.894×10^{12}

15. $(5x^3y^{-5}z^2)^4 = 5^4x^{12}y^{-20}z^8 = 625x^{12}y^{-20}z^8$

16. $(4x^2y^{-3})^{-4} = 4^{-4}x^{-8}y^{12} = \frac{1}{256}x^{-8}y^{12}$

CAUTION!   When raising a product such as $8x^2y^{-2}$ to a power, don't forget to raise *all* the factors to the power. For example,

$$(8x^2y^{-2})^3 = 8^3(x^2)^3(y^{-2})^3.$$

DO EXERCISES 27–33.

ii SCIENTIFIC NOTATION

Scientific notation is particularly useful for naming very large or very small numbers. It also has uses for our later study of logarithms. The following are examples of scientific notation:

$$7.8 \times 10^{13},$$
$$5.64 \times 10^{-8}.$$

Definition Scientific notation

Scientific notation for a number consists of exponential notation for a power of 10, and, if needed, decimal notation for a number *a* between 1 and 10 and a multiplication sign. The following are both scientific notation: $a \times 10^b$; 10^b.

We can convert to scientific notation by multiplying by 1, choosing an appropriate symbol $10^k/10^k$ for 1.

EXAMPLE 17   Convert 96,000 to scientific notation.

$$96{,}000 = 96{,}000 \times \frac{10^4}{10^4} \quad \text{We use } 10^4 \text{ in order to "move the decimal point" between 9 and 6.}$$

$$= \frac{96{,}000}{10^4} \times 10^4$$

$$= 9.6 \times 10^4$$

With practice such conversions can be done mentally, and you should try to do this as much as possible.

DO EXERCISES 34 AND 35.

EXAMPLE 18   Convert 0.00000478 to scientific notation.

$$0.00000478 = 0.00000478 \times \frac{10^6}{10^6} \quad \text{We use } 10^6 \text{ in order to "move the decimal point" between 4 and 7.}$$

$$= (0.00000478 \times 10^6) \times 10^{-6}$$

$$= 4.78 \times 10^{-6}$$

Again you should try to make conversions mentally as much as possible.

EXAMPLES   Convert to decimal notation.

19. $6.043 \times 10^5 = 604{,}300$

20. $4.7 \times 10^{-8} = 0.000000047$

DO EXERCISES 36–39.

iii   PROPERTIES OF ABSOLUTE VALUE

We now consider certain properties involving absolute value and use them to simplify certain expressions. In that way we can find equivalent expressions.

EXAMPLES

21. $|-3 \cdot 5| = |-15| = 15$ and $|-3| \cdot |5| = 3 \cdot 5 = 15$, so
$|-3 \cdot 5| = |-3| \cdot |5|$.

Similarly,

22. $|-4 \cdot (-3)| = |12| = 12$ and $|-4| \cdot |-3| = 4 \cdot 3 = 12$, so
$|-4 \cdot (-3)| = |-4| \cdot |-3|$.

The absolute value of a quotient is similarly the quotient of the absolute values.

EXAMPLE 23

$$\left| \frac{25}{-5} \right| = |-5| = 5 \quad \text{and} \quad \frac{|25|}{|-5|} = \frac{25}{5} = 5, \quad \text{so} \quad \left| \frac{25}{-5} \right| = \frac{|25|}{|-5|}.$$

The absolute value of an even power can be simplified by leaving off the absolute-value signs, because no even power can be negative. The absolute value of the additive inverse of a number is the same as the absolute value of the number. We can prove this using the definition of absolute value. Suppose a is positive. Then $-a$ is negative. Thus $|a| = a$ and $|-a| = -(-a) = a$, by Theorem 1. Now suppose a is negative or zero. Then $|a| = -a$ by the definition of absolute value and also $|-a| = -a$ since $-a$ is positive. Another way to think of the property $|a| = |-a|$ is that the distances of a and $-a$ from 0 are the same.

EXAMPLES

24. $|(-3)^2| = |9| = 9$ and $(-3)^2 = 9$,   so $|(-3)^2| = (-3)^2$.
25. $|-3| = 3$ and $|3| = 3$,   so $|-3| = |3|$.

Theorem 8 asserts the properties of absolute value that we have used. Each can be proven using the definition of absolute value.

Theorem 8

For any real numbers a and b and any nonzero number c:

1. $|ab| = |a| \cdot |b|$;
2. $\left| \dfrac{a}{c} \right| = \dfrac{|a|}{|c|}$;
3. $|a^n| = a^n$ if n is an even integer;
4. $|-a| = |a|$.

We can use Theorem 8 to find equivalent expressions.

Simplify.

40. $|-6ab|$

41. $|x^8|$

42. $|10m^2n^3|$

43. $\left|\dfrac{-2x^3}{y^2}\right|$

EXAMPLES Simplify, leaving as little as possible inside the absolute-value signs.

26. $|3x| = |3| \cdot |x| = 3|x|$

27. $|x^2| = x^2$

28. $|x^2y^3| = |x^2| \cdot |y^3| = x^2|y^3| = x^2y^2|y|$

29. $\left|\dfrac{x^2}{y}\right| = \dfrac{|x^2|}{|y|} = \dfrac{x^2}{|y|}$

30. $|-3x| = 3|x|$

DO EXERCISES 40–43.

EXERCISE SET 1.2

i Simplify.

1. $2^3 \cdot 2^{-4}$

2. $3^4 \cdot 3^{-5}$

3. $b^2 \cdot b^{-2}$

4. $c^3 \cdot c^{-3}$

5. $4^2 \cdot 4^{-5} \cdot 4^6$

6. $5^2 \cdot 5^{-4} \cdot 5^5$

7. $2x^3 \cdot 3x^2$

8. $3y^4 \cdot 4y^3$

9. $(5a^2b)(3a^{-3}b^4)$

10. $(4xy^2)(3x^{-4}y^5)$

11. $(2x)^3(3x)^2$

12. $(4y)^2(3y)^3$

13. $(6x^5y^{-2}z^3)(-3x^2y^3z^{-2})$

14. $(5x^4y^{-3}z^2)(-2x^2y^4z^{-1})$

15. $\dfrac{b^{40}}{b^{37}}$

16. $\dfrac{a^{39}}{a^{32}}$

17. $\dfrac{x^2y^{-2}}{x^{-1}y}$

18. $\dfrac{x^3y^{-3}}{x^{-1}y^2}$

19. $\dfrac{9a^2}{(-3a)^2}$

20. $\dfrac{16y^2}{(-4y)^2}$

21. $\dfrac{24a^5b^3}{8a^4b}$

22. $\dfrac{30x^6y^4}{5x^3y^2}$

23. $\dfrac{12x^2y^3z^{-2}}{21xy^2z^3}$

24. $\dfrac{15x^3y^4z^{-3}}{45xyz^5}$

25. $(2ab^2)^3$

26. $(4xy^3)^2$

27. $(-2x^3)^4$

28. $(-3x^2)^4$

29. $-(2x^3)^4$

30. $-(3x^2)^4$

31. $(6a^2b^3c)^2$

32. $(5x^3y^2z)^2$

33. $(-5c^{-1}d^{-2})^{-2}$

34. $(-4x^{-1}z^{-2})^{-2}$

35. $\dfrac{4^{-2}+2^{-4}}{8^{-1}}$

36. $\dfrac{3^{-2}+2^{-3}}{7^{-1}}$

37. $\dfrac{(-2)^4+(-4)^2}{(-1)^8}$

38. $\dfrac{(-3)^2+(-2)^4}{(-1)^6}$

39. $\dfrac{(3a^2b^{-2}c^4)^3}{(2a^{-1}b^2c^{-3})^2}$

40. $\dfrac{(2a^3b^{-3}c^3)^3}{(3a^{-1}b^{-3}c^{-5})^2}$

41. $\dfrac{6^{-2}x^{-3}y^2}{3^{-3}x^{-4}y}$

42. $\dfrac{5^{-2}x^{-4}y^3}{2^{-3}x^{-5}y}$

ii Convert to scientific notation.

43. 58,000,000

44. 27,000

45. 365,000

46. 3645

47. 0.0000027

48. 0.0000658

49. 0.027

50. 0.0038

51. \$910,000,000,000 (a recent national debt)

52. 93,000,000 (the distance, in miles, from the earth to the sun)

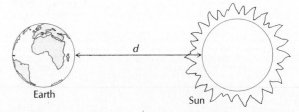

Convert to decimal notation.

53. 4×10^5

54. 5×10^{-4}

55. 6.2×10^{-3}

56. 7.8×10^6

57. 7.69×10^{12}

58. 8.54×10^{-7}

59. 9.46×10^{12} (the distance, in kilometers, that light travels in one year)

60. 1.7×10^{-24} (the mass, in grams, of a hydrogen atom)

Find $-x^2$ and $(-x)^2$, when:

61. $x = 5$.

62. $x = -7$.

63. $x = -1.08$.

64. $x = \sqrt{3}$.

iii Simplify.

65. $|9xy|$

66. $|y^4|$

67. $|3a^2b|$

68. $\left|\dfrac{4a}{b^2}\right|$

Simplify. Assume that all exponents are integers.

69. $(x^t \cdot x^{3t})^2$

70. $(x^y \cdot x^{-y})^3$

71. $(t^{a+x} \cdot t^{x-a})^4$

72. $(m^{x-y} \cdot m^{3y-x})^t$

73. $(x^a y^b \cdot x^b y^a)^c$

74. $(m^{x-b} \cdot n^{x+b})^x (m^b n^{-b})^x$

75. $\left[\dfrac{(3x^a y^b)^3}{(-3x^a y^b)^2}\right]^2$

76. $\left[\left(\dfrac{x^r}{y^t}\right)^2 \left(\dfrac{x^{2r}}{y^{4t}}\right)^{-2}\right]^{-3}$

The formula

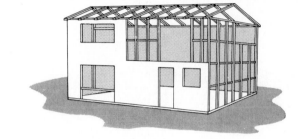

$$M = P\left[\dfrac{\dfrac{i}{12}\left(1 + \dfrac{i}{12}\right)^n}{\left(1 + \dfrac{i}{12}\right)^n - 1}\right]$$

gives the monthly mortgage payment M on a home loan of P dollars at interest rate i and n is the total number of payments (12 times the number of years).

77. ▦ The cost of a house is \$92,000. The down payment is \$14,000. The interest rate is $15\frac{3}{4}\%$. The loan period is 25 years. What is the monthly payment?

78. ▦ Repeat Exercise 77 for loan periods of 20 years and 30 years.

Find the error(s) in each of the following. Explain why each is an error. Then find the correct answer.

79. $x^4(x^3)^2 = x^9$

80. $\dfrac{x^4 y^{-7}}{x^{-2} y^5} = \dfrac{x^2}{y^2}$

81. $(2x^{-4} y^6 z^3)^3 = 6x^{-1} y^3 z^6$

1.3

ADDITION AND SUBTRACTION OF ALGEBRAIC EXPRESSIONS

i POLYNOMIALS

Expressions like the following are called *polynomials in one variable:**

$$-7x + 5, \qquad 3y^3 - 5y^2 + 7y - 4, \qquad 0, \qquad -5t^4, \qquad x^5 - 9.$$

---

*A *variable* is a symbol that can represent different numbers. Letters are generally used for variables. Letters used to represent numbers are not always variables, however. For example, if we choose to represent the distance to the moon by the letter d, then in that context d is not a variable, but a *constant*.

OBJECTIVES

You should be able to:

i Determine the degree of each term of a polynomial and the degree of the polynomial.

ii Add polynomials and other algebraic expressions.

iii Find the additive inverse of an algebraic expression.

iv Subtract polynomials and other algebraic expressions.

Determine the degree of each term and the degree of the polynomial.

1. $x^8 - 7x^6 + 2x^4 - 3x^9 + 2$

2. $8y^4 - 7xy^3 + 6x^2y^3 - 9x^5y - 1$

Definition

A *polynomial in one variable* is any expression of the type

$$a_n x^n + a_{n-1} x^{n-1} + \cdots + a_2 x^2 + a_1 x + a_0,$$

where *n* is a nonnegative integer and $a_n, \ldots, a_0$ are real numbers, called *coefficients*. Some or all of the coefficients may be 0. Each of the parts separated by plus signs is called a *term*.

The question might arise whether an expression such as

$$8x^3 - 6x^2 + 7x - 5$$

is a polynomial. It is indeed, since it is equivalent to

$$8x^3 + (-6x^2) + 7x + (-5).$$

Note that coefficients can be negative.

Expressions like the following are called *polynomials in several variables*:

$$5x^2y^3 + 17x^2y - 2, \qquad 14a^2b, \qquad \pi r^2 + 2\pi rh.$$

EXAMPLE 1   The polynomial $5x^3y - 7xy^2 + 2$ has three terms. They are

$$5x^3y, \qquad -7xy^2, \quad \text{and} \quad 2.$$

The coefficients of the terms are 5, −7, and 2.

The *degree of a term* is the sum of the exponents of the variables. The *degree* of a nonzero polynomial is the degree of the term of highest degree. The polynomial consisting only of the number 0 has no degree.

EXAMPLE 2   In the polynomial $5x^3y - 7xy^2 + 2$, the degrees of the terms are 4, 3, and 0. The polynomial is of degree 4.

A polynomial with just one term is called a *monomial*. If there are just two terms, a polynomial is called a *binomial*. If there are just three terms, it is called a *trinomial*.

DO EXERCISES 1 AND 2.

> CAUTION! In many applications, lower-case and capital letters are used to represent different numbers, as in
>
> $$R + r.$$
>
> In copying expressions, do *not* use a capital letter if a lower-case letter is given.

ii ADDITION

Much of the algebraic manipulation we do with polynomials can also be done with algebraic expressions that are not polynomials. Here are some examples of expressions that are not polynomials.

EXAMPLES

3. $3\sqrt{x} + 4y$

4. $\dfrac{3x^2 + 2}{x - 1}$

5. $4x^{1/2} - 5y^{3/2}$

If two terms of an expression have the same letters raised to the same powers, the terms are called *like*, or *similar*. Similar terms can be "combined" using the distributive laws.

EXAMPLES

6. $3x^2 - 4y + 2x^2 = 3x^2 + 2x^2 - 4y$ Rearranging using the commutative and associative laws

$\quad\quad\quad\quad\quad\quad\quad = (3 + 2)x^2 - 4y$ Using a distributive law

$\quad\quad\quad\quad\quad\quad\quad = 5x^2 - 4y$

7. $4x^{1/2}y + 7x^{1/2}y = 11x^{1/2}y$

8. $-2x^2\sqrt{y^3} + 5x^2\sqrt{y^3} = 3x^2\sqrt{y^3}$

DO EXERCISES 3–5.

The sum of two polynomials can be found by writing a plus sign between them and then combining similar terms. Ordinarily this can be done mentally.

EXAMPLE 9   Add $-3x^3 + 2x - 4$ and $4x^3 + 3x^2 + 2$.

$$(-3x^3 + 2x - 4) + (4x^3 + 3x^2 + 2) = x^3 + 3x^2 + 2x - 2$$

DO EXERCISES 6 AND 7.

iii ADDITIVE INVERSES

The additive inverse of an expression can be found by placing an inverse sign before the expression. An equivalent expression can be found using the following theorem.

Theorem 9

> The additive inverse of a polynomial can be found by replacing every term by its additive inverse.

EXAMPLE 10   Find two equivalent expressions for the additive inverse of $-3xy^2 + 4x^2y - 5x - 3$.

One expression is

$$-(-3xy^2 + 4x^2y - 5x - 3).$$

Another equivalent expression is

$$3xy^2 - 4x^2y + 5x + 3.$$

EXAMPLE 11   Find two equivalent expressions for the additive inverse of $7ab^2 - 6ab - 4b + 8$.

One expression is

$$-(7ab^2 - 6ab - 4b + 8).$$

Another equivalent expression is

$$-7ab^2 + 6ab + 4b - 8.$$

The preceding examples may bring to mind the following rule: To remove parentheses preceded by an additive inverse sign, change the sign of every term inside the parentheses.

DO EXERCISES 8 AND 9.

Combine similar terms.

3. $5x^3y^2 - 2x^2y^3 + 4x^3y^2$

4. $3xy^2 - 4x^2y + 4xy^2 + 2x^2y$

5. $5x^4\sqrt{y} - 2x^4\sqrt{y} + 2$

Add.

6. $3x^3 + 4x^2 - 7x - 2$ and $-7x^3 - 2x^2 + 3x + \frac{1}{2}$

7. $5p^2q^4 - 2p^2q^2 - 3q$ and $-6pq^2 + 3p^2q^2 + 5$

Find two equivalent expressions for the additive inverse of each expression.

8. $5x^2t^2 - 4xy^2t - 3xt + 6x - 5$

9. $-3x^2y + 5xy - 7x + 4y + 2$

Subtract.

10. $(5xy^4 - 7xy^2 + 4x^2 - 3)$
$- (-3xy^4 + 2xy^2 - 2y + 4)$

11. $\quad 5x^2y - 7x^3y^2 \qquad\qquad - x^2y^2 + 4y$
$\underline{-2x^2y + 2x^3y^2 - 5x^2y^3 \qquad\qquad - 5y}$

iv SUBTRACTION

By Theorem 2 we can subtract by adding an inverse. Thus to subtract one polynomial or other algebraic expression from another, we add its additive inverse. We change the sign of each term of the polynomial to be subtracted and then add. In simple cases this can be done mentally.

EXAMPLE 12   Subtract.

$(-9x^5 - x^3 + 2x^2 + 4) - (2x^5 - x^4 + 4x^3 - 3x^2)$
$\qquad\qquad = (-9x^5 - x^3 + 2x^2 + 4) + [-(2x^5 - x^4 + 4x^3 - 3x^2)]$
$\qquad\qquad = (-9x^5 - x^3 + 2x^2 + 4) + (-2x^5 + x^4 - 4x^3 + 3x^2)$
$\qquad\qquad = -11x^5 + x^4 - 5x^3 + 5x^2 + 4$

On occasion, it may be helpful to write polynomials to be subtracted with similar terms in columns.

EXAMPLE 13   Subtract the second polynomial from the first.

$\quad 4x^2y - 6x^3y^2 \qquad\qquad + x^2y^2 - \ 5y$
$\underline{\quad 4x^2y + \ \ x^3y^2 + 3x^2y^3 \qquad\quad + \ 6y} \qquad$ Mentally, change signs and add.
$\quad - 7x^3y^2 - 3x^2y^3 + x^2y^2 - 11y$

DO EXERCISES 10 AND 11.

EXERCISE SET 1.3

i Determine the degree of each term and the degree of the polynomial.

1. $-11x^4 - x^3 + x^2 + 3x - 9$

2. $t^3 - 3t^2 + t + 1$

3. $y^3 + 2y^6 + x^2y^4 - 8$

4. $u^2 + 3v^5 - u^3v^4 - 7$

5. $a^5 + 4a^2b^4 + 6ab + 4a - 3$

6. $8p^6 + 2p^4t^4 - 7p^3t + 5p^2 - 14$

ii Add.

7. $5x^2y - 2xy^2 + 3xy - 5$ and
$-2x^2y - 3xy^2 + 4xy + 7$

8. $6x^2y - 3xy^2 + 5xy - 3$ and
$-4x^2y - 4xy^2 + 3xy + 8$

9. $-3pq^2 - 5p^2q + 4pq + 3$ and
$-7pq^2 + 3pq - 4p + 2q$

10. $-5pq^2 - 3p^2q + 6pq + 5$ and
$-4pq^2 + 5pq - 6p + 4q$

11. $2x + 3y + z - 7$ and
$4x - 2y - z + 8$ and
$-3x + y - 2z - 4$

12. $2x^2 + 12xy - 11$ and
$6x^2 - 2x + 4$ and
$-x^2 - y - 2$

13. $7x\sqrt{y} - 3y\sqrt{x} + \frac{1}{5}$ and
$-2x\sqrt{y} - y\sqrt{x} - \frac{3}{5}$

14. $10x\sqrt{y} - 4y\sqrt{x} + \frac{4}{3}$ and
$-3x\sqrt{y} - y\sqrt{x} - \frac{1}{3}$

iii Find two equivalent expressions for the additive inverse of each expression.

15. $5x^3 - 7x^2 + 3x - 6$

16. $-4y^4 + 7y^2 - 2y - 1$

iv Subtract.

17. $(3x^2 - 2x - x^3 + 2) - (5x^2 - 8x - x^3 + 4)$

18. $(5x^2 + 4xy - 3y^2 + 2) - (9x^2 - 4xy + 2y^2 - 1)$

19. $(4a - 2b - c + 3d) - (-2a + 3b + c - d)$

20. $(5a - 3b - c + 4d) - (-3a + 5b + c - 2d)$

21. $(x^4 - 3x^2 + 4x) - (3x^3 + x^2 - 5x + 3)$

22. $(2x^4 - 5x^2 + 7x) - (5x^3 + 2x^2 - 3x + 5)$

23. $(7x\sqrt{y} - 4y\sqrt{x} + 7.5) - (-2x\sqrt{y} - y\sqrt{x} - 1.6)$

24. $(10x\sqrt{y} - 4y\sqrt{x} + \frac{4}{3}) - (-3x\sqrt{y} + y\sqrt{x} - \frac{1}{3})$

Simplify.

25. ▤ $(0.565p^2q - 2.167pq^2 + 16.02pq - 17.1)$
$+ (-1.612p^2q - 0.312pq^2 - 7.141pq - 87.044)$

26. ▤ $(5003.2xy^{-2} + 3102.4\sqrt{xy} - 5280)$
$- (2143.6xy^{-2} + 6153.8xy - 4141\sqrt{xy} + 4979.12)$

1.4

MULTIPLICATION OF ALGEBRAIC EXPRESSIONS

■i MULTIPLICATION OF ANY TWO POLYNOMIALS

Multiplication of polynomials is based on the distributive laws. To multiply two polynomials, we multiply each term of one by every term of the other and then add the results.

EXAMPLE 1   Multiply $4x^4y - 7x^2y + 3y$ by $2y - 3x^2y$.

$$
\begin{array}{r}
4x^4y - 7x^2y + 3y \\
2y - 3x^2y \\
\hline
8x^4y^2 - 14x^2y^2 + 6y^2 \\
-12x^6y^2 + 21x^4y^2 - 9x^2y^2 \\
\hline
-12x^6y^2 + 29x^4y^2 - 23x^2y^2 + 6y^2
\end{array}
$$

Multiplying by $2y$

Multiplying by $-3x^2y$

Adding

DO EXERCISES 1 AND 2.

The following methods allow us to multiply binomials more efficiently.

Products of Two Binomials

We can find a product of two binomials mentally. We multiply the first terms, then the outside terms, then the inside terms, then the last terms (this procedure is sometimes abbreviated FOIL), and then add the results. This procedure also works for multiplying algebraic expressions that are not polynomials.

EXAMPLES   Multiply.

$$\text{F} \quad \text{O} \quad \text{I} \quad \text{L}$$

2. $(3xy + 2x)(x^2 + 2xy^2) = 3x^3y + 6x^2y^3 + 2x^3 + 4x^2y^2$

3. $(x + \sqrt{2})(y - \sqrt{2}) = xy - \sqrt{2}x + \sqrt{2}y - 2$

4. $(2x - \sqrt{3})(y + 2) = 2xy + 4x - \sqrt{3}y - 2\sqrt{3}$

5. $(2x + 3y)(x - 4y) = 2x^2 - 5xy - 12y^2$

DO EXERCISES 3–5.

Squares of Binomials

Using FOIL to multiply a binomial $A + B$ by itself, we obtain the following:

$$(A + B)^2 = A^2 + 2AB + B^2$$

and

$$(A - B)^2 = A^2 - 2AB + B^2.$$

This gives us a way to square a binomial that is faster than FOIL. We square the first term, add twice the product of the terms, and then add the square of the second term.

OBJECTIVE

You should be able to:

■i Multiply any two polynomials, striving for speed and accuracy. Whenever possible you should write only the answer. In particular, you should be able to:

   a) Square a binomial, writing only the answer.
   b) Multiply the sum and difference of the same two terms, writing only the answer.
   c) Multiply any two binomials, writing only the answer.
   d) Cube a binomial.

Multiply.

1. $3x^2y - 2xy + 3y$ and $xy + 2y$

2. $p^2q + 2pq + 2q$ and $2p^2q - pq + q$

Multiply.

3. $(2xy + 3x)(x^2 - 2)$

4. $(3x - 2y)(5x + 3y)$

5. $(2x + \sqrt{2})(3y - \sqrt{2})$

Multiply.

6. $(4x - 5y)^2$

7. $(2y^2 + 6x^2y)^2$

Multiply.

8. $(4x + 7)(4x - 7)$

9. $(5x^2y + 2y)(5x^2y - 2y)$

10. $(4y^2 + \sqrt{3})(4y^2 - \sqrt{3})$

11. $(2x + 3 + 5y)(2x + 3 - 5y)$

12. $(-2x^3y^2 + 5t)(2x^3y^2 + 5t)$

EXAMPLES Multiply.

6. $(2x + 9y^2)^2 = (2x)^2 + 2(2x)(9y^2) + (9y^2)^2$
$\qquad\qquad = 4x^2 + 36xy^2 + 81y^4$

7. $(3x^2 - 5xy^2)^2 = (3x^2)^2 - 2(3x^2)(5xy^2) + (5xy^2)^2$
$\qquad\qquad = 9x^4 - 30x^3y^2 + 25x^2y^4$

The second term is $-5xy^2$ so twice the product of the terms is $-2(3x^2)(5xy^2)$.

CAUTION! The square of a sum is *not* the sum of squares; that is,

$$(A + B)^2 \neq A^2 + B^2.$$

DO EXERCISES 6 AND 7.

Products of Sums and Differences

We can also use FOIL to find the product of a sum and difference of the same two expressions:

$$(A + B)(A - B) = A^2 - AB + AB - B^2$$
$$= A^2 - B^2.$$

The product of a sum and a difference of the same two expressions is the difference of their squares. Thus to find such a product, we square the first expression, square the second expression, and write a minus sign between the results:

$$\boldsymbol{(A + B)(A - B) = A^2 - B^2}$$

EXAMPLES Multiply.

8. $(y + 5)(y - 5) = y^2 - 25$

9. $(3x + 2)(3x - 2) = (3x)^2 - 2^2$
$\qquad\qquad = 9x^2 - 4$

10. $(2xy^2 + 3x)(2xy^2 - 3x) = (2xy^2)^2 - (3x)^2$
$\qquad\qquad\qquad = 4x^2y^4 - 9x^2$

11. $(5x + \sqrt{2})(5x - \sqrt{2}) = (5x)^2 - (\sqrt{2})^2$
$\qquad\qquad\qquad = 25x^2 - 2$

12. $(5y + 4 + 3x)(5y + 4 - 3x) = (5y + 4)^2 - (3x)^2$
$\qquad\qquad\qquad = 25y^2 + 40y + 16 - 9x^2$

13. $(3xy^2 + 4y)(-3xy^2 + 4y) = -(3xy^2)^2 + (4y)^2$
$\qquad\qquad\qquad = 16y^2 - 9x^2y^4$

DO EXERCISES 8–12.

Cubing Binomials

The following multiplication gives another result to be remembered:

$$(A + B)^3 = (A + B)(A + B)^2$$
$$= (A + B)(A^2 + 2AB + B^2)$$
$$= (A + B)A^2 + (A + B)2AB + (A + B)B^2$$
$$= A^3 + A^2B + 2A^2B + 2AB^2 + AB^2 + B^3$$
$$= A^3 + 3A^2B + 3AB^2 + B^3.$$

The result to be remembered is as follows:

$$(A + B)^3 = A^3 + 3A^2B + 3AB^2 + B^3.$$

EXAMPLES Multiply.

14. $(x + 2)^3 = x^3 + 3x^2(2) + 3x(2)^2 + 2^3$
$= x^3 + 6x^2 + 12x + 8$

15. $(x - 2)^3 = [x + (-2)]^3$
$= x^3 + 3x^2(-2) + 3x(-2)^2 + (-2)^3$
$= x^3 - 6x^2 + 12x - 8$

16. $(5m^2 - 4n^3)^3 = (5m^2)^3 + 3(5m^2)^2(-4n^3) + 3(5m^2)(-4n^3)^2 + (-4n^3)^3$
$= 125m^6 - 300m^4n^3 + 240m^2n^6 - 64n^9$

Note in Examples 15 and 16 that a separate formula for $(A - B)^3$ need not be memorized. We can think of $(A - B)^3$ as $[A + (-B)]^3$.

DO EXERCISES 13–16.

In the following exercises you should do mentally as much of the calculating as you can. If possible, write only the answer. Work for speed with accuracy.

Multiply.

13. $(x + 1)^3$

14. $(x - 1)^3$

15. $(t^2 - 3b)^3$

16. $(2a^3 - 5b^2)^3$

EXERCISE SET 1.4

i Multiply.

1. $2x^2 + 4x + 16$ and $3x - 4$

2. $3y^2 - 3y + 9$ and $2y + 3$

3. $4a^2b - 2ab + 3b^2$ and $ab - 2b + 1$

4. $2x^2 + y^2 - 2xy$ and $x^2 - 2y^2 - xy$

5. $(a - b)(a^2 + ab + b^2)$

6. $(t + 1)(t^2 - t + 1)$

7. $(2x + 3y)(2x + y)$

8. $(2a - 3b)(2a - b)$

9. $(4x^2 - \frac{1}{2}y)(3x + \frac{1}{4}y)$

10. $(2y^3 + \frac{1}{5}x)(3y - \frac{1}{4}x)$

11. $(\sqrt{2}x^2 - y^2)(\sqrt{2}x - 2y)$

12. $(\sqrt{3}y^2 - 2)(\sqrt{3}y - x)$

13. $(2x + 3y)^2$

14. $(5x + 2y)^2$

15. $(2x^2 - 3y)^2$

16. $(4x^2 - 5y)^2$

17. $(2x^3 + 3y^2)^2$

18. $(5x^3 + 2y^2)^2$

19. $(\frac{1}{2}x^2 - \frac{3}{5}y)^2$

20. $(\frac{1}{4}x^2 - \frac{2}{5}y)^2$

21. $(0.5x + 0.7y^2)^2$

22. $(0.3x + 0.8y^2)^2$

23. $(3x - 2y)(3x + 2y)$

24. $(3x + 5y)(3x - 5y)$

25. $(x^2 + yz)(x^2 - yz)$

26. $(2x^2 + 5xy)(2x^2 - 5xy)$

27. $(3x^2 - \sqrt{2})(3x^2 + \sqrt{2})$

28. $(5x^2 - \sqrt{3})(5x^2 + \sqrt{3})$

29. $(2x + 3y + 4)(2x + 3y - 4)$

30. $(5x + 2y + 3)(5x + 2y - 3)$

31. $(x^2 + 3y + y^2)(x^2 + 3y - y^2)$

32. $(2x^2 + y + y^2)(2x^2 + y - y^2)$

33. $(x + 1)(x - 1)(x^2 + 1)$

34. $(y - 2)(y + 2)(y^2 + 4)$

35. $(2x + y)(2x - y)(4x^2 + y^2)$

36. $(5x + y)(5x - y)(25x^2 + y^2)$

37. ▦ $(0.051x + 0.04y)^2$

38. ▦ $(1.032x - 2.512y)^2$

39. ▦ $(37.86x + 1.42)(65.03x - 27.4)$

40. ▦ $(3.601x - 17.5)(47.105x + 31.23)$

41. $(y + 5)^3$

42. $(t - 7)^3$

43. $(m^2 - 2n)^3$

44. $(3t^2 + 4)^3$

Multiply. Assume that all exponents are natural numbers.

45. $(a^n + b^n)(a^n - b^n)$

46. $(t^a + 4)(t^a - 7)$

47. $(x^m - t^n)^3$

48. $y^3 z^n(y^{3n}z^3 - 4yz^{2n})$

49. $(x - 1)(x^2 + x + 1)(x^3 + 1)$

50. $(a^n + b^n)^2$

51. $[(2x - 1)^2 - 1]^2$

52. $[(a + b)(a - b)][5 - (a + b)][5 + (a + b)]$

53. $(x^{a-b})^{a+b}$

54. $(t^{m+n})^{m+n} \cdot (t^{m-n})^{m-n}$

55. $(a + b + c)^2$

56. $(a + b + c)^3$

57. $(a + b)^4$

58. $(x - y)(x^4 + x^3y + x^2y^2 + xy^3 + y^4)$

59. $(m + t)(m^4 - m^3t + m^2t^2 - mt^3 + t^4)$

60. $(a - b)(a^7 + a^6b + a^5b^2 + a^4b^3 + a^3b^4 + a^2b^5 + ab^6 + b^7)$

Find the error(s) in each of the following. Explain why each is an error. Then find the correct answer.

61. $(3a + b)^2 = 3a^2 + b^2$

62. $(2x - 3y)(2x - 3y) = 4x^2 - 9y^2$

63. $2x(x + 3) + 4(x^2 - 3) = 2x^2 + 3x + 4x^2 - 3$ (1)
$\qquad\qquad\qquad\qquad\qquad = 6x^2 + x$ (2)

64. $(2a - 3b)(3a + 2b) = 6a^2 - 6b^2$ (1)
$\qquad\qquad\qquad\qquad = a^2 - b^2$ (2)

65. Multiply. Assume that n is a natural number.

$$(x - y)(x^{n-1} + x^{n-2}y + x^{n-3}y^2 + \cdots + x^3y^{n-4} + x^2y^{n-3} + xy^{n-2} + y^{n-1})$$

OBJECTIVE

You should be able to:

i Determine the kind of factoring to try, when an expression is to be factored. Then factor expressions:

  a) by removing a common factor;
  b) that are differences of squares;
  c) that are trinomial squares;
  d) that are trinomials not squares;
  e) that are sums or differences of cubes.

1.5

FACTORING

i FACTORING POLYNOMIALS

To factor a polynomial we do the reverse of multiplying; that is, we find an equivalent expression that is a product. Facility in factoring is an important algebraic skill.

Terms with Common Factors

When an expression is to be factored, we should always look first for a possible factor that is common to all terms. We then "factor it out" using the distributive laws.

EXAMPLE 1   Factor $4x^2 + 8$.

$$4x^2 + 8 = 4(x^2 + 2)$$

Note that $4x^2$ and 8 are *terms* of the expression. The expression $4(x^2 + 2)$ is an equivalent expression that is a *product*. The expression $4 \cdot x^2 + 4 \cdot 2$ is not a correct answer. Although each term is factored, the entire expression has not been factored, that is, expressed as a product.

EXAMPLES   Factor.

2. $12x^2y - 20x^3y = 4x^2y(3 - 5x)$

3. $7x\sqrt{y} + 14x^2\sqrt{y} - 21\sqrt{y} = 7\sqrt{y}(x + 2x^2 - 3)$

4. $(a - b)(x + 5) + (a - b)(x - y^2) = (a - b)[(x + 5) + (x - y^2)]$
$\qquad\qquad\qquad\qquad\qquad\qquad\qquad = (a - b)(2x + 5 - y^2)$

In some polynomials pairs of terms have a common factor that can be removed, as in the following examples. This process is called *factoring by grouping*, and uses the distributive laws repeatedly.

EXAMPLES Factor.

5. $y^2 + 3y + 4y + 12 = y(y + 3) + 4(y + 3)$
$= (y + 4)(y + 3)$

6. $ax^2 + ay + bx^2 + by = a(x^2 + y) + b(x^2 + y)$
$= (a + b)(x^2 + y)$

DO EXERCISES 1–3.

Differences of Squares

Recall that $(A + B)(A - B) = A^2 - B^2$. We can use this equation in reverse to factor an expression that is a difference of two squares.

EXAMPLES Factor.

7. $x^2 - 9 = (x + 3)(x - 3)$
8. $y^2 - 2 = (y + \sqrt{2})(y - \sqrt{2})$
9. $9a^2 - 16x^4 = (3a)^2 - (4x^2)^2 = (3a + 4x^2)(3a - 4x^2)$
10. $9y^4 - 9x^4 = 9(y^4 - x^4)$ Remove the common factor first.
$= 9(y^2 + x^2)(y^2 - x^2)$
$= 9(y^2 + x^2)(y + x)(y - x)$

DO EXERCISES 4–7.

Trinomial Squares

You should recall that

$$A^2 + 2AB + B^2 = (A + B)^2 \quad \text{and} \quad A^2 - 2AB + B^2 = (A - B)^2.$$

We can use these equations to factor trinomials that are squares. To factor a trinomial, you should check to see if it is a square. For this to be the case, two of the terms must be squares and the other term must be twice the product of the square roots, or the additive inverse of that product.

EXAMPLES Factor.

11. $x^2 - 10x + 25 = (x - 5)^2$
12. $16y^2 + 56y + 49 = (4y + 7)^2$
13. $-4y^2 - 144y^8 + 48y^5 = -4y^2(1 + 36y^6 - 12y^3)$ We first removed
$= -4y^2(1 - 12y^3 + 36y^6)$ the common factor.
$= -4y^2(1 - 6y^3)^2$

DO EXERCISES 8–10.

Trinomials That Are Not Squares

Certain trinomials that are not squares can be factored into two binomials. To do this, we factor by trial and error using the equation

$$acx^2 + (ad + bc)x + bd = (ax + b)(cx + d).$$

Factor.

1. $20x^3y + 12x^2y$

2. $(p + q)(x + 2) + (p + q)(x + y)$

3. $4x^2 + 20x - 3x - 15$

Factor.

4. $x^2 - 16$

5. $25y^4 - 16x^2$

6. $2y^4 - 32x^4$

7. $x^2 - 3$

Factor.

8. $9y^2 - 30y + 25$

9. $16x^2 + 72xy + 81y^2$

10. $-12x^4y^2 + 60x^2y^5 - 75y^8$
(*Hint:* First remove a common factor.)

Factor.

11. $x^2 + 5x - 14$

12. $3x^2 + 5x + 2$

13. $6x^4y^6 - 9x^2y^3 - 60$

Factor.

14. $x^3 - 8$

15. $64 - t^3$

EXAMPLE 14   Factor $x^2 + 7x + 12$.

We look for factors of 12 whose sum is 7. By trial we determine the factorization to be as follows:

$$(x + 4)(x + 3).$$

EXAMPLE 15   Factor $3x^2 - 10x - 8$.

Method 1.   We look for binomials $ax + b$ and $cx + d$ for which the product of the first terms is $3x^2$. The product of the last terms must be -8. When we multiply the inside terms, then the outside terms, and add, we must get $-10x$. By trial, we determine the factorization to be as follows:

$$(3x + 2)(x - 4).$$

Method 2.   We multiply the leading coefficient 3 and the constant -8: $3(-8) = -24$. Then we try to factor -24 so that the sum of the factors is -10. By trial we find these numbers to be -12 and 2. We then write the middle term $-10x$ as a sum using -12 and 2. That is, we split the middle term as follows:

$$-10x = -12x + 2x.$$

Now we factor by grouping as follows:

$$3x^2 - 10x - 8 = 3x^2 - 12x + 2x - 8$$
$$= 3x(x - 4) + 2(x - 4)$$
$$= (3x + 2)(x - 4).$$

DO EXERCISES 11–13.

Sums or Differences of Cubes

We can use the following equations to factor a sum or a difference of two cubes:

$$A^3 + B^3 = (A + B)(A^2 - AB + B^2),$$

$$A^3 - B^3 = (A - B)(A^2 + AB + B^2).$$

Check them by multiplying the right-hand sides.

EXAMPLE 16   Factor $x^3 - 27$.

$$x^3 - 27 = x^3 - 3^3$$

In one set of parentheses we write the cube root of the first expression, x. Then we write the cube root of the second expression, -3. This gives us $x - 3$.

$$(x - 3)(\qquad\qquad)$$

To get the next factor we think of $x - 3$ and do the following.

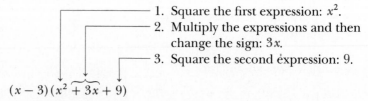

1. Square the first expression: x^2.
2. Multiply the expressions and then change the sign: $3x$.
3. Square the second expression: 9.

$$(x - 3)(x^2 + 3x + 9)$$

Note: We cannot factor $x^2 + 3x + 9$. (It is not a trinomial square.)

DO EXERCISES 14 AND 15.

EXAMPLE 17 Factor $125x^3 + y^3$.

$$125x^3 + y^3 = (5x)^3 + y^3$$

In one set of parentheses we write the cube root of the first expression, then a plus sign, and then the cube root of the second expression.

$$(5x + y)(\qquad\qquad)$$

To get the next factor, we think of $5x + y$ and do the following.

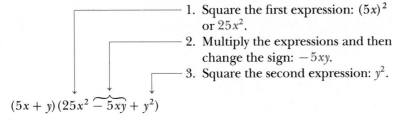

1. Square the first expression: $(5x)^2$ or $25x^2$.
2. Multiply the expressions and then change the sign: $-5xy$.
3. Square the second expression: y^2.

$$(5x + y)(25x^2 - 5xy + y^2)$$

DO EXERCISES 16 AND 17.

EXAMPLE 18 Factor $16x^7y + 54xy^7$.

We first look for a common factor.

$$2xy(8x^6 + 27y^6) = 2xy[(2x^2)^3 + (3y^2)^3]$$
$$= 2xy(2x^2 + 3y^2)(4x^4 - 6x^2y^2 + 9y^4)$$

DO EXERCISE 18.

Remember the following about factoring sums or differences of squares and cubes.

Sum of cubes:	$A^3 + B^3 = (A + B)(A^2 - AB + B^2)$
Difference of cubes:	$A^3 - B^3 = (A - B)(A^2 + AB + B^2)$
Difference of squares:	$A^2 - B^2 = (A + B)(A - B)$
Sum of squares:	$A^2 + B^2$ cannot be factored using real-number coefficients

---

Factor.

16. $27x^3 + y^3$

17. $8m^3 + 125t^3$

Factor.

18. $128y^7 - 250x^6y$

EXERCISE SET 1.5

i Factor.

1. $18a^2b - 15ab^2$

2. $4x^2y + 12xy^2$

3. $a(b - 2) + c(b - 2)$

4. $a(x^2 - 3) - 2(x^2 - 3)$

5. $x^2 + 3x + 6x + 18$

6. $3x^3 + x^2 - 18x - 6$

7. $9x^2 - 25$

8. $16x^2 - 9$

9. $4xy^4 - 4xz^2$

10. $5xy^4 - 5xz^4$

11. $y^2 - 6y + 9$

12. $x^2 + 8x + 16$

13. $1 - 8x + 16x^2$

14. $1 + 10x + 25x^2$

15. $4x^2 - 5$

16. $16x^2 - 7$

17. $x^2y^2 - 14xy + 49$

18. $x^2y^2 - 16xy + 64$

19. $4ax^2 + 20ax - 56a$

20. $21x^2y + 2xy - 8y$

21. $a^2 + 2ab + b^2 - c^2$

22. $x^2 - 2xy + y^2 - z^2$

23. $x^2 + 2xy + y^2 - a^2 - 2ab - b^2$
  (*Hint:* Factor $x^2 + 2xy + y^2$ and $-1(a^2 + 2ab + b^2)$.)

24. $r^2 + 2rs + s^2 - t^2 + 2tv - v^2$

25. $5y^4 - 80x^4$

26. $6y^4 - 96x^4$

27. $x^3 + 8$

28. $y^3 - 64$

29. $3x^3 - \frac{3}{8}$

30. $5y^3 + \frac{5}{27}$

31. $x^3 + 0.001$

32. $y^3 - 0.125$

33. $3z^3 - 24$

34. $4t^3 + 108$

35. $a^6 - t^6$

36. $64m^6 + y^6$

37. $16a^7b + 54ab^7$

38. $24a^2x^4 - 375a^8x$

39. ▤ $x^2 - 17.6$

40. ▤ $x^2 - 8.03$

41. ▤ $37x^2 - 14.5y^2$
     (*Hint:* First remove the common factor 37.)

42. ▤ $1.96x^2 - 17.4y^2$
     (*Hint:* First remove the common factor 1.96.)

Factor.

43. $(x + h)^3 - x^3$

44. $(x + 0.01)^2 - x^2$

45. $y^4 - 84 + 5y^2$

46. $11x^2 + x^4 - 80$

47. $y^2 - \frac{8}{49} + \frac{2}{7}y$

48. $x^2 + \frac{3}{5}x - \frac{4}{25}$

49. $t^2 - 0.27 + 0.6t$

50. $0.4m - 0.05 + m^2$

Factor. Assume that variables in exponents represent natural numbers.

51. $x^{2n} + 5x^n - 24$

52. $4x^{2n} - 4x^n - 3$

53. $x^2 + ax + bx + ab$

54. $bdy^2 + ady + bcy + ac$

55. $\frac{1}{4}t^2 - \frac{2}{5}t + \frac{4}{25}$

56. $\frac{4}{27}r^2 + \frac{5}{9}rs + \frac{1}{12}s^2 - \frac{1}{3}rs$

57. $25y^{2m} - (x^{2n} - 2x^n + 1)$

58. $4x^{4a} + 12x^{2a} + 10x^{2a} + 30$

59. $3x^{3n} - 24y^{3m}$

60. $x^{6a} - t^{3b}$

61. $(y - 1)^4 - (y - 1)^2$

62. $x^6 - 2x^5 + x^4 - x^2 + 2x - 1$

OBJECTIVES

You should be able to:

i   Determine sensible replacements in fractional expressions.

ii   Simplify fractional expressions.

iii   Multiply or divide fractional expressions, and simplify.

iv   Add or subtract fractional expressions, and simplify.

v   Simplify complex fractional expressions.

1.6

FRACTIONAL EXPRESSIONS

i   REPLACEMENTS IN FRACTIONAL EXPRESSIONS

Expressions like the following are called *fractional expressions*:

$$\frac{8}{5}, \quad \frac{x^2 - 9}{x - 3}, \quad \frac{3x^2 + 5\sqrt{x} - 2}{x^2 - y^2}, \quad \frac{x - 3}{x^2 - x - 2}.$$

Fractional expressions represent division. Certain substitutions are not sensible in such expressions. Since division by 0 is not defined, any number that makes a denominator zero is not a sensible replacement. For example, consider

$$\frac{x - 3}{x^2 - x - 2}.$$

To determine the sensible replacements, we set the denominator equal to 0 and solve:

$$x^2 - x - 2 = 0$$
$$(x + 1)(x - 2) = 0$$
$$x + 1 = 0 \quad \text{or} \quad x - 2 = 0$$
$$x = -1 \quad \text{or} \quad x = 2.$$

Thus -1 and 2 are not sensible replacements. All real numbers except -1 and 2 are sensible replacements.

DO EXERCISES 1 AND 2.

Multiplication and Division

To multiply two fractional expressions, we multiply their numerators and also their denominators. By Theorem 4, when we divide, we multiply by the reciprocal of the divisor. The latter can be obtained by inverting the divisor.

EXAMPLE 1 Multiply.

$$\frac{x+3}{y-4} \cdot \frac{x^3}{y+5} = \frac{(x+3)x^3}{(y-4)(y+5)}$$
$$= \frac{x^4 + 3x^3}{y^2 + y - 20}$$

EXAMPLE 2 Divide.

$$\frac{x-2}{x+1} \div \frac{x+5}{x-3} = \frac{x-2}{x+1} \cdot \frac{x-3}{x+5} \qquad \text{Inverting}$$
$$= \frac{(x-2)(x-3)}{(x+1)(x+5)} \qquad \text{Multiplying}$$
$$= \frac{x^2 - 5x + 6}{x^2 + 6x + 5}$$

DO EXERCISES 3 AND 4.

ii SIMPLIFYING

The basis of simplifying fractional expressions rests on the fact that certain expressions have a value of 1 for all sensible replacements. Such expressions have the same numerator and denominator.* Here are some examples:

$$\frac{x-2}{x-2}, \qquad \frac{3x^2 - 4x + 2}{3x^2 - 4x + 2}, \qquad \frac{4x-5}{4x-5}.$$

When we multiply by such an expression we obtain an equivalent expression. This means that the new expression will name the same number as the first for all sensible replacements. The set of sensible replacements may not be the same for the two expressions.

EXAMPLE 3 Multiply.

$$\frac{y+4}{y-3} \cdot \frac{y-2}{y-2} = \frac{(y+4)(y-2)}{(y-3)(y-2)}$$
$$= \frac{y^2 + 2y - 8}{y^2 - 5y + 6}$$

The expressions $(y+4)/(y-3)$ and $(y^2 + 2y - 8)/(y^2 - 5y + 6)$ are equivalent. That is, they will name the same number for all sensible replacements.

---
* By Theorem 4, $a \div a = a/a = a(1/a)$, and since a and $1/a$ are reciprocals, their product is 1.

Determine the sensible replacements.

1. $\dfrac{x^2 - 9}{x - 3}$

2. $\dfrac{x^3 - xy^2}{x^2 + 7x + 12}$

3. Multiply.

$$\frac{x+y}{2x^2 - 1} \cdot \frac{x+y}{7x}$$

4. Divide.

$$\frac{x-2}{x+2} \div \frac{x+2}{x+4}$$

5. Multiply

$$\frac{x+2}{x-5} \quad \text{by} \quad \frac{x+3}{x+3}$$

to obtain an equivalent expression. Name the sensible replacements for the two expressions.

Simplify. Name the sensible replacements in the original and the simplified expressions.

6. $\dfrac{6x^2 + 4x}{2x^2 + 4x}$

7. $\dfrac{y^2 + 3y + 2}{y^2 - 1}$

The only nonsensible replacement in the first expression is 3. For the second expression the nonsensible replacements are 2 and 3.

DO EXERCISE 5.

Simplification can be accomplished by reversing the procedure in the above example; that is, we try to factor the fractional expression in such a way that one of the factors is equal to 1 and then "remove" that factor. In this way we obtain an expression that is simpler, or less complicated, than the original, but equivalent to it.

EXAMPLE 4   Simplify.

$$\frac{15x^3y^2}{20x^2y} = \frac{(5x^2y)\,3xy}{(5x^2y)\,4} \qquad \text{Factoring numerator and denominator}$$

$$= \frac{5x^2y}{5x^2y} \cdot \frac{3xy}{4} \qquad \text{Factoring the expression}$$

$$= \frac{3xy}{4} \qquad \text{"Removing" a factor of 1}$$

Note that in the original expression neither x nor y can be 0. In the simplified expression, however, all replacements are sensible. The expressions are equivalent for all sensible replacements.

EXAMPLE 5   Simplify.

$$\frac{x^2 - 1}{2x^2 - x - 1} = \frac{(x-1)(x+1)}{(2x+1)(x-1)}$$

$$= \frac{x-1}{x-1} \cdot \frac{x+1}{2x+1}$$

$$= \frac{x+1}{2x+1}$$

In the original expression the sensible replacements are all real numbers except 1 and $-\frac{1}{2}$. In the simplified expression all real numbers except $-\frac{1}{2}$ are sensible replacements.

DO EXERCISES 6 AND 7.

iii   **MULTIPLYING, DIVIDING, AND SIMPLIFYING**

EXAMPLES

6. Multiply and simplify.

$$\frac{x+2}{x-2} \cdot \frac{x^2-4}{x^2+x-2} = \frac{(x+2)(x^2-4)}{(x-2)(x^2+x-2)} \qquad \text{Multiplying}$$

$$= \frac{(x+2)(x+2)(x-2)}{(x-2)(x+2)(x-1)} \qquad \text{Factoring}$$

$$= \frac{(x+2)(x-2)}{(x+2)(x-2)} \cdot \frac{x+2}{x-1} = \frac{x+2}{x-1} \qquad \text{"Removing" a factor of 1}$$

7. Divide and simplify.

$$\frac{a^2 - 1}{a + 1} \div \frac{a^2 - 2a + 1}{a + 1} = \frac{a^2 - 1}{a + 1} \cdot \frac{a + 1}{a^2 - 2a + 1}$$

$$= \frac{(a + 1)(a - 1)(a + 1)}{(a + 1)(a - 1)(a - 1)} = \frac{a + 1}{a - 1}$$

DO EXERCISES 8 AND 9.

iv ADDITION AND SUBTRACTION

When fractional expressions have the same denominator, we can add or subtract them by adding or subtracting the numerators and retaining the common denominator. If denominators are not the same, we then find equivalent expressions with the same denominator and add. If one denominator is the additive inverse of another, we can find a common denominator by multiplying by $-1/-1$.

EXAMPLE 8   Add.

$$\frac{3x^2 + 4x - 8}{x^2 + y^2} + \frac{-5x^2 + 5x + 7}{x^2 + y^2} = \frac{-2x^2 + 9x - 1}{x^2 + y^2}$$

In the following example, one denominator is the additive inverse of the other.

EXAMPLE 9   Add.

$$\frac{3x^2 + 4}{x - y} + \frac{5x^2 - 11}{y - x} = \frac{3x^2 + 4}{x - y} + \frac{-1}{-1} \cdot \frac{5x^2 - 11}{y - x}$$

We multiply by $-1/-1$ to convert the second denominator to its additive inverse.

$$= \frac{3x^2 + 4}{x - y} + \frac{-1(5x^2 - 11)}{-1(y - x)}$$

$$= \frac{3x^2 + 4}{x - y} + \frac{11 - 5x^2}{x - y} \qquad -1(y - x) = -y + x = x - y$$

$$= \frac{-2x^2 + 15}{x - y}$$

DO EXERCISES 10 AND 11.

When denominators are different, but not additive inverses of each other, we find a common denominator by factoring the denominators. Then we multiply each term by 1 appropriately to get the common denominator in each expression.

EXAMPLE 10   Add $\dfrac{1}{2x} + \dfrac{5x}{x^2 - 1} + \dfrac{3}{x + 1}$.

We first find the *least common multiple* (LCM) of the denominators. The denominators, when factored, are

$$2x, \qquad (x + 1)(x - 1), \qquad x + 1.$$

8. Multiply and simplify.

$$\frac{x^2 - 2xy + y^2}{x + y} \cdot \frac{3x + 3y}{x^2 - y^2}$$

9. Divide and simplify.

$$\frac{a^2 - b^2}{ab} \div \frac{a^2 - 2ab + b^2}{2a^2b^2}$$

Add.

10. $\dfrac{2x^2 + 5x - 9}{x - 5} + \dfrac{x^2 - x + 11}{x - 5}$

11. $\dfrac{3x^2 + 4}{x - 5} + \dfrac{x^2 - 7}{5 - x}$

Add.

12. $\dfrac{x^2 - 4xy + 4y^2}{2x^2 - 3xy + y^2} + \dfrac{x + 4y}{2x - 2y}$

The LCM is $2x(x + 1)(x - 1)$. Now we multiply each fractional expression by 1 appropriately.

$$\frac{1}{2x} \cdot \frac{(x + 1)(x - 1)}{(x + 1)(x - 1)} + \frac{5x}{(x + 1)(x - 1)} \cdot \frac{2x}{2x} + \frac{3}{(x + 1)} \cdot \frac{2x(x - 1)}{2x(x - 1)}$$

$$= \frac{(x + 1)(x - 1) + 10x^2 + 6x(x - 1)}{2x(x + 1)(x - 1)}$$

$$= \frac{17x^2 - 6x - 1}{2x(x + 1)(x - 1)} \quad \text{or} \quad \frac{17x^2 - 6x - 1}{2x^3 - 2x}$$

DO EXERCISE 12.

EXAMPLE 11 Subtract $\dfrac{x}{x^2 + 5x + 6} - \dfrac{2}{x^2 + 3x + 2}$.

$$\frac{x}{x^2 + 5x + 6} - \frac{2}{x^2 + 3x + 2}$$

$$= \frac{x}{(x + 2)(x + 3)} - \frac{2}{(x + 1)(x + 2)}$$

The LCM is $(x + 1)(x + 2)(x + 3)$.

$$= \frac{x}{(x + 2)(x + 3)} \cdot \frac{x + 1}{x + 1} - \frac{2}{(x + 1)(x + 2)} \cdot \frac{x + 3}{x + 3}$$

Subtract.

13. $\dfrac{x}{x^2 + 11x + 30} - \dfrac{5}{x^2 + 9x + 20}$

$$= \frac{x(x + 1) - [2(x + 3)]}{(x + 1)(x + 2)(x + 3)} \qquad \text{We use the colored brackets here to make sure we subtract the \textit{entire} numerator, not just part of it.}$$

$$= \frac{x^2 + x - [2x + 6]}{(x + 1)(x + 2)(x + 3)}$$

$$= \frac{x^2 + x - 2x - 6}{(x + 1)(x + 2)(x + 3)}$$

$$= \frac{x^2 - x - 6}{(x + 1)(x + 2)(x + 3)}$$

$$= \frac{(x - 3)(x + 2)}{(x + 1)(x + 2)(x + 3)}$$

$$= \frac{x - 3}{(x + 1)(x + 3)} \qquad \text{Always simplify at the end if possible.}$$

DO EXERCISE 13.

CAUTION! When subtracting one fractional expression from another, subtract numerators:

$$\frac{A}{C} - \frac{B}{C} = \frac{A - (B)}{C}.$$

Always be sure to subtract the *entire* numerator B and not just part of it. The use of parentheses or brackets helps. See Example 11.

---

v COMPLEX FRACTIONAL EXPRESSIONS

A complex fractional expression has a fractional expression in its numerator or denominator or both. To simplify such an expression, we can combine as necessary in numerator and denominator in order to obtain a single fractional expression for each. Then we divide the numerator by the denominator.

EXAMPLE 12 Simplify.

$$\frac{x + \dfrac{1}{5}}{x - \dfrac{1}{3}} = \frac{x \cdot \dfrac{5}{5} + \dfrac{1}{5}}{x \cdot \dfrac{3}{3} - \dfrac{1}{3}}$$

$$= \frac{\dfrac{5x + 1}{5}}{\dfrac{3x - 1}{3}} \qquad \text{Now we have a single fractional expression for both numerator and denominator.}$$

$$= \frac{5x + 1}{5} \cdot \frac{3}{3x - 1} \qquad \text{Here we divided by multiplying by the reciprocal of the denominator.}$$

$$= \frac{15x + 3}{15x - 5}$$

EXAMPLE 13 Simplify.

$$\frac{a^{-2} - b^{-2}}{a^{-1} + b^{-1}} = \frac{\dfrac{1}{a^2} - \dfrac{1}{b^2}}{\dfrac{1}{a} + \dfrac{1}{b}}$$

$$= \frac{\dfrac{b^2}{b^2} \cdot \dfrac{1}{a^2} - \dfrac{a^2}{a^2} \cdot \dfrac{1}{b^2}}{\dfrac{b}{b} \cdot \dfrac{1}{a} + \dfrac{a}{a} \cdot \dfrac{1}{b}}$$

$$= \frac{\dfrac{b^2 - a^2}{a^2 b^2}}{\dfrac{b + a}{ab}}$$

$$= \frac{b^2 - a^2}{a^2 b^2} \cdot \frac{ab}{b + a}$$

$$= \frac{(b - a)(b + a)\,ab}{(b + a)\,a^2 b^2}$$

$$= \frac{ab(b + a)}{ab(b + a)} \cdot \frac{b - a}{ab}$$

$$= \frac{b - a}{ab}$$

DO EXERCISES 14 AND 15.

Simplify.

14. $\dfrac{1 + \dfrac{x}{a}}{a - \dfrac{x^2}{a}}$

15. $\dfrac{\dfrac{1}{a} + \dfrac{1}{b}}{\dfrac{1}{a} - \dfrac{1}{b}}$

EXERCISE SET 1.6

i In Exercises 1–3, determine the sensible replacements.

1. $\dfrac{3x - 2}{x(x - 1)}$

2. $\dfrac{(x^2 - 4)(x + 1)}{(x + 2)(x^2 - 1)}$

3. $\dfrac{7y^2 - 2y + 4}{x(x^2 - x - 6)}$

ii   In Exercises 4–6, simplify. Then determine the replacements that are sensible in the simplified expression.

4. $\dfrac{25x^2y^2}{10xy^2}$

5. $\dfrac{x^2 - 4}{x^2 + 5x + 6}$

6. $\dfrac{x^2 - 3x + 2}{x^2 + x - 2}$

iii   Multiply or divide, and simplify.

7. $\dfrac{x^2 - y^2}{(x - y)^2} \cdot \dfrac{1}{x + y}$

8. $\dfrac{r - s}{r + s} \cdot \dfrac{r^2 - s^2}{(r - s)^2}$

9. $\dfrac{x^2 - 2x - 35}{2x^3 - 3x^2} \cdot \dfrac{4x^3 - 9x}{7x - 49}$

10. $\dfrac{x^2 + 2x - 35}{3x^3 - 2x^2} \cdot \dfrac{9x^3 - 4x}{7x + 49}$

11. $\dfrac{a^2 - a - 6}{a^2 - 7a + 12} \cdot \dfrac{a^2 - 2a - 8}{a^2 - 3a - 10}$

12. $\dfrac{a^2 - a - 12}{a^2 - 6a + 8} \cdot \dfrac{a^2 + a - 6}{a^2 - 2a - 24}$

13. $\dfrac{m^2 - n^2}{r + s} \div \dfrac{m - n}{r + s}$

14. $\dfrac{a^2 - b^2}{x - y} \div \dfrac{a + b}{x - y}$

15. $\dfrac{3x + 12}{2x - 8} \div \dfrac{(x + 4)^2}{(x - 4)^2}$

16. $\dfrac{a^2 - a - 2}{a^2 - a - 6} \div \dfrac{a^2 - 2a}{2a + a^2}$

17. $\dfrac{x^2 - y^2}{x^3 - y^3} \cdot \dfrac{x^2 + xy + y^2}{x^2 + 2xy + y^2}$

18. $\dfrac{c^3 + 8}{c^2 - 4} \div \dfrac{c^2 - 2c + 4}{c^2 - 4c + 4}$

19. $\dfrac{(x - y)^2 - z^2}{(x + y)^2 - z^2} \div \dfrac{x - y + z}{x + y - z}$

20. $\dfrac{(a + b)^2 - 9}{(a - b)^2 - 9} \cdot \dfrac{a - b - 3}{a + b + 3}$

iv   Add or subtract, and simplify.

21. $\dfrac{3}{2a + 3} + \dfrac{2a}{2a + 3}$

22. $\dfrac{a - 3b}{a + b} + \dfrac{a + 5b}{a + b}$

23. $\dfrac{y}{y - 1} + \dfrac{2}{1 - y}$

24. $\dfrac{a}{a - b} + \dfrac{b}{b - a}$

25. $\dfrac{x}{2x - 3y} - \dfrac{y}{3y - 2x}$

26. $\dfrac{3a}{3a - 2b} - \dfrac{2a}{2b - 3a}$

27. $\dfrac{3}{x + 2} + \dfrac{2}{x^2 - 4}$

28. $\dfrac{5}{a - 3} - \dfrac{2}{a^2 - 9}$

29. $\dfrac{y}{y^2 - y - 20} + \dfrac{2}{y + 4}$

30. $\dfrac{6}{y^2 + 6y + 9} - \dfrac{5}{y + 3}$

31. $\dfrac{3}{x + y} + \dfrac{x - 5y}{x^2 - y^2}$

32. $\dfrac{a^2 + 1}{a^2 - 1} - \dfrac{a - 1}{a + 1}$

33. $\dfrac{9x + 2}{3x^2 - 2x - 8} + \dfrac{7}{3x^2 + x - 4}$

34. $\dfrac{3y}{y^2 - 7y + 10} - \dfrac{2y}{y^2 - 8y + 15}$

35. $\dfrac{5a}{a - b} + \dfrac{ab}{a^2 - b^2} + \dfrac{4b}{a + b}$

36. $\dfrac{6a}{a - b} - \dfrac{3b}{b - a} + \dfrac{5}{a^2 - b^2}$

37. $\dfrac{7}{x + 2} - \dfrac{x + 8}{4 - x^2} + \dfrac{3x - 2}{4 - 4x + x^2}$

38. $\dfrac{6}{x + 3} - \dfrac{x + 4}{9 - x^2} + \dfrac{2x - 3}{9 - 6x + x^2}$

39. $\dfrac{1}{x + 1} - \dfrac{x}{x - 2} + \dfrac{x^2 + 2}{x^2 - x - 2}$

40. $\dfrac{x - 1}{x - 2} - \dfrac{x + 1}{x + 2} + \dfrac{x - 6}{x^2 - 4}$

v   Simplify.

41. $\dfrac{\dfrac{x^2 - y^2}{xy}}{\dfrac{x - y}{y}}$

42. $\dfrac{\dfrac{a - b}{b}}{\dfrac{a^2 - b^2}{ab}}$

43. $\dfrac{a - a^{-1}}{a + a^{-1}}$

44. $\dfrac{a - \dfrac{a}{b}}{b - \dfrac{b}{a}}$

45. $\dfrac{c + \dfrac{8}{c^2}}{1 + \dfrac{2}{c}}$

46. $\dfrac{x^{-1} + y^{-1}}{x^{-3} + y^{-3}}$

47. $\dfrac{x^2 + xy + y^2}{\dfrac{x^2}{y} - \dfrac{y^2}{x}}$

48. $\dfrac{\dfrac{a^2}{b} + \dfrac{b^2}{a}}{a^2 - ab + b^2}$

49. $\dfrac{\dfrac{x}{y} - \dfrac{y}{x}}{\dfrac{1}{y} + \dfrac{1}{x}}$

50. $\dfrac{\dfrac{a}{b} - \dfrac{b}{a}}{\dfrac{1}{a} - \dfrac{1}{b}}$

51. $\dfrac{x^2 y^{-2} - y^2 x^{-2}}{xy^{-1} + yx^{-1}}$

52. $\dfrac{a^2 b^{-2} - b^2 a^{-2}}{ab^{-1} - ba^{-1}}$

53. $\dfrac{\dfrac{a}{1-a} + \dfrac{1+a}{a}}{\dfrac{1-a}{a} + \dfrac{a}{1+a}}$

54. $\dfrac{\dfrac{1-x}{x} + \dfrac{x}{1+x}}{\dfrac{1+x}{x} + \dfrac{x}{1-x}}$

55. $\dfrac{\dfrac{1}{a^2} + \dfrac{2}{ab} + \dfrac{1}{b^2}}{\dfrac{1}{a^2} - \dfrac{1}{b^2}}$

56. $\dfrac{\dfrac{1}{x^2} - \dfrac{1}{y^2}}{\dfrac{1}{x^2} - \dfrac{2}{xy} + \dfrac{1}{y^2}}$

Simplify.

57. $\dfrac{(x+h)^2 - x^2}{h}$

58. $\dfrac{\dfrac{1}{x+h} - \dfrac{1}{x}}{h}$

59. $\dfrac{(x+h)^3 - x^3}{h}$

60. $\dfrac{\dfrac{1}{(x+h)^2} - \dfrac{1}{x^2}}{h}$

61. $\left[\dfrac{\dfrac{x+1}{x-1} + 1}{\dfrac{x+1}{x-1} - 1}\right]^5$

62. $1 + \dfrac{1}{1 + \dfrac{1}{1 + \dfrac{1}{1 + \dfrac{1}{x}}}}$

Find the error(s) in each of the following. Explain why each is an error. Then find the correct answer.

63.
$$\dfrac{a}{b} \div \left(\dfrac{a}{3} + \dfrac{b}{4}\right) = \dfrac{a}{b} \cdot \left(\dfrac{3}{a} + \dfrac{4}{b}\right) \quad (1)$$
$$= \dfrac{a}{b} \cdot \left(\dfrac{3b + 4a}{ab}\right) \quad (2)$$
$$= \dfrac{a(3b + 4a)}{ab^2} \quad (3)$$
$$= \dfrac{4a + 3b}{b} \quad (4)$$

64.
$$\dfrac{5x}{2y} + \dfrac{3y}{4x} = \dfrac{5x^2}{4xy} + \dfrac{3y^2}{4xy} \quad (1)$$
$$= \dfrac{5x^2 + 3y^2}{4xy} \quad (2)$$
$$= \dfrac{5x + 3y^2}{4y} \quad (3)$$
$$= \dfrac{5x + 3y}{4} \quad (4)$$

1.7

RADICAL NOTATION

A number a is said to be a square root of c if $a^2 = c$. Thus -3 is a square root of 9 because $(-3)^2 = 9$. Similarly, 3 is also a square root of 9 because $3^2 = 9$. A number a is said to be an nth root of c if $a^n = c$. For example, 5 is a third root (called the *cube root*) of 125 because $5^3 = 125$. Any real number has only one real-number cube root.

The symbol $\sqrt{a}$ denotes the nonnegative square root of the number a. The symbol $\sqrt[3]{a}$ denotes the cube root of a, and $\sqrt[4]{a}$ denotes the nonnegative fourth root of a. In general, $\sqrt[n]{a}$ denotes the nth root of a; that is, a number whose nth power is a. The symbol $\sqrt[n]{}$ is called a *radical* and the symbol under the radical is called the *radicand*. The number n (which is omitted when it is 2) is called the *index*.

OBJECTIVES

You should be able to:

i Determine whether a number is a sensible replacement in a radical expression.

ii Simplify radical expressions.

iii Solve applied problems involving radical expressions.

Determine whether the numbers are sensible replacements in the given expression.

1. $\sqrt{x - 2}$; 5, -1

2. $\sqrt{x + 3}$; -7, -3

3. $\sqrt{x^2}$; 18, $-\dfrac{1}{2}$

4. $\sqrt{x^2 + 1}$; -23, 47.2

ODD AND EVEN ROOTS

Any positive real number has two square roots, one positive and one negative. The same is true for fourth roots, or roots of any even index. The positive root is called the *principal* root. When a radical such as $\sqrt{4}$ or $\sqrt[4]{18}$ is used, it is understood to represent the principal (nonnegative) root. To denote a nonpositive root we use $-\sqrt{2}$, $-\sqrt[4]{18}$, and so on.

Definition

> A radical expression $\sqrt[n]{a}$, where *n* is even, represents the principal (nonnegative) *n*th root of *a*. The nonpositive root is denoted $-\sqrt[n]{a}$.

> CAUTION! Again, keep in mind that when the index is an even number E, then $\sqrt[E]{x}$ *never* represents a negative number. For example, $\sqrt{9}$ represents 3, and *not* -3!

■ i SENSIBLE REPLACEMENTS

Since negative numbers do not have even roots in the system of real numbers, any replacement that makes a radicand negative when the index is even is nonsensible.

Every real number, positive, negative, or zero, has just one cube root, and the same is true for any odd root. Thus $\sqrt[n]{a}$, where *n* is odd, represents the (only) *n*th root of *a*. In this case all real numbers are sensible replacements in the radicand.

EXAMPLE 1 Determine whether 0 and 3 are sensible replacements in $\sqrt{5x - 4}$.

We substitute 0 for x in $5x - 4$:

$$5(0) - 4 = 0 - 4 = -4.$$

Since the radicand is negative, 0 is not a sensible replacement. We substitute 3 for x in $5x - 4$:

$$5(3) - 4 = 15 - 4 = 11.$$

Since the radicand is not negative, 3 is a sensible replacement.

DO EXERCISES 1–4.

■ ii SIMPLIFYING RADICAL EXPRESSIONS

Consider the expression $\sqrt{(-3)^2}$. This is equivalent to $\sqrt{9}$, which simplifies to 3. Similarly, $\sqrt{3^2} = 3$. This illustrates an important general principle for simplifying radicals of even index.

Theorem 10

> For any radicand R, $\sqrt{R^2} = |R|$. Similarly, for any even index n, $\sqrt[n]{R^n} = |R|$.

EXAMPLES Simplify.

2. $\sqrt{x^2} = |x|$

3. $\sqrt{x^2 - 2ax + a^2} = \sqrt{(x - a)^2} = |x - a|$

4. $\sqrt{x^2 y^6} = \sqrt{(xy^3)^2} = |xy^3| = y^2|xy|$

If an index is odd, no absolute-value signs are necessary.

Theorem 11

For any radicand R and any odd index n, $\sqrt[n]{R^n} = R$.

DO EXERCISES 5–9.

A second property enables us to multiply radicals. We illustrate it with an example.

EXAMPLE 5   Compare $\sqrt{4} \cdot \sqrt{9}$ and $\sqrt{4 \cdot 9}$.

$\sqrt{4} \cdot \sqrt{9} = 2 \cdot 3 = 6$ and $\sqrt{4 \cdot 9} = \sqrt{36} = 6$, so $\sqrt{4} \cdot \sqrt{9} = \sqrt{4 \cdot 9}$.

Theorem 12

For any nonnegative real numbers a and b and any index n,

$$\sqrt[n]{a} \cdot \sqrt[n]{b} = \sqrt[n]{a \cdot b}.$$

EXAMPLES   Multiply.

6. $\sqrt{3} \cdot \sqrt{5} = \sqrt{3 \cdot 5} = \sqrt{15}$

7. $\sqrt{x+2} \cdot \sqrt{x-2} = \sqrt{(x+2)(x-2)} = \sqrt{x^2 - 4}$

8. $\sqrt[3]{4} \cdot \sqrt[3]{5} = \sqrt[3]{4 \cdot 5} = \sqrt[3]{20}$

DO EXERCISES 10–12.

Theorem 12 also enables us to simplify radical expressions. The idea is to factor the radicand, obtaining factors that are perfect nth powers.

EXAMPLES   Simplify.

9. $\sqrt{50} = \sqrt{25 \cdot 2} = \sqrt{25} \cdot \sqrt{2} = 5\sqrt{2}$

10. $\sqrt{5x^2} = \sqrt{x^2 \cdot 5} = \sqrt{x^2} \cdot \sqrt{5} = |x|\sqrt{5}$

11. $\sqrt[3]{32} = \sqrt[3]{8 \cdot 4} = \sqrt[3]{8} \cdot \sqrt[3]{4} = 2\sqrt[3]{4}$

12. $\sqrt{216x^5y^3} = \sqrt{36 \cdot 6 \cdot x^4 \cdot x \cdot y^2 \cdot y} = |6x^2y|\sqrt{6xy} = 6x^2|y|\sqrt{6xy}$

13. $\sqrt{2x^2 - 4x + 2} = \sqrt{2(x-1)^2} = |x-1|\sqrt{2}$

DO EXERCISES 13–17.

A third fundamental property of radicals is as follows.

Theorem 13

For any nonnegative number a and any positive number b, and any index n,

$$\sqrt[n]{\frac{a}{b}} = \frac{\sqrt[n]{a}}{\sqrt[n]{b}}.$$

This property can be used to divide and to simplify radical expressions.

Simplify.

5. $\sqrt{(x+2)^2}$

6. $\sqrt{x^2(y-2)^2}$

7. $\sqrt[4]{(x+2)^4}$

8. $\sqrt{x^2 + 8x + 16}$

9. $\sqrt[3]{(-4xy)^3}$

Multiply.

10. $\sqrt{19} \cdot \sqrt{7}$

11. $\sqrt{x+2y} \cdot \sqrt{x-2y}$

12. $\sqrt[4]{27} \cdot \sqrt[4]{3}$

Simplify.

13. $\sqrt{300}$

14. $\sqrt{36y^2}$

15. $\sqrt{2x^2 + 4x + 2}$

16. $\sqrt[3]{16}$

17. $\sqrt[3]{(a+b)^4}$

Simplify.

18. $\sqrt{\dfrac{49}{64}}$

19. $\sqrt{\dfrac{25}{y^2}}$

20. $\sqrt{\dfrac{7}{5}}$

21. $\sqrt[3]{\dfrac{7}{125}}$

22. $\dfrac{\sqrt{75}}{\sqrt{3}}$

23. $\dfrac{\sqrt{2x^3}}{\sqrt{50x}}$

24. $\dfrac{\sqrt[3]{24x^3y}}{\sqrt[3]{3y^4}}$

Simplify.

25. $\sqrt[3]{27^{10}}$

26. $(\sqrt{3})^8$

EXAMPLES Simplify.

14. $\sqrt{16x^3y^{-4}} = \sqrt{\dfrac{16x^3}{y^4}} = \dfrac{\sqrt{16x^3}}{\sqrt{y^4}} = \dfrac{\sqrt{16x^2 \cdot x}}{\sqrt{y^4}} = \dfrac{4|x|\sqrt{x}}{y^2}$

15. $\sqrt[3]{\dfrac{27y^5}{343x^3}} = \dfrac{\sqrt[3]{27y^5}}{\sqrt[3]{343x^3}} = \dfrac{\sqrt[3]{27y^3 \cdot y^2}}{\sqrt[3]{343x^3}} = \dfrac{3y\sqrt[3]{y^2}}{7x}$

Fractional expressions are often considered simpler when the denominator is free of radicals. Thus in simplifying, it is usual to remove the radicals in a denominator. This is called *rationalizing the denominator*, and it can be done by multiplying by 1.

EXAMPLES Simplify.

16. $\sqrt{\dfrac{1}{2}} = \sqrt{\dfrac{1}{2} \cdot \dfrac{2}{2}} = \sqrt{\dfrac{2}{4}} = \dfrac{\sqrt{2}}{\sqrt{4}} = \dfrac{\sqrt{2}}{2}$

17. $\sqrt[3]{\dfrac{7}{9}} = \sqrt[3]{\dfrac{7}{9} \cdot \dfrac{3}{3}} = \sqrt[3]{\dfrac{21}{27}} = \dfrac{\sqrt[3]{21}}{\sqrt[3]{27}} = \dfrac{\sqrt[3]{21}}{3}$

EXAMPLES Divide and simplify.

18. $\dfrac{18\sqrt{72}}{6\sqrt{6}} = 3\sqrt{\dfrac{72}{6}} = 3\sqrt{12} = 3\sqrt{4 \cdot 3} = 3 \cdot 2\sqrt{3} = 6\sqrt{3}$

19. $\dfrac{\sqrt[3]{32}}{\sqrt[3]{2}} = \sqrt[3]{\dfrac{32}{2}} = \sqrt[3]{16} = \sqrt[3]{8 \cdot 2} = \sqrt[3]{8} \cdot \sqrt[3]{2} = 2\sqrt[3]{2}$

DO EXERCISES 18–24.

A fourth fundamental principle of radicals involves an exponent under the radical. We illustrate with an example.

EXAMPLE 20 Compare $\sqrt{3^4}$ and $(\sqrt{3})^4$.

$$\sqrt{3^4} = \sqrt{81} = 9,$$
$$(\sqrt{3})^4 = \sqrt{3} \cdot \sqrt{3} \cdot \sqrt{3} \cdot \sqrt{3} = 3 \cdot 3 = 9$$

Thus $\sqrt{3^4} = (\sqrt{3})^4$.

The general principle is given in Theorem 14.

Theorem 14

For any nonnegative number a and any index n and any natural number m,
$$\sqrt[n]{a^m} = (\sqrt[n]{a})^m.$$

Theorem 14 sometimes facilitates radical simplification.

EXAMPLES Simplify.

21. $\sqrt[3]{8^5} = (\sqrt[3]{8})^5 = 2^5 = 32$

22. $(\sqrt{2})^6 = \sqrt{2^6} = 2^3 = 8$

DO EXERCISES 25 AND 26.

iii AN APPLICATION

EXAMPLE 23 (*Speed of a skidding car*). How do police determine the speed of a car that has skidded? The formula

$$r = 2\sqrt{5L}$$

can be used to approximate the speed r, in mph, of a car that has left a skid mark of length L, in feet. What was the speed of a car that left skid marks 306 ft long?

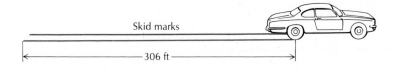

Skid marks

306 ft

We substitute 306 for L in the formula $r = 2\sqrt{5L}$ and use a calculator or Table 1 (at the back of the book) to approximate r:

$$r = 2\sqrt{5(306)} = 2\sqrt{1530}$$
$$\approx 2(39.115) \qquad \text{Using a calculator}$$
$$= 78.23$$

Table 1 contains powers, roots, and reciprocals of various numbers. We may have to simplify before using Table 1. Such is the case in this example:

$$r = 2\sqrt{5(306)} = 2\sqrt{1530}$$
$$= 2\sqrt{9 \cdot 170} \qquad \text{Simplifying}$$
$$= 2(3)\sqrt{170} \qquad \text{Taking the square root of 9}$$
$$\approx 6(13.038) \qquad \text{Using Table 1}$$
$$= 78.228$$

The speed of the car was about 78.23 mph.

DO EXERCISE 27.

27. What was the speed of a car that left skid marks 70 ft long?

EXERCISE SET 1.7

i Determine whether the specified numbers are sensible replacements in each expression.

1. $\sqrt{x-3}$; $-2, 5$ **2.** $\sqrt{2x-5}$; $3, 2$ **3.** $\sqrt{3-4x}$; $-1, 1$ **4.** $\sqrt{x^2+3}$; $0, 4.3$

5. $\sqrt{1-x^2}$; $1, 3$ **6.** $\sqrt{x^2+2x+1}$; $-3, 4$ **7.** $\sqrt[3]{2x+7}$; $-4, 5$ **8.** $\sqrt[4]{3-5x}$; $1, 2$

ii Simplify.

9. $\sqrt{(-11)^2}$ **10.** $\sqrt{(-1)^2}$ **11.** $\sqrt{16x^2}$ **12.** $\sqrt{36t^2}$

13. $\sqrt{(b+1)^2}$ **14.** $\sqrt{(2c-3)^2}$ **15.** $\sqrt[3]{-27x^3}$ **16.** $\sqrt[3]{-8y^3}$

17. $\sqrt{x^2-4x+4}$ **18.** $\sqrt{y^2+16y+64}$ **19.** $\sqrt[5]{32}$ **20.** $\sqrt[5]{-32}$

21. $\sqrt{180}$ **22.** $\sqrt{48}$ **23.** $\sqrt[3]{54}$ **24.** $\sqrt[3]{135}$

25. $\sqrt{128c^2d^{-4}}$ **26.** $\sqrt{162c^4d^{-6}}$ **27.** $\sqrt{3} \cdot \sqrt{6}$ **28.** $\sqrt{6} \cdot \sqrt{8}$

In the following exercises simplify, assuming that all letters represent positive numbers and that all radicands are positive. Thus no absolute-value signs will be needed.

29. $\sqrt{2x^3y}\sqrt{12xy}$ **30.** $\sqrt{3y^4z}\sqrt{20z}$ **31.** $\sqrt[3]{3x^2y}\sqrt[3]{36x}$ **32.** $\sqrt[5]{8x^3y^4}\sqrt[5]{4x^4y}$

33. $\sqrt[3]{2(x+4)}\,\sqrt[3]{4(x+4)^4}$

34. $\sqrt[3]{4(x+1)^2}\,\sqrt[3]{18(x+1)^2}$

35. $\dfrac{\sqrt{21ab^2}}{\sqrt{3ab}}$

36. $\dfrac{\sqrt{128ab^2}}{\sqrt{16a^2b}}$

37. $\dfrac{\sqrt[3]{40m}}{\sqrt[3]{5m}}$

38. $\dfrac{\sqrt{40xy}}{\sqrt{8x}}$

39. $\dfrac{\sqrt[3]{3x^2}}{\sqrt[3]{24x^5}}$

40. $\dfrac{\sqrt[3]{40xy^3}}{\sqrt[3]{8x}}$

41. $\dfrac{\sqrt{a^2-b^2}}{\sqrt{a-b}}$

42. $\dfrac{\sqrt{x^3-y^3}}{\sqrt{x-y}}$

43. $\sqrt{\dfrac{9a^2}{8b}}$

44. $\sqrt{\dfrac{5b^2}{12a}}$

45. $\sqrt[3]{\dfrac{2x^2 2y^3}{25z^4}}$

46. $\sqrt[3]{\dfrac{24x^3y}{3y^4}}$

47. $\dfrac{(\sqrt[3]{32x^4y})^2}{(\sqrt[3]{xy})^2}$

48. $\dfrac{(\sqrt[3]{16x^2y})^2}{(\sqrt[3]{xy})^2}$

49. $\dfrac{3\sqrt{a^2b^2}\,\sqrt{4xy}}{2\sqrt{a^{-1}b^{-2}}\,\sqrt{9x^{-3}y^{-1}}}$

50. $\dfrac{4\sqrt{xy^2}\,\sqrt{9ab}}{3\sqrt{x^{-1}y^{-2}}\,\sqrt{16a^{-5}b^{-1}}}$

iii

51. ▦ (*Pendulums*). The period T of a pendulum is the time it takes to make a move from one side to the other and back. A formula for the period is

$$T = 2\pi\sqrt{\frac{L}{32}},$$

where T is in seconds and L, the length of the pendulum, is in feet. Find the periods of pendulums of lengths 2 ft, 8 ft, 64 ft, and 100 ft. Use 3.14 for π.

52. A slow-pitch softball diamond is actually a square 65 ft on a side. How far is it from home to second base? (*Hint:* By the Pythagorean theorem, in right triangles $c^2 = a^2 + b^2$ and $c = \sqrt{a^2 + b^2}$, where a and b are the lengths of the legs and c is the length of the hypotenuse.)

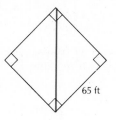

65 ft

Simplify, assuming that all letters represent positive numbers.

53. ▦ $\sqrt{8.2x^3y}\,\sqrt{12.5xy}$

54. ▦ $\sqrt{0.012y^4z}\,\sqrt{1.305z}$

55. ▦ $\sqrt{\dfrac{6.03a^2}{17.13b}}$

56. ▦ $\sqrt{\dfrac{3.2b^2}{82.1a}}$

An *equilateral* triangle is shown below.

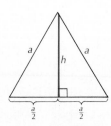

57. Find an expression for its height h in terms of a.

58. Find an expression for its area A in terms of a.

59. Figure *ABCD* is a square. Find the length of $\overline{AC}$.

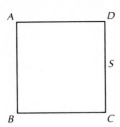

60. An isosceles right triangle has two sides of length *s*. Find a formula for the length of the third side.

61. The diagonal of a square has length $8\sqrt{2}$. Find the length of a side of the square.

62. The area of square *PQRS* is 100 ft^2 (square feet), and *A*, *B*, *C*, and *D* are the midpoints of the sides. Find the area of square *ABCD*.

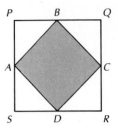

63. For what integer values of *n* is $|a^n| = |a|^n$ for all *a*?

64. For what values of *x* and *y* is $|x + y| = |x| + |y|$?

1.8

FURTHER CALCULATIONS WITH RADICAL NOTATION

i FURTHER SIMPLIFYING

Various calculations with radicals can be carried out using the properties of radicals and the properties of numbers, such as the distributive property. The following examples illustrate.

EXAMPLES Simplify.

1. $3\sqrt{8} - 5\sqrt{2} = 3\sqrt{4 \cdot 2} - 5\sqrt{2} = 3 \cdot 2\sqrt{2} - 5\sqrt{2}$

$\qquad = (6 - 5)\sqrt{2}$ Here we used a distributive law.

$\qquad = \sqrt{2}$

2. $(4\sqrt{3} + \sqrt{2})(\sqrt{3} - 5\sqrt{2}) = 4(\sqrt{3})^2 - 20\sqrt{3}\sqrt{2} + \sqrt{2}\sqrt{3} - 5(\sqrt{2})^2$

$\qquad = 4 \cdot 3 - 20\sqrt{6} + \sqrt{6} - 5 \cdot 2$

$\qquad = 12 - 19\sqrt{6} - 10$

$\qquad = 2 - 19\sqrt{6}$

DO EXERCISES 1–3.

ii RATIONALIZING DENOMINATORS OR NUMERATORS

When a fractional symbol contains radicals, we ordinarily rationalize the denominator, but on occasion we prefer to rationalize the numerator. In either case, we can accomplish the rationalization by multiplying by 1, as in the following examples.

OBJECTIVES

You should be able to:

i Simplify expressions containing radicals, using the distributive property and properties of radicals.

ii Rationalize numerators and denominators.

Simplify.

1. $7\sqrt{5} + 3\sqrt{5} - 8\sqrt{20}$

2. $5\sqrt[3]{16y^4} + 7\sqrt[3]{2y}$

3. $(\sqrt{3} - 5\sqrt{2})(2\sqrt{3} + \sqrt{2})$

EXAMPLES Rationalize the denominator.

3. $\dfrac{\sqrt{7}}{\sqrt{5}} = \dfrac{\sqrt{7}}{\sqrt{5}} \cdot \dfrac{\sqrt{5}}{\sqrt{5}} = \dfrac{\sqrt{35}}{\sqrt{25}} = \dfrac{\sqrt{35}}{5}$

4. $\dfrac{\sqrt{2a}}{\sqrt{5b}} = \dfrac{\sqrt{2a}}{\sqrt{5b}} \cdot \dfrac{\sqrt{5b}}{\sqrt{5b}} = \dfrac{\sqrt{10ab}}{\sqrt{(5b)^2}}$

$\qquad = \dfrac{\sqrt{10ab}}{|5b|} = \dfrac{\sqrt{10ab}}{5b}$ 　The absolute-value sign in the denominator is not necessary since $\sqrt{5b}$ would not exist at the outset unless $b > 0$.

5. $\dfrac{\sqrt[3]{54x^3}}{\sqrt[3]{4y^5}} = \sqrt[3]{\dfrac{54x^3}{4y^5} \cdot \dfrac{2y}{2y}}$

$\qquad = \sqrt[3]{\dfrac{54x^3 \cdot 2y}{8y^6}}$

$\qquad = \dfrac{\sqrt[3]{27x^3} \cdot \sqrt[3]{4y}}{\sqrt[3]{8y^6}}$

$\qquad = \dfrac{3x \cdot \sqrt[3]{4y}}{2y^2}$

When a numerator or denominator to be rationalized has two terms, we choose the symbol for 1 a little differently. The symbol for 1 will have two terms in its numerator and denominator. The following examples illustrate.

EXAMPLES Rationalize the denominator. Assume that all letters represent positive numbers.

6. $\dfrac{1}{\sqrt{2} + \sqrt{3}} = \dfrac{1}{\sqrt{2} + \sqrt{3}} \cdot \dfrac{\sqrt{2} - \sqrt{3}}{\sqrt{2} - \sqrt{3}}$

$\qquad = \dfrac{\sqrt{2} - \sqrt{3}}{(\sqrt{2} + \sqrt{3})(\sqrt{2} - \sqrt{3})}$

$\qquad = \dfrac{\sqrt{2} - \sqrt{3}}{(\sqrt{2})^2 - (\sqrt{3})^2}$

$\qquad = \dfrac{\sqrt{2} - \sqrt{3}}{2 - 3}$

$\qquad = \dfrac{\sqrt{2} - \sqrt{3}}{-1}$

$\qquad = \sqrt{3} - \sqrt{2}$

7. $\dfrac{\sqrt{x} + \sqrt{y}}{\sqrt{x} - \sqrt{y}} = \dfrac{\sqrt{x} + \sqrt{y}}{\sqrt{x} - \sqrt{y}} \cdot \dfrac{\sqrt{x} + \sqrt{y}}{\sqrt{x} + \sqrt{y}}$

$\qquad = \dfrac{(\sqrt{x} + \sqrt{y})^2}{(\sqrt{x})^2 - (\sqrt{y})^2}$

$\qquad = \dfrac{x + 2\sqrt{x}\sqrt{y} + y}{x - y}$

EXAMPLES Rationalize the numerator. Assume that all letters represent positive numbers.

8. $\dfrac{1 - \sqrt{2}}{5} = \dfrac{1 - \sqrt{2}}{5} \cdot \dfrac{1 + \sqrt{2}}{1 + \sqrt{2}}$

$= \dfrac{(1 - \sqrt{2})(1 + \sqrt{2})}{5(1 + \sqrt{2})}$

$= \dfrac{1 - 2}{5(1 + \sqrt{2})}$

$= \dfrac{-1}{5 + 5\sqrt{2}}$

9. $\dfrac{\sqrt{x + h} - \sqrt{x}}{h} = \dfrac{\sqrt{x + h} - \sqrt{x}}{h} \cdot \dfrac{\sqrt{x + h} + \sqrt{x}}{\sqrt{x + h} + \sqrt{x}}$

$= \dfrac{(x + h) - x}{h(\sqrt{x + h} + \sqrt{x})}$

$= \dfrac{h}{h(\sqrt{x + h} + \sqrt{x})}$

$= \dfrac{1}{\sqrt{x + h} + \sqrt{x}}$

DO EXERCISES 4–7.

Rationalize the denominator. Assume that all letters represent positive numbers.

4. $\dfrac{1}{\sqrt{3} - \sqrt{5}}$

5. $\dfrac{\sqrt{x} - 5}{\sqrt{x} + 2}$

Rationalize the numerator. Assume that all letters represent positive numbers.

6. $\dfrac{\sqrt{a + 2} - \sqrt{a}}{2}$

7. $\dfrac{\sqrt{x} - \sqrt{5}}{\sqrt{x} + \sqrt{5}}$

EXERCISE SET 1.8

In this exercise set, assume that all letters represent positive numbers and all radicands are positive. Thus, absolute-value signs will not be necessary.

i Simplify.

1. $8\sqrt{2} - 6\sqrt{20} - 5\sqrt{8}$

2. $\sqrt{12} - \sqrt{27} + \sqrt{75}$

3. $2\sqrt[3]{8x^2} + 5\sqrt[3]{27x^2} - 3\sqrt[3]{x^3}$

4. $5a\sqrt{(a + b)^3} - 2ab\sqrt{a + b} - 3b\sqrt{(a + b)^3}$

5. $3\sqrt{3y^2} - \dfrac{y\sqrt{48}}{\sqrt{2}} + \sqrt{\dfrac{12}{4y^{-2}}}$

6. $\sqrt[3]{x^5} - \dfrac{2\sqrt[3]{x}}{\sqrt[3]{x^{-1}}} + \sqrt[3]{\dfrac{8}{x^{-5}}}$

7. $(\sqrt{3} - \sqrt{2})(\sqrt{3} + \sqrt{2})$

8. $(\sqrt{8} + 2\sqrt{5})(\sqrt{8} - 2\sqrt{5})$

9. $(\sqrt{t} - x)^2$

10. $\left(\sqrt{a} + \dfrac{1}{\sqrt{a}}\right)^2$

11. $5\sqrt{7} + \dfrac{35}{\sqrt{7}}$

12. $(\sqrt{a^2b} + 3\sqrt{y})(2a\sqrt{b} - \sqrt{y})$

13. $(\sqrt{x + 3} - \sqrt{3})(\sqrt{x + 3} + \sqrt{3})$

14. $(\sqrt{x + h} - \sqrt{x})(\sqrt{x + h} + \sqrt{x})$

ii Rationalize the denominator.

15. $\dfrac{6}{3 + \sqrt{5}}$

16. $\dfrac{2}{\sqrt{3} - 1}$

17. $\sqrt[3]{\dfrac{16}{9}}$

18. $\dfrac{\sqrt[3]{3}}{\sqrt[3]{6}}$

19. $\dfrac{4\sqrt{x} - 3\sqrt{xy}}{2\sqrt{x} + 5\sqrt{y}}$

20. $\dfrac{5\sqrt{x} + 2\sqrt{xy}}{3\sqrt{x} - 2\sqrt{y}}$

ii   Rationalize the numerator.

21. $\dfrac{\sqrt{2} + \sqrt{5a}}{6}$

22. $\dfrac{\sqrt{3} + \sqrt{5y}}{4}$

23. $\dfrac{\sqrt{x+1} + 1}{\sqrt{x+1} - 1}$

24. $\dfrac{\sqrt{x+4} - 2}{\sqrt{x+4} + 2}$

25. $\dfrac{\sqrt{a+3} - \sqrt{3}}{3}$

26. $\dfrac{\sqrt{a+h} - \sqrt{a}}{h}$

Simplify.

27. $\sqrt{1 + x^2} + \dfrac{1}{\sqrt{1 + x^2}}$

28. $\sqrt{1 - x^2} - \dfrac{x^2}{2\sqrt{1 - x^2}}$

29. Show that $\sqrt{a + b} = \sqrt{a} + \sqrt{b}$ is false for positive real numbers a and b by finding two positive numbers a and b for which $\sqrt{a + b} \neq \sqrt{a} + \sqrt{b}$.

30. Show that $(\sqrt{5 + \sqrt{24}})^2 = (\sqrt{2} + \sqrt{3})^2$.

Prove the following.

31. For any positive real numbers a and b, there exist positive numbers c and d for which

$$\sqrt{a + b} = \sqrt{c} + \sqrt{d}.$$

Under what conditions does $a = c$ and $b = d$?

OBJECTIVES

You should be able to:

i   Convert from exponential notation to radical notation.

ii   Convert from radical notation to exponential notation.

iii   Simplify expressions using the properties of rational exponents and the arithmetic of rational numbers.

iv   Simplify certain expressions to expressions containing a single radical.

1.9

RATIONAL EXPONENTS

We are motivated to define fractional exponents so that the same rules, or laws, hold for them as for integer exponents. For example, if the laws of exponents are to hold, we must have

$$a^{1/2} \cdot a^{1/2} = a^{1/2 + 1/2} = a^1 = a.$$

Thus we are led to define $a^{1/2}$ to mean $\sqrt{a}$. Similarly, $a^{1/n}$ would mean $\sqrt[n]{a}$. Again, if the usual laws of exponents are to hold, we must have

$$(a^{1/n})^m = (a^m)^{1/n} = a^{m/n}.$$

Thus we are led to define $a^{m/n}$ to mean $(\sqrt[n]{a})^m$ or $\sqrt[n]{a^m}$.

Definition

> An expression $a^{m/n}$, where a is positive and m and n are natural numbers, is defined to mean $(\sqrt[n]{a})^m$ or $\sqrt[n]{a^m}$. An expression $a^{-m/n}$ is defined to mean $1/a^{m/n}$.

Note that in this definition we require a to be positive. Thus in manipulations with fractional exponents we assume that all letters represent positive numbers and that all radicands are positive. No absolute-value signs need be used.

Once the definition of rational exponents is made, the question arises whether the usual laws of exponents actually do hold. We shall not prove it, but the answer is that they do. Thus we can simplify or otherwise manipulate expressions containing rational exponents using those laws and the usual arithmetic of rational numbers.

i ■ CONVERTING TO RADICAL NOTATION

EXAMPLES Convert to radical notation and simplify, if possible.

1. $m^{2/3} = \sqrt[3]{m^2}$

2. $t^{-1/2} = \dfrac{1}{t^{1/2}} = \dfrac{1}{\sqrt{t}}$

3. $64^{5/2} = (64^{1/2})^5 = (\sqrt{64})^5 = 8^5 = 32{,}768$

4. $16^{-3/4} = \dfrac{1}{16^{3/4}} = \dfrac{1}{(16^{1/4})^3} = \dfrac{1}{(\sqrt[4]{16})^3} = \dfrac{1}{2^3} = \dfrac{1}{8}$

DO EXERCISES 1–4.

ii ■ CONVERTING TO EXPONENTIAL NOTATION

EXAMPLES Convert to exponential notation and simplify.

5. $(\sqrt[4]{7xy})^5 = (7xy)^{5/4}$

6. $\sqrt[3]{8^4} = 8^{4/3} = (8^{1/3})^4 = 2^4 = 16$

7. $\sqrt[6]{x^3} = x^{3/6} = x^{1/2}$ (or $\sqrt{x}$)

8. $\sqrt[6]{4} = 4^{1/6} = (4^{1/2})^{1/3} = 2^{1/3}$ (or $\sqrt[3]{2}$)

9. $\sqrt[3]{\sqrt{7}} = \sqrt[3]{7^{1/2}} = (7^{1/2})^{1/3} = 7^{1/6}$ (or $\sqrt[6]{7}$)

10. $\sqrt{6}\,\sqrt[3]{6} = 6^{1/2} \cdot 6^{1/3} = 6^{1/2+1/3} = 6^{5/6}$ (or $\sqrt[6]{6^5}$)

DO EXERCISES 5–9.

iii ■ SIMPLIFYING EXPRESSIONS WITH RATIONAL EXPONENTS

EXAMPLES Simplify and then write radical notation.

11. $x^{5/6} \cdot x^{2/3} = x^{5/6+2/3}$ Adding exponents
$= x^{9/6} = x^{3/2} = \sqrt{x^3}$
$= x\sqrt{x}$

12. $(a^5)^{-2/3} = a^{-10/3}$ Multiplying exponents
$= \dfrac{1}{a^{10/3}} = \dfrac{1}{\sqrt[3]{a^{10}}}$
$= \dfrac{1}{a^3 \cdot \sqrt[3]{a}}$

13. $(5^{1/3} - 5^{-5/3}) \cdot 5^{1/3} = 5^{1/3} \cdot 5^{1/3} - 5^{-5/3} \cdot 5^{1/3}$ Using a distributive law
$= 5^{2/3} - 5^{-4/3}$
$= \sqrt[3]{5^2} - \dfrac{1}{\sqrt[3]{5^4}}$
$= \sqrt[3]{25} - \dfrac{1}{5\sqrt[3]{5}}$

DO EXERCISES 10–12.

iv ■ WRITING SINGLE RADICALS

In certain expressions containing radicals or fractional exponents, it is possible to simplify in such a way that there is a single radical.

Convert to radical notation and simplify, if possible.

1. $n^{3/2}$

2. $y^{-6/7}$

3. $32^{4/5}$

4. $64^{-2/3}$

Convert to exponential notation and simplify.

5. $\sqrt[3]{(5ab)^4}$

6. $\sqrt[4]{16^3}$

7. $\sqrt[6]{a^4}$

8. $\sqrt{\sqrt[3]{4}}$

9. $\sqrt{5^3}\,\sqrt[3]{5}$

Simplify and then write radical notation.

10. $a^{3/4} \cdot a^{1/2}$

11. $(x^{-3})^{2/5}$

12. $(2^{1/4} + 2^{-3/4}) \cdot 2^{1/2}$

Write an expression containing a single radical.

13. $\sqrt[3]{5} \cdot \sqrt{2}$

14. $x^{2/3} y^{1/2} z^{5/6}$

15. $\dfrac{\sqrt[4]{(x+y)^3}}{\sqrt{x+y}}$

EXAMPLES Write an expression containing a single radical.

14. $a^{1/2} b^{-1/2} c^{5/6} = a^{3/6} b^{-3/6} c^{5/6}$
$$= (a^3 b^{-3} c^5)^{1/6} = \sqrt[6]{a^3 b^{-3} c^5}$$

15. $\dfrac{a^{1/4} b^{3/8}}{a^{1/2} b^{1/8}} = a^{-1/4} b^{1/4}$
$$= (a^{-1} b)^{1/4} = \sqrt[4]{a^{-1} b}$$

16. $\sqrt[4]{7} \sqrt{3} = 7^{1/4} \cdot 3^{1/2} = 7^{1/4} \cdot 3^{2/4}$
$$= (7 \cdot 3^2)^{1/4} = \sqrt[4]{63}$$

17. $\dfrac{\sqrt[4]{(x+2)^3} \sqrt[5]{x+2}}{\sqrt{x+2}} = \dfrac{(x+2)^{3/4} (x+2)^{1/5}}{(x+2)^{1/2}}$
$$= (x+2)^{3/4 + 1/5 - 1/2}$$
$$= (x+2)^{9/20}$$
$$= \sqrt[20]{(x+2)^9}$$

DO EXERCISES 13–15.

EXERCISE SET 1.9

i Convert to radical notation and simplify.

1. $x^{3/4}$ **2.** $y^{2/5}$ **3.** $16^{3/4}$ **4.** $4^{7/2}$

5. $125^{-1/3}$ **6.** $32^{-4/5}$ **7.** $a^{5/4} b^{-3/4}$ **8.** $x^{2/5} y^{-1/5}$

ii Convert to exponential notation and simplify.

9. $\sqrt[3]{20^2}$ **10.** $\sqrt[5]{17^3}$ **11.** $(\sqrt[4]{13})^5$ **12.** $(\sqrt[5]{12})^4$

13. $\sqrt[3]{\sqrt{11}}$ **14.** $\sqrt[3]{\sqrt[4]{7}}$ **15.** $\sqrt{5} \sqrt[3]{5}$ **16.** $\sqrt[3]{2} \sqrt{2}$

17. $\sqrt[5]{32^2}$ **18.** $\sqrt[3]{64^{-2}}$ **19.** $\sqrt[3]{8y^6}$ **20.** $\sqrt[5]{32 c^{10} d^{15}}$

21. $\sqrt[3]{a^2 + b^2}$ **22.** $\sqrt[4]{a^3 - b^3}$ **23.** $\sqrt[3]{27 a^3 b^9}$ **24.** $\sqrt[4]{81 x^8 y^8}$

25. $\sqrt[6]{\dfrac{m^{12} n^{24}}{64}}$ **26.** $\sqrt[8]{\dfrac{m^{16} n^{24}}{2^8}}$

iii Simplify and then write radical notation, unless inappropriate.

27. $(2a^{3/2})(4a^{1/2})$ **28.** $(3a^{5/6})(8a^{2/3})$ **29.** $\left(\dfrac{x^6}{9b^{-4}}\right)^{-1/2}$

30. $\left(\dfrac{x^{2/3}}{4y^{-2}}\right)^{-1/2}$ **31.** $\dfrac{x^{2/3} y^{5/6}}{x^{-1/3} y^{1/2}}$ **32.** $\dfrac{a^{1/2} b^{5/8}}{a^{1/4} b^{3/8}}$

iv Write an expression containing a single radical and simplify.

33. $\sqrt[3]{6} \sqrt{2}$ **34.** $\sqrt{2} \sqrt[4]{8}$

35. $\sqrt[4]{xy} \sqrt[3]{x^2 y}$ **36.** $\sqrt[3]{ab^2} \sqrt{ab}$

37. $\sqrt[3]{a^4} \sqrt{a^3}$ **38.** $\sqrt{a^3} \sqrt[3]{a^2}$

39. $\dfrac{\sqrt{(a+x)^3} \sqrt[3]{(a+x)^2}}{\sqrt[4]{a+x}}$ **40.** $\dfrac{\sqrt[4]{(x+y)^2} \sqrt[3]{(x+y)}}{\sqrt{(x+y)^3}}$

Simplify. (*Note:* Since $x^{1/4} = (x^{1/2})^{1/2}$ you can take a fourth root by taking a square root, and then the square root of the result. Or you can find decimal notation for the exponent, obtaining $x^{0.25}$, and use the power key, x^y. Remember, answers can vary depending on the type and readout of your calculator.) Round to three decimal places.

41. ▦ $(\sqrt[4]{13})^5$ **42.** ▦ $\sqrt[4]{17^3}$ **43.** ▦ $12.3^{3/2}$

44. ▦ $1.345^{5/2}$ **45.** ▦ $105.6^{3/4}$ **46.** ▦ $7.14^{5/4}$

In a psychological study, pavement signs were found to be most readable by a driver when the letters in the sign are of length L, given by

$$L = \frac{0.000169 d^{2.27}}{h},$$

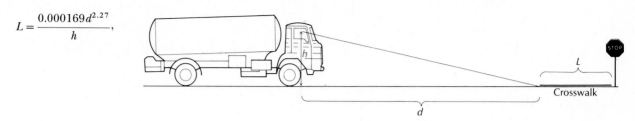

where d is the distance from the car to the lettering and h is the height of the eye above the road. All units are in feet. Find L, given the values of h and d.

47. ▦ $h = 4$ ft, $d = 180$ ft   **48.** ▦ $h = 4$ ft, $d = 100$ ft   **49.** ▦ $h = 4$ ft, $d = 200$ ft   **50.** ▦ $h = 4$ ft, $d = 300$ ft

Simplify.

51. $(\sqrt{a^{\sqrt{a}}})^{\sqrt{a}}$

52. $(2a^3 b^{5/4} c^{1/7})^4 \div (54a^{-2} b^{2/3} c^{6/5})^{-1/3}$

SUMMARY AND REVIEW: CHAPTER 1

The following contains a summary of what you should be able to do after completing this chapter. The review exercises are for practice. Answers are at the back of the book. If you miss an exercise, restudy the section indicated alongside the answer.

You should be able to:

Identify various kinds of real numbers.

Consider the numbers

$$-43.89, \quad 12, \quad -3, \quad -\frac{1}{5}, \quad \sqrt{7}, \quad \sqrt[3]{10}, \quad -1, \quad -\frac{4}{3}, \quad 7\frac{2}{3}, \quad -19, \quad 31, \quad 0.$$

1. Which are integers?   **2.** Which are natural numbers?   **3.** Which are rational numbers?

4. Which are real numbers?   **5.** Which are irrational numbers?   **6.** Which are whole numbers?

Compute with real numbers.

Compute.

7. $15 + (-19)$ **8.** $-12 + |-4|$ **9.** $-2.5 + (-2.5)$

10. $22 - (-8)$ **11.** $\dfrac{18}{-3}$ **12.** $(-17)(-9)$

13. $-10(20)(-5)(-3)$ **14.** $-\dfrac{15}{16} + \dfrac{3}{4}$ **15.** $\dfrac{5}{12} - \left(-\dfrac{7}{8}\right)$

Convert from decimal notation to scientific notation, and vice versa.

Convert to decimal notation.

16. 3.261×10^6 **17.** 4.1×10^{-4}

Convert to scientific notation.

18. 0.01432 **19.** $43,210$

Simplify an algebraic expression.

Simplify.

20. $(7a^2 b^4)(-2a^{-4} b^3)$ **21.** $\dfrac{54x^6 y^{-4} z^2}{9x^{-3} y^2 z^{-4}}$ **22.** $\sqrt[4]{81}$ **23.** $\sqrt[5]{-32}$

24. $\dfrac{b - a^{-1}}{a - b^{-1}}$

25. $\dfrac{\dfrac{x^2}{y} + \dfrac{y^2}{x}}{y^2 - xy + x^2}$

26. $(\sqrt{3} - \sqrt{7})(\sqrt{3} + \sqrt{7})$

27. $(5x^2 - \sqrt{2})^2$

28. $8\sqrt{5} + \dfrac{25}{\sqrt{5}}$

29. $(x + t)(x^2 - xt + t^2)$

30. $(5a + 4b)^3$

31. $(5xy^4 - 7xy^2 + 4x^2 - 3) - (-3xy^4 + 2xy^2 - 2y + 4)$

Factor a polynomial completely.

Factor.

32. $x^3 + 2x^2 - 3x - 6$

33. $12a^3 - 27ab^4$

34. $24x + 144 + x^2$

35. $9x^3 + 35x^2 - 4x$

36. $8x^3 - 1$

37. $27x^6 + 125y^6$

Simplify certain expressions to expressions containing a single radical.

Write an expression containing a single radical.

38. $\sqrt{y^5} \sqrt[3]{y^2}$

39. $\dfrac{\sqrt{(a + b)^3} \sqrt[3]{a + b}}{\sqrt[6]{(a + b)^7}}$

Convert from exponential notation to radical notation, and vice versa.

40. Convert to radical notation.

$$b^{7/5}$$

41. Convert to exponential notation and simplify.

$$\sqrt[8]{\dfrac{m^{32}n^{16}}{3^8}}$$

Add, subtract, multiply, and divide with fractional expressions.

42. Divide and simplify.

$$\dfrac{3x^2 - 12}{x^2 + 4x + 4} \div \dfrac{x - 2}{x + 2}$$

43. Subtract and simplify.

$$\dfrac{x}{x^2 + 9x + 20} - \dfrac{4}{x^2 + 7x + 12}$$

Rationalize the denominator of an algebraic expression.

44. Rationalize the denominator: $\dfrac{\sqrt{x} - \sqrt{y}}{\sqrt{x} + \sqrt{y}}$

What property is illustrated by each sentence?

45. $t + (-t) = 0$

46. $8(a + b) = 8a + 8b$

47. $-3(ab) = (-3a)b$

48. $tx = xt$

Multiply. Assume that all exponents are integers.

49. $(x^n + 10)(x^n - 4)$

50. $(t^a + t^{-a})^2$

51. $(y^b - z^c)(y^b + z^c)$

52. $(a^n - b^m)^3$

Factor.

53. $y^{2n} + 16y^n + 64$

54. $x^{2t} - 3x^t - 28$

55. $m^{6n} - m^{3n}$

TEST: CHAPTER 1

Consider the numbers

$$-14, \quad 23.77, \quad -5, \quad -\dfrac{4}{7}, \quad \sqrt{8}, \quad -\sqrt[3]{11}, \quad 56\dfrac{1}{4}, \quad \dfrac{9}{2}, \quad 0, \quad 233.$$

1. Which are whole numbers?

2. Which are irrational numbers?

3. Which are real numbers?

4. Which are rational numbers?

5. Which are natural numbers?

6. Which are integers?

Compute.

7. $-7 + |-7|$

8. $-7.4 + 9.4$

9. $(-6)(-2)$

10. $3 - (-5)$

11. $\dfrac{15}{-5}$

Convert to decimal notation.

12. 2.834×10^{-3}

13. 4.7×10^2

Convert to scientific notation.

14. 0.000816

15. 480.57

Simplify.

16. $(4x^4y^{-2})(-3x^{-5}y^{-4})$

17. $\dfrac{24pq^8r^{-4}}{36p^{-3}q^{-6}r^5}$

18. $\sqrt[3]{-27}$

19. $\sqrt[4]{625}$

20. $(\sqrt{8} - \sqrt{2})(\sqrt{8} + \sqrt{2})$

21. $\dfrac{\dfrac{x}{y^2} + \dfrac{y}{x^2}}{\dfrac{y^2 - yx + x^2}{y}}$

22. $(5a^2 - 2b)^2$

23. $(2x^2y - 3xy + y^2 - 4) - (-6x^2y + 2xy - y + 8)$

24. $(4y - 3)^3$

25. Write an expression containing a single radical.

$$\dfrac{\sqrt[4]{(c + d)^3} \cdot \sqrt{c + d}}{\sqrt[3]{(c + d)^4}}$$

26. Convert to radical notation.

$$t^{2/7}$$

Factor.

27. $16x^2 + 48x + 36$

28. $t^3 - 343$

29. $m^5 - 9m^3n^2$

30. $12p^4 + 9p^2 - 30$

31. Divide and simplify.

$$\dfrac{x^2 - 10x + 25}{x^2 + 10x + 25} \div \dfrac{x^2 - 25}{(x + 5)^3}$$

32. Subtract and simplify.

$$\dfrac{x}{x^2 - 4x - 12} - \dfrac{9}{x^2 - 36}$$

33. Rationalize the denominator.

$$\dfrac{7 - \sqrt{x}}{7 + \sqrt{x}}$$

34. Multiply.

$$(x^t + x^{-t})^3$$

35. Factor: $x^{16} - 16$.

How long will it take an object dropped from the top of the arch to reach the ground?

EQUATIONS, INEQUALITIES, AND PROBLEM SOLVING

In this chapter, we continue the review that was begun in Chapter 1. In particular, we review equations and how to solve them. We also begin to consider the payoff that results from the ability to solve equations— *problem solving*.

Problem solving is important in mathematics, as it is in life in general. From this introduction to problem solving you will go on, in later chapters, to consider problem solving again and again. The most important thing for you to learn here is that you must spend a good deal of time *thinking* about a problem situation, experimenting with it, making guesses, and so on. That is what will best enable you to *formulate* a problem mathematically, the hardest part of solving problems.

In this chapter you will also find an introduction to dimension symbols, such as ft, lb, hr, and mathematical ways to handle them.

2

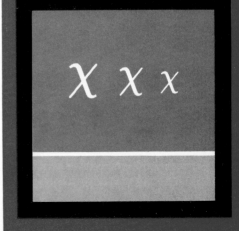

OBJECTIVES

You should be able to:

i Solve simple equations using the addition and multiplication principles and the principle of zero products.

ii Solve simple inequalities, and write set-builder notation for a set. Use inequalities to find sensible replacements in radical expressions.

Solve (find the solution set) by trial and error.

1. $x - 2 = 7$

2. $y^2 + y = 0$

2.1

SOLVING EQUATIONS AND INEQUALITIES

i SOLUTIONS AND SOLUTION SETS

Definition

A *solution* of an equation is any number that makes the equation true when that number is substituted for the variable.

The number 3 is a solution of $5x = 15$ because $5(3) = 15$ is true. The number -4 is not a solution of $5x = 15$ because $5(-4) = 15$ is false.

Definition

The set of all solutions of an equation is called its *solution set*. When we find all the solutions of an equation (find its solution set), we say that we have solved it.

The number 3 is the only solution of $5x = 15$, so the solution set consists of the number 3 and is denoted $\{3\}$. As another example consider

$$y^2 - y = 0.$$

The number 0 is a solution, and so is the number 1. There are no other solutions, so the solution set is $\{0, 1\}$.

DO EXERCISES 1 AND 2.

EQUATION-SOLVING PRINCIPLES

The Addition and Multiplication Principles

Two simple principles* allow us to solve many equations. The first of these is as follows:

The addition principle

For any real numbers a, b, and c, if an equation $a = b$ is true, then $a + c = b + c$ is true.

The second principle is similar to the first.

The multiplication principle

For any real numbers a, b, and c, if an equation $a = b$ is true, then $ac = bc$ is true.

Note that these principles also cover "subtracting on both sides" and "dividing on both sides" because subtracting c can be accomplished by

---

*These "principles" are actually very easy theorems. Suppose $a = b$ is true. Then a and b are the same number. If we add c to this number, the result is $a + c$. It is also $b + c$.

adding $-c$, and dividing by c can be accomplished by multiplying by $1/c$, provided that $c \neq 0$.

Now let us use the principles to solve some equations.

EXAMPLE 1 Solve $3x + 4 = 15$.

$$3x + 4 = 15$$
$$3x + 4 + (-4) = 15 + (-4) \qquad \text{Using the addition principle; adding } -4$$
$$3x = 11 \qquad \text{Simplifying}$$
$$\tfrac{1}{3} \cdot 3x = \tfrac{1}{3} \cdot 11 \qquad \text{Using the multiplication principle; multiplying by } \tfrac{1}{3}$$
$$x = \tfrac{11}{3}$$

Check:

$$\begin{array}{c|c} 3x + 4 = 15 \\ \hline 3 \cdot \tfrac{11}{3} + 4 & 15 \qquad \text{The only solution is the number } \tfrac{11}{3}. \\ 11 + 4 & \\ 15 & \end{array}$$

The solution set can be indicated as $\{\tfrac{11}{3}\}$, but for brevity we often omit the braces and just say "the solution is $\tfrac{11}{3}$."

EXAMPLE 2 Solve $3(7 - 2x) = 14 - 8(x - 1)$.

$$3(7 - 2x) = 14 - 8(x - 1)$$
$$21 - 6x = 14 - 8x + 8 \qquad \text{Multiplying, using the distributive laws, to remove parentheses}$$
$$21 - 6x = 22 - 8x \qquad \text{Collecting like terms}$$
$$8x - 6x = 22 - 21 \qquad \text{Adding } -21 \text{ and also } 8x$$
$$2x = 1 \qquad \text{Collecting like terms}$$
$$x = \tfrac{1}{2} \qquad \text{Multiplying by } \tfrac{1}{2}$$

Check:

$$\begin{array}{c|c} 3(7 - 2x) = 14 - 8(x - 1) \\ \hline 3(7 - 2 \cdot \tfrac{1}{2}) & 14 - 8(\tfrac{1}{2} - 1) \\ 3(7 - 1) & 14 - 8(-\tfrac{1}{2}) \\ 3 \cdot 6 & 14 + 4 \\ 18 & 18 \qquad \text{The number } \tfrac{1}{2} \text{ checks, so it is the solution.} \end{array}$$

EXAMPLE 3 Solve $x + 3 = x$.

$$x + 3 = x$$
$$-x + x + 3 = -x + x \qquad \text{Adding } -x$$
$$3 = 0 \qquad \text{Collecting like terms}$$

In Example 3 we get a false equation. No replacement for x will make the equation true. Since there are no solutions, the solution set is the *empty set*, denoted $\emptyset$.

DO EXERCISES 3–7.

The Principle of Zero Products

A third principle for solving equations is called the *principle of zero products*, as follows.

Solve.

3. $9x - 4 = 8$

4. $-4x + 2 + 5x = 3x - 15$

5. $3(y - 1) - 1 = 2 - 5(y + 5)$

6. $x - 7 = x$

7. $2x + 6 = 2x - 4$

Solve.

8. $(x - 7)(2x + 3) = 0$

9. $x^2 - x = 20$

10. $x^2 = 5x$

11. $25x^2 + 10x + 1 = 0$

12. $3x^3 - 11x^2 = 4x$

13. $5x^3 + x^2 - 5x - 1 = 0$
   [*Hint:* $x^2(5x + 1) - 1(5x + 1) = 0$.]

The principle of zero products

> For any numbers a and b, if $ab = 0$, then $a = 0$ *or* $b = 0$; and if $a = 0$ or $b = 0$, then $ab = 0$.

To solve an equation using this principle, there must be a 0 on one side of the equation and a product on the other. The solutions are then obtained by setting the factors equal to 0 separately.

EXAMPLE 4   Solve $x^2 + x - 12 = 0$.

$$x^2 + x - 12 = 0$$
$$(x + 4)(x - 3) = 0 \qquad \text{Factoring}$$
$$x + 4 = 0 \quad or \quad x - 3 = 0 \qquad \text{Using the principle of zero products}$$
$$x = -4 \quad or \qquad x = 3$$

The solutions are -4 and 3. The solution set is $\{-4, 3\}$.

EXAMPLE 5   Solve $2x^3 - x^2 = 3x$.

$$2x^3 - x^2 = 3x$$
$$2x^3 - x^2 - 3x = 0 \qquad \text{Addition principle}$$
$$x(2x^2 - x - 3) = 0 \qquad \text{Factoring}$$
$$x(2x - 3)(x + 1) = 0$$
$$x = 0 \quad or \quad 2x - 3 = 0 \quad or \quad x + 1 = 0 \qquad \text{Principle of zero products}$$
$$x = 0 \quad or \qquad x = \tfrac{3}{2} \quad or \qquad x = -1$$

The solution set is $\{0, \tfrac{3}{2}, -1\}$.

DO EXERCISES 8–13.

ii SOLVING INEQUALITIES

Principles for solving inequalities are similar to those for solving equations. We can add the same number on both sides of an inequality. We can also multiply on both sides by the same nonzero number, but if that number is negative, we must reverse the inequality sign.

EXAMPLE 6   Solve $3x < 11 - 2x$.

$$3x < 11 - 2x$$
$$3x + 2x < 11 \qquad \text{Adding } 2x$$
$$5x < 11 \qquad \text{Combining similar terms}$$
$$x < \tfrac{11}{5} \qquad \text{Multiplying by } \tfrac{1}{5}$$

Any number less than $\tfrac{11}{5}$ is a solution. The solution set is the set of all x such that $x < \tfrac{11}{5}$. We abbreviate this using *set-builder* notation as follows:

$$\{x \mid x < \tfrac{11}{5}\}.$$

For brevity we often write merely $x < \tfrac{11}{5}$.

EXAMPLE 7   Solve $16 - 7y \geqslant 10y - 4$.

$$16 - 7y \geqslant 10y - 4$$
$$-16 + 16 - 7y \geqslant -16 + 10y - 4 \qquad \text{Adding } -16$$
$$-7y \geqslant 10y - 20 \qquad \text{Simplifying}$$
$$-17y \geqslant -20 \qquad \text{Adding } -10y \text{ and simplifying}$$
$$y \leqslant \tfrac{20}{17} \qquad \text{Multiplying by } -\tfrac{1}{17} \text{ and reversing the inequality sign}$$

Any number less than or equal to $\frac{20}{17}$ is a solution. The solution set is $\{y | y \leqslant \frac{20}{17}\}$.

DO EXERCISES 14–19.

We can solve an inequality to find the sensible replacements in a radical expression.

EXAMPLE 8   Determine the sensible replacements in $\sqrt{5x - 4}$.

The sensible replacements are values of x for which $\sqrt{5x - 4}$ is a real number. Such replacements are those that make the radicand nonnegative; that is, numbers x for which

$$5x - 4 \geqslant 0$$
$$5x \geqslant 4$$
$$x \geqslant \tfrac{4}{5}.$$

Thus the sensible replacements are any numbers x for which $x \geqslant \frac{4}{5}$. These form the set $\{x | x \geqslant \frac{4}{5}\}$.

DO EXERCISES 20–22.

Solve.

14. $5x > 12 - 3x$

15. $17 - 5y \leqslant 8y - 5$

16. $12x - 6 < 10x + 4$

Write set-builder notation for each set.

17. The set of all x such that $x > \frac{5}{2}$.

18. The set of all y such that $y \geqslant -7$.

19. The set of all x such that $x^2 = 5$.

Determine the sensible replacements in each expression.

20. $\sqrt{x - 2}$

21. $\sqrt{x + 3}$

22. $\sqrt{22 - 4x}$

EXERCISE SET 2.1

i Solve.

1. $4x + 12 = 60$

2. $2y - 11 = 37$

3. $4 + \frac{1}{2}x = 1$

4. $4.1 - 0.2y = 1.3$

5. $y + 1 = 2y - 7$

6. $5 - 4x = x - 13$

7. $5x - 2 + 3x = 2x + 6 - 4x$

8. $5x - 17 - 2x = 6x - 1 - x$

9. $1.9x - 7.8 + 5.3x = 3.0 + 1.8x$

10. $2.2y - 5 + 4.5y = 1.7y - 20$

11. $7(3x + 6) = 11 - (x + 2)$

12. $4(5y + 3) = 3(2y - 5)$

13. $2x - (5 + 7x) = 4 - [x - (2x + 3)]$

14. $y - (9y - 8) = [5 - 2y - 3(2y - 3)] + 29$

15. $(2x - 3)(3x - 2) = 0$

16. $(5x - 2)(2x + 3) = 0$

17. $x(x - 1)(x + 2) = 0$

18. $x(x + 2)(x - 3) = 0$

19. $3x^2 + x - 2 = 0$

20. $10x^2 - 16x + 6 = 0$

21. $(x - 1)(x + 1) = 5(x - 1)$

22. $6(y - 3) = (y - 3)(y - 2)$

23. $x[4(x - 2) - 5(x - 1)] = 2$

24. $14[(x - 4) - \frac{1}{14}(x + 2)] = (x + 2)(x - 4)$

25. $(3x^2 - 7x - 20)(2x - 5) = 0$

26. $(8x + 11)(12x^2 - 5x - 2) = 0$

27. $16x^3 = x$

28. $9x^3 = x$

29. $2x^2 = 6x$

30. $18x + 9x^2 = 0$

31. $3y^3 - 5y^2 - 2y = 0$

32. $3t^3 + 2t = 5t^2$

33. $(2x - 3)(3x + 2)(x - 1) = 0$

34. $(y - 4)(4y + 12)(2y + 1) = 0$

35. $(2 - 4y)(y^2 + 3y) = 0$

36. $(y^2 - 9)(y^2 - 36) = 0$

ii Solve.

37. $x + 6 < 5x - 6$

38. $3 - x < 4x + 7$

39. $3x - 3 + 2x \geqslant 1 - 7x - 9$

40. $5y - 5 + y \leqslant 2 - 6y - 8$

41. $14 - 5y \leqslant 8y - 8$

42. $12x - 6 < 10x + 4$

43. $-\frac{3}{4}x \geqslant -\frac{5}{8} + \frac{2}{3}x$

44. $-\frac{5}{6}x \leqslant \frac{3}{4} + \frac{8}{3}x$

45. $4x(x - 2) < 2(2x - 1)(x - 3)$

46. $(x + 1)(x + 2) > x(x + 1)$

Write set-builder notation for each set.

47. The set of all x such that $x > 2.5$

48. The set of all y such that $y \leqslant -7$

49. The set of all t such that $2t^2 = 10$

50. The set of all m such that $m^3 + 3 = m^2 - 2$

Determine the sensible replacements in each of the following.

51. $\sqrt{x - 3}$ **52.** $\sqrt{2x - 5}$ **53.** $\sqrt{3 - 4x}$ **54.** $\sqrt{x^2 + 3}$

Solve.

55. ▤ $2.905x - 3.214 + 6.789x = 3.012 + 1.805x$

56. ▤ $(13.14x + 17.152)(15.15 - 7.616x) = 0$

57. ▤ $3.12x^2 - 6.715x = 0$

58. ▤ $9.25x^2 + 18.03x = 0$

59. ▤ $1.52(6.51x + 7.3) < 11.2 - (7.2x + 13.52)$

60. ▤ $4.73(5.16y + 3.62) \geqslant 3.005(2.75y - 6.31)$

Solve.

61. $7x^3 + x^2 - 7x - 1 = 0$
[*Hint:* $x^2(7x + 1) - 1(7x + 1) = 0.$]

62. $3x^3 + x^2 - 12x - 4 = 0$

63. $y^3 + 2y^2 - y - 2 = 0$

64. $t^3 + t^2 - 25t - 25 = 0$

65. $x^2 - x - 20 = x^2 - 25$

Solve.

66. $(x + 1)^3 = (x - 1)^3 + 26$

67. $(x - 2)^3 = x^3 - 2$

OBJECTIVES

You should be able to:

i Determine whether equations are equivalent.

ii Solve fractional equations.

2.2

FRACTIONAL EQUATIONS

i EQUIVALENT EQUATIONS

Definition

Equations that have the same solution set are called *equivalent equations.*

The three equations in Example 1 are equivalent.

EXAMPLE 1

$$3x = 6 \qquad 3x + 5 = 11 \qquad -12x = -24$$
Solution set $\{2\}$.   Solution set $\{2\}$.   Solution set $\{2\}$.

The pairs of equations in Examples 2 and 3 are *not* equivalent.

EXAMPLE 2

$$3x = 4x$$

Solution set $\{0\}$.

$$\frac{3}{x} = \frac{4}{x}$$

The solution set is $\emptyset$, the empty set (no solution, since division by 0 is not defined).

EXAMPLE 3

$$x = 1$$

Solution set $\{1\}$.

$$x^2 = x$$

Solution set $\{0, 1\}$.

DO EXERCISES 1–6.

Let us examine our equation-solving principles from the standpoint of equivalent equations. Consider the following example illustrating the addition principle:

$$x + 5 = 9$$
$$x + 5 + (-5) = 9 + (-5) \quad \text{Adding } -5 \text{ on both sides}$$
$$x = 4.$$

In this example the original equation and the last equation had exactly the same solutions. They were equivalent. Whenever the steps in an argument are reversible, such will be the case. When we use the addition principle, such will be the case, unless an expression is added having nonsensible replacements. To see that, note that we start with an equation $a = b$ and obtain $a + c = b + c$. By adding $-c$ we can always obtain $a = b$ again, so the steps are reversible.

DO EXERCISES 7 AND 8.

> The use of the addition principle produces an equation equivalent to the original, unless an expression is added having nonsensible replacements. Checking by substituting is therefore not necessary except to detect errors in solving.

Now let us consider the multiplication principle, which says that if $a = b$ is true, then $ac = bc$ is true. Does the multiplication principle yield equivalent equations? In other words, are the steps reversible? If the number c by which we multiply is not 0, then we can reverse the step by multiplying by $1/c$, but if c is 0, then $1/c$ does not exist and the step is not reversible.

> The use of the multiplication principle produces an equation equivalent to the original, if we multiply by a nonzero number.

Now let us look at the principle of zero products. According to this principle, if we start with an equation $ab = 0$, we obtain $a = 0$ or $b = 0$. Also, if we start with $a = 0$ or $b = 0$, we can obtain $ab = 0$. Thus when we use this principle, that step is reversible, so we have equivalent equations.

> The use of the principle of zero products yields the solutions of the original equation. Checking by substituting is not necessary except to detect errors in solving.

Determine whether the equations of each pair are equivalent.

1. $3x = 7$
   $-15x = -35$

2. $x = 7$
   $5x = 35$

3. $x = -5$
   $x^2 = 25$

4. $x = -5$
   $x + 1 = -4$

5. $\dfrac{(x-2)(x+8)}{x-2} = x + 8$

   $x + 8 = x + 8$

6. $3x^2 = 5x$
   $3x = 5$

7. Explain how you would derive $2x + 5x^2 = 5x^2$ from $2x = 0$.

8. Explain how you would derive $2x = 0$ from $2x + 5x^2 = 5x^2$.

Solve. Don't forget to check.

9. $\dfrac{-5}{x} = \dfrac{7}{x}$

10. $\dfrac{x+4}{x+5} = \dfrac{-1}{x+5}$

11. $\dfrac{2x-7}{x+4} = \dfrac{5}{x+4}$

ii FRACTIONAL EQUATIONS

Let us now consider equations containing fractional expressions. These are called *fractional equations*. Finding the solutions of such equations often involves multiplying by expressions with variables, as in the following example.

EXAMPLE 4 Solve $\dfrac{3}{x} = \dfrac{4}{x}$.

$$\frac{3}{x} = \frac{4}{x}$$

$$\frac{3}{x} \cdot x^2 = \frac{4}{x} \cdot x^2 \qquad \text{Multiplying by } x^2$$

$$3x = 4x$$

$$0 = x \qquad \text{Adding } -3x \text{ on both sides}$$

The number 0 is a solution of the last equation but is not a solution of the first. Thus we did not obtain equivalent equations. What can we do in practice?

> When we use the multiplication principle and multiply (or divide) by an expression with a variable, we may not obtain equivalent equations. We must check possible solutions by substituting in the original equation.

EXAMPLE 5 Solve $\dfrac{x-3}{x-7} = \dfrac{4}{x-7}$.

$$\frac{x-3}{x-7} = \frac{4}{x-7}$$

$$(x-7) \cdot \frac{x-3}{x-7} = (x-7) \cdot \frac{4}{x-7} \qquad \left\{ \begin{array}{l} \text{Caution! Here we multiplied} \\ \text{by an expression with a} \\ \text{variable. Thus we must check.} \end{array} \right.$$

$$x - 3 = 4$$

$$x = 7$$

The possible solution is 7. We check:

$$\frac{x-3}{x-7} = \frac{4}{x-7}$$

$$\begin{array}{c|c} \dfrac{7-3}{7-7} & \dfrac{4}{7-7} \\[2mm] \hline \dfrac{4}{0} & \dfrac{4}{0} \end{array}$$

Division by 0 is undefined; 7 is not a solution. The equation has no solutions. The solution set is $\emptyset$.

DO EXERCISES 9–11.

EXAMPLE 6   Solve $\dfrac{x^2}{x-3} = \dfrac{9}{x-3}$.

$$\frac{x^2}{x-3} = \frac{9}{x-3}$$

$$(x-3) \cdot \frac{x^2}{x-3} = (x-3) \cdot \frac{9}{x-3}$$

$\left\{\begin{array}{l}\text{Caution! Here we multiplied} \\ \text{by an expression with a} \\ \text{variable. Thus we must check.}\end{array}\right.$

$$x^2 = 9$$
$$x^2 - 9 = 0$$
$$(x+3)(x-3) = 0$$
$$x + 3 = 0 \quad \text{or} \quad x - 3 = 0$$
$$x = -3 \quad \text{or} \qquad x = 3$$

The possible solutions are 3 and -3. We must check since we have multiplied by an expression with a variable.

For 3:

$$\frac{x^2}{x-3} = \frac{9}{x-3}$$

$$\frac{3^2}{3-3} \;\bigg|\; \frac{9}{3-3}$$

$$\frac{9}{0} \;\bigg|\; \frac{9}{0}$$

3 does not check.

For -3:

$$\frac{x^2}{x-3} = \frac{9}{x-3}$$

$$\frac{(-3)^2}{-3-3} \;\bigg|\; \frac{9}{-3-3}$$

$$-\frac{9}{6} \;\bigg|\; -\frac{9}{6}$$

-3 checks.

Thus the solution set is $\{-3\}$.

DO EXERCISES 12 AND 13.

The usual procedure for solving fractional equations involves multiplying on both sides by the LCM of all the denominators. This procedure is called *clearing of fractions*.

EXAMPLE 7   Solve $\dfrac{14}{x+2} - \dfrac{1}{x-4} = 1$.

We multiply by the LCM of all the denominators: $(x+2)(x-4)$.

$$(x+2)(x-4) \cdot \frac{14}{x+2} - (x+2)(x-4) \cdot \frac{1}{x-4} = (x+2)(x-4) \cdot 1$$

$$14(x-4) - (x+2) = (x+2)(x-4)$$
$$14x - 56 - x - 2 = x^2 - 2x - 8$$
$$0 = x^2 - 15x + 50$$
$$0 = (x-10)(x-5)$$
$$x = 10 \quad \text{or} \quad x = 5$$

The possible solutions are 10 and 5. These check, so the solution set is $\{10, 5\}$.

DO EXERCISE 14.

Solve. Don't forget to check.

12. $\dfrac{y^2}{y+4} = \dfrac{16}{y+4}$

13. $\dfrac{x^2}{x-5} = \dfrac{36}{x-5}$

Solve.

14. $\dfrac{4}{x+5} + \dfrac{1}{x-5} = \dfrac{1}{x^2-25}$

EXERCISE SET 2.2

i Determine whether the equations of each pair are equivalent.

1. $3x + 5 = 12$
$3x = 7$

2. $x^2 = -7x$
$x = -7$

3. $x = 3$
$x^2 = 9$

4. $2y + 1 = -3$
$8y + 4 = -12$

5. $\dfrac{(x - 3)(x + 9)}{(x - 3)} = x + 9$
$x + 9 = x + 9$

6. $x^2 + x - 20 = 0$
$x^2 - 25 = 0$

ii Solve.

7. $\dfrac{1}{4} + \dfrac{1}{5} = \dfrac{1}{t}$

8. $\dfrac{1}{3} - \dfrac{5}{6} = \dfrac{1}{x}$

9. $\dfrac{3}{x - 8} = \dfrac{x - 5}{x - 8}$

10. $\dfrac{23}{y} = \dfrac{-5}{y}$

11. $\dfrac{x + 2}{4} - \dfrac{x - 1}{5} = 15$

12. $\dfrac{t + 1}{3} - \dfrac{t - 1}{2} = 1$

13. $x + \dfrac{6}{x} = 5$

14. $x = \dfrac{12}{x} + 1$

15. $\dfrac{x + 2}{2} + \dfrac{3x + 1}{5} = \dfrac{x - 2}{4}$

16. $\dfrac{2x - 1}{3} - \dfrac{x - 2}{5} = \dfrac{x}{2}$

17. $\dfrac{1}{2} + \dfrac{2}{x} = \dfrac{1}{3} + \dfrac{3}{x}$

18. $\dfrac{1}{t} + \dfrac{1}{2t} + \dfrac{1}{3t} = 5$

19. $\dfrac{4}{x^2 - 1} - \dfrac{2}{x - 1} = \dfrac{3}{x + 1}$

20. $\dfrac{3y + 5}{y^2 + 5y} + \dfrac{y + 4}{y + 5} = \dfrac{y + 1}{y}$

21. $\dfrac{1}{2t} - \dfrac{2}{5t} = \dfrac{1}{10t} - 3$

22. $\dfrac{3}{m + 2} + \dfrac{2}{m - 2} = \dfrac{4m - 4}{m^2 - 4}$

23. $1 - \dfrac{3}{x} = \dfrac{40}{x^2}$

24. $1 - \dfrac{15}{y^2} = \dfrac{2}{y}$

25. $\dfrac{11 - t^2}{3t^2 - 5t + 2} = \dfrac{2t + 3}{3t - 2} - \dfrac{t - 3}{t - 1}$

26. $\dfrac{1}{3y^2 - 10y + 3} = \dfrac{6y}{9y^2 - 1} + \dfrac{2}{1 - 3y}$

27. 🖩 $\dfrac{2.315}{y} - \dfrac{12.6}{17.4} = \dfrac{6.71}{7} + 0.763.$

28. 🖩 $\dfrac{6.034}{x} - 43.17 = \dfrac{0.793}{x} + 18.15$

Solve.

29. $\dfrac{(x - 3)^2}{x - 3} = x - 3$

30. $\dfrac{x^2 + 6x - 16}{x - 2} = x + 8$

31. $\dfrac{x^3 + 8}{x + 2} = x^2 - 2x + 4$

32. $\dfrac{x + 8}{x - 2} = \dfrac{8 + x}{-2 + x}$

33. $\dfrac{x + 3}{x} = 3$

34. Determine whether each equation is equivalent to the one that follows:

$$x^2 - x - 20 = x^2 - 25 \qquad (1)$$
$$(x - 5)(x + 4) = (x - 5)(x + 5) \qquad (2)$$
$$x + 4 = x + 5 \qquad (3)$$
$$4 = 5 \qquad (4)$$

Equations that are true for all sensible replacements of the variables are called *identities*. Determine which of the following equations are identities.

35. $\dfrac{x^2 + 6x - 16}{x - 2} = x + 8$

36. $x + 4 = 4 + x$

37. $(x - 1)(x^2 + x + 1) = x^3 - 1$

38. $\dfrac{x^3 + 8}{x^2 - 4} = \dfrac{x^2 - 2x + 4}{x - 2}$

39. $(x + 7)^2 = x^2 + 49$

40. $\sqrt{x^2 - 16} = x - 4$

41. Solve: $\dfrac{x + 3}{x + 2} - \dfrac{x + 4}{x + 3} = \dfrac{x + 5}{x + 4} - \dfrac{x + 6}{x + 5}$.

2.3

FORMULAS AND PROBLEM SOLVING

i FORMULAS

OBJECTIVES

You should be able to:

i Solve a formula for a given variable.

ii Use the problem-solving strategy to solve applied problems.

A formula is a "recipe" for doing a calculation. An example is $A = \pi r s + \pi r^2$, which gives the area A of a cone in terms of the slant height s and radius of the base r.

Suppose we want to find the slant height s when the area A and radius r are known. Our knowledge of equations allows us to get s alone on one side, or as we say, "solve the formula for s."

EXAMPLE 1 Solve $A = \pi r s + \pi r^2$ for s.

$$A - \pi r^2 = \pi r s$$

$$\frac{A - \pi r^2}{\pi r} = s$$

An equivalent expression for s is $\dfrac{A}{\pi r} - r$.

EXAMPLE 2 Solve

$$\frac{1}{R} = \frac{1}{r_1} + \frac{1}{r_2}$$

for R. (This is a formula from electricity.)

Resistance in parallel:

$$\frac{1}{R} = \frac{1}{r_1} + \frac{1}{r_2}$$

1. Solve

$$C = \frac{5}{9}(F - 32)$$

for F. (This is a formula for Celsius, or Centigrade, temperature in terms of Fahrenheit temperature F.)

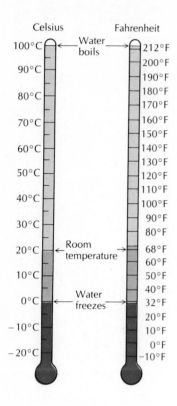

We first multiply by the LCM, which is $Rr_1 r_2$:

$$Rr_1 r_2 \cdot \frac{1}{R} = Rr_1 r_2 \cdot \left(\frac{1}{r_1} + \frac{1}{r_2}\right)$$

$$Rr_1 r_2 \cdot \frac{1}{R} = Rr_1 r_2 \cdot \frac{1}{r_1} + Rr_1 r_2 \cdot \frac{1}{r_2} \qquad \text{Using a distributive law}$$

$$r_1 r_2 = Rr_2 + Rr_1 \qquad \text{Simplifying}$$

$$r_1 r_2 = (r_2 + r_1)R \qquad \text{Factoring out the common factor } R$$

$$\frac{r_1 r_2}{r_2 + r_1} = R.$$

DO EXERCISES 1 AND 2.

ii PROBLEM SOLVING

By an *applied problem* we mean a problem in which mathematical techniques are used to answer some question. Problems like this may be posed orally. They may come about in the course of a conversation, or they can be hatched within the mind of one person. Thus to call them "word problems" or "story problems" is misleading.

There is no rule that will enable us to solve applied problems, because they are of many different kinds. We can, however, describe an overall, or general, strategy. The idea is to translate the problem situation to mathematical language and then calculate to find a solution.

The four-step problem-solving process

1. *Familiarize* yourself with the problem situation. If the problem is presented to you in written words, then of course this means to read carefully.

   a) Make a drawing, if it makes sense to do so. It is difficult to overemphasize the importance of this!
   b) Make a written list of the known facts and a list of what you wish to find out.
   c) Organize the information in a chart or table.

2. *Translate* the problem situation to mathematical language or symbolism. For most of the problems you will encounter in algebra this means to write one or more equations. You will of course also assign certain letters to represent unknown quantities.

3. *Carry out* some kind of mathematical manipulation. This means to use your mathematical knowledge to find a possible solution. In algebra this usually means to solve an equation or system of equations.

4. *Check* to see whether your possible solution actually fits the problem situation, and is thus really a solution of the problem.

---

2. Solve $\frac{1}{R} = \frac{1}{r_1} + \frac{1}{r_2}$, for r_2.

Whether you use all or part of the process depends on the difficulty of the problem. You may indeed guess a possible solution and check it in the problem. Usually in this text you will need to do more than that. We may sometimes illustrate all of the process, and sometimes not. In general, the more difficult the problem, the more of the process you will need to use.

Problems stated in textbooks are of necessity somewhat contrived. Problems that you encounter in nonclassroom situations will almost invariably contain insufficient information to obtain a firm, or exact, answer. They also usually contain a good deal of extraneous, useless information. Here is an example that illustrates this point.

EXAMPLE 3   There are 145 people in the junior class at Notown College. The prom committee wishes to know how much they will have to charge each person who attends the spring prom, to be held May 19 in the college gymnasium at 8:00 P.M.

The above problem is typical of problems encountered in real life. There is insufficient information to obtain an answer. To get an answer we need to know the following.

1. The cost of holding the prom. This would include such things as cost of music, refreshments, and hall rent.

2. The number of people who will attend.

The committee will need to find out information about costs, and in the course of finding it, they may revise their formulation of the problem. For example, if a band costs too much, they may have to settle for using records. In any event, they may find that they can determine costs only approximately; in other words, they must *estimate*.

The number 145 may help in estimating the number who will attend, but at best this number will be an estimate. Thus, on the basis of the best information and/or estimates available, the committee will do its arithmetic.

Note that the date, time, and place, while they are interesting information, do not contribute to the solution of the problem. Thus this is extraneous information.

In the following examples we illustrate how a problem situation can be translated to mathematical language. In certain simple situations, the translation is easy because certain words translate directly to mathematical symbols. Note how the word "is" translates to an equals sign, the word "what" translates to a variable, and the word "of" translates to a multiplication sign. Examples 4 and 5 are stated so explicitly that the familiarization step is not necessary.

EXAMPLE 4   What percent of 84 is 11.76?

2. *Translate:* $x\%$ $\cdot\ 84 = 11.76$

3. *Carry out:*   We solve the equation:

$$x \cdot 0.01 \cdot 84 = 11.76 \qquad \text{"\%" means "} \cdot 0.01\text{"}$$

$$x \cdot 0.84 = 11.76$$

$$x = \frac{11.76}{0.84} = 14.$$

4. *Check:* $14\% \cdot 84 = 0.14 \cdot 84 = 11.76$

The answer is 14%.

EXAMPLE 5   14% of what is 11.76?

2. *Translate:* $14\% \cdot$ y $= 11.76$

3. *Carry out:*   We solve the equation:

$$0.14 \cdot y = 11.76$$

$$y = \frac{11.76}{0.14} = 84.$$

4. *Check:* $14\% \cdot 84 = 0.14 \cdot 84 = 11.76$

The answer is 84.

DO EXERCISES 3–5.

3. What percent of 76 is 13.68?

4. 24% of what is 13.68?

5. 36% of 75 is what?

6. An investment is made at 14%, compounded annually. It grows to $2793 at the end of one year. How much was originally invested?

In the remainder of this section we consider applied problems of various types. Although there is no rule for solving applied problems because they can be so different, it does help somewhat to consider a few different types of problems. *The best way to learn to solve applied problems is to solve a lot of them and to use the process we have given you.*

Problems Involving Compound Interest

EXAMPLE 6 An investment is made at 16%, compounded annually. It grows to $1740 at the end of one year. How much was originally invested?

1. *Familiarize:* There is more than one way to translate the problem situation to mathematical language. The following is one method. We first restate the situation as follows:

The invested amount *plus* the interest is $1740.

2. *Translate:* Now the interest is 16% of the amount invested, so we have the following, which translates directly:

$$\underbrace{\text{Invested amount}}_{x} \text{ plus } \underbrace{16\%}_{16\%} \text{ of } \underbrace{\text{Invested amount}}_{x} \text{ is } \underbrace{\$1740}_{1740}.$$

3. *Carry out:* We solve the equation:

$$x + 16\%x = 1740$$
$$x + 0.16x = 1740$$
$$(1 + 0.16)x = 1740$$
$$1.16x = 1740$$
$$x = \frac{1740}{1.16} = 1500.$$

The number 1500 checks in the problem situation, so the answer is $1500.

DO EXERCISE 6.

Let us now consider an investment over a period longer than one year. If we invest P dollars at an interest rate i, compounded annually, we will have an amount in the account at the end of a year that we will call A_1.

Now $A_1 = P + Pi$, or

$$A_1 = P(1 + i), \quad \text{or}$$

$$A_1 = Pr, \quad \text{where we have let } r = 1 + i.$$

Going into the second year we have Pr dollars. By the end of the second year we will have A_2 dollars, given by

$$A_2 = A_1 \cdot r.$$

But $A_1 = Pr$, so $A_2 = (Pr)r$, or

$$A_2 = Pr^2.$$

Similarly, the amount A_3 in the account at the end of three years is given by

$$A_3 = Pr^3,$$

and so on. In general, the following applies.

Theorem 1

> If principal *P* is invested at an interest rate *i*, compounded annually, in *t* years it will grow to an amount *A* given by
>
> $$A = P(1 + i)^t.$$

▣ **EXAMPLE 7** Suppose $1000 is invested at 12%, compounded annually What amount will be in the account at the end of ten years?

We do not need the entire problem-solving strategy. We have a formula The solution to the problem comes from substituting into the formula. Using the equation $A = P(1 + i)^t$, we get

$$A = 1000(1 + 0.12)^{10} = 1000(1.12)^{10} \qquad \text{Use the } x^y \text{ key.}$$

$$\approx 3105.85.$$

The answer is $3105.85.

DO EXERCISE 7.

Interest may be compounded more often than once a year. Suppose it is compounded four times a year, or *quarterly*. The formula derived above can be altered to apply. We consider one-fourth of a year to be an *interest period*. The rate of interest for such a period is then $i/4$. The number of periods will be four times the number of years. This is shown in the following diagram.

$$A = P(1 + i)^t$$

The number of times interest is compounded (interest periods) goes from t to $4t$.

For $\frac{1}{4}$ year the interest rate will be $\frac{i}{4}$.

$$A = P\left(1 + \frac{i}{4}\right)^{4t}$$

7. ▣ Suppose $1000 is invested at 12.5%, compounded annually. What amount will be in the account at the end of eight years?

8. ▦ Suppose that $1000 is invested at 12.5% compounded semiannually ($n = 2$). How much will be in the account at the end of eight years?

Now suppose the number of interest periods per year is something other than 4, say n. Using the reasoning of the example above, we obtain a general formula.

Theorem 2

If principal P is invested at an interest rate i, compounded n times per year, in t years it will grow to an amount A given by

$$A = P\left(1 + \frac{i}{n}\right)^{nt}.$$

When problems involving compound interest are translated to mathematical language, the preceding formula is almost always used.

▦ **EXAMPLE 8**   Suppose $1000 is invested at 12%, compounded quarterly. How much will be in the account at the end of ten years?

In this case $n = 4$ and $t = 10$. We substitute into the formula:

$$A = P\left(1 + \frac{i}{n}\right)^{nt} = 1000\left(1 + \frac{0.12}{4}\right)^{4 \cdot 10}$$

$$= 1000(1.03)^{40} \qquad \text{Use the } x^y \text{ key.}$$

$$\approx 3262.04.$$

The answer is $3262.04.

DO EXERCISE 8.

Problems Involving Area

EXAMPLE 9   The radius of a circular swimming pool is 10 ft. A sidewalk of uniform width is constructed around the outside and has an area of 44π ft². How wide is the sidewalk?

1. *Familiarize:*   First make a drawing:

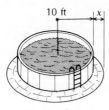

10 ft x

Let x represent the width of the walk. Then, recalling that a formula for the area of a circle is $A = \pi r^2$, we have

Area of pool = $\pi \cdot 10^2 = 100\pi$;

Area of sidewalk plus pool = $\pi \cdot (10 + x)^2 = \pi(100 + 20x + x^2)$.

2. *Translate:*   The translation is as follows.

$$\underbrace{\pi(100 + 20x + x^2)}_{\substack{\text{Area of sidewalk} \\ \text{plus pool}}} - \underbrace{100\pi}_{\substack{\text{Area of} \\ \text{pool}}} = \underbrace{44\pi}_{\substack{\text{Area of} \\ \text{sidewalk}}}$$

3. *Carry out:* We solve the equation:

$$100 + 20x + x^2 - 100 = 44 \qquad \text{Multiplying by } \frac{1}{\pi}$$

$$x^2 + 20x = 44$$

$$x^2 + 20x - 44 = 0$$

$$(x + 22)(x - 2) = 0$$

$$x = -22 \quad \text{or} \quad x = 2.$$

4. *Check:* We see that -22 ft is not a solution of the original problem since width has to be positive. The number 2 checks. The area of the pool is 100π ft². The area of the pool plus sidewalk is $\pi \cdot (10 + 2)^2$, or 144π ft², so the area of the sidewalk is 44π ft².

DO EXERCISE 9.

Problems Involving Motion

For problems that deal with distance, time, and speed, we almost always need to recall the definition of speed, or something equivalent to it.

> Speed = distance/time or $r = d/t$, where r = speed, d = distance, and t = time.

If you memorize the formula $r = d/t$, you can easily obtain either of the two equivalent equations $d = rt$ or $t = d/r$, as needed.

When translating these problems to mathematical language it is often helpful to look for some quantity that is constant in the problem. For example, two cars may travel for the same length of time, or two boats may travel the same distance. Such facts often provide the basis for setting up an equation.

EXAMPLE 10 A boat travels 246 km downstream in the same time that it takes to travel 180 km upstream. The speed of the current in the stream is 5.5 km/h. Find the speed of the boat in still water.

1. *Familiarize:* We first make a drawing and lay out the known facts and any other pertinent information. The time to go downstream is the same as the time to go upstream. We call it t. We know that the speed of the current is 5.5 mph, but we do not know the speed of the boat. Let's call it r.

t hours   Downstream   ⟶

$d_1 = 246$ km r_1 unknown, but equals speed of boat
                       plus speed of current: $r + 5.5$

t hours   Upstream

⟵ $d_2 = 180$ km r_2 unknown, but equals speed of boat
                       minus speed of current: $r - 5.5$

We can also organize the data in a table.

	Distance	Speed	Time
Downstream	246	$r + 5.5$	t
Upstream	180	$r - 5.5$	t

9. A rectangular garden is 60 ft by 80 ft. Part of the garden is torn up to install a sidewalk of uniform width around the garden. The new area of the garden is $\frac{1}{6}$ of the old area. How wide is the sidewalk?

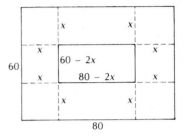

10. An airplane flies 1062 km with the wind in the same time that it takes to fly 738 km against the wind. The speed of the plane in still air is 200 km/h. Find the speed of the wind.

2. *Translate:* The table suggests that we use $t = d/r$, because we can get two different expressions for t. We get two equations:

$$t = \frac{246}{r + 5.5}$$

and

$$t = \frac{180}{r - 5.5}.$$

Thus,

$$\frac{246}{r + 5.5} = \frac{180}{r - 5.5}.$$

Solving for r we get 35.5 km/h. This checks. Thus the speed of the boat in still water is 35.5 km/h.

DO EXERCISE 10.

In the following example, although the distance is the same in two cases, an additional piece of information is given that is the key to the translation.

EXAMPLE 11 The speed of a boat in still water is 10 mph. It travels 24 mi upstream and 24 mi downstream in a total time of 5 hr. What is the speed of the current?

1. *Familiarize:* We first make a drawing and write out the pertinent information. We know that the speed of the boat is 10 mph, but we do not know the speed of the current. Let's call it c.

$$\xrightarrow{\hspace{6cm}} \text{Downstream}$$

$d_1 = 24$ mi t_1 unknown r_1 unknown, but is 10 mph plus speed of current: $10 + c$

$$\xleftarrow{\hspace{6cm}} \text{Upstream}$$

$d_2 = 24$ mi t_2 unknown, but $t_1 + t_2 = 5$ hr r_2 unknown, but is 10 mph minus speed of current: $10 - c$

We can also organize the information in a table.

	Distance	Speed	Time
Downstream	24	$10 + c$	t_1
Upstream	24	$10 - c$	t_2

2. *Translate:* This time the basis of our translation is the equation involving times:

$$t_1 + t_2 = 5.$$

We use $t = d/r$ with the information from the table. This leads us to

$$\frac{d_1}{r_1} + \frac{d_2}{r_2} = 5,$$

and then

$$\frac{24}{10+c} + \frac{24}{10-c} = 5,$$

where c is the speed of the current. Solving for c we get $c = -2$ or $c = 2$. Since speed cannot be negative in this problem, -2 cannot be a solution. But 2 checks, so the speed of the current is 2 mph.

DO EXERCISES 11 AND 12.

Problems Involving Work

EXAMPLE 12   Typist A can do a certain job in 3 hr. Typist B can do the same job in 5 hr. How long would it take both, working together, to do the same job?

1. *Familiarize:*   We list the facts:

A can do the typing job in 3 hr;

B can do the same typing job in 5 hr.

We want to know how long it will take them working together. Let's let t be that number of hours.

    We might try a guess. Suppose we guess 8 hr (we got that by adding 3 and 5). This is not a correct answer since either one of them can do the job alone in less than 8 hr.

    Let us make some kind of table. Thinking about A, we figure that that person should type $\frac{1}{3}$ of the job in 1 hr, since all of it can be done in 3 hr.

$\frac{1}{3}$ in 1 hr   All in 3 hr

Number of hours	1	2	3	4	5	6	
A	$\frac{1}{3}$	$\frac{2}{3}$	$\frac{3}{3}$	$\frac{4}{3}$	$\frac{5}{3}$		$\frac{1}{5}$ in 1 hr
B	$\frac{1}{5}$	$\frac{2}{5}$	$\frac{3}{5}$	$\frac{4}{5}$	$\frac{5}{5}$	$\frac{6}{5}$	All in 5 hr

In 2 hr, A can do $\frac{2}{3}$ of the job. In 3 hr, A can do it all, or $\frac{3}{3}$ of the job. Similarly, B can do $\frac{1}{5}$ of the job in 1 hr, $\frac{2}{5}$ in 2 hr, and so on. Thus B can do $\frac{5}{5}$, or all of it, in 5 hr.

    Since A can do $\frac{1}{3}$ of the typing in 1 hr and B can do $\frac{1}{5}$ of it in 1 hr, then working together they should do

$$\frac{1}{3} + \frac{1}{5} \quad \text{of the typing in 1 hr.}$$

11. A train leaves a station and travels north at 75 km/h. Two hours later a second train leaves on a parallel track traveling north at 125 km/h. How far from the station will the second train overtake the first train?

12. A train leaves Oldtown and travels 300 km to Newtown. On the return trip it travels 10 km/h faster. The total time for the round trip was 11 hr. How fast did the train travel on each part of the trip?

13. A can mow a lawn in 4 hr, while B can mow the same lawn in 5 hr. How long will it take them to mow the lawn if they work together?

We are assuming that it takes them t hours working together, so they should do $1/t$ of the typing in 1 hr.

2. *Translate:* They can do $1/t$ of the typing in 1 hr. They can also do $\frac{1}{3} + \frac{1}{5}$ of it in 1 hr. That gives us the following equation:

$$\frac{1}{3} + \frac{1}{5} = \frac{1}{t}.$$

3. *Carry out:* We solve the equation:

$$\frac{1}{3} + \frac{1}{5} = \frac{1}{t}$$

$$5t + 3t = 5 \cdot 3 \qquad \text{Multiplying by the LCM, } 3 \cdot 5 \cdot t$$

$$8t = 15$$

$$t = \frac{15}{8}, \quad \text{or} \quad 1\frac{7}{8} \text{ hr.}$$

And $t = 1\frac{7}{8}$ hr checks. Thus it takes them $1\frac{7}{8}$ hr to do the job if they work together.

DO EXERCISE 13.

The basic principle for translating work problems follows.

> If a job can be done in time t, then $1/t$ of it can be done in 1 unit of time.

14. Fran and Helen work together and get a certain job completed in 4 hr. It would take Helen 6 hr longer, working alone, to do the job than it would Fran. How long would each need to do the job working alone?

EXAMPLE 13 It takes Red 9 hr longer to build a wall than it takes Mort. If they work together they can build the wall in 20 hr. How long would it take each, working alone, to build the wall?

Let $t =$ the amount of time it takes Mort working alone. Then $t + 9 =$ the amount of time it takes Red working alone.

Then Mort can do $1/t$ of the work in 1 hr and Red can do $1/(t + 9)$ of it in 1 hr. Together they can do $[1/t] + [1/(t + 9)]$ of the work in 1 hr. We also know that they can do $\frac{1}{20}$ of the work in 1 hr. Thus,

$$\frac{1}{t} + \frac{1}{t + 9} = \frac{1}{20}.$$

Solving the equation, we get $t = -5$ or $t = 36$. Since negative time has no meaning in the problem, -5 is not a solution of the original problem. The number 36 checks in the original problem. Thus it would take Mort 36 hr and Red 45 hr.

DO EXERCISE 14.

EXERCISE SET 2.3

i

1. Solve $P = 2l + 2w$ for w.

2. Solve $F = ma$ for a.

3. Solve $E = IR$ for I.

4. Solve $F = \dfrac{km_1 m_2}{d^2}$ for m_2.

5. Solve $\dfrac{P_1 V_1}{T_1} = \dfrac{P_2 V_2}{T_2}$ for T_1.

6. Solve $\dfrac{P_1 V_1}{T_1} = \dfrac{P_2 V_2}{T_2}$ for V_2.

7. Solve $S = \dfrac{H}{m(v_1 - v_2)}$ for v_1.

8. Solve $S = \dfrac{H}{m(v_1 - v_2)}$ for v_2.

9. Solve $\dfrac{1}{F} = \dfrac{1}{m} + \dfrac{1}{p}$ for p.

10. Solve $\dfrac{1}{F} = \dfrac{1}{m} + \dfrac{1}{p}$ for F.

Solve for x.

11. $(x + a)(x - b) = x^2 + 5$

12. $(c + d)x + (c - d)x = c^2$

13. $10(a + x) = 8(a - x)$

14. $4(a + b + x) + 3(a + b - x) = 8a$

ii Problem Solving

15. 79.2 is what percent of 180?

16. 6% of what number is 480?

17. What percent of 28 is 1.68?

18. What is 7% of 45.3?

19. A person gets an 11% raise that is $1595. What was the old salary? the new salary?

20. A person gets a 12% raise that is $2520. What was the old salary? the new salary?

21. An investment is made at 13%, compounded annually. It grows to $734.50 at the end of one year. How much was originally invested?

22. An investment is made at 14%, compounded annually. It grows to $912 at the end of one year. How much was originally invested?

23. In triangle ABC, angle B is five times as large as angle A. The measure of angle C is $2°$ less than that of angle A. Find the measures of the angles. (*Hint:* The sum of the angle measures is $180°$.)

24. In triangle ABC, angle B is twice as large as angle A. Angle C measures $20°$ more than angle A. Find the measures of the angles.

25. The perimeter of a rectangle is 322 m. The length is 25 m more than the width. Find the dimensions.

26. The length of a rectangle is twice the width. The perimeter is 39 m. Find the dimensions.

27. A student's scores on three tests are 87%, 64%, and 78%. What must the student score on the fourth test so that the average will be 80%?

28. A student's scores on three tests are 74%, 55%, and 68%. What must the student score on the fourth test so that the average will be 70%?

29. An open box is made from a 10-cm by 20-cm piece of tin by cutting a square from each corner and folding up the edges. The area of the resulting base is 96 cm². What is the length of the sides of the squares?

30. The frame of a picture is 28 cm by 32 cm outside and is of uniform width. What is the width of the frame if 192 cm² of the picture shows?

31. After a 2% increase, the population of a city is 826,200. What was the former population?

32. After a 3% increase, the population of a city is 741,600. What was the former population?

33. A boat goes 50 km downstream in the same time that it takes to go 30 km upstream. The speed of the stream is 3 km/h. Find the speed of the boat in still water.

34. A boat goes 50 km downstream in the same time that it takes to go 30 km upstream. The speed of the boat in still water is 16 km/h. Find the speed of the stream.

35. The speed of train A is 12 mph slower than the speed of train B. Train A travels 230 mi in the same time it takes train B to travel 290 mi. Find the speed of each train.

36. The speed of a passenger train is 14 mph faster than the speed of a freight train. The passenger train travels 400 mi in the same time it takes the freight train to travel 330 mi. Find the speed of each train.

37. An airplane leaves Chicago at a speed of 475 mph. Twenty minutes later a plane leaves Cleveland, which is 350 miles from Chicago, at 500 mph. When they meet, how far are they from Cleveland? (*Hint:* It is usually best to make the units consistent. That is, consider 20 min as $\frac{1}{3}$ hr.)

38. A private airplane leaves an airport and flies due east at 180 km/h. Two hours later a jet leaves the same airport and flies due east at 900 km/h. How far from the airport will the jet overtake the private plane?

39. A can do a certain job in 3 hr, B can do the same job in 5 hr, and C can do the same job in 7 hr. How long would the job take with all working together?

40. Pipe A can fill a tank in 4 hr, pipe B can fill it in 10 hr, and pipe C can fill it in 12 hr. The pipes are connected to the same tank. How long does it take to fill the tank if all three are running together?

41. 🖩 A can do a certain job, working alone, in 3.15 hr. A working with B can do the same job in 2.09 hr. How long would it take B, working alone, to do the job?

42. 🖩 At a factory, smokestack A pollutes the air 2.13 times as fast as smokestack B. When both stacks operate together they yield a certain amount of pollution in 16.3 hr. Find the amount of time it would take each to yield the same amount of pollution if it operated alone.

43. 🖩 Suppose $1000 is invested at $13\frac{3}{4}\%$. How much is in the account at the end of one year, if interest is compounded (a) annually? (b) semiannually? (c) quarterly? (d) daily (use 365 days per yr)? (e) hourly?

44. 🖩 Suppose $1000 is invested at 15.5%. How much is in the account at the end of five years, if interest is compounded (a) annually? (b) semiannually? (c) quarterly? (d) daily (use 365 days per yr)? (e) hourly?

45. A car is driven 144 mi. If it had gone 4 mph faster it could have made the trip in $\frac{1}{2}$ hr less time. What was the speed?

46. A car is driven 280 mi. If it had gone 5 mph faster it could have made the trip in 1 hr less time. What was the speed?

Average speed is defined as total distance divided by total time.

47. A student drove 3 hr on a freeway at 55 mph and then drove 10 mi in the city at 35 mph. What was the average speed?

48. For the first 100 km of a 200-km trip, a student drove at a speed of 40 km/h. For the second half of the trip the student drove at a speed of 60 km/h. What was the average speed for the entire trip? (It is not 50 km/h.)

49. A driver drove half the distance of a trip at 40 mph. At what speed would the driver have to drive for the rest of the distance so that the average speed for the entire trip would be 45 mph?

50. At what time after 4:00 will the minute hand and the hour hand of a clock first be in the same position?

51. At what time after 10:30 will the hands of a clock first be perpendicular?

52. Three trucks, A, B, and C, working together, can move a load of sand in t hours. When working alone, it takes A 1 extra hour to move the sand; B, 6 extra hours; and C, t extra hours. Find t.

53. An airplane is flying from Los Angeles to Hawaii with a tailwind of 50 mph. The distance, in statute miles, from Los Angeles to Honolulu is 2574 mi.
  a) Find the point at which it takes the same amount of time to go back to Los Angeles as it does to go on to Honolulu.
  b) After traveling 1187 mi, the pilot determines that it is necessary to make an emergency landing. Would it require less time to continue to Honolulu or return to Los Angeles?

54. A commuter drives to work at 45 mph and arrives one minute early. At 40 mph, the commuter would arrive one minute late. How far is it to work?

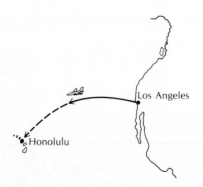

55. Suppose that b is 20% more than a, c is 25% more than b, and d is $k\%$ less than c. Find k such that $a = d$.

2.4

QUADRATIC EQUATIONS

[i] SOLVING QUADRATIC EQUATIONS

Solving $ax^2 + c = 0$

Definition

A *quadratic equation,* also known as an equation of *second degree,* is any equation equivalent to the following:

$$ax^2 + bx + c = 0, \quad \text{where } a \neq 0.$$

The above is the *standard form* of a quadratic equation. It has a second-degree polynomial on one side and 0 on the other. You have solved such equations by factoring. We now consider other methods.

We first consider equations in which the first-degree term is missing; that is, $ax^2 + c = 0$. Such equations can be solved by taking square roots on both sides. To see how that works, we shall proceed by factoring:

$$ax^2 + c = 0$$
$$ax^2 = -c$$
$$x^2 = -\frac{c}{a}.$$

Here we have x^2 on one side and some number on the other. To simplify writing we will write k for the constant. Now we solve by factoring:

$$x^2 = k$$
$$x^2 - k = 0$$
$$x^2 - (\sqrt{k})^2 = 0 \qquad \text{Expressing } k \text{ as } (\sqrt{k})^2$$
$$(x + \sqrt{k})(x - \sqrt{k}) = 0 \qquad \text{Factoring}$$
$$x + \sqrt{k} = 0 \quad \text{or} \quad x - \sqrt{k} = 0 \qquad \text{Principle of zero products}$$
$$x = -\sqrt{k} \quad \text{or} \qquad x = \sqrt{k}$$

There are two solutions. They are the principal square root $\sqrt{k}$ and its additive inverse $-\sqrt{k}$.

EXAMPLE 1 Solve $5x^2 = 15$.

$$5x^2 = 15$$
$$x^2 = 3 \qquad \text{Multiplying by } \tfrac{1}{5}$$
$$x = \sqrt{3} \quad \text{or} \quad x = -\sqrt{3} \qquad \text{Taking square roots}$$

These numbers check, so the solutions are $\sqrt{3}$ and $-\sqrt{3}$.

In general, the equation $x^2 = k$, where k is nonnegative, has two solutions, $\sqrt{k}$ and $-\sqrt{k}$, if $k > 0$. If $k = 0$, there is just one solution, 0. If $k < 0$, there are no real-number solutions.

DO EXERCISES 1–4.

OBJECTIVES

You should be able to:

[i] Solve quadratic equations using factoring, completing the square, or the quadratic formula.

[ii] Use the discriminant to determine the nature of the solutions of a given quadratic equation.

[iii] Write a quadratic equation having specified solutions.

Solve for x.

1. $x^2 = 3$

2. $5x^2 = 0$

3. $3x^2 = \pi$

4. $mx^2 = n$

Solve.

5. $(x + 4)^2 = 7$

6. $(x - 5)^2 = 3$

7. $(x + 5)^2 = 4$

We now use this method in another example.

EXAMPLE 2   Solve $(x + 5)^2 = 3$.

Taking the principal square root, we have

$$|x + 5| = \sqrt{3}.$$

Then

$$x + 5 = \pm\sqrt{3}$$
$$x = -5 \pm \sqrt{3}.$$

The solution set is $\{-5 - \sqrt{3}, -5 + \sqrt{3}\}$.

DO EXERCISES 5–7.

Completing the Square

If the equation is not in the form $(x - h)^2 = k$, we can put it into that form by *completing the square*.

EXAMPLE 3   Solve by completing the square: $x^2 - 6x - 12 = 0$.

We consider first the x^2- and x-terms.

$$x^2 - 6x \qquad - 12 = 0$$

We construct a trinomial square. First note that $x^2 - 6x + 9$ is a perfect square: $(x - 3)^2$. So if we add 9 to $x^2 - 6x$ we will have a perfect square. But, to get an equivalent equation we have to subtract 9 as well. This is the same as adding $9 - 9$, or 0. Then we proceed as follows:

$$x^2 - 6x + 9 - 9 - 12 = 0$$
$$(x - 3)^2 - 21 = 0$$
$$(x - 3)^2 = 21$$
$$x - 3 = \pm\sqrt{21}$$
$$x = 3 \pm \sqrt{21}.$$

The solution set is $\{3 + \sqrt{21}, 3 - \sqrt{21}\}$, which we often abbreviate as $\{3 \pm \sqrt{21}\}$.

The following is a general procedure we use called *completing the square*.

To solve $ax^2 + bx + c = 0$, by *completing the square*:

1. If $a \neq 1$, multiply both sides of the equation by $1/a$ to get the equation in the form $x^2 + Bx + C = 0$.
2. To complete the square on $x^2 + Bx$, take half the coefficient of x and square it. Then add and subtract that number: $(B/2)^2$.
3. Take the square roots and solve for x.

EXAMPLE 4   Solve by completing the square: $x^2 + 3x - 5 = 0$.

1. Since $a = 1$, no multiplication by $1/a$ is necessary to get the equation in proper form: $x^2 + 3x - 5 = 0$.

2. To complete the square on $x^2 + 3x$, we take half the coefficient of x and square it. Then we add and subtract that number: $(\frac{3}{2})^2$, or $\frac{9}{4}$:

$$x^2 + 3x + \frac{9}{4} - \frac{9}{4} - 5 = 0$$

$$\left(x + \frac{3}{2}\right)^2 - \frac{29}{4} = 0$$

$$\left(x + \frac{3}{2}\right)^2 = \frac{29}{4}.$$

3. We now take the square roots and solve for x:

$$x + \frac{3}{2} = \pm\sqrt{\frac{29}{4}} = \pm\frac{\sqrt{29}}{2}$$

$$x = -\frac{3}{2} \pm \frac{\sqrt{29}}{2}$$

$$= \frac{-3 \pm \sqrt{29}}{2}.$$

The solutions are

$$\frac{-3 + \sqrt{29}}{2} \quad \text{and} \quad \frac{-3 - \sqrt{29}}{2}, \quad \text{or} \quad \frac{-3 \pm \sqrt{29}}{2}.$$

DO EXERCISES 8–16.

EXAMPLE 5 Solve by completing the square: $2x^2 - 3x - 1 = 0$.

1. Since $a \neq 1$, we multiply on both sides by $1/a$, which is $\frac{1}{2}$:

$$2x^2 - 3x - 1 = 0$$
$$x^2 - \tfrac{3}{2}x - \tfrac{1}{2} = 0. \qquad \text{Multiplying by } \tfrac{1}{2}$$

2. To complete the square on $x^2 - \frac{3}{2}x$, we take half the coefficient of x and square it. Then we add and subtract that number: $(\frac{1}{2} \cdot \frac{3}{2})^2$, or $\frac{9}{16}$:

$$x^2 - \frac{3}{2}x + \frac{9}{16} - \frac{9}{16} - \frac{1}{2} = 0$$

$$\left(x - \frac{3}{4}\right)^2 - \frac{17}{16} = 0$$

$$\left(x - \frac{3}{4}\right)^2 = \frac{17}{16}.$$

3. We now take the square roots and solve for x:

$$x - \frac{3}{4} = \pm\sqrt{\frac{17}{16}} = \pm\frac{\sqrt{17}}{4}$$

$$x = \frac{3}{4} \pm \frac{\sqrt{17}}{4} = \frac{3 \pm \sqrt{17}}{4}.$$

The solution set is $\left\{\dfrac{3 + \sqrt{17}}{4}, \dfrac{3 - \sqrt{17}}{4}\right\}$.

DO EXERCISES 17 AND 18.

Find the term that completes the square, then fill in the second expression.

8. $x^2 + 4x + \underline{\hspace{1cm}} = (\hspace{1cm})^2$

9. $x^2 - 6x + \underline{\hspace{1cm}} = (\hspace{1cm})^2$

10. $x^2 + 5x + \underline{\hspace{1cm}} = (\hspace{1cm})^2$

11. $x^2 - 7x + \underline{\hspace{1cm}} = (\hspace{1cm})^2$

12. $x^2 + \dfrac{3}{4}x + \underline{\hspace{1cm}} = (\hspace{1cm})^2$

13. $x^2 - x + \underline{\hspace{1cm}} = (\hspace{1cm})^2$

Solve by completing the square.

14. $x^2 + 4x - 3 = 0$

15. $x^2 - 6x + 8 = 0$

16. $x^2 - 5x + 6 = 0$

Solve by completing the square.

17. $2x^2 + 2x - 3 = 0$

18. $4x^2 + 3x - 1 = 0$

The Quadratic Formula

We studied completing the square for two reasons. The most important is that it is a useful tool in other places in mathematics. The second reason is that it can be used to derive a general formula for solving quadratic equations, called the quadratic formula. You will be asked to find one such derivation in the exercises (see Exercise 64). In the following discussion we give another shorter proof of the quadratic formula. Both proofs use completing the square. We consider the standard form of the quadratic equation, with unspecified coefficients. Solving that equation gives us the formula we seek. We begin with

$$ax^2 + bx + c = 0.$$

We multiply on both sides by $4a$, to obtain

$$4a^2x^2 + 4abx + 4ac = 0.$$

Now we complete the square on the left. Note that $4a^2x^2$ is a square. It is $(2ax)^2$. Now the middle term $4abx$ is equivalent to $2(2a \cdot b)x$, so if the last term were b^2, we would have a trinomial square. We complete the square by adding $b^2 - b^2$ on the left, and then rearrange as follows:

$$4a^2x^2 + 4abx + b^2 - b^2 + 4ac = 0,$$

or

$$(4a^2x^2 + 4abx + b^2) - (b^2 - 4ac) = 0,$$

or

$$(2ax + b)^2 - (b^2 - 4ac) = 0.$$

The left side can be factored as a difference of squares, after which we use the principle of zero products:

$$[(2ax + b) - \sqrt{b^2 - 4ac}][(2ax + b) + \sqrt{b^2 - 4ac}] = 0$$

$$2ax + b - \sqrt{b^2 - 4ac} = 0 \quad \text{or} \quad 2ax + b + \sqrt{b^2 - 4ac} = 0$$

$$x = \frac{-b + \sqrt{b^2 - 4ac}}{2a} \quad \text{or} \quad x = \frac{-b - \sqrt{b^2 - 4ac}}{2a}.$$

We can abbreviate the last line using the sign $\pm$. This gives us the quadratic formula.

Theorem 3   The Quadratic Formula

If a quadratic equation $ax^2 + bx + c = 0$ has solutions, they are given by

$$x = \frac{-b \pm \sqrt{b^2 - 4ac}}{2a}.$$

When using the quadratic formula it is helpful to first find the standard form so that the coefficients a, b, and c can be determined.

EXAMPLE 6   Solve $3x^2 + 2x = 7$.

First find the standard form and determine a, b, and c:

$$3x^2 + 2x - 7 = 0,$$
$$a = 3, \quad b = 2, \quad c = -7.$$

Then use the quadratic formula:

$$x = \frac{-b \pm \sqrt{b^2 - 4ac}}{2a} = \frac{-2 \pm \sqrt{2^2 - 4 \cdot 3 \cdot (-7)}}{2 \cdot 3}$$

$$x = \frac{-2 \pm \sqrt{4 + 84}}{6} \qquad \text{To prevent careless mistakes it helps to write out } all \text{ the steps.}$$

$$x = \frac{-2 \pm \sqrt{88}}{6} = \frac{-2 \pm \sqrt{4 \cdot 22}}{6}$$

$$x = \frac{-2 \pm 2\sqrt{22}}{6} = \frac{2(-1 \pm \sqrt{22})}{2 \cdot 3}$$

$$x = \frac{-1 \pm \sqrt{22}}{3}.$$

Should such irrational solutions arise in an applied problem, we can find approximations using a calculator or Table 1:

$$\frac{-1 + \sqrt{22}}{3} \approx \frac{-1 + 4.69}{3}$$

$$\approx \frac{3.69}{3} \approx 1.2; \qquad \text{Rounded to the nearest tenth}$$

$$\frac{-1 - \sqrt{22}}{2} \approx \frac{-1 - 4.69}{3}$$

$$\approx \frac{-5.69}{3} \approx -1.9. \qquad \text{Rounded to the nearest tenth}$$

The solutions of a quadratic equation can *always* be found using the quadratic formula. Solutions that are irrational are difficult to find by factoring. A general strategy for solving quadratic equations is as follows.

1. Try factoring.

2. If factoring seems difficult, use the quadratic formula; it *always works*!

DO EXERCISES 19 AND 20.

ii THE DISCRIMINANT

The expression $b^2 - 4ac$, called the *discriminant*, determines the nature of the solutions. If x_1 and x_2 represent the solutions, then

$$x_1 = \frac{-b + \sqrt{b^2 - 4ac}}{2a}, \qquad x_2 = \frac{-b - \sqrt{b^2 - 4ac}}{2a}.$$

Note that if $b^2 - 4ac = 0$, then x_1 and x_2 are the same, $-b/2a$. If $b^2 - 4ac > 0$, then the expression $\sqrt{b^2 - 4ac}$ represents a positive real number that is to be added to or subtracted from $-b$. In this case the solutions are different. If $b^2 - 4ac < 0$, the expression $\sqrt{b^2 - 4ac}$ does not represent a real number. Thus there are no real-number solutions. In a later chapter we will consider a number system in which such solutions exist.

19. a) Solve $2x^2 + 7x = 4$, by factoring.

b) Solve the equation in (a) using the quadratic formula. Compare your answers.

20. Solve $5x^2 - 8x = 3$ using the quadratic formula.

Determine the nature of the solutions of each equation. Do not solve.

21. $9x^2 - 12x + 4 = 0$

22. $x^2 + 5x + 8 = 0$

23. $x^2 + 5x + 6 = 0$

For each pair of numbers, write a quadratic equation having those numbers as solutions.

24. $-4, \dfrac{5}{3}$

25. $-7, 8$

26. m, n

Theorem 4

If $b^2 - 4ac = 0$, $ax^2 + bx + c = 0$ has just one real-number solution.

If $b^2 - 4ac > 0$, $ax^2 + bx + c = 0$ has two real-number solutions.

If $b^2 - 4ac < 0$, $ax^2 + bx + c = 0$ has no real-number solution.

DO EXERCISES 21–23.

■iii WRITING EQUATIONS FROM SOLUTIONS

We can use the principle of zero products to write a quadratic equation whose solutions are known.

EXAMPLE 7 Write a quadratic equation whose solutions are 3 and $-\frac{2}{5}$.

$$x = 3 \quad \text{or} \quad x = -\tfrac{2}{5}$$
$$x - 3 = 0 \quad \text{or} \quad x + \tfrac{2}{5} = 0$$
$$(x - 3)\left(x + \tfrac{2}{5}\right) = 0 \qquad \text{Multiplying}$$
$$x^2 - \tfrac{13}{5}x - \tfrac{6}{5} = 0 \quad \text{or} \quad 5x^2 - 13x - 6 = 0$$

DO EXERCISES 24–26.

EXERCISE SET 2.4

■i Solve for x.

1. $2x^2 = 14$ **2.** $6x^2 = 36$ **3.** $9x^2 = 5$ **4.** $4x^2 = 3$

5. $ax^2 = b$ **6.** $\pi x^2 = k$ **7.** $(x - 7)^2 = 5$ **8.** $(x + 3)^2 = 2$

9. $(x - h)^2 = a$ **10.** $y = a(x - h)^2 + k$

■i Solve by completing the square. (It is important to practice this method since we will need it later.)

11. $x^2 + 6x + 4 = 0$ **12.** $x^2 - 6x - 4 = 0$ **13.** $y^2 + 7y - 30 = 0$ **14.** $y^2 - 7y - 30 = 0$

15. $5x^2 - 4x - 2 = 0$ **16.** $12y^2 - 14y + 3 = 0$ **17.** $2x^2 + 7x - 15 = 0$ **18.** $9x^2 - 30x = -25$

■i Solve using the quadratic formula.

19. $x^2 + 4x = 5$ **20.** $x^2 = 2x + 15$ **21.** $2y^2 - 3y - 2 = 0$ **22.** $5m^2 + 3m - 2 = 0$

23. $3t^2 + 8t + 5 = 0$ **24.** $3u^2 = 18u - 6$ **25.** $3 + u^2 = 12u$ **26.** $40 + 30p + 5p^2 = 0$

■ii Determine the nature of the solutions of each equation. Do not solve.

27. $x^2 - 2x + 10 = 0$ **28.** $4x^2 - 4\sqrt{3}x + 3 = 0$ **29.** $9x^2 - 6x = 0$ **30.** $2x^2 - 12x + 1 = 0$

■iii Write quadratic equations whose solutions are as follows.

31. $-11, 9$ **32.** $-4, 4$ **33.** 7, only solution   **34.** $-\dfrac{2}{3}$, only solution

35. $-\dfrac{2}{5}, \dfrac{6}{5}$ **36.** $-\dfrac{1}{4}, -\dfrac{1}{2}$ **37.** $\dfrac{c}{2}, \dfrac{d}{2}$ **38.** $\dfrac{k}{3}, \dfrac{m}{4}$

39. $\sqrt{2}, 3\sqrt{2}$ **40.** $-\sqrt{3}, 2\sqrt{3}$

Solve.

41. ▦ $x^2 - 0.75x - 0.5 = 0$ **42.** ▦ $5.33x^2 - 8.23x - 3.24 = 0$

Solve using any method. (In general, try factoring first. Then use the quadratic formula if factoring is not possible.)

43. $2x^2 - x - 6 = 0$

44. $3x^2 + 5x + 2 = 0$

45. $x^2 + 3x = 8$

46. $x^2 - 6x = 4$

47. $x + \dfrac{1}{x} = \dfrac{13}{6}$

48. $\dfrac{3}{x} + \dfrac{x}{3} = \dfrac{5}{2}$

49. $t^2 + 0.2t - 0.3 = 0$

50. $p^2 + 0.3p - 0.2 = 0$

51. $x^2 + x - \sqrt{2} = 0$

52. $x^2 - x - \sqrt{3} = 0$

53. $x^2 + \sqrt{5}x - \sqrt{3} = 0$

54. $2x^2 + \sqrt{3}x - \pi = 0$

55. $\sqrt{2}x^2 - \sqrt{3}x - \sqrt{5} = 0$

56. $\sqrt{2}x^2 + 5x + \sqrt{2} = 0$

57. $(2t - 3)^2 + 17t = 15$

58. $2y^2 - (y + 2)(y - 3) = 12$

59. $(x + 3)(x - 2) = 2(x + 11)$

60. $9t(t + 2) - 3t(t - 2) = 2(t + 4)(t + 6)$

61. $2x^2 + (x - 4)^2 = 5x(x - 4) + 24$

62. $(c + 2)^2 + (c - 2)(c + 2) = 44 + (c - 2)^2$

---

63. Prove that the solutions of $ax^2 + bx + c = 0$ are the reciprocals of the solutions of the equation

$$cx^2 + bx + a = 0, \qquad c \neq 0, \ a \neq 0.$$

64. Derive the quadratic formula by completing the square.

2.5

FORMULAS AND PROBLEM SOLVING

i FORMULAS

To solve a formula for a certain letter, we use the principles of equation solving we have developed until we have an equation with the letter alone on one side. In most formulas the letters represent nonnegative numbers, so we usually do not need to use absolute values when taking principal square roots.

EXAMPLE 1 Solve $V = \pi r^2 h$, for r.

$$V = \pi r^2 h$$

$$\frac{V}{\pi h} = r^2$$

$$\sqrt{\frac{V}{\pi h}} = r$$

DO EXERCISE 1.

EXAMPLE 2 Solve $A = \pi r s + \pi r^2$ for r.

$$\pi r^2 + \pi r s - A = 0.$$

OBJECTIVES

You should be able to:

i Solve a formula for a given letter.

ii Solve applied problems involving quadratic equations.

1. Solve $V = \frac{1}{3}\pi r^2 h$, for r.

2. Solve $S = 16t^2 + v_0 t$ for t.

Then $a = \pi$, $b = \pi s$, $c = -A$, and we use the quadratic formula:

$$r = \frac{-b \pm \sqrt{b^2 - 4ac}}{2a}$$

$$= \frac{-\pi s \pm \sqrt{(\pi s)^2 - 4 \cdot \pi \cdot (-A)}}{2\pi}$$

$$= \frac{-\pi s \pm \sqrt{\pi^2 s^2 + 4\pi A}}{2\pi}$$

or just

$$r = \frac{-\pi s + \sqrt{\pi^2 s^2 + 4\pi A}}{2\pi},$$

since the negative square root would result in a negative solution.

DO EXERCISE 2.

ii PROBLEM SOLVING

EXAMPLE 3 (*An application: Compound interest*). An investment of $2560 is made at interest rate i, compounded annually. In 2 years it grows to $3240. What is the interest rate?

3. An investment of $2560 is made at interest rate i, compounded annually. In 2 years it grows to $3610. What is the interest rate?

We substitute 2560 for P, 3240 for A, and 2 for t in the formula $A = P(1 + i)^t$, and solve for i:

$$A = P(1 + i)^t$$
$$3240 = 2560(1 + i)^2$$

$$\frac{3240}{2560} = (1 + i)^2$$

$$\sqrt{\frac{324}{256}} = |1 + i| \qquad \text{Taking the principal square root}$$

$$\pm\frac{18}{16} = 1 + i$$

$$-1 + \frac{18}{16} = i \quad \text{or} \quad -1 - \frac{18}{16} = i$$

$$\frac{2}{16} = i \quad \text{or} \quad -\frac{34}{16} = i.$$

Since the interest rate cannot be negative, $i = \frac{2}{16} = \frac{1}{8} = 0.125 = 12.5\%$.

DO EXERCISE 3.

EXAMPLE 4 A ladder 10 ft long leans against a wall. The bottom of the ladder is 6 ft from the wall. The bottom of the ladder is then pulled out 3 ft farther. How much does the top end move down the wall?

1. *Familiarize:* The first thing to do is to make a drawing and label it with the known and unknown data. The dotted line in the figure shows the ladder in its original position, with the lower end 6 ft from the wall.

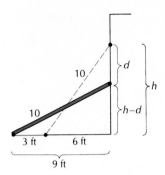

4. A 13-ft ladder leans against a wall. The bottom of the ladder is 5 ft from the wall. The bottom is then pulled out 4 ft farther. How much does the top end move down the wall?

2. *Translate:* In the figure there are right triangles. This is a clue that we may wish to use the Pythagorean theorem. We use that theorem and begin to write equations. From the taller triangle, we get

$$h^2 + 6^2 = 10^2. \qquad (1)$$

From the other triangle, we get

$$9^2 + (h - d)^2 = 10^2. \qquad (2)$$

3. *Carry out:* We solve Equation (1) for h and get $h = 8$ ft. Thus we know that $h = 8$ in Equation (2), so we have

$$9^2 + (8 - d)^2 = 10^2$$

or

$$d^2 - 16d + 45 = 0.$$

Using the quadratic formula we get

$$d = \frac{16 \pm \sqrt{76}}{2} = \frac{16 \pm 2\sqrt{19}}{2} = 8 \pm \sqrt{19}.$$

4. *Check:* The length $8 + \sqrt{19}$ is not a solution since it exceeds the original length. The number $8 - \sqrt{19} \approx 8 - 4.359 = 3.641$ checks and is the solution. Therefore, the top moves down 3.641 ft when the bottom is moved out 3 ft.

DO EXERCISE 4.

We use the following in Example 5.

> When an object is dropped or thrown downward, the distance, in meters, that it falls in t seconds is given by the following formula:
>
> $$s = 4.9t^2 + v_0 t.$$
>
> In this formula v_0 is the initial velocity.

▓ **EXAMPLE 5**

a) An object is dropped from the top of the Gateway Arch in St. Louis, which is 195 m high. How long does it take to reach the ground?

5. a) An object is dropped from the top of the Statue of Liberty, which is 92 m tall. How long does it take to reach the ground?

   b) An object is thrown downward from the statue at an initial velocity of 40 m/sec. How long does it take to reach the ground?

   c) How far will an object fall in 1 sec, thrown downward from the statue at an initial velocity of 40 m/sec?

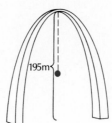

195m

Since the object was *dropped* its initial velocity was 0. So we substitute 0 for v_0 and 195 for s and then solve for t:

$$195 = 4.9t^2 + 0 \cdot t$$
$$195 = 4.9t^2$$
$$t^2 = 39.8$$
$$t = \sqrt{39.8} \approx 6.31. \qquad \text{Use the square root key.}$$

Thus it takes about 6.31 sec to reach the ground.

b) An object is thrown downward from the arch at an initial velocity of 16 m/sec. How long does it take to reach the ground?
   We substitute 195 for s and 16 for v_0 and solve for t:

$$195 = 4.9t^2 + 16t$$
$$0 = 4.9t^2 + 16t - 195.$$

By the quadratic formula and a calculator we obtain

$$t = -8.15 \quad \text{or} \quad t = 4.88.$$

The negative answer is meaningless in this problem, so the answer is about 4.88 sec.

c) How far will an object fall in 3 sec if it is thrown downward from the arch at an initial velocity of 16 m/sec?
   We substitute 16 for v_0 and 3 for t and solve for s:

$$s = 4.9t^2 + v_0 t = 4.9(3)^2 + 16 \cdot 3 = 92.1.$$

Thus the object falls 92.1 m in 3 sec.

DO EXERCISE 5.

EXERCISE SET 2.5

i Solve each formula for the given letter. Assume that all letters represent positive numbers.

1. $F = \dfrac{kM_1 M_2}{d^2}$, for d

2. $E = mc^2$, for c

3. $S = \dfrac{1}{2}at^2$, for t

4. $V = 4\pi r^2$, for r

5. $s = -16t^2 + v_0 t$, for t

6. $A = 2\pi r^2 + 3\pi rh$, for r

7. $d = \dfrac{n^2 - 3n}{2}$, for n

8. $\sqrt{2}\,t^2 + 3k = \pi t$, for t

9. $A = P(1 + i)^2$, for i

10. $A = P\left(1 + \dfrac{i}{2}\right)^2$, for i

ii Problem Solving

What is the interest rate if interest is compounded annually?

11. $5120 grows to $7220 in 2 years

12. $1000 grows to $1210 in 2 years

13. ▦ $8000 grows to $9856.80 in 2 years

14. ▦ $1000 grows to $1271.26 in 2 years

The number of diagonals, d, of a polygon of n sides is given by

$$d = \frac{n^2 - 3n}{2}.$$

15. A polygon has 27 diagonals. How many sides does it have?

16. A polygon has 44 diagonals. How many sides does it have?

17. A ladder 10 ft long leans against a wall. The bottom of the ladder is 6 ft from the wall. How much would the lower end of the ladder have to be pulled away so that the top end would be pulled down the same amount?

18. A ladder 13 ft long leans against a wall. The bottom of the ladder is 5 ft from the wall. How much would the lower end of the ladder have to be pulled away so that the top end would be pulled down the same amount?

19. The area of a triangle is 18 cm². The base is 3 cm longer than the height. Find the height.

20. A baseball diamond is a square 90 ft on a side. How far is it directly from second base to home?

21. Trains A and B leave the same city at right angles at the same time. Train B travels 5 mph faster than train A. After 2 hr they are 50 mi apart. Find the speed of each train.

22. Trains A and B leave the same city at right angles at the same time. Train A travels 14 km/h faster than train B. After 5 hr they are 130 km apart. Find the speed of each train.

For Exercises 23 and 24 use the formula $s = 4.9t^2 + v_0 t$.

23. ▦ a) An object is dropped 75 m from an airplane. How long does it take to reach the ground?
b) An object is thrown downward from the plane at an initial velocity of 30 m/sec. How long does it take to reach the ground?
c) How far will an object fall in 2 sec, thrown downward at an initial velocity of 30 m/sec?

24. ▦ a) An object is dropped 500 m from an airplane. How long does it take to reach the ground?
b) An object is thrown downward from the plane at an initial velocity of 30 m/sec. How long does it take to reach the ground?
c) How far will an object fall in 5 sec, thrown downward at an initial velocity of 30 m/sec?

25. ▦ The diagonal of a square is 1.341 cm longer than a side. Find the length of the side.

26. ▦ The hypotenuse of a right triangle is 8.312 cm long. The sum of the lengths of the legs is 10.23 cm. Find the lengths of the legs.

The following formula will be helpful in Exercises 27–30:

$$T = c \cdot N.$$

Total cost = (Cost per item) · (Number of items)

Total cost = (Cost per person) · (Number of persons)

27. A group of students share equally in the $140 cost of a boat. At the last minute 3 students drop out and this raises the share of each remaining student $15. How many students were in the group at the outset?

28. An investor bought a group of lots for $8400. All but 4 of them were sold for $8400. The selling price for each lot was $350 greater than the cost. How many lots were bought?

29. An investor buys some stock for $720. If each share had cost $15 less, 4 more shares could have been bought for the same $720. How many shares of stock were bought?

30. A sorority is going to spend $112 for a party. When 14 new pledges join the sorority, this reduces each student's cost by $4. How much did it cost each student before?

Solve for x.

31. $kx^2 + (3 - 2k)x - 6 = 0$

32. $x^2 - 2x + kx + 1 = kx^2$

33. $(m + n)^2 x^2 + (m + n)x = 2$

34. Solve $x^2 - 3xy - 4y^2 = 0$ (a) for x; (b) for y.

35. ▦ For interest compounded annually, what is the interest rate when $9826 grows to $13,704 in 3 years?

36. ▦ An equilateral triangle is inscribed in a circle whose circumference is 6π. Find the area of the triangle.

37. Two equilateral triangles have sides of respective lengths a_1 and a_2. Find the length of a side a_3 of a third equilateral triangle whose area is the sum of the areas of the two triangles.

38. The sides of triangle A are 25 ft, 25 ft, and 30 ft. The sides of triangle B are 25 ft, 25 ft, and 40 ft. Which triangle has the greatest area?

39. A rectangle of 12-cm² area is inscribed in the right triangle ABC as shown in the drawing. What are its dimensions?

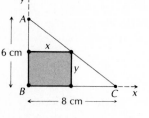

40. The world record for free-fall to the earth without a parachute by a woman is 175 ft and is held by Kitty O'Neill. Approximately how long did the fall take?

175 ft

OBJECTIVE

You should be able to:

■i Solve radical equations using the principle of powers.

2.6

RADICAL EQUATIONS

■i SOLVING RADICAL EQUATIONS

A *radical equation* is an equation in which variables occur in one or more radicands. For example, $\sqrt[3]{x} + \sqrt[3]{4x - 7} = 2$. To solve such equations we need a new principle.

Theorem 5 **The principle of powers**

For any number n, if an equation $a = b$ is true, then $a^n = b^n$ is true.

It is important to check when using the principle of powers. This principle may *not* produce equivalent equations. For example, when we square both sides of an equation, the new equation may have solutions that the first equation does not. Consider, for example, the equation

$$x = 3.$$

This equation has just one solution, the number 3. When we square both sides we get

$$x^2 = 9,$$

which has two solutions, 3 and -3. Thus the equations $x = 3$ and $x^2 = 9$ are *not* equivalent.

As another example, consider $\sqrt{x} = -3$. At the outset we should note that this equation has no solution because, by definition, $\sqrt{x}$ must be a *nonnegative* number. Suppose, though, that we try to solve by squaring both sides. We would get $(\sqrt{x})^2 = (-3)^2$, or $x = 9$. The number 9 does not check.

CAUTION!   It is imperative when using the principle of powers to check possible solutions in the original equation.

EXAMPLE 1   Solve $x - 5 = \sqrt{x + 7}$.

$$x - 5 = \sqrt{x + 7}$$
$$(x - 5)^2 = (\sqrt{x + 7})^2 \qquad \text{Principle of powers; squaring both sides}$$
$$x^2 - 10x + 25 = x + 7$$
$$x^2 - 11x + 18 = 0$$
$$(x - 9)(x - 2) = 0$$
$$x = 9 \quad \text{or} \quad x = 2$$

The possible solutions are 9 and 2. We check, as follows. For 9, we substitute 9 for x on each side of the equation and simplify each side separately:

$$\frac{x - 5 = \sqrt{x + 7}}{\begin{array}{c|c} 9 - 5 & \sqrt{9 + 7} \\ 4 & 4 \end{array}}$$

Since the results, 4, are the same, 9 checks. Thus it is a solution.

For 2, we substitute 2 for x on each side and simplify each side separately:

$$\frac{x - 5 = \sqrt{x + 7}}{\begin{array}{c|c} 2 - 5 & \sqrt{2 + 7} \\ -3 & 3 \end{array}}$$

Since the results are not the same, 2 is not a solution. The solution is 9.

DO EXERCISES 1 AND 2.

EXAMPLE 2   Solve $\sqrt[3]{4x^2 + 1} = 5$.

$$\sqrt[3]{4x^2 + 1} = 5$$
$$(\sqrt[3]{4x^2 + 1})^3 = 5^3 \qquad \text{Principle of powers; cubing both sides}$$
$$4x^2 + 1 = 125$$
$$4x^2 = 124$$
$$x^2 = 31$$
$$x = \pm\sqrt{31}$$

Both $\sqrt{31}$ and $-\sqrt{31}$ check. These are the solutions.

DO EXERCISE 3.

Solve. Don't forget to check!

1. $\sqrt{2x} = -5$

2. $x - 1 = \sqrt{x + 5}$

Solve. Don't forget to check.

3. $\sqrt[4]{3x - 1} = 2$

Solve. Don't forget to check.

4. $\sqrt{3x + 1} - \sqrt{x + 4} = 1$

5. Solve

$$A = \sqrt{1 + \frac{a^2}{b^2}}$$

for b. Assume that the variables represent positive numbers.

Equations with Two Radical Terms

A general strategy for solving equations with two radical terms is as follows.

> **1.** Isolate one of the radical terms.
> **2.** Use the principle of powers.
> **3.** If a radical remains, perform steps (1) and (2) again.
> **4.** Check possible solutions.

EXAMPLE 3 Solve $\sqrt{2x - 5} - \sqrt{x - 3} = 1$.

$$\sqrt{2x - 5} - \sqrt{x - 3} = 1$$

$$\sqrt{2x - 5} = 1 + \sqrt{x - 3} \qquad \text{Adding } \sqrt{x - 3} \text{ to isolate one of the radical terms}$$

$$(\sqrt{2x - 5})^2 = (1 + \sqrt{x - 3})^2 \qquad \text{Principle of powers; squaring both sides}$$

$$2x - 5 = 1 + 2\sqrt{x - 3} + (x - 3) \qquad \text{On the right we squared a binomial. We can also use FOIL.}$$

$$x - 3 = 2\sqrt{x - 3} \qquad \text{Isolating the remaining radical term}$$

$$(x - 3)^2 = (2\sqrt{x - 3})^2 \qquad \text{Squaring both sides}$$

$$x^2 - 6x + 9 = 4(x - 3)$$

$$x^2 - 6x + 9 = 4x - 12$$

$$x^2 - 10x + 21 = 0$$

$$(x - 7)(x - 3) = 0$$

$$x = 7 \quad \text{or} \quad x = 3$$

The numbers 7 and 3 check. Thus the solutions are 7 and 3.

DO EXERCISE 4.

EXAMPLE 4 Solve

$$A = \sqrt{1 + \frac{a^2}{b^2}}$$

for a. Assume that the variables represent positive numbers.

$$A = \sqrt{1 + \frac{a^2}{b^2}}$$

$$A^2 = 1 + \frac{a^2}{b^2}$$

$$b^2 A^2 = b^2 + a^2$$

$$b^2 A^2 - b^2 = a^2$$

$$\sqrt{b^2 A^2 - b^2} = a$$

$$\sqrt{b^2(A^2 - 1)} = a$$

$$b\sqrt{A^2 - 1} = a$$

DO EXERCISE 5.

EXERCISE SET 2.6

i Solve. Don't forget to check!

1. $\sqrt{3x-4} = 1$

2. $\sqrt[3]{2x+1} = -5$

3. $\sqrt[4]{x^2-1} = 1$

4. $\sqrt{m+1} - 5 = 8$

5. $\sqrt{y-1} + 4 = 0$

6. $5 + \sqrt{3x^2 + \pi} = 0$

7. $\sqrt{x-3} + \sqrt{x+5} = 4$ (*Hint:* You are squaring a binomial.)

8. $\sqrt{x} - \sqrt{x-5} = 1$

9. $\sqrt{3x-5} + \sqrt{2x+3} + 1 = 0$

10. $\sqrt{2m-3} = \sqrt{m+7} - 2$

11. $\sqrt[3]{6x+9} + 8 = 5$

12. $\sqrt[5]{3x+4} = 2$

13. $\sqrt{6x+7} = x+2$

14. $\sqrt{6x+7} - \sqrt{3x+3} = 1$

15. $\sqrt{20-x} = \sqrt{9-x} + 3$

16. $\sqrt{n+2} + \sqrt{3n+4} = 2$

17. ▦ $\sqrt{7.35x + 8.051} = 0.345x + 0.067$

18. ▦ $\sqrt{1.213x + 9.333} = 5.343x + 2.312$

19. $x^{1/3} = -2$

20. $t^{1/5} = 2$

21. $t^{1/4} = 3$

22. $m^{1/2} = -7$

23. $8 = \dfrac{1}{\sqrt{x}}$

24. $3 = \dfrac{1}{\sqrt{y}}$

25. $\sqrt[3]{m} = -5$

26. $\sqrt[4]{t} = -5$

For Exercises 27 and 28 assume that the variables represent positive numbers.

27. Solve $T = 2\pi\sqrt{L/g}$ for L; for g.

28. Solve $H = \sqrt{c^2 + d^2}$ for c.

(*An application: Distance to the horizon*). The formula $V = 1.2\sqrt{h}$ can be used to approximate the distance V, in miles, that a person can see to the horizon from a height h, in feet.

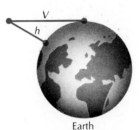

Earth

29. ▦ How far can you see to the horizon through an airplane window at a height of 30,000 ft?

30. ▦ How far can a sailor see to the horizon from the top of a 72-ft mast?

31. ▦ A person can see 144 mi to the horizon from an airplane window. How high is the airplane?

32. ▦ A sailor can see 11 mi to the horizon from the top of a mast. How high is the mast?

Solve.

33. $(x-5)^{2/3} = 2$

34. $(x-3)^{2/3} = 2$

35. $\dfrac{x + \sqrt{x+1}}{x - \sqrt{x+1}} = \dfrac{5}{11}$

36. $\sqrt{\sqrt{x+25}} - \sqrt{x} = 5$ (*Hint:* You are squaring a binomial.)

37. $\sqrt{x+2} - \sqrt{x-2} = \sqrt{2x}$

38. $2\sqrt{x+3} = \sqrt{x} + \sqrt{x+8}$

39. $\sqrt[4]{x+2} = \sqrt{3x+1}$

40. $\sqrt[3]{2x-1} = \sqrt[6]{x+1}$

41. Prove Theorem 5.

OBJECTIVES

You should be able to:

i Solve equations that are reducible to quadratic.

ii Solve applied problems involving equations reducible to quadratic.

1. a) Solve $x + \sqrt{x} - 12 = 0$.

     b) Can you think of another procedure for solving this equation? See the previous section. Which seemed easier?

2.7

EQUATIONS REDUCIBLE TO QUADRATIC

i **SOLVING EQUATIONS REDUCIBLE TO QUADRATIC**

Certain equations that are not really quadratic can be thought of in such a way that they can be solved as quadratic. For example,

$$x + 3\sqrt{x} - 10 = 0$$

$$(\sqrt{x})^2 + 3\sqrt{x} - 10 = 0 \qquad \text{Thinking of } x \text{ as } (\sqrt{x})^2$$

$$u^2 + 3u - 10 = 0 \qquad \text{To make this clearer, write } u \text{ instead of } \sqrt{x}.$$

The equation $u^2 + 3u - 10 = 0$ can be solved by factoring or by using the quadratic formula. After that, we can find x by remembering that $\sqrt{x} = u$. Equations that can be solved in this way are said to be *reducible to quadratic*.

> To solve equations reducible to quadratic we first make a substitution, solve for the new variable, and then solve for the original variable.

EXAMPLE 1 Solve $x + 3\sqrt{x} - 10 = 0$.

Let $u = \sqrt{x}$. Then we solve the equation resulting from substituting u for $\sqrt{x}$:

$$u^2 + 3u - 10 = 0$$
$$(u + 5)(u - 2) = 0$$
$$u = -5 \quad \text{or} \quad u = 2$$

We have solved for u. Now we solve for x. We substitute $\sqrt{x}$ for u and solve these equations:

$$\sqrt{x} = -5 \qquad \text{or} \quad \sqrt{x} = 2$$
$$(\text{no solution}) \quad \text{or} \quad x = 4.$$

Check:

$$\begin{array}{c|c} x + 3\sqrt{x} - 10 = 0 & \\ \hline 4 + 3\sqrt{4} - 10 & 0 \\ 4 + 6 - 10 & \\ 0 & \\ \end{array}$$

The solution is 4.

DO EXERCISE 1.

EXAMPLE 2 Solve $x^4 - 6x^2 + 7 = 0$.

Let $u = x^2$. Then we solve the equation resulting from substituting u for x^2.

We have

$$u^2 - 6u + 7 = 0$$
$$a = 1, b = -6, c = 7$$
$$u = \frac{-b \pm \sqrt{b^2 - 4ac}}{2a} = \frac{-(-6) \pm \sqrt{(-6)^2 - 4 \cdot 1 \cdot 7}}{2 \cdot 1}$$
$$u = \frac{6 \pm \sqrt{8}}{2}$$
$$u = \frac{2 \cdot 3 \pm 2\sqrt{2}}{2 \cdot 1} = 3 \pm \sqrt{2}.$$

We have solved for u. Now we solve for x. We substitute x^2 for u and solve for x:

$$x^2 = 3 + \sqrt{2} \qquad \text{or} \quad x^2 = 3 - \sqrt{2}$$
$$x = \pm\sqrt{3 + \sqrt{2}} \quad \text{or} \quad x = \pm\sqrt{3 - \sqrt{2}}.$$

Thus we have the four solutions $\sqrt{3 + \sqrt{2}}$, $-\sqrt{3 + \sqrt{2}}$, $\sqrt{3 - \sqrt{2}}$, and $-\sqrt{3 - \sqrt{2}}$.

DO EXERCISE 2.

EXAMPLE 3   Solve $(x^2 - x)^2 - 14(x^2 - x) + 24 = 0$.

Let $u = x^2 - x$. Then we solve the equation resulting from substituting u for $x^2 - x$:

$$u^2 - 14u + 24 = 0$$
$$(u - 12)(u - 2) = 0$$
$$u = 12 \quad \text{or} \quad u = 2.$$

We have solved for u. Now we solve for x. We substitute $x^2 - x$ for u and solve:

$$x^2 - x = 12 \qquad \text{or} \qquad x^2 - x = 2$$
$$x^2 - x - 12 = 0 \qquad \text{or} \qquad x^2 - x - 2 = 0$$
$$(x - 4)(x + 3) = 0 \qquad \text{or} \quad (x - 2)(x + 1) = 0$$
$$x = 4 \quad \text{or} \quad x = -3 \quad \text{or} \quad x = 2 \quad \text{or} \quad x = -1.$$

The solutions are $4, -3, 2, -1$.

DO EXERCISE 3.

EXAMPLE 4   Solve $t^{2/5} - t^{1/5} - 2 = 0$.

Let $u = t^{1/5}$. Then we solve the equation resulting from substituting u for $t^{1/5}$:

$$u^2 - u - 2 = 0$$
$$(u - 2)(u + 1) = 0$$
$$u = 2 \quad \text{or} \quad u = -1.$$

Now we substitute $t^{1/5}$ for u and solve:

$$t^{1/5} = 2 \quad \text{or} \quad t^{1/5} = -1$$
$$t = 32 \quad \text{or} \quad t = -1 \qquad \text{Principle of powers;}$$
$$\text{raising to the 5th power}$$

The solutions are 32 and -1.

DO EXERCISE 4.

2. Solve.

$$2x^4 - 10x^2 + 11 = 0$$

3. Solve.

$$(x^2 - 1)^2 - (x^2 - 1) - 2 = 0$$

4. Solve.

$$t^{2/3} - 3t^{1/3} - 10 = 0$$

ii PROBLEM SOLVING

EXAMPLE 5   (*A well problem*). An object is dropped into a well. Two seconds later the sound of the splash is heard at the top. The speed of sound is 1100 ft/sec. How deep is the well?

1. *Familiarize:* We first make a drawing and label it with known and unknown information. We can picture the situation as follows. We have a well s feet deep. The time it takes for the object to fall to the bottom of the well can be represented by t_1. The time it takes for the sound to get back to the top of the well can be represented by t_2. The total amount of time is the sum of t_1 and t_2. This gives us the equation

$$t_1 + t_2 = 2. \tag{1}$$

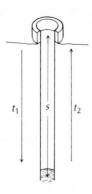

2. *Translate:* Now can we find any relationship between the two times and the distance s? Often in problem solving you may need to look up related formulas in perhaps a physics book, an encyclopedia, or maybe another mathematics book. It turns out that the formula for falling objects, which we considered in Section 2.5, becomes

$$s = 16t^2 + v_0 t,$$

when the distance is in feet. Since the object is dropped, $v_0 = 0$. The time t_1 that it takes for the object to reach the bottom of the well can be found as follows:

$$s = 16t_1^2, \quad \text{or} \quad t_1 = \frac{\sqrt{s}}{4}. \tag{2}$$

We now have an expression for t_1. What about t_2? To find how long it takes for the sound to get back to the top of the well we can use the formula $d = rt$, which we considered in Section 2.3. We want to find s, so $d = s$. The speed r is the speed of sound. (You might need to look this up.) It is 1100 ft/sec. Then

$$s = 1100 \cdot t_2, \quad \text{or} \quad t_2 = \frac{s}{1100}. \tag{3}$$

We have an expression for t_1 in Eq. (2) and an expression for t_2 in Eq. (3). We substitute these into Eq. (1) and obtain

$$t_1 + t_2 = 2, \quad \text{or} \quad \frac{\sqrt{s}}{4} + \frac{s}{1100} = 2. \tag{4}$$

3. *Carry out:* We solve Eq. (4) for *s*. Multiplying by 1100, we get

$$275\sqrt{s} + s = 2200, \quad \text{or} \quad s + 275\sqrt{s} - 2200 = 0.$$

This equation is reducible to quadratic with $u = \sqrt{s}$. Substituting, we get

$$u^2 + 275u - 2200 = 0.$$

Using the quadratic formula we can solve for *u*:

$$u = \frac{-275 + \sqrt{275^2 + 8800}}{2} \qquad \text{We want the positive solution.}$$

$$= \frac{-275 + \sqrt{84{,}425}}{2}$$

$$\approx \frac{-275 + 290.56}{2} \qquad \text{Use your calculator or factor and approximate.}$$

$$u \approx \frac{15.56}{2} = 7.78.$$

Thus $u \approx 7.78 \approx \sqrt{s}$, so $s \approx 60.5284$; that is, the well is about 60.53 ft deep.

DO EXERCISE 5.

5. An object is dropped into a well. In 5 sec the sound of the splash reaches the top of the well. If we assume that the speed of sound is 1100 ft/sec, how deep is the well?

EXERCISE SET 2.7

i Solve.

1. $x - 10\sqrt{x} + 9 = 0$

2. $2x - 9\sqrt{x} + 4 = 0$

3. $x^4 - 10x^2 + 25 = 0$

4. $x^4 - 3x^2 + 2 = 0$

5. $t^{2/3} + t^{1/3} - 6 = 0$

6. $w^{2/3} - 2w^{1/3} - 8 = 0$

7. $z^{1/2} = z^{1/4} + 2$

8. $6 = m^{1/3} - m^{1/6}$

9. $(x^2 - 6x)^2 - 2(x^2 - 6x) - 35 = 0$

10. $(1 + \sqrt{x})^2 + (1 + \sqrt{x}) - 6 = 0$

11. $(y^2 - 5y)^2 + (y^2 - 5y) - 12 = 0$

12. $(2t^2 + t)^2 - 4(2t^2 + t) + 3 = 0$

13. $w^4 - 4w^2 - 2 = 0$

14. $t^4 - 5t^2 + 5 = 0$

15. $x^{-2} - x^{-1} - 6 = 0$

16. $4x^{-2} - x^{-1} - 5 = 0$

17. $2x^{-2} + x^{-1} = 1$

18. $10 - 9m^{-1} = m^{-2}$

19. $\left(\dfrac{x^2 - 2}{x}\right)^2 - 7\left(\dfrac{x^2 - 2}{x}\right) - 18 = 0$

20. $\left(\dfrac{x^2 + 1}{x}\right)^2 - 8\left(\dfrac{x^2 + 1}{x}\right) + 15 = 0$

21. $\dfrac{x}{x-1} - 6\sqrt{\dfrac{x}{x-1}} - 40 = 0$

22. $\dfrac{2x+1}{x} + 30 = 7\sqrt{\dfrac{2x+1}{x}} \quad \left(\textit{Hint: } u = \dfrac{2x+1}{x}.\right)$

23. $5\left(\dfrac{x+2}{x-2}\right)^2 = 3\left(\dfrac{x+2}{x-2}\right) + 2$

24. $\left(\dfrac{x+1}{x+3}\right)^2 + \left(\dfrac{x+1}{x+3}\right) - 6 = 0$

25. $\left(\dfrac{x^2 - 1}{x}\right)^2 - \left(\dfrac{x^2 - 1}{x}\right) - 2 = 0$

26. $5\left(\dfrac{x+2}{x-2}\right)^2 = 3\left(\dfrac{x+2}{x-2}\right) + 2$

ii **Problem Solving**

27. A stone is dropped from a cliff. In 3 sec the sound of the stone striking the ground reaches the top of the cliff. If we assume that the speed of sound is 1100 ft/sec, how high is the cliff?

28. A stone is dropped from a cliff. In 4 sec the sound of the stone striking the ground reaches the top of the cliff. If we assume that the speed of sound is 1100 ft/sec, how high is the cliff?

Solve. Check possible solutions by substituting into the original equation.

29. ▤ $6.75x = \sqrt{35x} + 5.36$

30. ▤ $\pi x^4 - \sqrt{99.3} = \pi^2 x^2$

Solve.

31. $9x^{3/2} - 8 = x^3$

32. $\sqrt[3]{2x + 3} = \sqrt[6]{2x + 3}$

33. $\sqrt{x - 3} - \sqrt[4]{x - 3} = 2$

34. $a^3 - 26a^{3/2} - 27 = 0$

35. Solve.

$$\frac{2x + 1}{x} = 3 + 7\sqrt{\frac{2x + 1}{x}}$$

OBJECTIVE

You should be able to:

i Find equations of variation and solve applied problems involving variation.

1. Find an equation of variation where y varies directly as x, and $y = 32$ when $x = 0.2$.

2.8

VARIATION

i TYPES OF VARIATION

Direct Variation

There are many situations that yield linear equations like $y = kx$, where k is some positive constant. Note that as x increases, y increases. In such a situation we say that we have *direct variation*, and k is called the *variation constant*. Usually only positive values of x and y are considered.

Definition

> If two variables x and y are related as in the equation $y = kx$, where k is a positive constant, we say that "y varies directly as x," or that "y is directly proportional to x."

For example, the circumference C of a circle varies directly as its diameter D: $C = \pi D$. A car moves at a constant speed of 65 mph. The distance d it travels varies directly as the time t: $d = 65t$.

EXAMPLE 1 Find an equation of variation where y varies directly as x, and $y = 5.6$ when $x = 8$.

We know that $y = kx$, so $5.6 = k \cdot 8$ and $0.7 = k$. Thus $y = 0.7x$.

DO EXERCISE 1.

Suppose y varies directly as x. Then we have $y = kx$. When (x_1, y_1) and (x_2, y_2) are solutions of the equation, we have $y_1 = kx_1$ and $y_2 = kx_2$, and

$$\frac{y_2}{y_1} = \frac{kx_2}{kx_1} = \frac{k}{k} \cdot \frac{x_2}{x_1} = \frac{x_2}{x_1}.$$

The equation $y_2/y_1 = x_2/x_1$ is called a *proportion*. Such equations are helpful in solving applied problems.

EXAMPLE 2   (*A spring problem*). *Hooke's law* states that the distance *d* that an elastic object such as a spring is stretched by placing a certain weight on it varies directly as the weight *w* of the object. If the distance is 40 cm when the weight is 3 kg, what is the distance stretched when a 2-kg weight is attached?

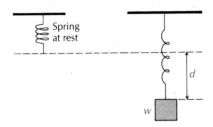

Method 1.   First find *k* using the first set of data. Then solve for d_2 using the second set of data.

$$d_1 = kw_1 \qquad\qquad d_2 = \frac{40}{3} \cdot w_2$$

$$40 = k \cdot 3 \qquad\qquad d_2 = \frac{40}{3} \cdot 2$$

$$\frac{40}{3} = k \qquad\qquad d_2 = \frac{80}{3}, \quad \text{or } 26.7 \text{ cm}$$

Method 2.   Use a proportion to solve for d_2 without first finding the variation constant.

$$\frac{d_2}{d_1} = \frac{w_2}{w_1}$$

$$\frac{d_2}{40} = \frac{2}{3}$$

$$3 \cdot d_2 = 40 \cdot 2$$

$$d_2 = 26.7 \text{ cm}$$

DO EXERCISES 2–4.

Inverse Variation

There are also situations that yield equations of the type $y = k/x$, where *k* is a positive constant. Note that as positive values of *x* increase, *y* decreases. In such a situation, we say that we have *inverse variation*, and *k* is called the *variation constant*. In applications we are usually considering only positive values of *x* and *y*.

Definition

If two variables *x* and *y* are related as in the equation $y = k/x$, where *k* is a positive constant, we say that "*y* varies inversely as *x*," or that "*y* is inversely proportional to *x*."

DO EXERCISE 5.

2. Under the conditions of Example 2, what weight would be required to stretch the spring 60 cm?

3. *Ohm's law* states that the voltage *V* in an electric circuit varies directly as the number of amperes *I* of electric current in the circuit. If the voltage is 10 volts when the current is 3 amperes, what is the voltage when the current is 15 amperes?

4. The amount of garbage *G* produced in the United States varies directly as the number of people *N* who produce the garbage. It is known that 50 tons of garbage is produced by 200 people in 1 yr. The population of San Francisco is 705,000. How much garbage is produced by San Francisco in 1 yr?

5. Find an equation of variation where *y* varies inversely as *x*, and $y = 32$ when $x = 0.2$.

Suppose that y varies inversely as x. Then we have $y = k/x$. When (x_1, y_1) and (x_2, y_2) are solutions of the equation, we have $y_1 = k/x_1$ and $y_2 = k/x_2$. Then

$$\frac{y_2}{y_1} = \frac{\dfrac{k}{x_2}}{\dfrac{k}{x_1}} = \frac{k}{x_2} \cdot \frac{x_1}{k} = \frac{k}{k} \cdot \frac{x_1}{x_2} = \frac{x_1}{x_2}.$$

The equation $y_2/y_1 = x_1/x_2$ is a proportion that also can be helpful in solving applied problems. Compare it to the proportion for direct variation. Note how the variables are interchanged.

EXAMPLE 3 (*Stocks and gold*). Certain economists theorize that stock prices are inversely proportional to the price of gold. That is, when the price of gold goes up, the prices of stock go down; and when the price of gold goes down, the prices of stock go up. Let us assume that the Dow-Jones Industrial Average, D, an index of the overall price of stock, is inversely proportional to the price of gold, G, in dollars per ounce. One day the Dow-Jones Industrial Average was 846 and the price of gold was $528 per ounce. What will the Dow-Jones average be if the price of gold drops to $480 per ounce?

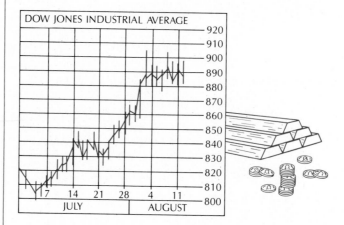

Method 1. Find k using the first set of data. Then solve for D_2.

$$D_1 = \frac{k}{G_1}$$

$$846 = \frac{k}{528}$$

$$k = 846(528)$$

$$= 446{,}688$$

$$D_2 = \frac{k}{G_2}$$

$$= \frac{446{,}688}{480}$$

$$= 930.6$$

Method 2. Use a proportion to solve for D_2 without first finding the variation constant.

$$\frac{D_2}{D_1} = \frac{G_1}{G_2}$$

$$\frac{D_2}{846} = \frac{528}{480}$$

$$480(D_2) = 846(528)$$

$$D_2 = 930.6$$

Warning! Do not put too much "stock" in the equation of Example 3. It is meant to give us an idea of economic relationships. An equation to predict the stock market accurately has not been found!

DO EXERCISE 6.

Other Kinds of Variation

There are many other types of variation.

Definition

y varies directly as the square of *x* if there is some positive constant *k* such that $y = kx^2$.

For example, the area A of a circle varies directly as the square of the radius r: $A = \pi r^2$.

DO EXERCISES 7 AND 8.

Definition

y varies inversely as the square of *x* if there is some positive constant *k* such that $y = k/x^2$.

For example, the weight W of a body varies inversely as the square of the distance d from the center of the earth: $W = k/d^2$.

DO EXERCISES 9 AND 10.

Definition

y varies *jointly* as *x* and *z* if there is some positive constant *k* such that $y = kxz$.

For example, consider the equation for the area of a triangle: $A = \frac{1}{2}bh$. The area varies jointly as b and h. The variation constant is $\frac{1}{2}$.

DO EXERCISES 11 AND 12.

Several kinds of variation can occur together. For example, if $y = k \cdot xz^3/w^2$, then y varies jointly as x and the cube of z and inversely as the square of w.

DO EXERCISE 13.

6. The time t required to drive a fixed distance varies inversely as the speed r. It takes 5 hr at 60 km/h to drive a fixed distance. How long would it take to drive the fixed distance at 40 km/h?

7. Find a proportion involving $\dfrac{A_2}{A_1}$, where A_1 and A_2 are areas of circles.

8. Find an equation of variation where y varies directly as the square of x, and $y = 12$ when $x = 2$.

9. Find a proportion involving $\dfrac{W_2}{W_1}$.

10. Find an equation of variation where y varies inversely as the square of x, and $y = \frac{1}{4}$ when $x = 6$.

11. Find a proportion involving $\dfrac{A_2}{A_1}$.

12. Find an equation of variation where y varies jointly as x and z, and $y = 42$ when $x = 2$ and $z = 3$.

13. Find an equation of variation where y varies jointly as x and z and inversely as the square of w, and $y = 105$ when $x = 3$, $z = 20$, and $w = 2$.

14. The distance S that an object falls from some point above the ground varies directly as the square of the time t that it falls. If the object falls 4 ft in 0.5 sec, how long will it take the object to fall 64 ft?

EXAMPLE 4 (*The volume of a tree trunk*). The volume of wood V in a tree trunk varies jointly as the height h and the square of the girth g (girth is distance around). If the volume is 7750 ft^3 when the height is 100 ft and the girth is 5 ft, what is the height when the volume is 37,975 ft^3 and the girth is 7 ft?

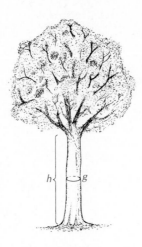

Method 1. First find k using the first set of data. Then solve for h_2 using the second set of data.

$$V_1 = k \cdot h_1 \cdot g_1^2$$
$$7750 = k \cdot 100 \cdot 5^2$$
$$3.1 = k$$

Then

$$37{,}975 = 3.1 \cdot h_2 \cdot 7^2$$
$$250 = h_2$$
$$h_2 = 250 \text{ ft.}$$

Method 2. Use a proportion to solve for h_2 without first finding the variation constant.

$$\frac{V_2}{V_1} = \frac{h_2 \cdot g_2^2}{h_1 \cdot g_1^2}$$

$$\frac{37{,}975}{7750} = \frac{h_2 \cdot 7^2}{100 \cdot 5^2}$$

$$h_2 = 250 \text{ ft}$$

DO EXERCISE 14.

EXAMPLE 5 (*The weight of an astronaut*). The weight W of an object varies inversely as the square of the distance d from the object to the center of the earth. At sea level (4000 mi from the center of the earth) an astronaut weighs 200 lb. Find his weight when he is 100 mi above the surface of the earth, and the spacecraft is not in motion.

We use the proportion

$$\frac{W_2}{W_1} = \frac{d_1^2}{d_2^2}$$

$$\frac{W_2}{200} = \frac{4000^2}{4100^2}$$

$$16{,}810{,}000 \cdot W_2 = 200(16{,}000{,}000)$$

$$W_2 = 190 \text{ lb (to the nearest pound)}.$$

DO EXERCISE 15.

15. a) How much will the 200-lb astronaut weigh when he is 1000 mi from earth?

b) How far is he from earth when his weight is reduced to 50 lb?

EXERCISE SET 2.8

i Find an equation of variation where:

1. y varies directly as x, and $y = 0.6$ when $x = 0.4$.

2. y varies inversely as x, and $y = 0.4$ when $x = 0.8$.

3. y varies inversely as the square of x, and $y = 0.15$ when $x = 0.1$.

4. y varies jointly as x and z, and $y = 56$ when $x = 7$ and $z = 10$.

5. y varies jointly as x and z and inversely as w, and $y = \frac{3}{2}$ when $x = 2$, $z = 3$, and $w = 4$.

6. y varies jointly as x and the square of z, and $y = 105$ when $x = 14$ and $z = 5$.

7. y varies jointly as x and z and inversely as the square of w, and $y = \frac{12}{5}$ when $x = 16$, $z = 3$, and $w = 5$.

8. y varies jointly as x and z and inversely as the product of w and p, and $y = \frac{3}{28}$ when $x = 3$, $z = 10$, $w = 7$, and $p = 8$.

9. Suppose y varies directly as x and x is doubled. What is the effect on y?

10. Suppose y varies inversely as x and x is tripled. What is the effect on y?

11. Suppose y varies inversely as the square of x and x is multiplied by n. What is the effect on y?

12. Suppose y varies directly as the square of x and x is multiplied by n. What is the effect on y?

13. The amount of pollution A entering the atmosphere varies directly as the amount of people N living in an area. If 60,000 people cause 42,600 tons of pollutants, how many tons entered the atmosphere in a city with a population of 750,000?

14. The volume V of a given mass of gas varies directly as the temperature T and inversely as the pressure P. If $V = 231$ in^3 when $T = 420°$ and $P = 20$ lb/in^2, what is the volume when $T = 300°$ and $P = 15$ lb/in^2?

15. (*The strength of a beam*). The safe load (amount it supports without breaking) L of a beam varies jointly as its width w and the square of its height h and inversely as its length l. If the width and height are doubled at the same time the length is halved, what is the effect on L?

16. For a chord $\overline{PQ}$ through a fixed point A in a circle, the length of $\overline{PA}$ is inversely proportional to the length of $\overline{AQ}$. If the length of $\overline{PA} = 64$ when the length of $\overline{AQ} = 16$, what is the length of $\overline{AQ}$ when the length of $\overline{PA} = 4$?

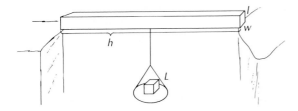

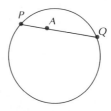

17. ▦ (*A sighting problem*). The distance d that one can see to the horizon varies directly as the square root of the height above sea level. If a person 19.5 m above sea level can see 28.97 km, how high above sea level must one be to see 54.32 km?

18. ▦ (*Electrical resistance*). At a fixed temperature and chemical composition, the resistance of a wire varies directly as the length l and inversely as the square of the diameter d. If the resistance of a certain kind of wire is 0.112 ohm when the diameter is 0.254 cm and the length is 15.24 m, what is the resistance of a wire whose length is 608.7 m and whose diameter is 0.478 cm?

19. (*Area of a cube*). The area of a cube varies directly as the square of the length of a side. If a cube has an area of 168.54 m^2 when the length of a side is 5.3 m, what will the area be when the length of a side is 10.2 m?

20. (*Intensity of light*). The intensity of light l from a light bulb varies inversely as the square of the distance d from the bulb. Suppose l is 90 W/m^2 when the distance is 5 m. Find the intensity at a distance of 10 m.

21. (*Earned run average*). A pitcher's earned run average *A* varies directly as the number of earned runs *R* allowed and inversely as the number of innings pitched *I*. In a recent year a pitcher had an earned run average of 2.92. He gave up 85 earned runs in 262 innings. How many earned runs would he have given up had he pitched 300 innings? Round to the nearest whole number.

22. (*Volume of a gas*). The volume *V* of a given mass of a gas varies directly as the temperature *T* and inversely as the pressure *P*. If $V = 231$ cm³ when $T = 42°$ and $P = 20$ kg/cm², what is the volume when $T = 30°$ and $P = 15$ kg/cm²?

23. Show that if *p* varies directly as *q*, then *q* varies directly as *p*.

24. Show that if *u* varies inversely as *v*, then *v* varies inversely as *u*, and $1/u$ varies directly as *v*.

25. The area of a circle varies directly as the square of the length of a diameter. What is the variation constant?

26. *P* varies directly as the square of *t*. How does *t* vary in relationship to *P*?

27. (*The gravity model in sociology*). It has been determined that the average number of telephone calls in a day *N*, between two cities, is directly proportional to the populations P_1 and P_2 of the cities and inversely proportional to the square of the distance *d* between the cities. That is,

$$N = \frac{kP_1P_2}{d^2}.$$

This theory is known as the "gravity model" because the equation is similar to the one used in Newton's theory of gravity. Use a calculator to find solutions to these problems.

  a) The population of Indianapolis is about 744,624, the population of Cincinnati is about 452,524, and the distance between the cities is 174 km. The average number of daily phone calls between the two cities is 11,153. Find the value *k* and write the equation of variation.

  b) The population of Detroit is about 1,511,482 and it is 446 km from Indianapolis. Find the average number of daily phone calls between Detroit and Indianapolis.

  c) The average number of daily phone calls between Indianapolis and New York City is 4270 and the population of New York City is about 7,895,563. Find the distance between Indianapolis and New York City.

  d) Why is this model not appropriate for adjoining cities such as Minneapolis and St. Paul? (Sociologists say that as the communication between two cities increases, the cities tend to merge.)

OBJECTIVES

You should be able to:

i Perform a given calculation involving dimension symbols and simplify, if possible, without making any unit changes.

ii Perform a given change of dimension symbols, using substitution or "multiplying by one."

2.9

HANDLING DIMENSION SYMBOLS

In this section we learn to manipulate dimension symbols. We make calculations, simplifications, and changes of unit. The algebraic skills we have reviewed up to now are quite useful for this. The following table contains abbreviations for some dimension symbols.

Dimension symbol	Unit
m	meter
cm	centimeter (0.01 m)
km	kilometer (1000 m)
g	gram
cg	centigram (0.01 g)
kg	kilogram (1000 g)
s or sec	second
h or hr	hour
L	liter
mL	milliliter (0.001 L)

SPEED

Speed is often determined by measuring a distance and a time and then dividing the distance by the time (this is *average* speed):

$$\text{Speed} = \frac{\text{Distance}}{\text{Time}}.$$

If a distance is measured in kilometers and the time required to travel that distance is measured in hours, the speed will be computed in *kilometers per hour* (km/h). For example, if a car travels 100 km in 2 h, the average speed is

$$\frac{100 \text{ km}}{2 \text{ h}}, \quad \text{or} \quad 50 \frac{\text{km}}{\text{h}}.$$

DO EXERCISES 1 AND 2.

■ DIMENSION SYMBOLS

The symbol 100 km/2 h makes it look as though we are dividing 100 km by 2 h. It may be argued that we cannot divide 100 km by 2 h (we can only divide 100 by 2). Nevertheless, it is convenient to treat dimension symbols such as *kilometers, hours, feet, seconds,* and *pounds* as though they were numerals or variables, for the reason that correct results can thus be obtained mechanically. Compare, for example,

$$\frac{100x}{2y} = \frac{100}{2} \cdot \frac{x}{y} = 50 \frac{x}{y}$$

with

$$\frac{100 \text{ km}}{2 \text{ h}} = \frac{100}{2} \cdot \frac{\text{km}}{\text{h}} = 50 \frac{\text{km}}{\text{h}}.$$

The analogy holds in other situations, as shown in the following examples.

EXAMPLE 1 Compare

$$3 \text{ ft} + 2 \text{ ft} = (3 + 2) \text{ ft} = 5 \text{ ft}$$

with

$$3x + 2x = (3 + 2)x = 5x.$$

This looks like a distributive law in use.

DO EXERCISES 3–5.

EXAMPLE 2 Compare

$$4 \text{ m} \cdot 3 \text{ m} = 3 \cdot 4 \cdot \text{m} \cdot \text{m} = 12 \text{ m}^2 \text{ (sq m)}$$

with

$$4x \cdot 3x = 4 \cdot 3 \cdot x \cdot x = 12x^2.$$

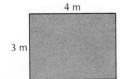

4 m

3 m

What is the speed in m/sec?

1. 186,000 m, 10 sec

2. 8 m, 16 sec

Add these measures.

3. 45 ft, 17 ft

4. $\frac{3}{4}$ kg, $\frac{2}{5}$ kg

5. $70 \frac{\text{cm}}{\text{sec}}$, $35 \frac{\text{cm}}{\text{sec}}$

Perform these calculations and simplify if possible. Do not make any unit changes.

6. $36 \text{ ft} \cdot \dfrac{1 \text{ yd}}{3 \text{ ft}}$

7. $5 \text{ lb} \cdot \dfrac{16 \text{ oz}}{1 \text{ lb}}$

8. $\dfrac{4 \text{ kg}}{5 \text{ ft}} \cdot \dfrac{7 \text{ ft}}{8 \text{ kg}}$

9. $\dfrac{5 \text{ in.} \cdot 9 \text{ lb/hr}}{4 \text{ hr}}$

10. $\dfrac{10 \text{ lb}}{7 \text{ m}} \cdot \dfrac{14 \text{ lb}}{5 \text{ m}}$

Perform the following changes of unit. Use substitution.

11. 34 yd, change to in.

12. 11 mi, change to ft.

13. 5 hr, change to sec.

Perform the following changes of unit. Use multiplying by one.

14. 720 in., change to yd.

15. 36,960 m, change to km.

16. 360,000 sec, change to hr.

EXAMPLE 3   Compare

$$5 \text{ men} \cdot 8 \text{ hr} = 5 \cdot 8 \cdot \text{man-hr} = 40 \text{ man-hr}$$

with

$$5x \cdot 8y = 5 \cdot 8 \cdot x \cdot y = 40xy.$$

In each of the above examples, dimension symbols are treated as though they were variables or numerals, and as though a symbol such as 3 m represents a product 3 times m. A symbol like km/h is treated as though it represents a division of km by h (*kilometers* by *hours*). Any two measures can be "multiplied" or "divided."

DO EXERCISES 6–10.

ii CHANGES OF UNIT

Changes of unit can be accomplished by substitutions.

EXAMPLE 4   Change to inches: 25 yd.

$$
\begin{aligned}
25 \text{ yd} &= 25 \cdot 1 \text{ yd} \\
&= 25 \cdot 3 \text{ ft} \qquad \text{Substituting 3 ft for 1 yd} \\
&= 25 \cdot 3 \cdot 1 \text{ ft} \\
&= 25 \cdot 3 \cdot 12 \text{ in.} \qquad \text{Substituting 12 in. for 1 ft} \\
&= 900 \text{ in.}
\end{aligned}
$$

DO EXERCISES 11–13.

The notion of "multiplying by one" can also be used to change units.

EXAMPLE 5   Change 7.2 in. to yd.

$$7.2 \text{ in.} = 7.2 \text{ in.} \cdot \frac{1 \text{ ft}}{12 \text{ in.}} \cdot \frac{1 \text{ yd}}{3 \text{ ft}}$$

Both of these are equal to 1.

$$= \frac{7.2}{12 \cdot 3} \cdot \frac{\text{in.}}{\text{in.}} \cdot \frac{\text{ft}}{\text{ft}} \text{ yd}$$

$$= 0.2 \text{ yd}$$

In Example 5, we first used the following symbol for 1:

$$\frac{1 \text{ ft}}{12 \text{ in.}}.$$

"ft" in the numerator is the unit we are changing *to*.

"in." in the denominator is the unit we are changing *from*.

In the final multiplication we were converting from ft to yd, so 1 yd was in the numerator and 3 ft was in the denominator.

DO EXERCISES 14–16.

EXAMPLE 6   Change $60 \dfrac{km}{h}$ to $\dfrac{m}{sec}$.

$$60\,\frac{km}{h} = 60\,\frac{km}{h} \cdot \frac{1000\ m}{1\ km} \cdot \frac{1\ h}{60\ min} \cdot \frac{1\ min}{60\ sec}$$

$$= \frac{60 \cdot 1000}{60 \cdot 60} \cdot \frac{km}{km} \cdot \frac{h}{h} \cdot \frac{min}{min} \cdot \frac{m}{sec}$$

$$= 16.67\,\frac{m}{sec}$$

$$= 16.67\ \text{m/s.*}$$

EXAMPLE 7   Change $55 \dfrac{mi}{hr}$ to $\dfrac{ft}{sec}$.

$$55\,\frac{mi}{hr} = 55\,\frac{mi}{hr} \cdot \frac{5280\ ft}{1\ mi} \cdot \frac{1\ hr}{60\ min} \cdot \frac{1\ min}{60\ sec}$$

$$= \frac{55 \cdot 5280}{60 \cdot 60} \cdot \frac{mi}{mi} \cdot \frac{hr}{hr} \cdot \frac{min}{min} \cdot \frac{ft}{sec}$$

$$= 80\frac{2}{3}\,\frac{ft}{sec}$$

DO EXERCISES 17–20.

Perform the following changes of unit. Use multiplying by one.

17. $120 \dfrac{mi}{hr}$, change to $\dfrac{ft}{sec}$.

18. $3600\ cm^2$, change to m^2.

19. $50 \dfrac{kg}{L}$, change to $\dfrac{g}{cm^3}$.
(*Hint:* $1\ L = 1000\ cm^3$.)

20. $\dfrac{\$72}{day}$, change to $\dfrac{\cent}{hr}$.

EXERCISE SET 2.9

i Perform these calculations and simplify if possible. Do not make any unit changes.

1. $36\ ft \cdot \dfrac{1\ yd}{3\ ft}$

2. $6\ lb \cdot \dfrac{16\ oz}{1\ lb}$

3. $6\ kg \cdot 8\,\dfrac{hr}{kg}$

4. $9\,\dfrac{km}{h} \cdot 3\ h$

5. $3\ cm \cdot \dfrac{2\ g}{2\ cm}$

6. $\dfrac{9\ km}{3\ days} \cdot 6\ days$

7. $6\ m + 2\ m$

8. $10\ tons + 6\ tons$

9. $5\ ft^3 + 7\ ft^3$

10. $10\ yd^3 + 17\ yd^3$

11. $\dfrac{3\ kg}{5\ m} \cdot \dfrac{7\ kg}{6\ m}$

12. $3\ acres \times 60\,\dfrac{1}{acre}$

13. $\dfrac{2000\ lb \cdot (6\ mi/hr)^2}{100\ ft}$

14. $\dfrac{7\ m \cdot 8\ kg/sec}{4\ sec}$

15. $\dfrac{6\ cm^2 \cdot 5\ cm/sec}{2\ sec^2/cm^2 \cdot 2\,\dfrac{1}{kg}}$

16. $\dfrac{320\ lb \cdot (5\ ft/sec)^2}{2 \cdot 32\,\dfrac{ft}{sec^2}}$

ii Perform the following changes of unit, using substitution or multiplying by one.

17. 72 in., change to ft

18. 17 hr, change to min

19. 2 days, change to sec

20. 360 sec, change to hr

21. $60 \dfrac{kg}{m}$, change to g/cm

22. $44 \dfrac{ft}{sec}$, change to mi/hr

23. $216\ m^2$, change to cm^2

24. $60 \dfrac{lb}{ft^3}$, change to ton/yd^3

25. $\dfrac{\$36}{day}$, change to $\dfrac{\cent}{hr}$

26. 1440 man-hr, change to man-days

---

* The standard abbreviation for meters per second is m/s and for kilometers per hour is km/h.

27. $1.73 \dfrac{mL}{sec}$, change to $\dfrac{L}{hr}$
(*Hint:* 1 liter = 1 L = 1000 mL = 1000 milliliters.)

28. $1800 \dfrac{g}{L}$, change to $\dfrac{cg}{mL}$

29. $186{,}000 \dfrac{mi}{sec}$ (speed of light), change to $\dfrac{mi}{yr}$.
(Let 365 days = 1 yr.)

30. $1100 \dfrac{ft}{sec}$ (speed of sound), change to $\dfrac{mi}{yr}$.
(Let 365 days = 1 yr.)

Use Table 6 at the back of the book to do the following unit changes.

31. ▦ $89.2 \dfrac{ft}{sec}$, change to m/min

32. ▦ 1013 yd^3, change to m^3

33. ▦ 640 mi^2, change to km^2

34. ▦ $312.2 \dfrac{kg}{m}$, change to lb/ft

35. If a steel rod 2 cm long weighs 5 g, how much does a rod of the same type weigh whose length is 3 cm? 5 m?

36. In Exercise 35, how long is a rod that weighs 4.3 g? 20 cg?

In chemistry, 1 *mole* of a substance is that mass of the substance, in grams, equal to its molecular weight. For example, 1 mole of oxygen is 32 grams because its molecular weight is 32, and 1 mole of neon is 20.2 grams.

Convert to grams.

37. 50 moles of oxygen

38. 44 moles of neon

Convert to moles.

39. 303 grams of neon

40. 377.6 grams of oxygen

SUMMARY AND REVIEW: CHAPTER 2

The following contains a summary of what you should be able to do after completing this chapter. The review exercises are for practice. Answers are at the back of the book. If you miss an exercise, restudy the section indicated alongside the answer.

You should be able to:

Solve equations using the addition and multiplication principles and the principle of zero products. Solve inequalities using the addition and multiplication principles. Solve quadratic equations using the quadratic formula, and solve radical equations using the principle of powers. Solve equations reducible to quadratic.

Solve.

1. $\dfrac{3x + 2}{7} - \dfrac{5x}{3} = \dfrac{32}{21}$

2. $(p - 3)(3p + 2)(p + 2) = 0$

3. $3x^2 + 2x - 8 = 0$

4. $r^2 + 3r - 1 = 0$

5. $3t^2 = 4 + 3t$

6. $\dfrac{5}{2x + 3} + \dfrac{1}{x - 6} = 0$

7. $y^4 - 3y^2 + 1 = 0$

8. $x = 2\sqrt{x} - 1$

9. $(x^2 - 1)^2 - (x^2 - 1) - 2 = 0$

10. $t^{2/3} - 10 = 3t^{1/3}$

11. $\sqrt{x - 1} - \sqrt{x - 4} = 1$

12. $\sqrt{5x + 1} - 1 = \sqrt{3x}$

13. Solve by completing the square: $y^2 - 6y = 16$.
Show your work.

14. $y^2 - 3y = 18$

15. $3[x - 5(4 + 2x)] = 7x - 10(3x - 2)$

16. $(z^2 - 1) + z = 14 - z$

17. $(x - 2)(x + 3) + 4 = 0$

18. $14 - 4y < 22$

19. $(x - 5)(x + 5) \geqslant (x - 5)(x + 4)$

Determine the sensible replacements in a radical expression.

20. Determine the sensible replacements in the radical expression $\sqrt{24 - 6x}$.

Determine the nature of the solutions of a quadratic equation and write a quadratic equation whose solutions are two given numbers.

Determine the nature of the solutions of each equation.

21. $4y^2 + 5y + 10 = 0$

22. $3z^2 + 2z - 1 = 0$

23. Write a quadratic equation whose solutions are -3 and $\frac{1}{2}$.

Solve a formula for a given letter.

24. Solve $v = \sqrt{2gh}$ for h.

25. Solve $\dfrac{1}{a} + \dfrac{1}{b} = \dfrac{1}{t}$ for t.

Solve problems involving the solution of the types of equations we have studied in this chapter.

Solve.

26. A student scores 73% and 79% on two tests. If the third test counts as though it were two tests, what score must the student make on the third test so the average will be 85%?

27. A can mow a lawn in 4 hr. B can mow it in 2 hr. How long would it take if they worked together?

28. A can mow a lawn in 3 hr. Working together, A and B can mow the lawn in 1 hr. How long would it take B, working alone, to mow the lawn?

29. 18 is 30% of what?

30. What is 9 percent of 50?

31. Two trains leave the same city at right angles. The first train travels 60 km/h. In 1 hr the trains are 100 km apart. How fast is the second train traveling?

32. In a right triangle the perimeter is 40 and the sum of the squares of the sides is 578. Find the lengths of the sides.

33. A boat travels 2 km upstream and 2 km downstream. The total time for both parts of the trip is 1 hr. The speed of the stream is 2 km/h. What is the speed of the boat in still water? Round to the nearest tenth.

Find equations of variation and solve applied problems involving variation.

Solve.

34. Find an equation of variation where y varies inversely as the square of x, and $y = 0.005$ when $x = 10$.

35. Find an equation of variation where T varies directly as the square of x and inversely as p, and $T = 0.01$ when $x = 6$ and $p = 20$.

36. It is theorized that the dividends paid on utilities stocks are inversely proportional to the prime (interest) rate. Recently, the dividends D on the stock of Indianapolis Power and Light were $2.09 per share and the prime rate was 19%. The prime rate, R, dropped to 17.5%. What dividends were then paid?

37. For a body falling freely from rest, the distance (s ft) that the body falls varies directly as the square of the time (t sec). Given that $s = 64$ when $t = 2$, find a formula for s in terms of t. How long will it take the body to fall 900 ft?

Perform a change of dimension symbols, using substitution or "multiplying by one."

38. Change $10\,\dfrac{\text{km}}{\text{h}}$ to $\dfrac{\text{m}}{\text{min}}$.

Determine whether equations are equivalent.

Determine whether the equations of each pair are equivalent.

39. $x = -5$
$x^2 = 25$

40. $x - 7 = \dfrac{x^2 - 49}{x + 7}$

$x - 7 = x - 7$

41. Suppose a, b, c, and d are nonzero real numbers such that c and d are solutions of

$$x^2 + ax + b = 0$$

and a and b are solutions of

$$x^2 + cx + d = 0.$$

Find $a + b + c + d$.

42. Determine whether

$$x - 7 = \frac{x^2 - 49}{x + 7}$$

is an identity.

43. The area of a circle varies directly as the square of its circumference. Write an equation of variation and find the variation constant.

44. Solve:

$$\sqrt{\sqrt{\sqrt{\sqrt{x}}}} = 2.$$

45. Determine whether $\dfrac{x+1}{x+2} = \dfrac{1}{2}$ is an identity.

TEST: CHAPTER 2

Solve.

1. $(2y - 9)(y + 4)(y - 5) = 0$

2. $x - 7\sqrt{x} + 6 = 0$

3. $2t^2 - t - 21 = 0$

4. $4t^2 - 3t - 5 = 0$

5. $\dfrac{8}{3x - 5} = \dfrac{4}{x + 3}$

6. $5x^2 + 12x - 2 = 0$

7. $\sqrt{y - 11} = \sqrt{y + 10} - 3$

8. $\dfrac{2x - 3}{3} = \dfrac{x + 4}{4} - 2$

9. $(a + 3)(a - 7) - 11 = 0$

10. $22 - 8y < 38$

11. $x^2 + x = 42$

12. $3x - 4(x + 6) = 2[x - 3(4 - x)]$

13. 18 is what percent of 9?

14. Grain flows through spout A five times faster than through spout B. When grain flows through both spouts, a grain bin is filled in 4 hr. How many hours would it take to fill the grain bin if grain flows through spout B alone?

15. The speed of train A is 8 mph slower than the speed of train B. Train A travels 144 mi in the same time it takes train B to travel 240 mi. Find the speed of train A.

16. Solve by completing the square. Show your work.

$$3x^2 - 12x - 6 = 0$$

17. Determine the nature of the solutions of $5z^2 + 8z - 4 = 0$.

18. Write a quadratic equation whose solutions are -6 and $\frac{2}{5}$.

19. Solve $\dfrac{W_1}{S_1 T_1} = \dfrac{W_2}{S_2 T_2}$ for T_2.

20. The hypotenuse of a right triangle is 50 ft. One leg is 10 ft longer than the other. What are the lengths of the legs?

21. Find an equation of variation where y varies jointly as x and w and inversely as the square of z, and $y = 10$ when $x = \frac{4}{3}$, $w = 5$, and $z = 2$.

22. The length l of rectangles of *fixed area* is inversely proportional to the width w. Suppose the length is 64 cm when the width is 3 cm. Find the length when the width is 12 cm.

23. Determine whether these equations are equivalent.

$$2x + 3 = -4$$
$$3x + 3 = x - 4$$

24. Determine the sensible replacements in the radical expression $\sqrt{10 - 5x}$.

25. Change $1200 \dfrac{m}{min}$ to $\dfrac{km}{hr}$.

Solve.

26. $x = \dfrac{1}{1 + x}$

27. $\sqrt{7x} - \sqrt{3x} = 7 - 3$

How can we determine which of these letters have a line of symmetry?

As the title indicates, this chapter deals with *relations*, *functions*, and *transformations*. We start the chapter by considering what we mean, or *should* mean, mathematically, by a *relation*. A relation is defined as a set of ordered pairs. An example is a solution set of an equation in two variables, since that kind of solution set is a set of ordered pairs.

A *function* is defined to be a special kind of relation. That kind of relation is probably the most important kind in all of mathematics, so we will learn well what a function is and how to deal with functions.

When we draw a picture of a set of ordered pairs, we say that we have *graphed* a relation. If we alter a relation by moving the graph to a new location, stretching it or shrinking it, we say that we have performed a *transformation*. We also begin a study of transformations in this chapter.

3

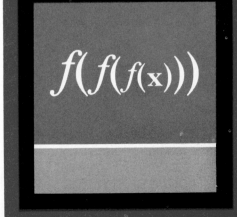

OBJECTIVES

You should be able to:

i Determine whether an ordered pair is in a relation.

ii Find Cartesian products of small sets.

iii Indicate or find certain relations.

iv Find the domain and range of a relation.

1. Consider the set in Example 1.

   a) List the ordered pairs in the relation "is a brother of."

   b) Determine whether the pair (Father, Chuck) is in the relation. Explain why or why not.

   c) Determine whether the pair (Deron, Vanessa) is in the relation. Explain why or why not.

3.1

RELATIONS AND ORDERED PAIRS

i **RELATIONS AND ORDERED PAIRS**

We begin with a discussion of ordered pairs, Cartesian products, and relations. We start out with some examples and define terms as we go along.

EXAMPLE 1   The following set consists of the six members of a family: the father, the mother, two daughters, and two sons.

$$\{\text{Father, Mother, Elaine, Vanessa, Chuck, Deron}\}.$$

We want to consider a particular kind of relationship, or *relation*, among members of the family. We shall define "relation" precisely later. The relation is "is a sister of." We have six facts concerning this relation:

Elaine is a sister of Chuck,
Elaine is a sister of Deron,
Elaine is a sister of Vanessa,
Vanessa is a sister of Chuck,
Vanessa is a sister of Deron,
Vanessa is a sister of Elaine.

Instead of listing the relations we have, we can write down a set of ordered pairs as follows:

$$\{(\text{Elaine, Chuck}), (\text{Elaine, Deron}), (\text{Elaine, Vanessa}),$$
$$(\text{Vanessa, Chuck}), (\text{Vanessa, Deron}), (\text{Vanessa, Elaine})\}.$$

a)   Determine whether the ordered pair (Elaine, Deron) is in the relation. Explain why or why not.

b)   Determine whether the ordered pair (Chuck, Elaine) is in the relation. Explain why or why not.

We solve as follows.

a)   The ordered pair (Elaine, Deron) is in the relation because the statement "Elaine is the sister of Deron" is true.

b)   The ordered pair (Chuck, Elaine) is not in the relation because the statement "Chuck is a sister of Elaine" is not true.

DO EXERCISE 1.

EXAMPLE 2   The following table gives a relation between years and the cost of first-class postage.

Year	Cost of first-class postage
1964	5¢
1974	10¢
1978	15¢
1983	20¢
1984	20¢

We say that the ordered pair (a, b) is in the relation if "the cost of first-class postage in year a is b."

a)   List five ordered pairs in the relation.

b) Determine whether the ordered pair (1985, 22¢) is in the relation.

c) Determine whether the ordered pair (15¢, 1983) is in the relation.

We solve as follows.

a) Five ordered pairs of the relation are (1964, 5¢), (1974, 10¢), (1978, 15¢), (1983, 20¢), and (1984, 20¢). There are others.

b) The ordered pair (1985, 22¢) is in the relation. The price of first-class postage was raised to 22¢ in February of 1985.

c) The ordered pair (15¢, 1978) is *not* in the relation. The year is to be listed first.

DO EXERCISE 2.

ii CARTESIAN PRODUCTS

The ordered pairs of a relation are formed by taking members of a set or two sets. We now consider sets consisting of all possible such ordered pairs.

Consider the sets $A = \{1, 2, 3\}$ and $B = \{a, b\}$. From these sets we can pick numbers and form ordered pairs: for example,

$$(1, a), \qquad (2, a), \quad \text{and} \quad (3, b).$$

Definition

The *Cartesian product* of two sets A and B, symbolized $A \times B$ and read "A cross B," is defined as the set of *all* ordered pairs having the first member from set A and the second member from set B.

EXAMPLE 3 Find the Cartesian product $A \times B$, where $A = \{1, 2, 3\}$ and $B = \{a, b\}$.

The Cartesian product, $A \times B$, is as follows:

$$(1, a), \qquad (2, a), \qquad (3, a),$$
$$(1, b), \qquad (2, b), \qquad (3, b).$$

CAUTION! Be sure in forming Cartesian products that in each pair of $A \times B$ the first member is taken from A and the second member is taken from B.

DO EXERCISE 3.

We now consider an example in which A and B are the same set.

EXAMPLE 4 Consider the set $\{2, 3, 4, 5\}$. Find the Cartesian product of this set by itself.

The Cartesian product of this set by itself is called a *Cartesian square* and is as follows:

```
5 | (2, 5)   (3, 5)   (4, 5)   (5, 5)
4 | (2, 4)   (3, 4)   (4, 4)   (5, 4)
3 | (2, 3)   (3, 3)   (4, 3)   (5, 3)
2 | (2, 2)   (3, 2)   (4, 2)   (5, 2)
   ------------------------------------
      2        3        4        5
```

2. Consider the relation in Example 2.

a) Determine whether the ordered pair (1938, 3¢) is in the relation. Explain why or why not.

b) Determine whether the ordered pair (1986, 22¢) is in the relation. Explain why or why not.

3. Let $A = \{d, e, f\}$ and $B = \{1, 2\}$.

a) List all the ordered pairs in $A \times B$.

b) List all the ordered pairs in $B \times A$.

4. List all the ordered pairs in the Cartesian square of {1, 2, 3, 4}. Save the list for later use.

The headings at the bottom and at the left are for reference only. The Cartesian square consists only of the ordered pairs.

DO EXERCISES 4 AND 5.

iii RELATIONS

In a Cartesian product we can pick out ordered pairs that make up common relations, such as $=$ or $<$ as in the following examples.

EXAMPLE 5 In the Cartesian square of Example 4 indicate all ordered pairs for which the first member is the same as the second. This set of ordered pairs is the relation $=$ (equals).

We know that an ordered pair (a, b) is in the relation if $a = b$. The members of the Cartesian square that are in this relation are in color below.

$$(2, 5) \quad (3, 5) \quad (4, 5) \quad (5, 5)$$
$$(2, 4) \quad (3, 4) \quad (4, 4) \quad (5, 4)$$
$$(2, 3) \quad (3, 3) \quad (4, 3) \quad (5, 3)$$
$$(2, 2) \quad (3, 2) \quad (4, 2) \quad (5, 2)$$

5. List all the ordered pairs in the Cartesian square $A \times A$, where A is the set of family members in Example 1.

EXAMPLE 6 In the Cartesian square of Example 4 indicate all ordered pairs for which the first member is less than the second. This set of ordered pairs is the relation $<$ (is less than).

We know that an ordered pair (a, b) is in the relation if $a < b$. The members of the Cartesian square that are in this relation are in color below.

$$(2, 5) \quad (3, 5) \quad (4, 5) \quad (5, 5)$$
$$(2, 4) \quad (3, 4) \quad (4, 4) \quad (5, 4)$$
$$(2, 3) \quad (3, 3) \quad (4, 3) \quad (5, 3)$$
$$(2, 2) \quad (3, 2) \quad (4, 2) \quad (5, 2)$$

6. In the Cartesian square of Exercise 4, indicate all ordered pairs for which the first member is the same as the second. This is the relation $=$.

DO EXERCISES 6 AND 7.

There are also many relations that do not have common names and relations with which we are not already familiar. Any time we select a set of ordered pairs from a Cartesian product, we have selected some relation. This is true even if we make the selection at random.

EXAMPLE 7 The following set of ordered pairs is a relation, but is not a familiar one. It has no common name.

7. In the Cartesian square of Exercise 4, indicate all ordered pairs for which the first member is greater than the second. This is the relation $>$ (is greater than). An ordered pair (a, b) is in the relation if $a > b$.

$$(2, 5) \quad (3, 5) \quad (4, 5) \quad (5, 5)$$
$$(2, 4) \quad (3, 4) \quad (4, 4) \quad (5, 4)$$
$$(2, 3) \quad (3, 3) \quad (4, 3) \quad (5, 3)$$
$$(2, 2) \quad (3, 2) \quad (4, 2) \quad (5, 2)$$

We shall now make our definition of relation.

Definition

A *relation* from a set *A* to a set *B* is defined to be any set of ordered pairs in *A* × *B*.

iv DOMAIN AND RANGE

Definition

The set of all first members in a relation is called its *domain*. The set of all second members in a relation is called its *range*.

EXAMPLE 8 Find the domain and range of the relation in Example 1.

Domain: {Elaine, Vanessa}

Range: {Chuck, Deron, Elaine, Vanessa}

EXAMPLE 9 Find the domain and range of the relation = in Example 5.

Domain: {2, 3, 4, 5}; Range: {2, 3, 4, 5}

EXAMPLE 10 Find the domain and range of the relation < in Example 6.

Domain: {2, 3, 4}; Range: {3, 4, 5}

EXAMPLE 11 Find the domain and range of the relation in Example 7.

Domain: {2, 4, 5}; Range: {2, 3, 5}

DO EXERCISES 8–11.

8. Find the domain and range of the relation "is a brother of" in Exercise 1(a).

9. Find the domain and range of the relation = in Exercise 6.

10. Find the domain and range of the relation > in Exercise 7.

11. Consider the relation whose ordered pairs are (2, 2), (1, 1), (1, 2), and (1, 3). Find the domain and range.

EXERCISE SET 3.1

i

Consider the set in Example 1 for Exercises 1 and 2.

1. a) List the ordered pairs in the relation "is a father of."
  b) Determine whether the ordered pair (Father, Chuck) is in the relation.
  c) Determine whether the ordered pair (Vanessa, Father) is in the relation.

2. a) List the ordered pairs in the relation "is a mother of."
  b) Determine whether the ordered pair (Mother, Deron) is in the relation.
  c) Determine whether the ordered pair (Elaine, Vanessa) is in the relation.

3. The following table gives a relation between years that have passed and the average cash value of a $10,000 whole-life insurance policy.

Year	Average value of a $10,000 whole-life policy
0	$ 0
5	490
10	1270
15	2000
20	2790

We say that the ordered pair (a, b) is in the relation if "after a years the average value of a $10,000 whole-life policy is b."

a) Determine whether the ordered pair (10, $1270) is in the relation.
b) Determine whether the ordered pair (20, $3000) is in the relation.

4. The following table gives the relation between years and the estimated population of the USSR, in millions.

Year	Population, in millions
1959	209
1969	231
1979	255
1989	282

We say that the ordered pair (a, b) is in the relation if "the population of the USSR in the year a is b, in millions."

a) Determine whether the ordered pair (282, 1989) is in the relation.
b) Determine whether the ordered pair (1979, 255) is in the relation.

ii

5. Let $A = \{0, 2\}$ and $B = \{a, b, c\}$.

a) List all the ordered pairs in $A \times B$. Remember that first coordinates come from set A and second coordinates come from set B.

b) List all the ordered pairs in $B \times A$.

c) List all the ordered pairs in the Cartesian square $A \times A$.

d) List all the ordered pairs in the Cartesian square $B \times B$.

6. Let $A = \{1, 3, 5, 9\}$ and $B = \{d, e, f\}$.

a) List all the ordered pairs in $A \times B$.

b) List all the ordered pairs in $B \times A$.

c) List all the ordered pairs in $A \times A$.

d) List all the ordered pairs in $B \times B$.

iii For Exercises 7–12, let $A = \{-1, 0, 1, 2\}$ and consider the Cartesian square $A \times A$.

7. Find the set of ordered pairs in the relation $<$ (is less than).

8. Find the set of ordered pairs in the relation $>$ (is greater than).

9. Find the set of ordered pairs in the relation $\leqslant$ (is less than or equal to).

10. Find the set of ordered pairs in the relation $\geq$ (is greater than or equal to).

11. Find the set of ordered pairs in the relation $=$.

12. Find the set of ordered pairs in the relation $\neq$.

iv

13. Find the domain and range of the relation in Exercise 3.

14. Find the domain and range of the relation in Exercise 4.

15. Find the domain and range of the relation in Exercise 7.

16. Find the domain and range of the relation in Exercise 8.

17. Find the domain and range of the relation in Exercise 9.

18. Find the domain and range of the relation in Exercise 10.

19. a) List all the ordered pairs in the Cartesian square $D \times D$, where $D = \{-1, 0, 1, 2\}$.

b) Consider the relation whose ordered pairs are $(0, 0)$, $(1, 1)$, $(0, 1)$, and $(1, 2)$. Circle the ordered pairs of this relation on the answer to part (a).

c) List the domain and range of the relation in (b).

20. a) List all the ordered pairs in the Cartesian square $E \times E$, where $E = \{-1, 1, 3, 5\}$.

b) Consider the relation whose ordered pairs are $(-1, 1)$, $(1, 1)$, $(-1, 3)$, and $(1, 3)$. Circle the ordered pairs of this relation on the answer to part (a).

c) List the domain and range of the relation in (b).

OBJECTIVES

You should be able to:

i Determine whether an ordered pair of numbers is a solution of an equation with two variables.

ii Graph certain equations.

iii Given the graph of a relation, describe the domain and the range.

3.2

GRAPHS OF EQUATIONS

RELATIONS IN REAL NUMBERS

We are most interested in relations involving $R \times R$, where R is the set of real numbers. The set R is infinite. Thus relations involving R may be infinite, and therefore cannot be indicated by listing the ordered pairs one at a time. We usually indicate such relations with some sort of picture in $R \times R$. This kind of picture is called a *graph*.

POINTS IN THE PLANE AND ORDERED PAIRS

On a number line each point corresponds to a number. On a plane each point corresponds to a number pair from $R \times R$. To represent $R \times R$ we draw an x-axis and a y-axis perpendicular to each other. Their intersection is called the *origin* and is labeled 0. The arrows show the positive directions.

This is called a Cartesian coordinate system. The first member of an ordered pair is called the *first coordinate*. The second member is called the

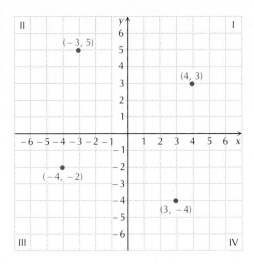

1. a) Plot the points in the relation $(3, 2)$, $(-5, -2)$, $(-4, 3)$.

  b) Find the domain and range of the relation.

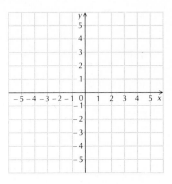

*second coordinate.** Together these are called the *coordinates of a point*. The axes divide the plane into four *quadrants*, indicated by the Roman numerals.

DO EXERCISE 1.

i ▌ SOLUTIONS OF EQUATIONS

If an equation has two variables, its solutions are ordered pairs of numbers. A *solution* is an ordered pair that when substituted alphabetically for the variables produces a true equation.

EXAMPLE 1 Determine whether the following ordered pairs are solutions of the equation $y = 3x - 1$: $(-1, -4)$, $(7, 5)$.

$$\begin{array}{c|c} \multicolumn{2}{c}{y = 3x - 1} \\ \hline -4 & 3(-1) - 1 \\ -4 & -3 - 1 \\ & -4 \end{array}$$ We substitute -1 for x and -4 for y (alphabetical order of variables).

The equation becomes true: $(-1, -4)$ is a solution.

$$\begin{array}{c|c} \multicolumn{2}{c}{y = 3x - 1} \\ \hline 5 & 3 \cdot 7 - 1 \\ 5 & 21 - 1 \\ & 20 \end{array}$$ We substitute.

The equation becomes false: $(7, 5)$ is not a solution.

DO EXERCISES 2–5.

2. Determine whether $(1, 7)$ is a solution of $y = 2x + 5$.

3. Determine whether $(-1, 4)$ is a solution of $y = 2x + 5$.

4. Determine whether $(-2, 5)$ is a solution of $y = x^2$.

5. Determine whether $(4, -5)$ is a solution of $x^3 - y^2 = 39$.

---

* The first coordinate is sometimes called the *abscissa* and the second coordinate the *ordinate*.

6. Graph $y = -3x + 1$.

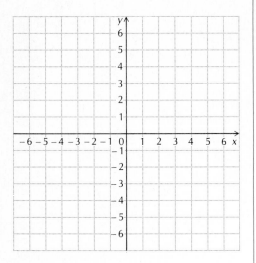

ii GRAPHS OF EQUATIONS

The solutions of an equation are ordered pairs and thus constitute a relation. To *graph* an equation, or a relation, means to make a drawing of its solutions. Some general suggestions for graphing are as follows.

Graphing suggestions

a) Use graph paper.

b) Label axes with symbols for the variables.

c) Use arrows to indicate positive directions.

d) Scale the axes; that is, mark numbers on the axes.

e) Plot solutions and complete the graph. When finished, write down the equation or relation being graphed.

EXAMPLE 2   Graph $y = 3x - 1$.

We find some ordered pairs that are solutions. To find an ordered pair, we choose *any* number that is a sensible replacement for x and then determine y. For example, if we choose 2 for x, then $y = 3(2) - 1$, or 5. We have found the solution $(2, 5)$. We continue making choices for x, and finding the corresponding values for y. We make some negative choices for x, as well as positive ones. We keep track of the solutions in a table.

x	y
0	-1
1	2
2	5
-1	-4
-2	-7

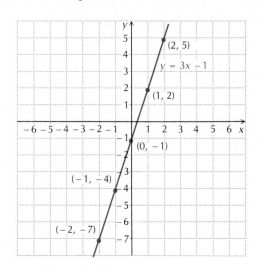

The table gives us the ordered pairs $(0, -1)$, $(1, 2)$, $(2, 5)$, and so on. Next, we plot these points. If we had enough of them, they would make a solid line. We can draw the line with a ruler, and label it $y = 3x - 1$.

Note that the equation $y = 3x - 1$ has an infinite (unending) set of solutions. The *graph of the equation* is a drawing of the relation that is made up of its solutions. Thus the relation consists of all pairs (x, y) such that $y = 3x - 1$ is true. That is, $\{(x, y) \mid y = 3x - 1\}$.

DO EXERCISE 6.

EXAMPLE 3   Graph $y = x^2 - 5$.

We select numbers for x and find the corresponding values for y. The table gives us the ordered pairs $(0, -5)$, $(-1, -4)$, and so on.

x	y
0	−5
−1	−4
1	−4
−2	−1
2	−1
−3	4
3	4

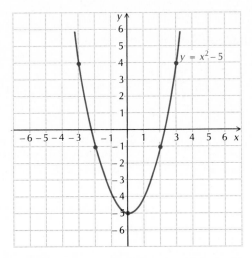

7. Graph $y = 3 - x^2$.

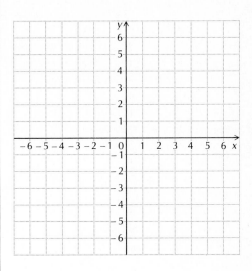

Next we plot these points. We note that as the absolute value of x increases, $x^2 - 5$ also increases. Thus the graph is a curve that rises gradually on either side of the y-axis, as shown above.

This graph shows the relation $\{(x, y) \mid y = x^2 - 5\}$.

DO EXERCISE 7.

EXAMPLE 4 Graph the equation $x = y^2 + 1$.

Here we shall select numbers for y and then find the corresponding values for x. This time we arrange them in a horizontal table.

x	2	2	1	5	5
y	−1	1	0	−2	2

We must remember that x is the first coordinate and y is the second coordinate. Thus the table gives us the ordered pairs $(2, -1)$, $(2, 1)$, $(1, 0)$, and so on. We note that as the absolute value of y gradually increases, the value of x also gradually increases. Thus the graph is a curve that stretches farther and farther to the right as it gets farther from the x-axis.

8. Graph $x = y^2 - 5$. (*Hint:* Select values for y and then find the corresponding values of x. When you plot, be sure to find x (horizontally) first.) Compare it with the graph of Example 3.

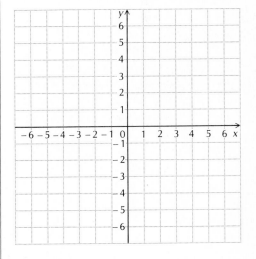

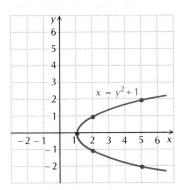

This graph shows the relation $\{(x, y) \mid x = y^2 + 1\}$.

DO EXERCISE 8.

You can always use a calculator to find as many values as desired. This can be especially helpful when you are uncertain about the shape of a graph.

9. Graph $xy = 1$.

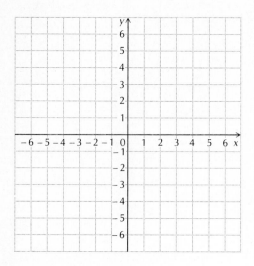

10. Graph the equation $x = |y|$. Compare it with the graph of Example 6.

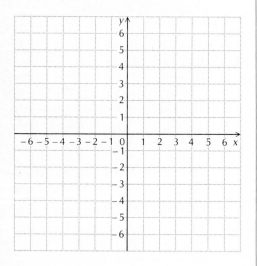

EXAMPLE 5 Graph the equation $xy = 12$.

We find numbers that satisfy the equation.

x	1	-1	2	-2	3	-3	4	-4	6	-6	12	-12
y	12	-12	6	-6	4	-4	3	-3	2	-2	1	-1

We plot these points and connect them. To see how to do this, note that neither x nor y can be 0. Thus the graph does not cross either axis. As the absolute value of x gets small, the absolute value of y must get large, and vice versa. Thus the graph consists of two curves, as follows.

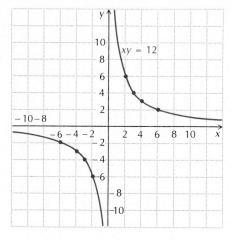

DO EXERCISE 9.

EXAMPLE 6 Graph the equation $y = |x|$.

We find numbers that satisfy the equation.

x	0	1	-1	2	-2	3	-3	4	-4
y	0	1	1	2	2	3	3	4	4

We plot these points and connect them. To see how to do this, note that as we get farther from the origin, to the left or right, the absolute value of x increases. Thus the graph is a curve that rises to the left and right of the y-axis. It actually consists of parts of two straight lines, as follows.

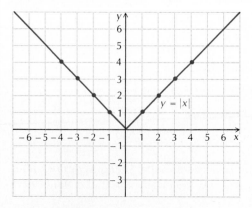

DO EXERCISE 10.

iii DOMAINS AND RANGES

Recall that the domain of a relation is the set of all first coordinates and the range is the set of all second coordinates.

EXAMPLE 7 For the relation in part (a) of the figure, locate or shade the domain on the *x*-axis and the range on the *y*-axis.

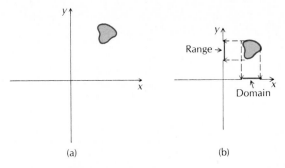

(a)   (b)

From a point (x, y) on the graph, we find y by drawing or thinking of a horizontal line back to the *y*-axis. To locate the range we want all such numbers on the *y*-axis. For this relation, we determine the outermost upper or lower extremities of the relation, and draw horizontal lines from them to the *y*-axis. The range consists of the numbers where the lines cross the *y*-axis and all numbers in between. From a point (x, y) on the graph, we find x by drawing or thinking of a vertical line back to the *x*-axis. To locate the domain we want all such numbers on the *x*-axis. For this relation, we determine the outermost left and right extremities of the relation, and draw vertical lines back to the *x*-axis. The domain consists of the numbers where the lines cross the *x*-axis and all numbers in between. The domain and range are shown in part (b) of the figure.

EXAMPLE 8 Graph $x = y^2 + 1$. Locate the domain on the *x*-axis and the range on the *y*-axis.

The graph is shown in part (a) of the figure. It is the relation we considered in Example 4. The left extremity of the graph yields the number 1 on the *x*-axis. There is no right extremity. The domain is the set of all real numbers greater than or equal to 1. The graph extends up and to the right, and down and to the right. There are no upper–lower extremities. The range consists of the entire set of real numbers. The domain and range are shown in part (b) of the figure.

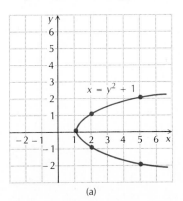

(a)

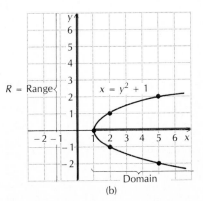

(b)

DO EXERCISES 11–13.

On each diagram in Exercises 11 and 12:

a) shade (on the *x*-axis) the domain;

b) shade (on the *y*-axis) the range.

11.

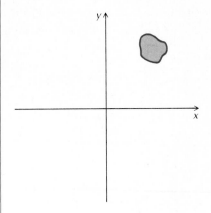

12.

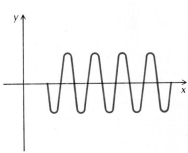

13. Graph $y = x^2$. Then shade and describe the domain on the *x*-axis and the range on the *y*-axis.

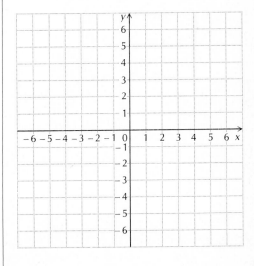

EXERCISE SET 3.2

i

1. Determine whether $(-1, -3)$ is a solution of $y = 5x + 2$.
2. Determine whether $(-2, 8)$ is a solution of $y = 3x - 7$.
3. Determine whether $(-2, 7)$ is a solution of $4x + 3y = 12$.
4. Determine whether $(0, -2)$ is a solution of $2x - 3y = 6$.
5. Determine whether $(3, 0)$ is a solution of $x^2 - 2y = 6$.
6. Determine whether $(-2, -1)$ is a solution of $x^2 - 2y = 6$.

ii Graph.

7. $y = x + 3$
8. $y = x - 2$
9. $y = 3x - 2$
10. $y = -4x + 1$
11. $y = x^2$
12. $y = -x^2$
13. $y = x^2 + 2$
14. $y = x^2 - 2$
15. $x = y^2 + 2$
16. $x = y^2 - 2$
17. $y = |x + 1|$
18. $y = |x - 1|$
19. $x = |y + 1|$
20. $x = |y - 1|$
21. $xy = 10$
22. $xy = -18$
23. $y = \dfrac{1}{x}$
24. $y = -\dfrac{2}{x}$
25. $y = \dfrac{1}{x^2}$
26. $y = x^3$
27. $y = \sqrt{x}$
28. $y = x^{1/3}$
29. $y = 8 - x^2$
30. $x = 4 - y^2$

Graph and compare.

31. $y = x^2 + 1$ and $y = (-x)^2 + 1$
32. $y = x^2 - 2$ and $y = 2 - x^2$

iii

33. Graph a relation as follows.
   a) Draw a circle with radius of length 2, centered at $(4, 3)$. Shade the circle and its interior.
   b) Shade (on the x-axis) the domain. Describe the domain.
   c) Shade (on the y-axis) the range. Describe the range.

34. Graph a relation as follows.
   a) Draw a triangle with vertices at $(1, 1)$, $(4, 2)$, and $(3, 6)$. Shade the triangle and its interior.
   b) Shade (on the x-axis) the domain. Describe the domain.
   c) Shade (on the y-axis) the range. Describe the range.

Use the graphs found in Exercises 7–32. Describe the domain and range of each of the following relations.

35. $y = -x^2$
36. $y = x - 2$
37. $x = |y + 1|$
38. $y = \dfrac{1}{x}$

In each of the following, all relations to be considered are in $R \times R$, where R is the set of real numbers.

39. Graph the relation in which the second coordinate is always 2 and the first coordinate may be any real number.

40. Graph the relation in which the first coordinate is always -3, and the second coordinate may be any real number.

41. Graph the relation in which the second coordinate is always 1 more than the first coordinate, and the first coordinate may be any real number.

42. Graph the relation in which the second coordinate is always 1 less than the first coordinate and the first coordinate may be any real number.

43. Graph the relation in which the second coordinate is always twice the first coordinate, and the first coordinate may be any real number.

44. Graph the relation in which the second coordinate is always half the first coordinate, and the first coordinate may be any real number.

45. Graph the relation in which the second coordinate is always the square of the first coordinate, and the first coordinate may be any real number.

46. Graph the relation in which the first coordinate is always the square of the second coordinate, and the second coordinate may be any real number.

Graph the following equations. The definition of absolute value in Section 1.1 may be helpful for some of these exercises.

47. $y = |x| + x$
48. $y = x|x|$
49. $y = |x^2 - 4|$
50. $y = x^{2/3}$
51. $|y| = x + 1$
52. $y = |x^3|$
53. $|y| = |x|$
54. $|x| + |y| = 0$
55. $|xy| = 1$
56. Graph the relation $\{(x, y) \mid 1 < x < 4 \text{ and } -3 < y < -1\}$.

57. Graph $|x| + |y| = 1$.
58. Graph the relation $\{(x, y) \mid |x| \leqslant 1 \text{ and } |y| \leqslant 2\}$.

📱 Graph. In each case use your calculator and substitute at least twenty values of x between -10 and 10.

59. $y = \dfrac{1}{3}x^3 - x + \dfrac{2}{3}$

60. $y = \dfrac{1}{3}x^3 - \dfrac{1}{2}x^2 - 2x + 1$

61. $y = x + \dfrac{4}{x}$

3.3

FUNCTIONS

i   RECOGNIZING GRAPHS OF FUNCTIONS

A *function* is a special kind of relation. It is defined as follows.

Definition

> A *function* is a relation in which no two ordered pairs have the same first coordinate and different second coordinates.

In a function, given a member of the domain (a first coordinate) there is one and only one member of the range that goes with it (the second coordinate). Thus each member of the domain *determines* exactly one member of the range. It is easy to recognize the graph of a relation that is a function. If there are two or more points of the graph on the same vertical line, then the relation is not a function. Otherwise it is. Here are some graphs of functions. In graph (c), the solid dot indicates that $(-1, 1)$ belongs to the graph. The open dot indicates that $(-1, -2)$ does not belong to the graph. Thus no vertical line crosses the graph more than once.

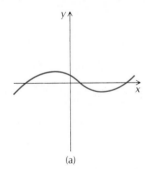

(a)

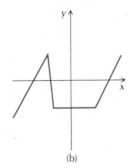

(b)

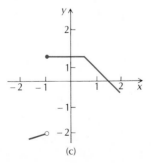

(c)

The following are not graphs of functions, because they fail the so-called *vertical-line test*. That is, we can find a vertical line that meets the graph in more than one point.

OBJECTIVES

You should be able to:

i   Recognize a graph of a function.

ii Use a formula to find function values.

iii Find the domain of a function, given by a formula.

iv Compose pairs of functions f and g, finding formulas for $f \circ g$ and $g \circ f$.

1. Which of the following are graphs of functions?

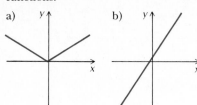

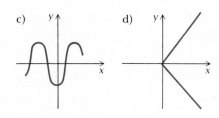

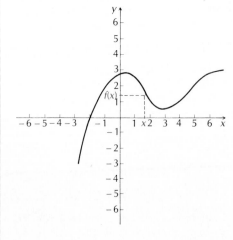

(a)   (b)   (c)

> *The vertical-line test.*   If it is possible for a vertical line to meet a graph more than once, the graph is not the graph of a function.

DO EXERCISE 1.

ii NOTATION FOR FUNCTIONS

Functions are often named by letters, such as f or g. A function f is thus a set of ordered pairs. If we represent the first coordinate of a pair by x, then we may represent the second coordinate by $f(x)$. The symbol $f(x)$ is read "f of x" or "f at x." The number represented by $f(x)$ is called the "value" of the function at x. [*Note:* "$f(x)$" does *not* mean "f times x."]

EXAMPLE 1   Below let us call the function in color g.

$$(1, 4) \quad (2, 4) \quad (3, 4) \quad (4, 4)$$
$$(1, 3) \quad (2, 3) \quad (3, 3) \quad (4, 3)$$
$$(1, 2) \quad (2, 2) \quad (3, 2) \quad (4, 2)$$
$$(1, 1) \quad (2, 1) \quad (3, 1) \quad (4, 1)$$

Here $g(1) = 4$, $g(2) = 3$, $g(3) = 2$, and $g(4) = 4$.

2. In this function, what is $f(1)$? $f(2)$? $f(3)$? $f(4)$?

$$(1, 4) \quad (2, 4) \quad (3, 4) \quad (4, 4)$$
$$(1, 3) \quad (2, 3) \quad (3, 3) \quad (4, 3)$$
$$(1, 2) \quad (2, 2) \quad (3, 2) \quad (4, 2)$$
$$(1, 1) \quad (2, 1) \quad (3, 1) \quad (4, 1)$$

What is the domain of this function? What is the range?

EXAMPLE 2   Let us call the function graphed below f. To find function values, we locate x on the x-axis, and then find $f(x)$ on the y-axis. From the graph we can see that $f(-2) = 0$, that $f(0)$ is about 2.7, and that $f(3)$ is about 0.6.

3. From the graph, find the following function values approximately: $f(2)$, $f(0)$, $f(-3)$.

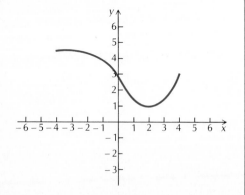

DO EXERCISES 2 AND 3.

Some functions in real numbers can be defined by formulas, or equations. Here are some examples:

$$g(s) = 3, \qquad p(t) = \frac{1}{t}, \qquad f(x) = 3x^2 + 4x - 5, \qquad u(y) = |y| + 3.$$

Function values can be obtained by making substitutions for the variables.

EXAMPLE 3 For the function f given by $f(x) = 1/x$, find $f(2)$, $f(-\frac{1}{4})$, and $f(0)$, if possible.

We find function values by making substitutions for the variables as follows:

$$f(2) = \frac{1}{2},$$

$$f(-\tfrac{1}{4}) = \frac{1}{-\frac{1}{4}} = -4,$$

$$f(0) = \frac{1}{0}, \text{ but this function value } does\ not\ exist \text{ since we cannot divide by } 0.$$

For this function $f(-\frac{1}{4}) = -4$. This means that "f of $-\frac{1}{4}$" or "f at $-\frac{1}{4}$" is -4. The value of the function at $-\frac{1}{4}$ is -4.

DO EXERCISE 4.

A function g given by $g(s) = 3$ is called a *constant function* because all its function values are the same. Thus $g(5) = 3$, $g(-7) = 3$, and so on. The range contains only one number, 3.

DO EXERCISE 5.

EXAMPLE 4 For the function f given by $f(x) = 2x^2 - 3$, find $f(0)$, $f(-1)$, $f(5a)$, and $f(a - 4)$.

One way to find function values when a formula is given is to think of the formula as follows:

$$f(\ \ ) = 2(\ \ )^2 - 3.$$

Then whatever goes in the blank on the left between parentheses goes in the blank on the right between parentheses.

$$f(0) = 2 \cdot (0)^2 - 3 = -3;$$
$$f(-1) = 2(-1)^2 - 3 = 2 \cdot 1 - 3 = -1;$$
$$f(5a) = 2(5a)^2 - 3 = 2 \cdot 25a^2 - 3 = 50a^2 - 3;$$
$$f(a - 4) = 2(a - 4)^2 - 3 = 2(a^2 - 8a + 16) - 3$$
$$= 2a^2 - 16a + 29.$$

DO EXERCISE 6.

Simplifying expressions like

$$\frac{f(a + h) - f(a)}{h}$$

is a valuable skill in calculus.

4. For the function f given by $f(x) = \sqrt{x}$, find $f(16)$, $f(3)$, and $f(-4)$, if possible.

5. For the function G given by $G(x) = 7$, find $G(0)$, $G(-3)$, and $G(\frac{1}{2})$. What is the range?

6. Using $f(x) = 3x^2 + 1$, find:

a) $f(0)$;
b) $f(1)$;
c) $f(-1)$;
d) $f(2a)$;
e) $f(a + 1)$;

For each function f given as follows, construct the expression

$$\frac{f(a + h) - f(a)}{h}.$$

Then simplify.

7. $f(x) = 3x^2 + 1$

EXAMPLE 5 For the function f given by $f(x) = 2x^2 - 3$, construct the expression

$$\frac{f(a + h) - f(a)}{h}.$$

Then simplify.

We find $f(a + h)$ and $f(a)$:

$$f(a + h) = 2(a + h)^2 - 3, \qquad f(a) = 2(a)^2 - 3.$$

Then

$$\frac{f(a + h) - f(a)}{h} = \frac{[2(a + h)^2 - 3] - [2a^2 - 3]}{h}.$$

Now we simplify:

$$\frac{f(a + h) - f(a)}{h} = \frac{2a^2 + 4ah + 2h^2 - 3 - 2a^2 + 3}{h}$$

$$= \frac{4ah + 2h^2}{h}$$

$$= 4a + 2h.$$

DO EXERCISE 7.

8. $f(x) = x^2 - x$

EXAMPLE 6 For the function f given by $f(x) = x^3 + x$, construct the expression

$$\frac{f(a + h) - f(a)}{h}.$$

Then simplify.

We find $f(a + h)$ and $f(a)$:

$$f(a + h) = (a + h)^3 + (a + h) \qquad \text{Replacing each occurrence of } x \text{ by } a + h$$
$$= a^3 + 3a^2h + 3ah^2 + h^3 + a + h;$$
$$f(a) = a^3 + a.$$

Then

$$\frac{f(a + h) - f(a)}{h} = \frac{[a^3 + 3a^2h + 3ah^2 + h^3 + a + h] - [a^3 + a]}{h}.$$

Now we simplify:

$$\frac{f(a + h) - f(a)}{h} = \frac{a^3 + 3a^2h + 3ah^2 + h^3 + a + h - a^3 - a}{h}$$

$$= \frac{3a^2h + 3ah^2 + h^3 + h}{h}$$

$$= \frac{h(3a^2 + 3ah + h^2 + 1)}{h}$$

$$= 3a^2 + 3ah + h^2 + 1$$

DO EXERCISE 8.

Some comment is in order regarding the notation $f(x)$ and f. When we refer to a function f, we are referring to a set of ordered pairs. The notation $f(x)$ refers to the second coordinate of an ordered pair that has x as its first coordinate. Nevertheless, it is a fact of life in the literature of mathematics to be "sloppy" about this notation and say things like "the function $f(x)$" and "the function $f(x) = x^2$."

iii FINDING DOMAINS OF FUNCTIONS

When a function f in $R \times R$ is given by a formula, the domain is understood to be the set of all real numbers that are sensible replacements for x. For example, consider the function f given by $f(x) = 1/x$. The number 0 is not a sensible replacement because division by 0 is not possible. All other real numbers are sensible replacements, so the domain of f consists of all non-zero real numbers, $\{x \mid x \neq 0\}$. Consider the function g given by $g(x) = \sqrt{x}$. The negative numbers are not sensible replacements. Thus the domain of g consists of all nonnegative numbers, $\{x \mid x \geqslant 0\}$.

EXAMPLE 7 Find the domain of the function g given by

$$g(x) = \frac{x}{x^2 + 2x - 3}.$$

The formula makes sense as long as a replacement for x does not make the denominator 0. To find those replacements that do make the denominator 0, we solve $x^2 + 2x - 3 = 0$:

$$x^2 + 2x - 3 = 0$$
$$(x - 1)(x + 3) = 0$$
$$x - 1 = 0 \quad \text{or} \quad x + 3 = 0$$
$$x = 1 \quad \text{or} \quad x = -3.$$

Thus the domain consists of the set of all real numbers except 1 and -3. We can name this set $\{x \mid x \neq 1 \text{ and } x \neq -3\}$.

EXAMPLE 8 Find the domain of the function f given by $f(x) = \sqrt{5x - 3}$.

The formula makes sense as long as a replacement makes the radicand nonnegative (no negative number has a real square root). Thus to find the domain we solve the inequality $5x - 3 \geqslant 0$:

$$5x - 3 \geqslant 0$$
$$5x \geqslant 3$$
$$x \geqslant \tfrac{3}{5}.$$

The domain is $\{x \mid x \geqslant \tfrac{3}{5}\}$.

EXAMPLE 9 Find the domain of the function t given by $t(x) = x^3 + |x|$.

There are no restrictions on the numbers we can substitute into this formula. We can cube any real number, we can take the absolute value of any real number, and we can add the results. Thus the domain is the entire set of real numbers.

DO EXERCISES 9–11.

Find the domain of the function given by each formula.

9. $f(x) = \dfrac{x + 1}{3x^2 + 10x + 8}$

10. $g(x) = \sqrt{10x + 25}$

11. $p(x) = x^3 - 4x^2 + 2x + 8$

FUNCTIONS, MAPPINGS, AND MACHINES

Functions can be thought of as mappings. A function f *maps* the set of first coordinates (the domain) to the set of second coordinates (the range).

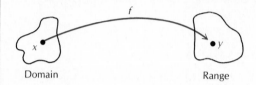

The definition of function at the beginning of this section says that in a function no two ordered pairs have the same first coordinate. Thus a number x in the domain corresponds (or is mapped onto) *just one y* in the range. That y is the (unique) second coordinate of the ordered pair (x, y). A situation like the following *cannot* happen for a function.

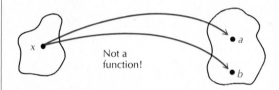

No x in the domain can be mapped to two or more different members of the range.

EXAMPLE 10   Consider the function f given by $f(x) = 2x + 3$.

Since $f(0) = 3$, this function maps 0 to 3.

Since $f(3) = 9$, this function maps 3 to 9.

The concept of function is illuminated somewhat by considering a so-called "function machine." In this drawing we see a function machine designed, or programmed, to do the mapping (function) f. The inputs acceptable to the machine are the members of the domain of f. The outputs are, of course, members of the range of f.

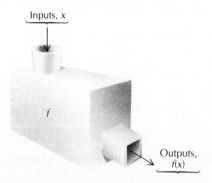

We may sometimes refer to an ordered pair $(x, f(x))$ as an input–output pair. The graph of a function consists of all the input–output pairs.

iv COMPOSITION OF FUNCTIONS

Functions can be combined in a way called *composition* of functions. Consider, for example,

$$f(x) = x^2 \qquad \text{This function squares each input.}$$

and

$$g(x) = x + 1. \qquad \text{This function adds 1 to each input.}$$

We define a new function that first does what g does (adds 1) and then does what f does (squares). The new function is called the *composite* of f and g, and is symbolized $f \circ g$. Let us think of hooking two function machines f and g together to get the resultant function machine $f \circ g$.

Definition

The *composed* function $f \circ g$ (the *composite* of f and g) is defined as follows:

$$f \circ g(x) = f(g(x)).$$

To find $f \circ g(x)$, first find $g(x)$. Then substitute $g(x)$ for x in $f(x)$ to find $f \circ g(x)$.

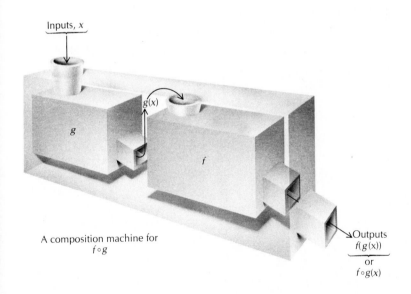

A composition machine for
$f \circ g$

Outputs
$f(g(x))$
or
$f \circ g(x)$

Now let us see how to find a formula for $f \circ g$, and also the reverse composition $g \circ f$.

EXAMPLE 11 For functions f and g given by $f(x) = x^2$ and $g(x) = x + 1$, find formulas for $f \circ g(x)$ and $g \circ f(x)$.

By definition of $\circ$,

$$f \circ g(x) = f(g(x)).$$

12. For functions f and g given by $f(x) = x - 2$ and $g(x) = 2x^2$, find $f \circ g(x)$ and $g \circ f(x)$.

Now

$$f(g(x)) = f(x + 1) \qquad \text{Substituting } x + 1 \text{ for } g(x)$$

$$= (x + 1)^2 \qquad \text{Substituting } x + 1 \text{ for } x \text{ in the formula for } f(x)$$

$$= x^2 + 2x + 1.$$

Similarly, $g \circ f(x) = g(f(x))$. Now

$$g \circ f(x) = g(x^2) \qquad \text{Substituting } x^2 \text{ for } f(x)$$

$$= x^2 + 1. \qquad \text{Substituting } x^2 \text{ for } x \text{ in the formula for } g(x)$$

Note that the function $f \circ g$ is not the same as the function $g \circ f$.

In order for functions to be composable, such as f and g above, the outputs of g must be acceptable as inputs for f. In other words, if any number $g(x)$ is not in the domain of f, then x is not in the domain of $f \circ g$. Note also the order of happenings in $f \circ g$. The function $f \circ g$ does *first* what g does, *then* what f does.

DO EXERCISE 12.

13. For functions u and v given by $u(x) = 2x^2$ and $v(x) = 3x + 2$, find $u \circ v(x)$ and $v \circ u(x)$.

EXAMPLE 12 For functions f and g given by $f(x) = 2x$ and $g(x) = x^2 + 1$, find $f \circ g(x)$ and $g \circ f(x)$.

By definition of $\circ$,

$$f \circ g(x) = f(g(x))$$

$$= f(x^2 + 1) \qquad \text{Substituting } x^2 + 1 \text{ for } g(x)$$

$$= 2(x^2 + 1) \qquad \text{Substituting } x^2 + 1 \text{ for } x \text{ in the formula for } f(x)$$

$$= 2x^2 + 2.$$

Similarly,

$$g \circ f(x) = g(f(x))$$

$$= g(2x) \qquad \text{Substituting } 2x \text{ for } f(x)$$

$$= (2x)^2 + 1 \qquad \text{Substituting } 2x \text{ for } x \text{ in the formula for } g(x)$$

$$= 4x^2 + 1.$$

DO EXERCISE 13.

EXERCISE SET 3.3

1. Which of the following are graphs of functions?

a)
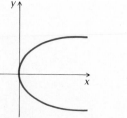

b)
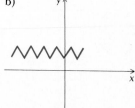

c)
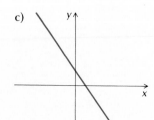

d)
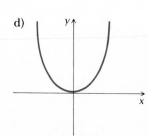

2. Which of the following are graphs of functions? An open circle indicates that the point does not belong to the graph.

a)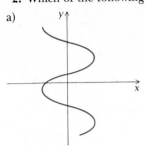

b)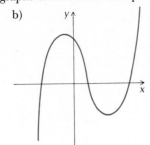

c)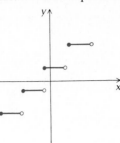

d)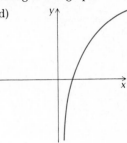

ii In Exercises 3–10, functions are given by formulas. Find the indicated function values and construct and simplify the expressions like $\dfrac{f(a+h)-f(a)}{h}$.

3. $f(x) = 5x^2 + 4x$. Find:

   a) $f(0)$;   b) $f(-1)$;

   c) $f(3)$;   d) $f(t)$;

   e) $f(t-1)$;   f) $\dfrac{f(a+h)-f(a)}{h}$.

4. $g(x) = 3x^2 - 2x + 1$. Find:

   a) $g(0)$;   b) $g(-1)$;

   c) $g(3)$;   d) $g(t)$;

   e) $g(a+h)$;   f) $\dfrac{g(a+h)-g(a)}{h}$.

5. $f(x) = 2|x| + 3x$. Find:

   a) $f(1)$;   b) $f(-2)$;

   c) $f(-4)$;   d) $f(2y)$;

   e) $f(a+h)$;   f) $\dfrac{f(a+h)-f(a)}{h}$.

6. $g(x) = x^3 - 2x$. Find:

   a) $g(1)$;   b) $g(-2)$;

   c) $g(-4)$;   d) $g(3y)$;

   e) $g(2+h)$;   f) $\dfrac{g(2+h)-g(2)}{h}$.

7. ▦ $f(x) = 4.3x^2 - 1.4x$. Find:

   a) $f(1.034)$;   b) $f(-3.441)$;

   c) $f(27.35)$;   d) $f(-16.31)$.

8. ▦ $g(x) = \sqrt{2.2|x| + 3.5}$. Find:

   a) $g(17.3)$;   b) $g(-64.2)$;

   c) $g(0.095)$;   d) $g(-6.33)$.

9. $f(x) = \dfrac{x^2 - x - 2}{2x^2 - 5x - 3}$. Find:

   a) $f(0)$;   b) $f(4)$;

   c) $f(-1)$;   d) $f(3)$.

10. $s(x) = \sqrt{\dfrac{3x-4}{2x+5}}$. Find:

   a) $s(10)$;   b) $s(2)$;

   c) $s(1)$;   d) $s(-1)$.

iii In Exercises 11–20, functions are given by formulas. Find the domain of each function.

11. $f(x) = 7x + 4$

12. $f(x) = |3x - 2|$

13. $f(x) = 4 - \dfrac{2}{x}$

14. $f(x) = \sqrt{x - 3}$

15. $f(x) = \sqrt{7x + 4}$

16. $f(x) = \dfrac{1}{9 - x^2}$

17. $f(x) = \dfrac{1}{x^2 - 4}$

18. $f(x) = \dfrac{2x + 6}{x^3 - 4x}$

19. $f(x) = \dfrac{4x^3 + 4}{4x^2 - 5x - 6}$

20. $f(x) = \dfrac{x^3 + 8}{x^2 - 4}$

iv In Exercises 21–26, find $f \circ g(x)$ and $g \circ f(x)$.

21. $f(x) = 3x^2 + 2$, $g(x) = 2x - 1$

22. $f(x) = 4x + 3$, $g(x) = 2x^2 - 5$

23. $f(x) = 4x^2 - 1$, $g(x) = \dfrac{2}{x}$

24. $f(x) = \dfrac{3}{x}$, $g(x) = 2x^2 + 3$

25. $f(x) = x^2 + 1$, $g(x) = x^2 - 1$

26. $f(x) = \dfrac{1}{x^2}$, $g(x) = x + 2$

27. From this graph, find approximately $g(-2)$, $g(-3)$, $g(0)$, and $g(2)$.

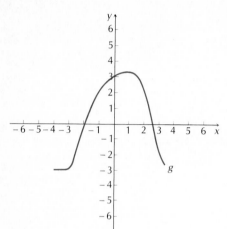

28. From this graph, find approximately $h(-2)$, $h(0)$, $h(3)$, and $h(-3)$.

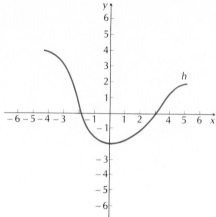

For each function, construct and simplify $\dfrac{f(x + h) - f(x)}{h}$.

29. $f(x) = \dfrac{1}{x}$

30. $f(x) = \dfrac{1}{x^2}$

31. $f(x) = \sqrt{x}$ (*Hint:* Rationalize the numerator.)

Find the domain of each function.

32. $f(x) = \dfrac{\sqrt{x}}{2x^2 - 3x - 5}$

33. $f(x) = \dfrac{\sqrt{x + 3}}{x^2 - x - 2}$

34. $f(x) = \dfrac{\sqrt{x + 1}}{x + |x|}$

35. $f(x) = \sqrt{x^2 + 1}$

36. For $f(x) = \dfrac{1}{1 - x}$, find $f \circ f(x)$ and $f \circ [f \circ f(x)]$.

37. In Exercise 23, find the domain of $f \circ g$ and $g \circ f$.

38. Determine whether the relation $\{(x, y) \mid xy = 0\}$ is a function.

39. The *greatest integer function* $f(x) = [x]$ is defined as follows: $[x]$ is the greatest integer that is less than or equal to x. For example, if $x = 3.74$, then $[x] = 3$; and if $x = -0.98$, then $[x] = -1$. Graph the greatest integer function for values of x such that $-5 \leqslant x \leqslant 5$.

40. Graph the equation $[y] = [x]$. See Exercise 39. Is this the graph of a function?

OBJECTIVES

You should be able to:

i Given an equation defining a relation, determine whether a graph is symmetric with respect to a coordinate axis.

ii Given an equation defining a relation, determine whether a graph is symmetric with respect to the origin.

3.4

SYMMETRY

SYMMETRY WITH RESPECT TO A LINE

In the figure points P and P_1 are said to be *symmetric* with respect to line l. They are the same distance from l.

Definition

Two points P and P_1 are *symmetric with respect to a line l* if and only if l is the perpendicular bisector of the segment $\overline{PP_1}$. The line l is known as the *line of symmetry*.

We also say that the two points P and P_1 are *reflections* of each other across the line. The line is therefore also known as a *line of reflection*.

Now consider a set of points (geometric figure) as in the color curve below. This figure is said to be symmetric with respect to the line l, because if you pick any point Q in the set, you can find another point Q_1 in the set such that Q and Q_1 are symmetric with respect to l.

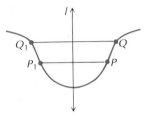

Definition

A figure, or set of points, is *symmetric with respect to a line l* if and only if for each point Q in the set there exists another point Q_1 in the set for which Q and Q_1 are symmetric with respect to line l.

Imagine picking up the preceding figure and flipping it about the line l. Points P and P_1 would be interchanged. Points Q and Q_1 would be interchanged. These are, then, pairs of symmetric points. The entire figure would look exactly like it did before flipping. This means that *each* point of the figure is symmetric with *some* point of the figure. Thus the figure is symmetric with respect to the line. A point and its reflection are known as *images* of each other. Thus P_1 is the image of P, for example. The line l is known as an *axis* of symmetry.

i SYMMETRY WITH RESPECT TO THE AXES

There are special and interesting kinds of symmetry in which a coordinate axis is a line of symmetry. The following example shows figures that are symmetric with respect to an axis and a figure that is not.

EXAMPLE 1 In graph (a), flipping the figure about the y-axis would not change the figure. In graph (b), flipping the graph about the x-axis would not change the figure. In graph (c), flipping about either axis would change the figure.

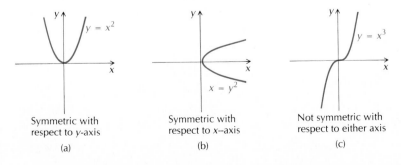

Symmetric with respect to y-axis

(a)

Symmetric with respect to x-axis

(b)

Not symmetric with respect to either axis

(c)

1. a) Plot the point $(3, 2)$. Let the y-axis be a line of symmetry. Plot the point symmetric to $(3, 2)$. What are its coordinates?

  b) Plot the point $(-4, -5)$. Let the y-axis be a line of reflection. Plot the image of $(-4, -5)$. What are its coordinates?

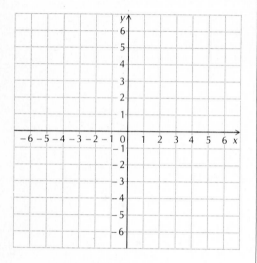

2. Let the x-axis be a line of symmetry.

  a) Plot the point $(4, 3)$. Plot the point symmetric to it. What are its coordinates?

  b) Plot the point $(3, -5)$. Plot its image after reflection across the x-axis. What are its coordinates?

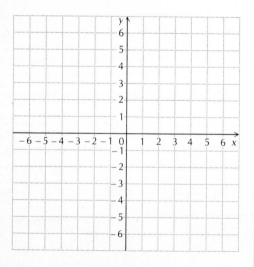

Let us consider a figure like graph (a), symmetric with respect to the y-axis. For every point of the figure there is another point the same distance across the y-axis. The first coordinates of such a pair of points are additive inverses of each other.

EXAMPLE 2   In the relation $y = x^2$ there are points $(2, 4)$ and $(-2, 4)$. The first coordinates, 2 and -2, are additive inverses of each other, while the second coordinates are the same. For every point of the figure (x, y), there is another point $(-x, y)$.

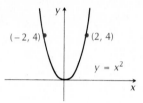

DO EXERCISE 1.

Let us consider a figure symmetric with respect to the x-axis. For every point of such a figure there is another point the same distance across the x-axis. The second coordinates of such a pair of points are additive inverses of each other.

EXAMPLE 3   In the relation $x = y^2$ there are points $(4, 2)$ and $(4, -2)$. The second coordinates, 2 and -2, are additive inverses of each other, while the first coordinates are the same. For every point of the figure (x, y), there is another point $(x, -y)$.

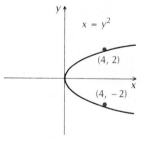

DO EXERCISE 2.

Let us consider a figure symmetric with respect to the y-axis, as in Example 2. Suppose it is defined by an equation. If in this equation we replace x by $-x$, we obtain a new equation, but we get the same figure. This is true because any number x gives us the same y-value as its additive inverse, $-x$.

Let us consider a figure symmetric with respect to the x-axis, as in Example 3. Suppose it is defined by an equation. If we replace y by $-y$ in the equation, we obtain a new equation, but we get the same figure. This is true because any number y gives us the same x-value as $-y$. We thus have a means of testing a relation for symmetry, when it is defined by an equation.

> **When a relation is defined by an equation:**
>
> 1. If replacing x by $-x$ produces an equivalent equation, then the graph is symmetric with respect to the y-axis.
> 2. If replacing y by $-y$ produces an equivalent equation, then the graph is symmetric with respect to the x-axis.

EXAMPLE 4   Test $y = x^2 + 2$ for symmetry with respect to the y-axis.

a)   Replace x by $-x$.

$$y = x^2 + 2 \qquad (1)$$
$$y = (-x)^2 + 2$$

b)   Simplify, if possible.

$$y = (-x)^2 + 2 = x^2 + 2 \qquad (2)$$

c)   Is the resulting equation (2) equivalent to the original (1)?

Since the answer is yes, the graph is symmetric with respect to the y-axis.

EXAMPLE 5   Test $y = x^2 + 2$ for symmetry with respect to the x-axis.

a)   Replace y by $-y$.

$$y = x^2 + 2 \qquad (1)$$
$$-y = x^2 + 2$$

b)   Simplify, if possible.

The equation is simplified for the most part, although we could multiply on both sides by -1, obtaining

$$y = -(x^2 + 2). \qquad (2)$$

c)   Is the resulting equation (2) equivalent to the original (1)?

This answer may be obvious to you and is no. To be sure one might use some trial-and-error reasoning as follows. Suppose we substitute 0 for x in Eq. (1). Then

$$y = 0^2 + 2 = 2,$$

so $(0, 2)$ is a solution of Eq. (1). For Eqs. (1) and (2) to be equivalent, $(0, 2)$ must also be a solution of Eq. (2). We substitute to find out:

$$
\begin{array}{c|c}
y = & -(x^2 + 2) \\
\hline
2 & -(0^2 + 2) \\
& -2
\end{array}
$$

Thus $(0, 2)$ is not a solution of Eq. (2); hence the equations are not equivalent and the graph is not symmetric with respect to the x-axis.

Test the following for symmetry with respect to the coordinate axes.

3. $y = x^2 - 3$

4. $y^2 = x^3$

5. $x^4 = y^2 + 2$

6. $3y^2 + 4x^2 = 12$

7. $a + 3b = 5$

8. $2p^3 + 4q^3 = 1$

EXAMPLE 6 Test $a^2 + b^4 + 5 = 0$ for symmetry with respect to the a-axis.

a) Replace b by $-b$.

$$a^2 + b^4 + 5 = 0 \qquad (1)$$
$$a^2 + (-b)^4 + 5 = 0$$

b) Simplify, if possible.

$$a^2 + (-b)^4 + 5 = a^2 + b^4 + 5 = 0 \qquad (2)$$

c) Is the resulting equation equivalent to the first?

Since the answer is yes, the graph is symmetric with respect to the a-axis.

DO EXERCISES 3–8.

SYMMETRY WITH RESPECT TO A POINT

Two points are symmetric with respect to a point when they are situated as shown in the following figure. That is, the points are the same distance from that point, and all three points are on a line.

Definition

Two points P and P_1 are *symmetric with respect to a point* Q if and only if Q (the point of symmetry) is the midpoint of segment $\overline{PP_1}$.

A *set* of points is symmetric with respect to a point when each point in the set is symmetric with some point in the set. This is illustrated here. Imagine sticking a pin in this figure at O and then rotating the figure $180°$. Points P and P_1 would be interchanged. Points Q and Q_1 would be interchanged. These are pairs of symmetric points. The entire figure would look exactly as it did before rotating. This means that *each* point of the figure is symmetric with *some* point of the figure. Thus the figure is symmetric with respect to the point O.

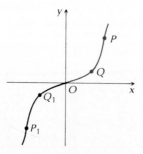

Definition

A set of points is *symmetric with respect to a point B* if and only if for every point *P* in the set there exists another point P_1 in the set for which *P* and P_1 are symmetric with respect to *B*.

ⅱ SYMMETRY WITH RESPECT TO THE ORIGIN

A special kind of symmetry with respect to a point is symmetry with respect to the origin.

EXAMPLE 7   In graphs (a) and (b), rotating the figure $180°$ about the origin would not change the figure. In graph (c) such a rotation would change the figure.

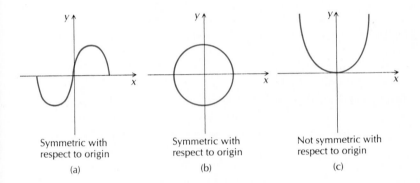

Symmetric with respect to origin	Symmetric with respect to origin	Not symmetric with respect to origin
(a)	(b)	(c)

Let us consider figures like (a) and (b) above, symmetric with respect to the origin. For every point of the figure there is another point the same distance across the origin. The first coordinates of such a pair are additive inverses of each other, and the second coordinates are additive inverses of each other.

DO EXERCISE 9.

EXAMPLE 8   The relation $y = x^3$ is symmetric with respect to the origin. In this relation are the points $(2, 8)$ and $(-2, -8)$. The first coordinates are additive inverses of each other. The second coordinates are additive inverses of each other. For every point of the figure (x, y), there is another point $(-x, -y)$.

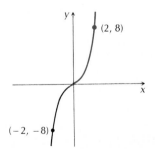

In a relation symmetric with respect to the origin, as in Example 8, if we replace *x* by $-x$ and *y* by $-y$, we obtain a new equation, but we will get

9. Draw coordinate axes. Let the origin be a point of symmetry.

a) Plot the point $(3, 2)$. Plot the point symmetric to it. What are its coordinates?

b) Plot the point $(-4, 3)$. Plot the point symmetric to it. What are its coordinates?

c) Plot the point $(-5, -7)$. Plot the point symmetric to it. What are its coordinates?

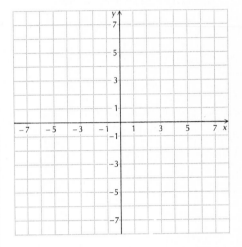

Test the following for symmetry with respect to the origin.

10. $x^2 + 3y^2 = 4$

11. $x = y$ (*Hint:* After you substitute, multiply on both sides by -1.)

12. $x = -y$

13. $xy = 5$

14. $ab = -5$

15. $u = |v|$

the same figure. This is true because whenever a point (x, y) is in the relation, the point $(-x, -y)$ is also in the relation. This gives us a means of testing a relation for symmetry when it is defined by an equation.

> When a relation is defined by an equation, if replacing x by $-x$ and replacing y by $-y$ produces an equivalent equation, then the graph is symmetric with respect to the origin.

EXAMPLE 9 Test $x^2 = y^2 + 2$ for symmetry with respect to the origin.

a) Replace x by $-x$ and y by $-y$.

$$x^2 = y^2 + 2$$
$$(-x)^2 = (-y)^2 + 2$$

b) Simplify, if possible.

Since $(-x)^2 = x^2$ and $(-y)^2 = y^2$, $(-x)^2 = (-y)^2 + 2$ simplifies to

$$x^2 = y^2 + 2.$$

c) Is the resulting equation equivalent to the original?

Since the answer is yes, the graph is symmetric with respect to the origin.

EXAMPLE 10 Test $2a + 3b = 8$ for symmetry with respect to the origin.

a) Replace a by $-a$ and b by $-b$.

$$2a + 3b = 8$$
$$2(-a) + 3(-b) = 8$$

b) Simplify, if possible.

$$2(-a) + 3(-b) = -2a - 3b$$

so

$$-(2a + 3b) = 8 \quad \text{and} \quad 2a + 3b = -8$$

c) Is the resulting equation equivalent to the original?

The equation $2a + 3b = -8$ is *not* equivalent to $2a + 3b = 8$, so the graph is not symmetric with respect to the origin.

DO EXERCISES 10–15.

EXERCISE SET 3.4

i, **ii** In Exercises 1–12, test for symmetry with respect to the coordinate axes and the origin.

1. $3y = x^2 + 4$ **2.** $5y = 2x^2 - 3$ **3.** $y^3 = 2x^2$ **4.** $3y^3 = 4x^2$

5. $2x^4 + 3 = y^2$ **6.** $3y^2 = 2x^4 - 5$ **7.** $2y^2 = 5x^2 + 12$ **8.** $3x^2 - 2y^2 = 7$

9. $2x - 5 = 3y$ **10.** $5y = 4x + 5$ **11.** $3b^3 = 4a^3 + 2$ **12.** $p^3 - 4q^3 = 12$

ii In Exercises 13–24, test for symmetry with respect to the origin.

13. $3x^2 - 2y^2 = 3$ **14.** $5y^2 = -7x^2 + 4$ **15.** $5x - 5y = 0$ **16.** $3x = 3y$

17. $3x + 3y = 0$ **18.** $7x = -7y$ **19.** $3x = \dfrac{5}{y}$ **20.** $3y = \dfrac{7}{x}$

21. $y = |2x|$ **22.** $3x = |y|$ **23.** $3a^2 + 4a = 2b$ **24.** $5v = 7u^2 - 2u$

Consider the figure on the right for Exercises 25–28.

25. Graph the reflection across the *x*-axis.

26. Graph the reflection across the *y*-axis.

27. Graph the reflection across the line $y = x$.

28. Graph the figure formed by reflecting each point through the origin.

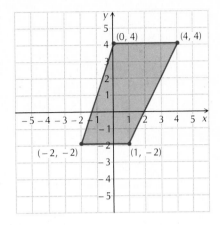

29. Consider symmetries with respect to the *x*-axis, the *y*-axis, and the origin. Prove that symmetry with respect to any two of these implies symmetry with respect to the other.

3.5

TRANSFORMATIONS

i TRANSFORMATIONS OF RELATIONS

Given a relation, we can find various ways of altering it to obtain another relation. Such an alteration is called a *transformation*. If such an alteration consists merely of moving the graph without changing its shape or orientation, the transformation is called a *translation*.

Vertical Translations

Consider the relations $y = x^2$ and $y = 1 + x^2$ whose graphs are shown below. The graph of $y = 1 + x^2$ has the same shape as that of $y = x^2$, but is moved upward a distance of 1 unit. Consider any equation $y = f(x)$. Adding a constant *a* to produce $y = a + f(x)$ changes each function value by the same amount, *a*. Thus it produces no change in the shape of the graph, but merely translates it upward if the constant is positive. If *a* is negative, the graph will be moved downward.

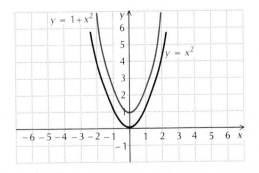

DO EXERCISE 1.

OBJECTIVE

You should be able to:

i Given the graph of a relation, graph its transformation under translations, reflection, stretchings, and shrinkings.

1. a) Graph $y = x^2$. Then graph $y = 2 + x^2$ and compare.

   b) Graph $y = -2 + x^2$ and compare.

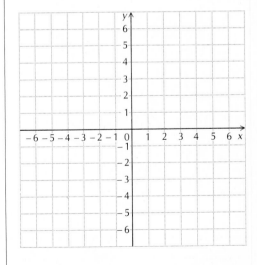

2. Sketch a graph of $y = -3 + x^2$.

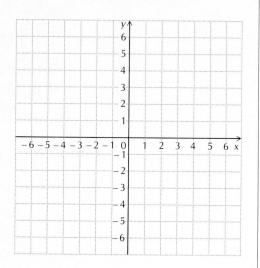

Note that $y = 1 + x^2$ is equivalent to $y - 1 = x^2$. Thus the transformation described above amounts to replacing y by $y - 1$ in the original equation.

Theorem 1

In an equation of a relation, replacing y by $y - a$, where a is a constant, translates the graph vertically a distance of $|a|$. If a is positive, the translation is upward. If a is negative, the translation is downward.*

If in an equation we replace y by $y + 3$, this is the same as replacing it by $y - (-3)$. In this case the constant a is -3 and the translation is downward. If we replace y by $y - 5$, the constant a is 5 and the translation is upward.

EXAMPLE 1 Sketch a graph of $y = |x|$ and then one of $y = -2 + |x|$.

The graph of $y = |x|$ is shown below. Now consider $y = -2 + |x|$. Note that $y = -2 + |x|$ is equivalent to $y + 2 = |x|$ or $y - (-2) = |x|$. This shows that the new equation can be obtained by replacing y in $y = |x|$ by $y - (-2)$, so the translation is downward, 2 units, as shown below.

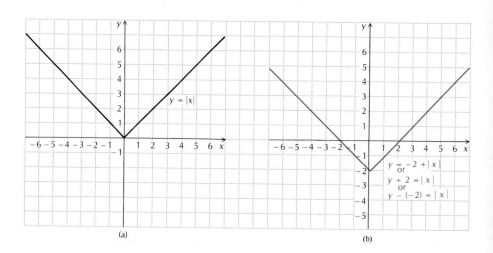

(a)

(b)

3. Here is a graph of $y = f(x)$. There is no formula for it. Sketch a graph of $y = 3 + f(x)$.

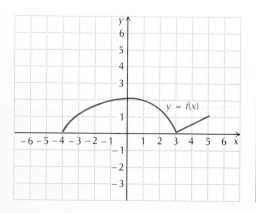

DO EXERCISES 2 AND 3.

Horizontal Translations

Replacing y by $y - a$ in an equation translates vertically a distance of $|a|$. The translation is in the positive direction (upward) if a is positive. A similar thing happens in the horizontal direction. If we replace x by $x - b$ everywhere x occurs in the equation, we translate a distance of $|b|$ horizontally. If b is positive, we translate in the positive direction (to the right). If b is negative, we translate in the negative direction (to the left).

---

*In practice, when working with functions, we are more inclined to write $y = a + f(x)$ instead of $y - a = f(x)$, but for relations in general we are more likely to replace y by $y - a$.

Theorem 2

In an equation of a relation, replacing x by $x - b$, where b is a constant, translates the graph horizontally a distance of $|b|$. If b is positive, the translation is to the right. If b is negative, the translation is to the left.

EXAMPLE 2 Given a graph of $y = |x|$, sketch a graph of $y = |x + 2|$.

Here we note that x is replaced by $x + 2$, or $x - (-2)$. Thus $b = -2$, and the graph will be moved 2 units in the negative direction (to the left).

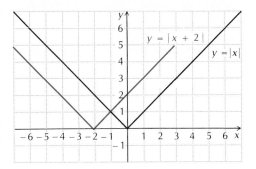

EXAMPLE 3 A circle centered at the origin with radius of length 1 has an equation $x^2 + y^2 = 1$. If we replace x by $x - 1$ and y by $y + 2$, we translate the circle so that the center is at the point $(1, -2)$.

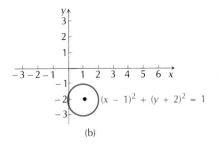

(a)   (b)

DO EXERCISES 4 AND 5.

Vertical Stretchings and Shrinkings

Consider the function $y = |x|$. We will compare its graph with that of $y = 2|x|$ and $y = \frac{1}{2}|x|$. The graph of $y = 2|x|$ looks like that of $y = |x|$ but every output is doubled, so the graph is stretched in a vertical direction. The graph of $y = \frac{1}{2}|x|$ is flattened or shrunk in a vertical direction since every output is cut in half.

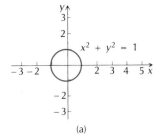

x	$2|x|$	$\frac{1}{2}|x|$
-3	6	$\frac{3}{2}$
-2	4	1
0	0	0
2	4	1
3	6	$\frac{3}{2}$

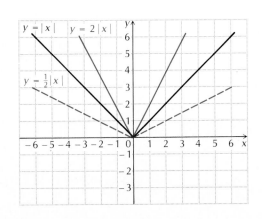

4. Sketch a graph of $y = (x + 3)^2$.

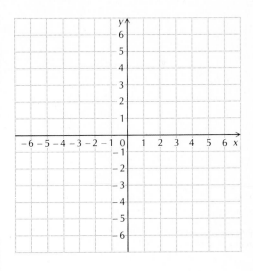

5. Here is a graph of $y = g(x)$. There is no formula for it. Sketch a graph of $y = g(x - 4)$.

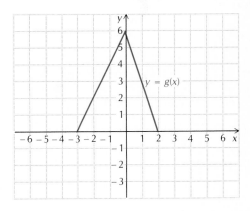

6. Graph $y = x^2$. Graph $y = 2x^2$ and compare.

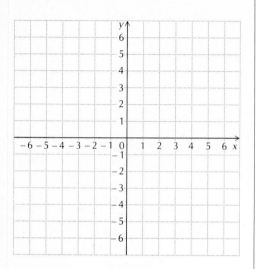

7. Graph $y = \frac{1}{2}x^2$. Compare with $y = x^2$.

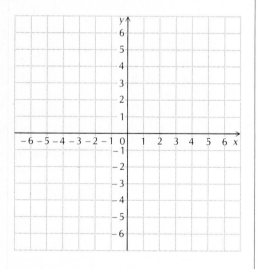

Consider any equation such as $y = f(x)$. Multiplying on the right by the constant 2 will double every function value, thus stretching the graph both ways from the horizontal axis. A similar thing is true for any constant greater than 1. If the constant is between 0 and 1, then the graph will be flattened or shrunk vertically.

DO EXERCISES 6 AND 7.

When we multiply by a negative constant, the graph is reflected across the x-axis as well as being stretched or shrunk.

EXAMPLE 4 Compare the graphs of $y = |x|$, $y = -2|x|$, and $y = -\frac{1}{2}|x|$.

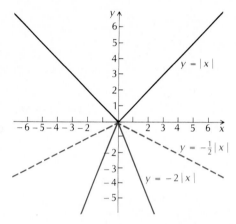

Multiplying by c on the right is, of course, equivalent to dividing by c on the left.

Theorem 3

> In an equation of a relation, dividing y by a constant c does the following to the graph.
>
> 1. If $|c| > 1$, the graph is stretched vertically.
> 2. If $|c| < 1$, the graph is shrunk vertically.
> 3. If c is negative, the graph is also reflected across the x-axis.*

Note that if $c = -1$, this has the effect of replacing y by $-y$ and we obtain a reflection without stretching or shrinking.

EXAMPLE 5 The following page shows a graph of $y = f(x)$. Sketch a graph of $y/2 = f(x)$ or $y = 2f(x)$.

---

* Again, with functions, we are more inclined to write $y = cf(x)$ rather than $y/c = f(x)$, but for relations in general we are more likely to replace y by y/c.

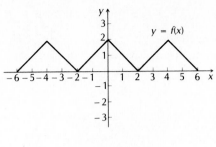

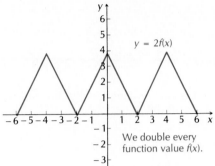

8. Graph $y = -2x^2$. Compare with $y = x^2$.

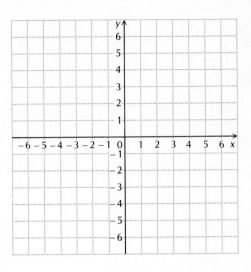

EXAMPLE 6 The following figure shows a graph of $y = g(x)$. Sketch a graph of $\dfrac{y}{-\frac{1}{2}} = g(x)$ or $y = -\frac{1}{2}g(x)$.

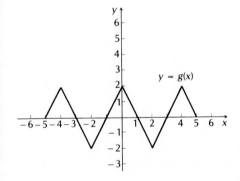

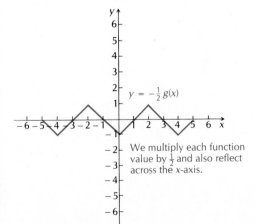

9. Graph $y = -\frac{1}{2}x^2$. Compare with $y = x^2$.

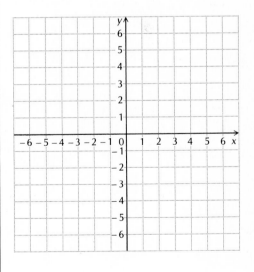

DO EXERCISES 8 AND 9.

Here is a graph of $y = t(x)$. Use graph paper to sketch the following.

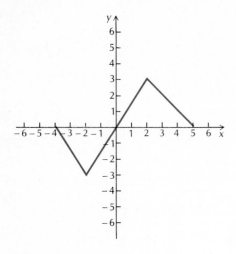

10. Sketch a graph of $y = \frac{1}{2}t(x)$.

11. Sketch a graph of $y = -2t(x)$.

12. Sketch a graph of $y = t(2x)$.

13. Sketch a graph of $y = t(-\frac{1}{2}x)$.

Horizontal Stretchings and Shrinkings

For vertical stretchings and shrinkings we divided y by a constant c. Similarly, if we divide x by a constant wherever it occurs, we will get a horizontal stretching or shrinking.

Theorem 4

In an equation of a relation, dividing x wherever it occurs by d does the following to the graph.

1. If $|d| < 1$, the graph is shrunk horizontally.
2. If $|d| > 1$, the graph is stretched horizontally.
3. If d is negative, the graph is also reflected across the y-axis.*

Note that if $d = -1$, this has the effect of replacing x by $-x$ and we obtain a reflection without stretching or shrinking.

EXAMPLE 7   Here is a graph of $y = f(x)$. Sketch a graph of $y = f\left(\dfrac{x}{\frac{1}{2}}\right)$ or $y = f(2x)$, a graph of $y = f\left(\dfrac{x}{2}\right)$ or $y = f\left(\dfrac{1}{2}x\right)$, and a graph of $y = f\left(\dfrac{x}{-2}\right)$ or $y = f\left(-\dfrac{1}{2}x\right)$.

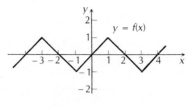

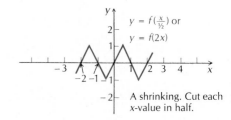

A shrinking. Cut each x-value in half.

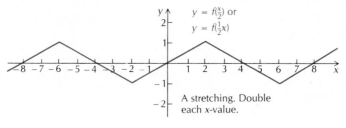

A stretching. Double each x-value.

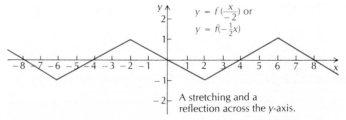

A stretching and a reflection across the y-axis.

DO EXERCISES 10–13.

---

* Again, with functions, we are more inclined to write $y = f(kx)$ rather than $y = f(x/d)$, but for relations in general we usually consider replacing x by x/d.

EXERCISE SET 3.5

i In Exercises 1–14, sketch graphs by transforming the graph of $y = |x|$.

1. $y + 3 = |x|$ **2.** $y = 2 + |x|$ **3.** $y = |x - 1|$ **4.** $y = |x + 2|$

5. $y = -4|x|$ **6.** $\dfrac{y}{3} = |x|$ **7.** $y = \dfrac{1}{3}|x|$ **8.** $y = -\dfrac{1}{4}|x|$

9. $y = |2x|$ **10.** $y = \left|\dfrac{x}{3}\right|$ **11.** $y = |x - 2| + 3$ **12.** $y = 2|x + 1| - 3$

13. $y = -3|x - 2|$ **14.** $y = \dfrac{1}{3}|x + 2| + 1$

Here is a graph of $y = f(x)$. No formula will be given for this function. In Exercises 15–33, sketch graphs by transforming this one.

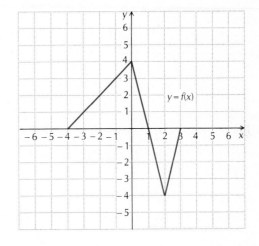

15. $y = 2 + f(x)$ **16.** $y + 1 = f(x)$ **17.** $y = f(x - 1)$ **18.** $y = f(x + 2)$

19. $\dfrac{y}{-2} = f(x)$ **20.** $y = 3f(x)$ **21.** $y = \dfrac{1}{3}f(x)$ **22.** $y = -\dfrac{1}{2}f(x)$

23. $y = f(2x)$ **24.** $y = f(3x)$ **25.** $y = f(-2x)$ **26.** $y = f(-3x)$

27. $y = f\left(\dfrac{x}{-2}\right)$ **28.** $y = f\left(\dfrac{1}{3}x\right)$ **29.** $y = f(x - 2) + 3$ **30.** $y = -3f(x - 2)$

31. $y = 2 \cdot f(x + 1) - 2$ **32.** $y = \dfrac{1}{2}f(x + 2) - 1$ **33.** $y = -\dfrac{1}{2}f(x - 3) + 2$

Here is a graph of $y = f(x)$. No formula will be given for this function. In Exercises 34–38, sketch graphs by transforming this one.

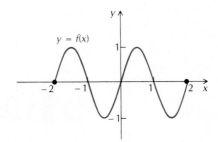

34. $y = -2f(x + 1) - 1$ **35.** $y = 3f(x + 2) + 1$ **36.** $y = \dfrac{5}{2}f(x - 3) - 2$

37. $y = -\sqrt{2}f(x + 1.8)$ **38.** $y = \dfrac{\sqrt{3}}{2} \cdot f(x - 2.5) - 5.3$

39. A *linear* transformation of a coordinatized line is one that takes any point x to the point x', where $x' = ax + b$ (a and b constants). For example, the transformation $x' = 3x + 5$ takes the point 2 to the point $3 \cdot 2 + 5$, or 11. Suppose for a particular linear transformation the point 2 goes to 5 and the point 3 goes to 7.

a) Describe the transformation as $x' = ax + b$.
b) This transformation leaves one point fixed. What point is it?
c) The set $\{x \mid -2 \leqslant x \leqslant 1\}$ is mapped to what set under this transformation?

41. Show that any linear transformation $x' = ax + b$, $a \neq 1$, leaves one point fixed.

40. How does the linear transformation $x' = x - 3$ transform each of the following sets?

a) $\{x \mid 0 < x < 1\}$
b) $\{x \mid -4 \leqslant x < -1\}$
c) $\{x \mid 8 \leqslant x \leqslant 20\}$

42. A professor gives a test with 80 possible points. She decides that a score of 55 should be passing. She performs a linear transformation on the scores so that a perfect paper gets 100 and 70 is passing.

a) Describe the linear transformation that the professor used.
b) What grade remains unchanged under the transformation?

43. Suppose the average score on the test of Exercise 42 was 60. What will be the average of the transformed scores?

Given the graph of $y = f(x)$ for Exercises 34–38, graph each of the following.

44. $\dfrac{y}{3} = f\left(2x + \dfrac{1}{2}\right)$

45. $y = -4 \cdot f(5x + 10)$

OBJECTIVES

You should be able to:

i Given the graph of a function or a formula, determine whether the function is even, odd, or neither even nor odd.

ii Given the graph of a function, determine whether it is periodic, and if it is periodic, determine its period.

iii Write interval notation for certain sets.

iv Given the graph of a function, determine whether it is continuous over a specified interval, and indicate discontinuities.

v Given the graph of a function, determine whether it is increasing, decreasing, or neither increasing nor decreasing.

vi Graph functions defined piecewise.

3.6

SOME SPECIAL CLASSES OF FUNCTIONS

i EVEN AND ODD FUNCTIONS

If the graph of a function is symmetric with respect to the y-axis, it is an *even* function. Recall (Section 3.4) that a function will be symmetric to the y-axis if in its equation we can replace x by $-x$ and obtain an equivalent equation. Thus if we have a function given by $y = f(x)$, then $y = f(-x)$ will give the same function if the function is even. In other words, an even function is one for which $f(x) = f(-x)$ for all x in its domain. This is the definition of an even function.

Definition

A function f is an *even* function if $f(x) = f(-x)$ for all x in the domain of f.

EXAMPLE 1 Determine whether the function $f(x) = x^2 + 1$ is even.

a) Find $f(-x)$ and simplify.

$$f(-x) = (-x)^2 + 1 = x^2 + 1$$

b) Compare $f(x)$ and $f(-x)$.

Since $f(x) = f(-x)$ for all x in the domain, f is an even function.

Note that the graph is symmetric with respect to the y-axis.

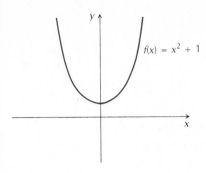

$f(x) = x^2 + 1$

a)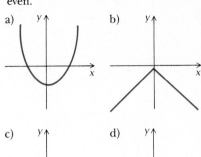

b)

c)

d)

EXAMPLE 2 Determine whether the function $f(x) = x^2 + 8x^3$ is even.

a) Find $f(-x)$ and simplify.

$$f(-x) = (-x)^2 + 8(-x)^3 = x^2 - 8x^3$$

b) Compare $f(x)$ and $f(-x)$.

Since $f(x)$ and $f(-x)$ are *not* the same for all x in the domain, f is *not* an even function.

DO EXERCISES 1 AND 2.

If the graph of a function is symmetric with respect to the origin, it is an *odd* function. Recall that a function will be symmetric with respect to the origin if in its equation we can replace x by $-x$ and y by $-y$ and obtain an equivalent equation. Thus if we have a function given by $y = f(x)$, then $-y = f(-x)$ will be equivalent if f is an odd function. In other words, an odd function is one for which $f(-x) = -f(x)$ for all x in the domain. Let us make this our definition.

Definition

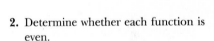

A function f is an *odd* function when $f(-x) = -f(x)$ for all x in the domain of f.

EXAMPLE 3 Determine whether $f(x) = x^3$ is an odd function.

a) Find $f(-x)$ and $-f(x)$ and simplify.

$$f(-x) = (-x)^3 = -x^3,$$
$$-f(x) = -x^3$$

b) Compare $f(-x)$ and $-f(x)$.

Since $f(-x) = -f(x)$ for all x in the domain, f is odd.

Note that the graph is symmetric with respect to the origin.

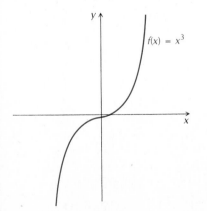

$f(x) = x^3$

2. Determine whether each function is even.

a) $f(x) = x^2 + 3x$

b) $f(x) = |x|$

c) $f(x) = 3x^2 - x^4$

d) $f(x) = 2x^2 + 1$

3. Determine whether each function is even, odd, or neither even nor odd.

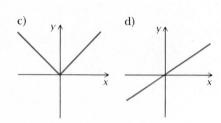

EXAMPLE 4 Determine whether $f(x) = x^2 - 4x^3$ is even, odd, or neither even nor odd.

a) Find $f(-x)$ and $-f(x)$ and simplify.

$$f(-x) = (-x)^2 - 4(-x)^3 = x^2 + 4x^3,$$
$$-f(x) = -x^2 + 4x^3$$

b) Compare $f(x)$ and $f(-x)$ to determine whether f is even.

Since $f(x)$ and $f(-x)$ are *not* the same for all x in the domain, f is *not* even.

c) Compare $f(-x)$ and $-f(x)$ to determine whether f is odd.

Since $f(-x)$ and $-f(x)$ are *not* the same for all x in the domain, f is *not* odd. Thus f is *neither* even nor odd.

DO EXERCISES 3 AND 4.

ii PERIODIC FUNCTIONS

Certain functions with a repeating pattern are called *periodic*. Here are some examples.

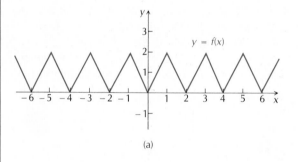

(a)

4. Determine whether each function is even, odd, or neither even nor odd.

a) $f(x) = x^3 + 2$

b) $f(x) = x^4 - x^6$

c) $f(x) = x^3 + x$

d) $f(x) = 3x^2 + 3x^5$

e) $f(x) = x^2 - \dfrac{1}{x}$

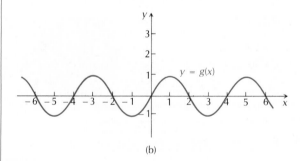

(b)

The function values of the function f repeat themselves every two units as we move from left to right. In other words, for any x, we have $f(x) = f(x + 2)$. To see this another way, think of the part of the graph between 0 and 2 and note that the rest of the graph consists of copies of it. In terms of translations, if we translate the graph 2 units to the left or right, the original graph will be obtained.

In the function g, the function values repeat themselves every 4 units. Hence $g(x) = g(x + 4)$ for any x, and if the graph is translated 4 units to the left, or right, it will coincide with itself. Or, think of the part of the graph between 0 and 4. The rest of the graph consists of copies of it.

We say that f has a *period* of 2, and that g has a period of 4.

Definition

If a function f has the property that $f(x + p) = f(x)$ whenever x and $x + p$ are in the domain, where p is a positive constant, then f is said to be **periodic**. The smallest positive number p (if there is one) for which $f(x + p) = f(x)$ for all x is called the **period** of the function.

The period p can be thought of as the length of the shortest recurring interval.

DO EXERCISES 5–7.

iii INTERVAL NOTATION

For a and b real numbers such that $a < b$, we define the *open interval* (a, b) as follows:

(a, b) is the set of all numbers x such that $a < x < b$,

or

$$\{x \,|\, a < x < b\}.$$

Its graph is as follows:

(a, b) ——○————————○————
$\quad\quad\quad a \quad\quad\quad\quad\quad b$

Note that the endpoints are not included. Be careful not to confuse this notation with that of an ordered pair. The context of the writing should make the meaning clear. If not, we might say "the interval $(-2, 3)$." When we mean an ordered pair, we might say "the pair $(-2, 3)$."

DO EXERCISES 8 AND 9.

The *closed interval* $[a, b]$ is defined as follows:

$[a, b]$ is the set of all x such that $a \leqslant x \leqslant b$,

or

$$\{x \,|\, a \leqslant x \leqslant b\}.$$

Its graph is as follows:

$[a, b]$ ——●————————●————
$\quad\quad\quad a \quad\quad\quad\quad\quad b$

Note that the endpoints are included. For example, the graph of $[-2, 3]$ is as follows:

$[-2, 3]$ ——●——|——|——|——●——
$\quad\quad\quad -2 \ -1 \ \ 0 \ \ 1 \ \ 2 \ \ 3$

There are two kinds of *half-open intervals* defined as follows:

1. $(a, b] = \{x \,|\, a < x \leqslant b\}$.

This is open on the left. Its graph is as follows:

$(a, b]$ ——○————————●————
$\quad\quad\quad a \quad\quad\quad\quad\quad b$

5. For the function f whose graph was just considered, how does $f(x)$ compare with $f(x + 4)$? with $f(x + 6)$?

6. a) Determine whether this function is periodic.

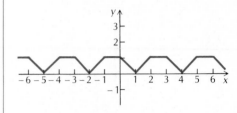

  b) If so, what is the period?

7. (*Optional*). For the function t, where $t(x) = 3$, how does $t(x)$ compare with $t(x + 1)$? By the definition, is t periodic? If so, does it have a period?

8. Write interval notation for each set pictured.
  a)

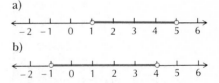

  b)

9. Write interval notation for each set.
  a) $\{x \,|\, -2 < x < 3\}$
  b) $\{x \,|\, 0 < x < 1\}$
  c) $\{x \,|\, -\frac{1}{4} < x < \sqrt{2}\}$

10. Write interval notation for each set pictured.

a)

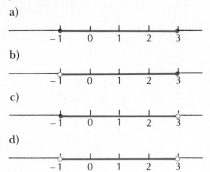

b)

c)

d)

11. Write interval notation for each set.

a) $\{x \mid 4 \leqslant x \leqslant 5\frac{1}{2}\}$

b) $\{x \mid -3 < x \leqslant 0\}$

c) $\{x \mid -\frac{1}{2} \leqslant x < \frac{1}{2}\}$

d) $\{x \mid -\pi < x < \pi\}$

12. Is this function continuous

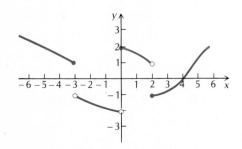

a) in the interval $(-1, 1)$?

b) in the interval $[-2, -1]$?

c) in the interval $(1, 2]$?

d) in the interval $(1, 2)$?

e) in the interval $(-5, 0)$?

13. Where are the discontinuities of the above function?

2. $[a, b) = \{x \mid a \leqslant x < b\}$.

This is open on the right. Its graph is as follows:

DO EXERCISES 10 AND 11.

iv CONTINUOUS FUNCTIONS

Some functions have graphs that are continuous curves, without breaks or holes in them. Such functions are called *continuous functions.*

EXAMPLE 5   The function f below is continuous because it has no breaks, jumps, or holes in it. The function g has *discontinuities* where $x = -2$ and $x = 5$. The function g is continuous on the interval $[-2, 5)$ or on any interval contained therein. It is also continuous on other intervals, but it is not continuous on an interval such as $[-3, 1]$ or $(3, 7)$.

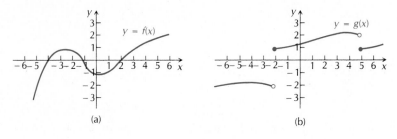

(a)

(b)

DO EXERCISES 12 AND 13.

v INCREASING AND DECREASING FUNCTIONS

If the graph of a function rises from left to right, it is said to be an *increasing* function. If the graph of a function drops from left to right, it is said to be a *decreasing* function. This can be stated more formally.

Definition

1. A function f is an *increasing* function when for all a and b in the domain of f, if $a < b$, then $f(a) < f(b)$.

2. A function f is a *decreasing* function when for all a and b in the domain of f, if $a < b$, then $f(a) > f(b)$.

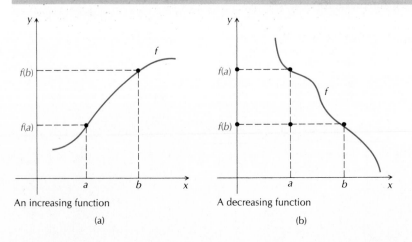

An increasing function

(a)

A decreasing function

(b)

EXAMPLES Determine whether each function is increasing, decreasing, or neither.

6.

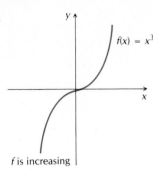

$f(x) = x^3$

f is increasing

7.

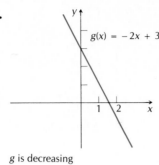

$g(x) = -2x + 3$

g is decreasing

8.

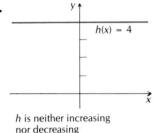

$h(x) = 4$

h is neither increasing
nor decreasing

9.

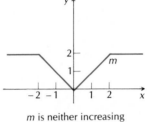

m

m is neither increasing
nor decreasing

14. Determine whether each function is increasing, decreasing, or neither increasing nor decreasing.

a)

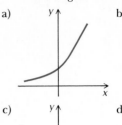

b)

c)

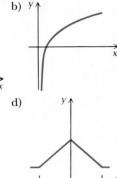

d)

−3   3

In Example 9, the function m is neither increasing nor decreasing on the entire real line. But it is increasing on the interval $[0, 2]$ and decreasing on the interval $[-2, 0]$.

DO EXERCISES 14 AND 15.

vi **FUNCTIONS DEFINED PIECEWISE**

Sometimes functions are defined piecewise. That is, we have different output formulas for different parts of the domain.

15. For the function in Exercise 14(d), find an interval on which the function is (a) increasing; (b) decreasing.

EXAMPLE 10 Graph the function defined as follows.

$$f(x) = \begin{cases} 4 & \text{for } x \leqslant 0 \\ & \text{(This means that for any input } x \text{ less than or} \\ & \text{equal to 0 the output is 4.)} \\ 4 - x^2 & \text{for } 0 < x \leqslant 2 \\ & \text{(This means that for any input } x \text{ greater than 0} \\ & \text{and less than or equal to 2, the output is } 4 - x^2.) \\ 2x - 6 & \text{for } x > 2 \\ & \text{(This means that for any input } x \text{ greater than 2,} \\ & \text{the output is } 2x - 6.) \end{cases}$$

See the graph below.

a) We graph $f(x) = 4$ for inputs less than or equal to 0 (that is, $x \leqslant 0$). Note that $f(x) = 4$ *only* for numbers x less than or equal to 0.

16. Graph the function defined as follows:

$$f(x) = \begin{cases} x + 3 & \text{for } x \leqslant -2, \\ 1 & \text{for } -2 < x \leqslant 3, \\ x^2 - 10 & \text{for } 3 < x \end{cases}$$

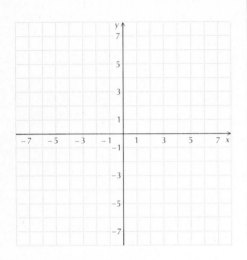

b) We graph $f(x) = 4 - x^2$ for inputs greater than 0 and less than or equal to 2 (that is, $0 < x \leqslant 2$). Note that $f(x) = 4 - x^2$ *only* on the interval $(0, 2]$.

c) We graph $f(x) = 2x - 6$ for inputs greater than 2 (that is, $x > 2$). Note that $f(x) = 2x - 6$ *only* for numbers x greater than 2.

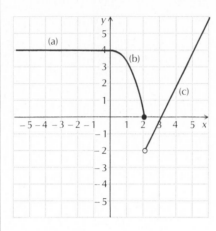

DO EXERCISE 16.

EXERCISE SET 3.6

1. Determine whether each function is even, odd, or neither even nor odd.

a)

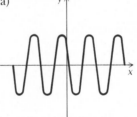

b)

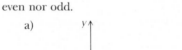

c)

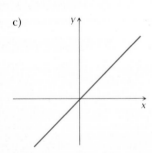

d)

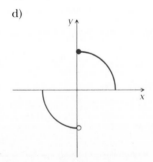

2. Determine whether each function is even, odd, or neither even nor odd.

a)

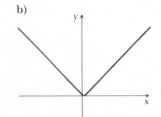

b)

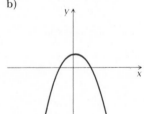

c)

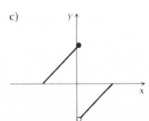

d)

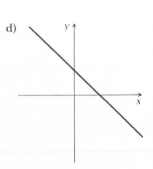

Determine whether each function is even, odd, or neither even nor odd.

3. $f(x) = 2x^2 + 4x$

4. $f(x) = -3x^3 + 2x$

5. $f(x) = 3x^4 - 4x^2$

6. $f(x) = 5x^2 + 2x^4 - 1$

7. $f(x) = 7x^3 + 4x - 2$

8. $f(x) = 4x$

9. $f(x) = |3x|$

10. $f(x) = x^{24}$

11. $f(x) = x^{17}$

12. $f(x) = x + \dfrac{1}{x}$

13. $f(x) = x - |x|$

14. $f(x) = \sqrt{x}$

15. $f(x) = \sqrt[3]{x}$

16. $f(x) = 7$

17. $f(x) = 0$

18. $f(x) = \sqrt[3]{x - 2}$

19. $f(x) = \sqrt{x^2 + 1}$

20. $f(x) = \dfrac{x^2 + 1}{x^3 - x}$

ii

21. Determine whether each function is periodic.

a)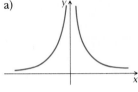

b)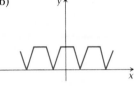

c)

d)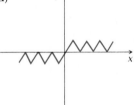

22. Determine whether each function is periodic.

a)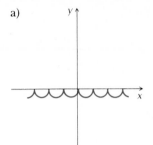

b)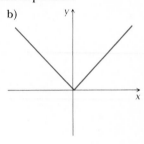

c)

d)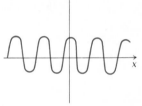

23. What is the period of this function?

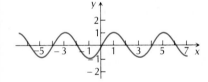

24. What is the period of this function?

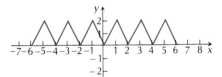

iii

25. Write interval notation for each set pictured.

a)

b)

c)

d)

26. Write interval notation for each set pictured.

a)

b)

c)

d)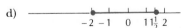

27. Write interval notation for each set.

a) $\{x \mid -2 < x < 4\}$

b) $\left\{x \mid -\dfrac{1}{4} < x \leqslant \dfrac{1}{4}\right\}$

c) $\{x \mid 7 \leqslant x < 10\pi\}$

d) $\{x \mid -9 \leqslant x \leqslant -6\}$

28. Write interval notation for each set.

a) $\{x \mid -5 < x < 0\}$

b) $\{x \mid -\sqrt{2} \leqslant x < \sqrt{2}\}$

c) $\left\{x \mid -\dfrac{\pi}{2} < x \leqslant \dfrac{\pi}{2}\right\}$

d) $\left\{x \mid -12 \leqslant x \leqslant -\dfrac{1}{2}\right\}$

iv

29. Is this function continuous

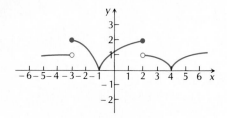

a) in the interval [0, 2]?
b) in the interval (−2, 0)?
c) in the interval [0, 3)?
d) in the interval [−3, −1]?
e) in the interval (−3, −1]?

30. Is this function continuous

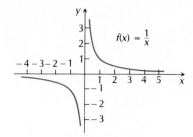

$f(x) = \frac{1}{x}$

a) in the interval [−3, −1]?
b) in the interval (1, 4)?
c) in the interval [−1, 1]?
d) in the interval [−2, 4]?
e) in the interval (0, 1]?

31. Where are the discontinuities of the function in Exercise 29?

32. Where are the discontinuities of the function in Exercise 30?

v

33. Determine whether each function is increasing, decreasing, or neither increasing nor decreasing.

a)
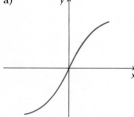

b)
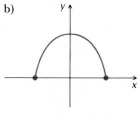

c)
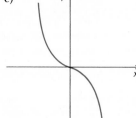

d)
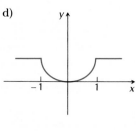

34. Determine whether each function is increasing, decreasing, or neither increasing nor decreasing.

a)
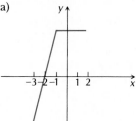

b)
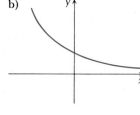

c)
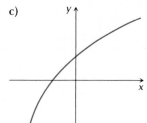

d)
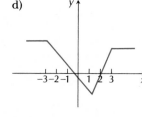

vi Graph.

35. $f(x) = \begin{cases} 1 & \text{for } x < 0, \\ -1 & \text{for } x \geqslant 0 \end{cases}$

37. $f(x) = \begin{cases} 3 & \text{for } x \leqslant -3, \\ |x| & \text{for } -3 < x \leqslant 3, \\ -3 & \text{for } x > 3 \end{cases}$

36. $f(x) = \begin{cases} 2 & \text{for } x \text{ an integer}, \\ -2 & \text{for } x \text{ not an integer} \end{cases}$

38. $f(x) = \begin{cases} -2x - 6 & \text{for } x \leqslant -2, \\ 2 - x^2 & \text{for } -2 < x < 2, \\ 2x - 6 & \text{for } x \geqslant 2 \end{cases}$

39. (*The postage function*). Postage rates are as follows: 22¢ for the first ounce plus 17¢ for each additional ounce or fraction thereof. Thus if x is the weight of a letter in ounces, then $p(x)$ is the cost of mailing the letter, where

$$p(x) = \begin{cases} 22\text{¢} & \text{if } 0 < x \leqslant 1, \\ 39\text{¢} & \text{if } 1 < x \leqslant 2, \\ 56\text{¢} & \text{if } 2 < x \leqslant 3, \end{cases}$$

and so on, up to 12 ounces, after which postal cost also depends on distance. Graph this function for x such that $0 < x \leqslant 12$.

40. Graph.

$$f(x) = \begin{cases} 3 + x & \text{for } x \leqslant 0, \\ \sqrt{x} & \text{for } 0 < x < 4, \\ x^2 - 4x - 1 & \text{for } x \geqslant 4 \end{cases}$$

41. For the function in Exercise 33 (d), find an interval on which the function is (a) increasing; (b) decreasing.

42. For the function in Exercise 34 (d), find an interval on which the function is (a) increasing; (b) decreasing.

43. Graph each function. Then determine whether it is increasing, decreasing, or neither increasing nor decreasing.

  a) $f(x) = -2x - 3$ b) $f(x) = 2x - 3$
  c) $f(x) = 3x^2$ d) $f(x) = \sqrt{3}$
  e) $f(x) = |x| + 2$ f) $f(x) = x^3 - 2$

44. Graph each function. Then determine whether it is increasing, decreasing, or neither increasing nor decreasing.

  a) $f(x) = 3x + 4$ b) $f(x) = -3x + 4$
  c) $f(x) = x^2 + 1$ d) $f(x) = -2$
  e) $f(x) = |x| + 2$ f) $f(x) = x^3 - 2$

45. To what interval is each of the following mapped under the linear transformation $x' = x + 2$?

  a) $[0, 1]$ b) $(-2, 7]$ c) $(-8, -1)$

46. To what interval is each of the following mapped under the linear transformation $x' = 0.4x - 3$?

  a) $[0, 1]$ b) $(-4, -20)$ c) $[6, 11)$

SUMMARY AND REVIEW: CHAPTER 3

The following contains a summary of what you should be able to do after completing this chapter. The review exercises are for practice. Answers are at the back of the book. If you miss an exercise, restudy the section indicated alongside the answer.

You should be able to:

Find Cartesian products of small sets.

  1. List all the ordered pairs in the Cartesian square $G \times G$, where $G = \{1, 3, 5, 7\}$.

Find the domain and range of a relation.

  2. List the domain and the range of the relation whose ordered pairs are $(3, 1)$, $(5, 3)$, $(7, 7)$, and $(3, 5)$.

Graph relations and functions given by equations and formulas.

Graph.

  3. $x = |y|$ **4.** $y = (x + 1)^2$ **5.** $y = |x| - 2$ **6.** $f(x) = \sqrt{x}$
  7. $f(x) = \sqrt{x - 2}$ **8.** $f(x) = 2\sqrt{x + 3}$ **9.** $f(x) = \frac{1}{4}\sqrt{x - 1} + 2$ **10.** $|x - y| = 1$

Given an equation defining a relation, determine whether a graph is symmetric with respect to a coordinate axis or the origin.

Consider the following relations for Exercises 11–13.

  a) $y = 7$ b) $x^2 + y^2 = 4$ c) $x^3 = y^3 - y$ d) $y^2 = x^2 + 3$
  e) $x + y = 3$ f) $x = 3$ g) $y = x^2$ h) $y = x^3$

11. Which are symmetric with respect to the x-axis?

12. Which are symmetric with respect to the y-axis?

13. Which are symmetric with respect to the origin?

Recognize a graph of a function.

14. Which of the following are graphs of functions?

a)

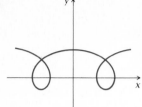

b)

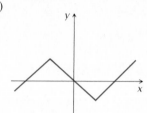

c)

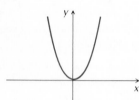

d)

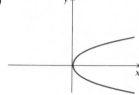

For a function given by a formula, find function values and construct and simplify expressions involving function values.

Use $f(x) = x^2 - x - 3$ to answer Exercises 15–17. Find:

15. $f(0)$.

16. $f(-3)$.

17. $\dfrac{f(a + h) - f(a)}{h}$.

Use $g(x) = 2\sqrt{x - 1}$ to answer Exercises 18–20. Find:

18. $g(1)$.

19. $g(5)$.

20. $g(a + 2)$.

Find the domain of a function given by a formula.

Find the domain of each function.

21. $f(x) = \sqrt{7 - 3x}$

22. $f(x) = \dfrac{1}{x^2 - 6x + 5}$

23. $f(x) = (x - 9x^{-1})^{-1}$

24. $f(x) = \dfrac{\sqrt{1 - x}}{x - |x|}$

Compose pairs of functions f and g, finding formulas for $f \circ g$ and $g \circ f$.

Find $f \circ g(x)$ and $g \circ f(x)$ in Exercises 25 and 26.

25. $f(x) = \dfrac{4}{x^2}$; $g(x) = 3 - 2x$

26. $f(x) = 3x^2 + 4x$; $g(x) = 2x - 1$

Given the graph of a relation, graph its transformation under translations, reflections, stretchings, and shrinkings.

27. Here is a graph of $y = f(x)$.

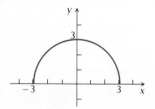

Sketch graphs of the following.

a) $y = 1 + f(x)$

b) $y = \frac{1}{2}f(x)$

c) $y = f(x + 1)$

Given the graph of a function or a formula, determine whether the function is even, odd, or neither even nor odd; periodic and if it is periodic, determine its period; continuous over a specified interval, and indicate discontinuities; increasing, decreasing, or neither increasing nor decreasing.

Use the following to answer Exercises 28–30.

a)

b)

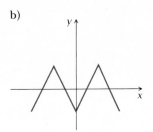

c) $f(x) = 3x^2 - 2$
e) $f(x) = 3x^3$

d) $f(x) = x + 3$
f) $f(x) = x^5 - x^3$

28. Which of the previous are even?

29. Which of the previous are odd?

30. Which of the previous are neither even nor odd?

31. Which of the following functions are periodic?

a)

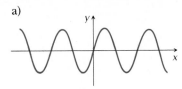

b)

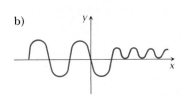

c)

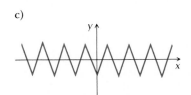

d)

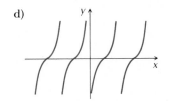

32. What is the period of this function?

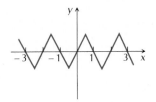

33. Is this function continuous (a) in the interval $[-3, -1]$? (b) in the interval $[-1, 1]$?

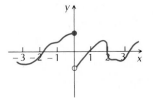

Use the following for Exercises 34–36.

a)

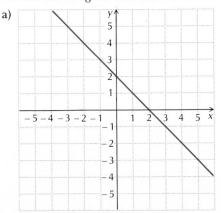

b)

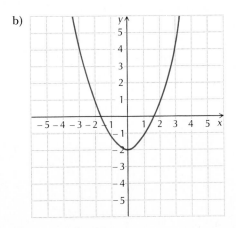

c)

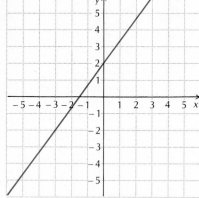

d)

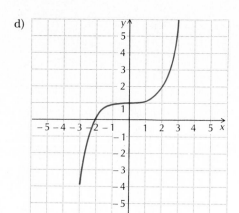

34. Which of the above are increasing?

35. Which of the above are decreasing?

36. Which of the above are neither increasing nor decreasing?

Write interval notation for certain sets.

Write interval notation for the following.

37. $\{x \mid -\pi \leqslant x \leqslant 2\pi\}$

38. $\{x \mid 0 < x \leqslant 1\}$

Graph functions defined piecewise.

39. Graph

$$f(x) = \begin{cases} x^2 + 2 & \text{for } x < 0, \\ x^3 & \text{for } 0 \leqslant x < 2, \\ -4x + 5 & \text{for } x \geqslant 2. \end{cases}$$

40. Graph several functions of the type $y = |f(x)|$. Describe a procedure, involving transformations, for graphing such functions.

TEST: CHAPTER 3

1. List all the ordered pairs in the Cartesian product $A \times B$, where $A = \{1, 3, 7\}$ and $B = \{3, 6\}$.

2. List the range of the relation whose ordered pairs are $(2, 7)$, $(-2, -7)$, $(7, -2)$, $(3, 7)$, and $(0, 2)$.

3. Graph $y = (x - 2)^2$.

4. Graph $x = |y + 1|$.

Consider the following relations for Questions 5 and 6.

a) $2y - \dfrac{5}{x} = 0$

b) $x = -3$

c) $y = x^2 - 2$

d) $x^3 - x = y^3$

e) $x^2 - 1 = y^2$

f) $3y = |x|$

5. Which are symmetric with respect to the origin?

6. Which are symmetric with respect to the x-axis?

7. Find the domain:

$$f(x) = \frac{1}{16 - x^2}.$$

8. Find $f \circ g(x)$ and $g \circ f(x)$:

$$f(x) = 2x - 1, \qquad g(x) = x^2 + 6.$$

9. Which of the following are graphs of functions?

a)

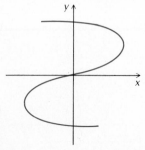

b)

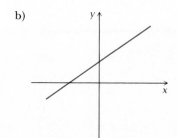

c)

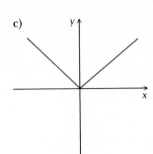

d)

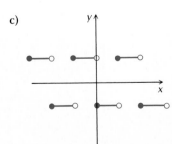

Use $g(x) = x^2 - x + 2$ to answer Questions 10–13. Find:

10. $g(-1)$.

11. $g(0)$.

12. $g(a-1)$.

13. $\dfrac{g(a+h) - g(a)}{h}$.

14. Here is a graph of $y = f(x)$.

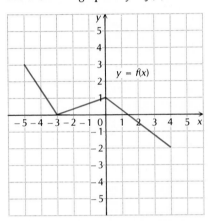

Sketch graphs of each of the following.

a) $y = f(x - 1)$

b) $y = f(2x)$

c) $y = 3 + f(x)$

Use the following to answer Questions 15 and 16.

a)

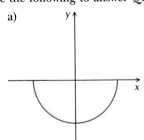

b)

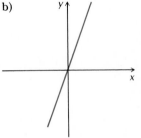

c) $f(x) = x^2 - 9$

e) $f(x) = |x|$

d) $f(x) = x^3 - 2x + 4$

f) $f(x) = x^7 - x^5$

15. Which are even?

16. Which are neither even nor odd?

17. Write interval notation for $\{x \mid -7 < x < 2\}$.

18. Which of the following functions are periodic?

a)

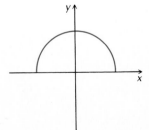

b)

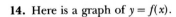

c)

19. What is the period of this function?

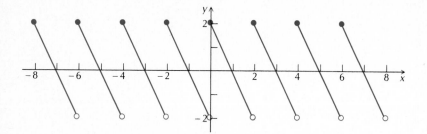

20. Is this function continuous (a) in the interval $[-1, 1)$?   (b) in the interval $(-2, -1)$?

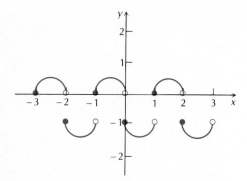

Use the following for Questions 21 and 22.

a)

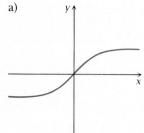

b)

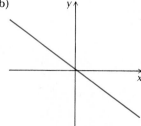

c)

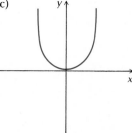

21. Which are decreasing?

22. Which are neither increasing nor decreasing?

23. Graph:

$$f(x) = \begin{cases} x^3 & \text{for } x < -2, \\ |x| & \text{for } -2 \leqslant x < 2, \\ \sqrt{x-1} & \text{for } x \geqslant 2. \end{cases}$$

What does it mean to say that the grade of a road is three percent?

A *linear function* is a function whose graph is a straight line. Such a function can always be described by a linear, or first-degree, polynomial or equation, $f(x) = mx + b$. A *quadratic function* is a function that can be described by a quadratic, or second-degree, polynomial or equation $f(x) = ax^2 + bx + c$, $a \neq 0$. In this chapter we study these kinds of functions in some detail, looking at their graphs and their properties and using them in problem solving.

While we are studying linear and quadratic equations, we shall also look at linear and quadratic inequalities as well as some functions that involve the use of absolute value.

LINEAR AND QUADRATIC FUNCTIONS AND INEQUALITIES

4

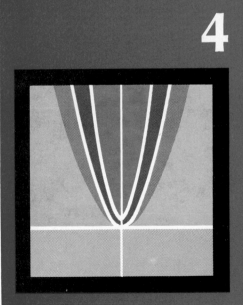

1. Determine whether each equation is
    linear.

    a) $3x = 2y + 4$

    b) $7y = 11$

    c) $5y^2 x = 13$

    d) $x = \dfrac{4}{y}$

2. Graph $6x - 4y = 12$.

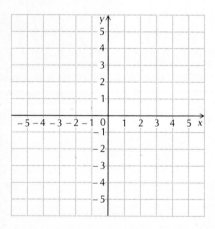

4.1

LINES AND LINEAR FUNCTIONS

i LINEAR EQUATIONS

An equation of the type $Ax + By = C$ is called a *linear equation* because its
graph is a straight line. In the above, A, B, and C are constants, but A and
B cannot both be 0. Any equation that is equivalent to one of this form has
a straight-line graph.

EXAMPLE 1   Determine whether each equation is linear.

a) The equation $5x + 8y - 3 = 0$ is linear because it is equivalent to
    $5x + 8y = 3$. Here, $A = 5$, $B = 8$, and $C = 3$.

b) The equation $3x^2 - 4y + 5 = 0$ is not linear because x is squared.

c) The equation $4x + 3xy = 25$ is not linear because the product xy occurs.

DO EXERCISE 1.

ii GRAPHS OF LINEAR EQUATIONS

Since two points determine a line, we can graph a linear equation by finding
two of its points. Then we draw a line through those points.
    A third point should always be used as a check. The easiest points to
find are often the intercepts (the points where the line crosses the axes).

EXAMPLE 2   Graph $4x + 5y = 20$.

    We set $x = 0$ and find that $y = 4$. Thus $(0, 4)$ is a point of the graph
(the *y*-intercept).
    We set $y = 0$ and find that $x = 5$. Thus $(5, 0)$ is a point of the graph
(the *x*-intercept). The graph is shown below. The point $(-2, 5\frac{3}{5})$ was used
as a check.

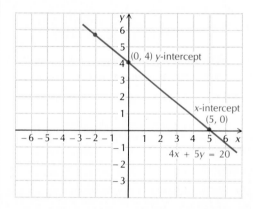

    If a graph, such as $y = 7x$, goes through the origin, it has only one
intercept, and other points will be needed for graphing.
    If an equation has a missing variable ($A = 0$ or $B = 0$), then its graph
is parallel to one of the axes.

DO EXERCISE 2.

EXAMPLE 3

a) Graph $y = 3$.

The graph is shown below.

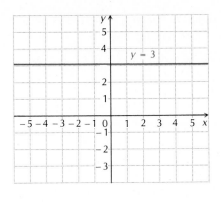

b) Graph $x = -2$.

The graph is shown below.

Graph.

3. $3x + 2y = 6$

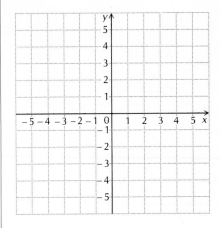

DO EXERCISES 3–5.

4. $x = 4$

iii SLOPE

The graph of a linear equation may slant upward or downward in various ways. Let us see how this relates to equations.

Suppose points P_1 and P_2 with coordinates (x_1, y_1) and (x_2, y_2) are two different points on a line not parallel to an axis. Consider a right triangle as shown with legs parallel to the axes. The point P_3 with coordinates (x_2, y_1) is the third vertex of the triangle. As we move from P_1 to P_2, y changes from y_1 to y_2. The change in y is $y_2 - y_1$. Similarly, the change in x is $x_2 - x_1$. The ratio of these changes is called the *slope*.

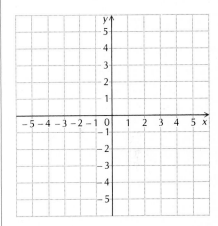

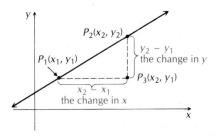

5. $y = -3$

Definition

The *slope m* of a line containing two points (x_1, y_1) and (x_2, y_2) is defined by

$$m = \frac{y_2 - y_1}{x_2 - x_1}.$$

Note that when $x_1 = x_2$, $x_1 - x_2 = 0$, and the slope is not defined.

EXAMPLE 4 Graph the line through the points $(1, 2)$ and $(3, 6)$ and find its slope.

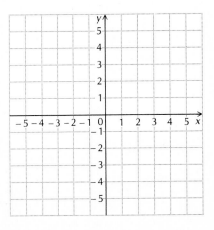

Use graph paper. Graph the lines through these points and find their slopes.

6. $(1, 3)$ and $(2, 5)$

7. $(3, 7)$ and $(5, 3)$

8. $(1, 1)$ and $(2, 3)$

9. $(3, 5)$ and $(2, -1)$

Use graph paper. Graph the lines through these points and find their slopes.

10. $(-1, -1)$ and $(2, -4)$

11. $(0, 2)$ and $(3, 1)$

Find the slopes, if they exist, of the lines containing these points.

12. $(4, 6)$ and $(-2, 6)$

13. $(-3, 5)$ and $(-3, 7)$

14. $(9, 0)$ and $(6, 0)$

Let us call the slope m and let $(1, 2)$ be (x_1, y_1) and $(3, 6)$ be (x_2, y_2). Applying the definition, we obtain

$$m = \frac{y_2 - y_1}{x_2 - x_1} = \frac{6 - 2}{3 - 1} = \frac{4}{2} = 2.$$

Note that we can also use the points in the opposite order, so long as we are consistent. We get the same slope:

$$m = \frac{y_2 - y_1}{x_2 - x_1} = \frac{2 - 6}{1 - 3} = \frac{-4}{-2} = 2.$$

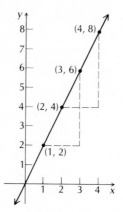

From Example 4 we see that it does not matter in which order we choose the points, so long as we take differences in the same order. From Example 4 we can also see that it does not matter which two points of a line we choose to determine the slope. No matter what points we choose we get the same number for the slope. For example, if we choose $(2, 4)$ and $(4, 8)$ we get for the slope,

$$\frac{8 - 4}{4 - 2} = \frac{4}{2} = 2.$$

DO EXERCISES 6–9.

If a line slants upward from left to right, it has a positive slope, as in Example 4. If a line slants downward from left to right, the change in x and the change in y have opposite signs, so the line has a negative slope.

> **If a line is horizontal, the change in y for any two points is 0. Thus a horizontal line has zero slope.**
>
> **If a line is vertical, the change in x for any two points is 0. Thus the slope is not defined, because we cannot divide by zero.**
>
> **A vertical line does not have a slope.**

DO EXERCISES 10–14.

iv POINT–SLOPE EQUATIONS OF LINES

Suppose we have a nonvertical line and that the coordinates of one point P_1 are (x_1, y_1). We think of point P_1 as fixed. Suppose, also, that we have a movable point P on the line with coordinates (x, y). Thus the slope would be given by

$$\frac{(y - y_1)}{(x - x_1)} = m. \qquad (1)$$

Note that this is true only when (x, y) is a point different from (x_1, y_1). If we use the multiplication principle, we get*

$$(y - y_1) = m(x - x_1). \qquad \textit{Point–slope equation} \qquad (2)$$

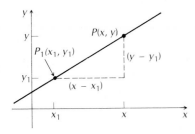

Equation (2) will be true even if $(x, y) = (x_1, y_1)$. Equation (2) is called the *point–slope equation* of a line. Thus if we know the slope of a line and the coordinates of a point on the line, we can find an equation of the line.

EXAMPLE 5   Find an equation of the line containing the point $(\frac{1}{2}, -1)$ with slope 5.

If we substitute in $(y - y_1) = m(x - x_1)$, we get $y - (-1) = 5(x - \frac{1}{2})$, which simplifies to

$$y + 1 = 5(x - \tfrac{1}{2})$$

or

$$y = 5x - \tfrac{5}{2} - 1$$

or

$$y = 5x - \tfrac{7}{2}.$$

DO EXERCISES 15–17.

v. TWO-POINT EQUATIONS OF LINES

Suppose a nonvertical line contains the points $P_1(x_1, y_1)$ and $P_2(x_2, y_2)$. The slope of the line is

$$\frac{y_2 - y_1}{x_2 - x_1}.$$

If we substitute $(y_2 - y_1)/(x_2 - x_1)$ for m in the point–slope equation,

$$y - y_1 = m(x - x_1),$$

---

* *Hint:* It may be easier to remember Eq. (1), since it relates to the slope formula. We can easily get Eq. (2) from it.

Find an equation of the line.

15. Containing the point $(-2, \frac{1}{4})$ with slope -3

16. With y-intercept $(0, -9)$ and slope $\frac{1}{4}$

17. With x-intercept $(5, 0)$ and slope $-\frac{1}{2}$

Find an equation of the line containing the given pair of points.

18. $(1, 4)$ and $(3, -2)$

we have*

$$y - y_1 = \frac{y_2 - y_1}{x_2 - x_1}(x - x_1). \qquad \textit{Two-point equation}$$

This is known as the *two-point equation* of a line. Note that either of the two given points can be called P_1 or P_2 and (x, y) is any point on the line.

EXAMPLE 6   Find an equation of the line containing the points $(2, 3)$ and $(1, -4)$.

19. $(3, -6)$ and $(0, 4)$

If we take $(2, 3)$ as P_1 and use the two-point equation, we get

$$y - 3 = \frac{-4 - 3}{1 - 2}(x - 2),$$

which simplifies to $y = 7x - 11$.

DO EXERCISES 18 AND 19.

vi SLOPE-INTERCEPT EQUATIONS OF LINES

Suppose a nonvertical line has slope m and y-intercept $(0, b)$. We sometimes, for brevity, refer to the number b as the y-intercept. Let us substitute m and $(0, b)$ in the point–slope equation. We get

$$y - y_1 = m(x - x_1)$$
$$y - b = m(x - 0).$$

Find the slope and y-intercept.

20. $y = -7x + 11$

This simplifies to

$$y = mx + b. \qquad \textit{Slope–intercept equation}$$

This is called the *slope–intercept equation* of a line. The advantage of such an equation is that we can read the slope m and the y-intercept b from the equation.

EXAMPLE 7   Find the slope and the y-intercept of $y = 5x - \frac{1}{4}$.

$$y = 5x - \tfrac{1}{4}$$

Slope: 5 y-intercept: $-\frac{1}{4}$

21. $y = -4$

EXAMPLE 8   Find the slope and y-intercept of $y = 8$.

We can rewrite this equation as $y = 0x + 8$. We then see that the slope is 0 and the y-intercept is 8.

DO EXERCISES 20 AND 21.

---

* *Hint:* It may be easier to remember

$$\frac{y - y_1}{x - x_1} = \frac{y_2 - y_1}{x_2 - x_1},$$

interpreting each side as slope.

To find the slope–intercept equation of a line, when given another equation, we solve for y.

EXAMPLE 9 Find the slope and y-intercept of the line having the equation $3x - 6y - 7 = 0$.

We solve for y, obtaining $y = \frac{1}{2}x - \frac{7}{6}$. Thus the slope is $\frac{1}{2}$ and the y-intercept is $-\frac{7}{6}$.

If a line is vertical it has no slope. Thus it has no slope–intercept equation. Such a line does have a simple equation, however. All vertical lines have equations $x = c$, where c is some constant.

DO EXERCISE 22.

LINEAR FUNCTIONS

Any nonvertical straight line is the graph of a function. Such a function is called a *linear function*. Any nonvertical line has an equation $y = mx + b$. Thus a function f is a linear function if and only if it has an equation $f(x) = mx + b$. If the slope m is zero, the equation simplifies to $f(x) = b$. A function like this is called a *constant function*.

22. a) Find the slope–intercept equation of the line whose equation is $-2x + 3y - 6 = 0$.

b) Find the slope and y-intercept of this line.

EXERCISE SET 4.1

i

1. Determine whether each equation is linear.
   a) $3y = 2x - 5$
   b) $5x + 3 = 4y$
   c) $3y = x^2 + 2$
   d) $y = 3$
   e) $xy = 5$
   f) $3x^2 + 2y = 4$
   g) $3x + \dfrac{1}{y} = 4$
   h) $5x - 2 = 4y$

2. Determine whether each equation is linear.
   a) $5y = 3x - 4$
   b) $3x + 5 = 7y$
   c) $4y = 3x^2 - 4$
   d) $4x - \dfrac{2}{y} = 3$
   e) $2xy = 4$
   f) $5x^2 + 3y = -4$
   g) $x = -4$
   h) $6x - 7 = 3y$

ii Graph.

3. $8x - 3y = 24$ **4.** $5x - 10y = 50$ **5.** $3x + 12 = 4y$ **6.** $4x - 20 = 5y$
7. $y = -2$ **8.** $2y - 3 = 9$ **9.** $5x + 2 = 17$ **10.** $19 = 5 - 2x$

iii Find the slopes of the lines containing these points.

11. $(6, 2)$ and $(-2, 1)$ **12.** $(-2, 1)$ and $(-4, -2)$
13. $(2, -4)$ and $(4, -3)$ **14.** $(5, -3)$ and $(-5, 8)$
15. $(2\pi, 5)$ and $(\pi, 4)$ **16.** $(\sqrt{2}, -4)$ and $(\pi, -4)$

iv Find equations of the following lines.

17. Through $(3, 2)$ with $m = 4$ **18.** Through $(4, 7)$ with $m = -2$
19. With y-intercept -5 and $m = 2$ **20.** With y-intercept π and $m = \frac{1}{4}$

v Find equations of the following lines.

21. Containing $(1, 4)$ and $(5, 6)$ **22.** Containing $(-2, 0)$ and $(2, 3)$

vi Find the slope and *y*-intercept of each line.

23. $y = 2x + 3$

24. $y = 6 - x$

25. $2y = -6x + 10$

26. $-3y = -12x + 9$

27. $3x - 4y = 12$

28. $5x + 2y = -7$

29. $3y + 10 = 0$

30. $y = 7$

Find equations of the following lines.

31. ▦ Through $(3.014, -2.563)$ with slope 3.516

32. ▦ Through $(-173.4, -17.6)$ with slope -0.00014

33. ▦ Through the points $(1.103, 2.443)$ and $(8.114, 11.012)$

34. ▦ Through the points $(473.78, 910.2)$ and $(993.55, 171.43)$

Suppose *f* is a linear function. Then $f(x) = mx + b$. Find a formula for $f(x)$ given that:

35. $f(3x) = 3f(x)$.

36. $f(kx) = kf(x)$, for some number k.

37. $f(x + 2) = f(x) + 2$.

38. $[f(x)]^2$ is linear.

Suppose *f* is a linear function. Then $f(x) = mx + b$. Determine whether the following are true or false.

39. $f(c + d) = f(c) + f(d)$

40. $f(cd) = f(c)f(d)$

41. $f(kx) = kf(x)$

42. $f(c - d) = f(c) - f(d)$

43. Determine whether these three points are on a line. [*Hint:* Compare the slopes of $\overline{AB}$ and $\overline{BC}$. ($\overline{AB}$ refers to the segment from *A* to *B*.)]

$$A(9, 4), \quad B(-1, 2), \quad C(4, 3)$$

44. Determine whether these three points are on a line. (See the hint for Exercise 43.)

$$A(-1, -1), \quad B(2, 2), \quad C(-3, -4)$$

45. Use graph paper. Plot the points $A(0, 0)$, $B(8, 2)$, $C(11, 6)$, and $D(3, 4)$. Draw $\overline{AB}$, $\overline{BC}$, $\overline{CD}$, and $\overline{DA}$. Find the slopes of these four segments. Compare the slopes of $\overline{AB}$ and $\overline{CD}$. Compare the slopes of $\overline{BC}$ and $\overline{DA}$. (Figure *ABCD* is a parallelogram and its opposite sides are parallel.)

46. Use graph paper. Plot the points $E(-2, -5)$, $F(2, -2)$, $G(7, -2)$, and $H(3, -5)$. Draw $\overline{EF}$, $\overline{FG}$, $\overline{GH}$, $\overline{HE}$, $\overline{EG}$, and $\overline{FH}$. Compare the slopes of $\overline{EG}$ and $\overline{FH}$. (Figure *EFGH* is a rhombus and its diagonals are perpendicular.)

47. (*Fahrenheit temperature as a function of Celsius temperature*). Fahrenheit temperature *F* is a linear function of Celsius (or Centigrade) temperature *C*. When *C* is 0, *F* is 32. When *C* is 100, *F* is 212. Use these data to express *F* as a linear function of *C*.

48. (*Celsius temperature as a function of Fahrenheit temperature*). Celsius (Centigrade) temperature *C* is a linear function of Fahrenheit temperature *F*. When *F* is 32, *C* is 0. When *F* is 212, *C* is 100. Use these data to express *C* as a linear function of *F*.

49. Suppose *P* is a nonconstant linear function of *Q*. Show that *Q* is a linear function of *P*.

50. Suppose *y* is directly proportional to *x*. Show that *y* is a linear function of *x*.

(*Road grade*). Numbers like 2%, 3%, and 6% are often used to represent the *grade* of a road. Such a number is meant to tell how steep a road up a hill or mountain is. For example, a 3% grade means that for every horizontal distance of 100 ft, the road rises 3 ft. In each case find the road grade and an equation giving the height *y* of a vehicle in terms of a horizontal distance *x*.

51.

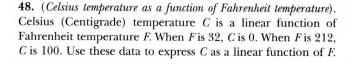
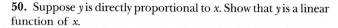

52.

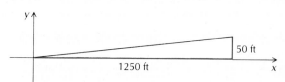

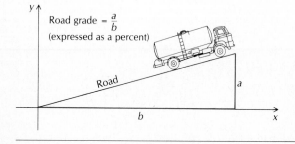

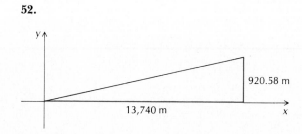

4.2

PARALLEL AND PERPENDICULAR LINES; THE DISTANCE FORMULA

i PARALLEL AND PERPENDICULAR LINES

If two lines are vertical, they are parallel. Thus equations such as $x = c_1$ and $x = c_2$ (where c_1 and c_2 are constants) have graphs that are parallel lines. Now we consider nonvertical lines. In order that such lines be parallel, they must have the same slope but different y-intercepts. Thus equations such as $y = mx + b_1$ and $y = mx + b_2$, $b_1 \neq b_2$, have graphs that are parallel lines.

Theorem 1

> Vertical lines are parallel. Nonvertical lines are parallel if and only if they have the same slope and different y-intercepts.

If two equations are equivalent, they represent the same line. Thus if two lines have the same slope and the same y-intercept, they are not really two different lines. They are the same line. In such a case we sometimes speak of *coincident lines*.

If one line is vertical and the other is horizontal, such as $x = c_1$ and $y = c_2$, they are perpendicular. Otherwise, how can we tell whether two lines are perpendicular? Let's look at the following figure.

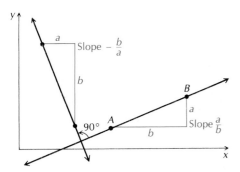

Consider a line $\overleftrightarrow{AB}$ as shown, with slope a/b. Then think of rotating the figure 90° to get a line perpendicular to $\overleftrightarrow{AB}$. For the new line the change in y and the change in x are interchanged, but the change in x is now the additive inverse of what it was. Thus the slope of the new line is $-b/a$. Let us multiply the slopes:

$$\frac{a}{b}\left(-\frac{b}{a}\right) = -1.$$

This is the condition under which lines will be perpendicular.

Theorem 2

> Two lines with slopes m_1 and m_2 are perpendicular if and only if $m_1 m_2 = -1$.

If one line has slope m_1, the slope m_2 of a line perpendicular to it is $-1/m_1$.

DO EXERCISES 1 AND 2.

OBJECTIVES

You should be able to:

i Given equations of two lines, tell whether they are parallel, perpendicular, or neither.

ii Given an equation of a line and the coordinates of a point, find equations of lines parallel or perpendicular to that line and containing the given point.

iii Given the coordinates of two points, find the distance between them.

iv Determine whether three points with given coordinates are vertices of a right triangle.

v Given the coordinates of the endpoints of a segment, find the coordinates of its midpoint.

In each situation, slopes of the lines are given. Determine whether they are perpendicular.

1.

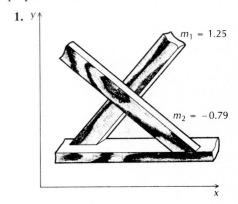

$m_1 = 1.25$

$m_2 = -0.79$

2.

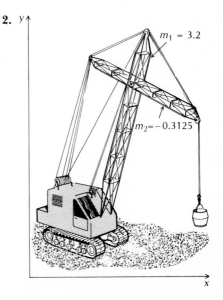

$m_1 = 3.2$

$m_2 = -0.3125$

Determine whether each pair of lines is parallel, perpendicular, or neither.

3. $2y - x = 2$,
$y + 2x = 4$

4. $3y = 2x + 15$,
$2y = 3x + 10$

5. $5y = 3 - 4x$,
$8x + 10y = 1$

6. Find equations of the lines parallel and perpendicular to the line $4 - y = 2x$ and containing the point $(3, 4)$.

EXAMPLE 1 Determine whether the following lines are parallel, perpendicular, or neither.

a) $y + 2 = 5x$, $5y + x = -15$.

We solve for y:

$$y = 5x - 2, \quad y = -\tfrac{1}{5}x - 3.$$

The slopes are 5 and $-\tfrac{1}{5}$. Their product is -1, so the lines are perpendicular.

b) $2y + 4x = 8$, $5 + 2x = -y$.

By solving for y we determine that $m_1 = -2$ and $m_2 = -2$ and the y-intercepts are different, so the lines are parallel.

c) $2x + 1 = y$, $y + 3x = 4$.

By solving for y we determine that $m_1 = 2$ and $m_2 = -3$, so the lines are neither parallel nor perpendicular.

DO EXERCISES 3–5.

ii PARALLEL OR PERPENDICULAR LINES THROUGH A POINT

EXAMPLE 2 Write equations of the lines perpendicular and parallel to the line $4y - x = 20$ and containing the point $(2, -3)$.

We first solve for y: $y = \tfrac{1}{4}x + 5$, so the slope is $\tfrac{1}{4}$. The slope of the perpendicular line is -4.

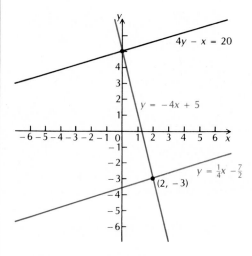

Now we use the point–slope equation to write an equation with slope -4 and containing the point $(2, -3)$:

$$y - y_1 = m(x - x_1)$$
$$y - (-3) = -4(x - 2).$$

This simplifies to $y = -4x + 5$.

The line parallel to the given line will have slope $\tfrac{1}{4}$. The equation is

$$y - (-3) = \tfrac{1}{4}(x - 2).$$

This simplifies to $y = \tfrac{1}{4}x - \tfrac{7}{2}$.

DO EXERCISE 6.

Vertical Lines

If a line is vertical, it has no slope. Every vertical line has an equation $x = c$, where c is a constant. Any line parallel to a vertical line must also be vertical, hence must also have an equation $x = k$, where k is a constant.

A line will be perpendicular to a vertical line if and only if that line is horizontal. Every horizontal line has zero slope and has an equation $y = c$, where c is a constant. Thus if a line is vertical, it is simple to determine whether another line is parallel to it or perpendicular to it.

EXAMPLE 3 Find equations of the lines parallel and perpendicular to the line $x = 4$ and containing the point $(-2, 3)$.

The line $x = 4$ is vertical, so any line parallel to it must be vertical. The line we seek has one x-coordinate, which is -2, so all x-coordinates on the line must be -2. The equation is $x = -2$.

The perpendicular line has a y-coordinate, which is 3, so all y-coordinates must be 3. The equation is $y = 3$.

DO EXERCISE 7.

iii THE DISTANCE FORMULA

We develop a formula for finding the distance between two points whose coordinates are known. Suppose the points are on a horizontal line, thus having the same second coordinates. We can find the distance between them by subtracting their first coordinates. This difference may be negative, depending on the order in which we subtract. So to make sure we get a positive number, we take the absolute value of this difference. The distance between two points on a horizontal line (x_1, y) and (x_2, y) is thus $|x_1 - x_2|$. Similarly, the distance between two points on a vertical line (x, y_1) and (x, y_2) is $|y_1 - y_2|$.

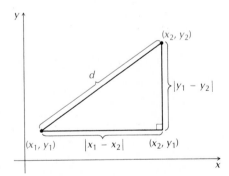

Now consider any two points not on a horizontal or vertical line (x_1, y_1) and (x_2, y_2). These points are vertices of a right triangle, as shown. The other vertex is (x_2, y_1). The legs of this triangle have the lengths $|x_1 - x_2|$ and $|y_1 - y_2|$. Now by the Pythagorean theorem we obtain a relation between the length of the hypotenuse d and the lengths of the legs:

$$d^2 = |x_1 - x_2|^2 + |y_1 - y_2|^2.$$

We may now dispense with the absolute-value signs because squares of numbers are never negative. Thus we have

$$d^2 = (x_1 - x_2)^2 + (y_1 - y_2)^2.$$

By taking the square root we obtain the distance between the two points.

7. Find equations of the lines parallel and perpendicular to the line $x = -3$ and containing the point $(5, -4)$.

Find the distance between the points of each pair.

8. $(-5, 3)$ and $(2, -7)$

9. $(3, 3)$ and $(-3, -3)$

10. $(9, -5)$ and $(9, 11)$

11. $(0, \pi)$ and $(8, \pi)$

Determine whether the points are vertices of a right triangle.

12. $(11, 1)$, $(6, 6)$, $(2, 2)$

13. $(10, -4)$, $(3, 5)$, $(0, 0)$

Theorem 3 The distance formula

> The distance between any two points (x_1, y_1) and (x_2, y_2) is given by
> $$d = \sqrt{(x_1 - x_2)^2 + (y_1 - y_2)^2}.$$

Although we derived the distance formula by considering two points not on a horizontal or vertical line, the formula holds for *any* two points. The subtraction of the *x*-coordinates can be done in any order, as can the subtraction of the *y*-coordinates.

EXAMPLE 4 Find the distance between the points $A(-2, 2)$ and $B(4, -3)$ on the islands in the figure.

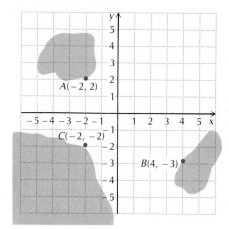

$$d = \sqrt{(-2 - 4)^2 + [2 - (-3)]^2}$$
$$= \sqrt{(-6)^2 + 5^2}$$
$$= \sqrt{36 + 25}$$
$$= \sqrt{61}$$

An approximation for this distance is 7.8.

DO EXERCISES 8–11.

iv VERTICES OF RIGHT TRIANGLES

EXAMPLE 5 Determine whether the points $A(-2, 2)$, $B(4, -3)$, and $C(-2, -2)$ in the figure of Example 4 are vertices of a right triangle.

First we find the squares of the distances between the points:

$$d_1^2 = (-2 - 4)^2 + [2 - (-3)]^2 = (-6)^2 + 5^2 = 61,$$
$$d_2^2 = [-2 - (-2)]^2 + [2 - (-2)]^2 = 0^2 + 4^2 = 16,$$
$$d_3^2 = [4 - (-2)]^2 + [-3 - (-2)]^2 = 6^2 + (-1)^2 = 37.$$

Since no sum of any two squares is another, it follows by the converse of the Pythagorean theorem that the points are not vertices of a right triangle.

DO EXERCISES 12 AND 13.

MIDPOINTS OF SEGMENTS

The distance formula can be used to verify or derive a formula for finding the coordinates of the midpoint of a segment when the coordinates of its endpoints are known. We shall not derive this formula but simply state it.

Theorem 4 The midpoint formula

If the endpoints of a segment are (x_1, y_1) and (x_2, y_2), then the coordinates of the midpoint are

$$\left(\frac{x_1 + x_2}{2}, \frac{y_1 + y_2}{2}\right).$$

Note that we obtain the coordinates of the midpoint by averaging the coordinates of the endpoints. This is an easy way to remember this formula.

EXAMPLE 6 Find the midpoint of the segment in Example 4 with endpoints $A(-2, 2)$ and $B(4, -3)$.

Using the midpoint formula, we obtain

$$\left(\frac{-2 + 4}{2}, \frac{2 + (-3)}{2}\right), \quad \text{or} \quad \left(1, -\frac{1}{2}\right).$$

DO EXERCISES 14 AND 15.

Find the midpoints of the segments having endpoints as given.

14. $(-2, 1)$ and $(5, -6)$

15. $(9, -6)$ and $(9, -4)$

EXERCISE SET 4.2

i In Exercises 1–4, determine whether the lines are parallel, perpendicular, or neither.

1. $2x - 5y = -3$, $\quad$ **2.** $x + 2y = 5$, $\quad$ **3.** $y = 4x - 5$, $\quad$ **4.** $y = -x + 7$,
$2x + 5y = 4$ $\qquad\quad$ $2x + 4y = 8$ $\qquad\quad$ $4y = 8 - x$ $\qquad\quad$ $y = x + 3$

ii In Exercises 5–10, find an equation of the line containing the given point and parallel to the given line.

5. $(0, 3)$, $3x - y = 7$ $\qquad$ **6.** $(-4, -5)$, $2x + y = -4$ $\qquad$ **7.** $(3, 8)$, $x = 2$

8. $(3, -3)$, $x = -1$ $\qquad$ **9.** $(-2, -3)$, $y = 4$ $\qquad$ **10.** $(-3, 2)$, $y = -3$

In Exercises 11–16, find an equation of the line containing the given point and perpendicular to the given line.

11. $(-3, -5)$, $5x - 2y = 4$ $\qquad$ **12.** $(3, -2)$, $3x + 4y = 5$ $\qquad$ **13.** $(0, 3)$, $x = 1$

14. $(-2, -2)$, $x = 3$ $\qquad$ **15.** $(-3, -7)$, $y = 2$ $\qquad$ **16.** $(4, -5)$, $y = -1$

17. ▦ Find an equation of the line parallel to the one given, and containing the given point.

$4.323x - 7.071y = 16.61$, $\quad (-2.603, 1.818)$

18. ▦ Find an equation of the line containing the given point and perpendicular to the given line.

$6.232x + 4.001y = 4.881$, $\quad (3.149, -2.908)$

iii Find the distance between the points of each pair.

19. $(-3, -2)$ and $(1, 1)$ $\qquad\qquad\qquad$ **20.** $(5, 9)$ and $(-1, 6)$

21. $(0, -7)$ and $(3, -4)$ $\qquad\qquad\quad$ **22.** $(2, 2)$ and $(-2, -2)$

23. $(a, -3)$ and $(2a, 5)$ $\qquad\qquad\quad$ **24.** $(5, 2k)$ and $(-3, k)$

25. $(0, 0)$ and (a, b) $\qquad\qquad\qquad$ **26.** $(\sqrt{2}, \sqrt{3})$ and $(0, 0)$

27. $(\sqrt{a}, \sqrt{b})$ and $(-\sqrt{a}, \sqrt{b})$ $\qquad\quad$ **28.** $(c - d, c + d)$ and $(c + d, d - c)$

29. ▦ $(7.3482, -3.0991)$ and $(18.9431, -17.9054)$ $\qquad$ **30.** ▦ $(-25.414, 175.31)$ and $(275.34, -95.144)$

iv Determine whether the points are vertices of a right triangle.

31. $(9, 6)$, $(-1, 2)$, and $(1, -3)$

32. $(-5, -8)$, $(1, 6)$, and $(5, -4)$

v Find the midpoints of the segments having the following endpoints.

33. $(-4, 7)$ and $(3, -9)$

34. $(4, 5)$ and $(6, -7)$

35. (a, b) and $(a, -b)$

36. $(-c, d)$ and (c, d)

37. ▦ $(-3.895, 8.1212)$ and $(2.998, -8.6677)$

38. ▦ $(4.1112, 6.9898)$ and $(5.1928, -6.9143)$

---

39. Find an equation of the line containing the point $(4, -2)$ and parallel to the line containing $(-1, 4)$ and $(2, -3)$.

40. Find an equation of the line containing the point $(-1, 3)$ and parallel to the line containing $(3, -5)$ and $(-2, -7)$.

41. Find the point on the x-axis that is equidistant from the points $(1, 3)$ and $(8, 4)$.

42. Find the point on the y-axis that is equidistant from the points $(-2, 0)$ and $(4, 6)$.

---

43. Consider any right triangle with base b and height h, situated as shown. Show that the midpoint of the hypotenuse P is equidistant from the three vertices of the triangle.

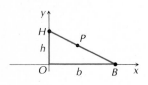

44. Consider any quadrilateral situated as shown. Show that the segments joining the midpoints of the sides, in order as shown, form a parallelogram.

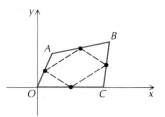

45. Prove that the distance formula holds when two points are on either a vertical line or a horizontal line.

46. The distance formula can be interpreted as a function whose inputs are ordered pairs (P, Q), where P is a pair (x_1, y_1) and Q is a pair (x_2, y_2).

    a) What is the domain of the function?

    b) What is the range of the function?

    c) Under what conditions is an output 0?

OBJECTIVES

You should be able to:

i Given a quadratic function, find the vertex of its graph, the line of symmetry, and the maximum or minimum value.

ii Graph a quadratic function.

iii Find the x-intercepts of the graph of a quadratic function.

4.3

QUADRATIC FUNCTIONS

i PROPERTIES OF QUADRATIC FUNCTIONS

If a function can be described by a second-degree polynomial, it is called *quadratic*. The following is a more precise definition.

Definition

A *quadratic* function is a function that can be described as follows:

$$f(x) = ax^2 + bx + c, \qquad \text{where } a \neq 0.$$

In this definition we insist that $a \neq 0$; otherwise the polynomial would not be of degree two. One or both of the constants b and c can be 0.

Consider $f(x) = x^2$. This is an even function, so the y-axis is a line of symmetry. The graph opens upward, as shown on the following page. If we multiply by a constant a to get $f(x) = ax^2$, we obtain a vertical stretching or shrinking, and a reflection if a is negative. Examples of this are shown in the figure. Bear in mind that if a is negative the graph opens downward.

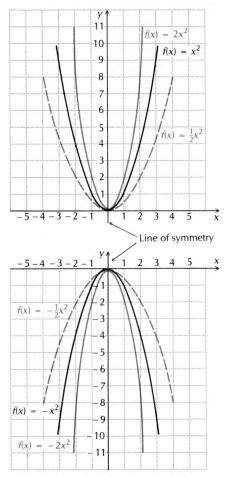

Graphs of quadratic functions. Parabolas.

Graphs of quadratic functions are called *parabolas*. In each parabola shown, the point $(0, 0)$ is the *vertex*.

DO EXERCISES 1 AND 2.

Let us consider $f(x) = a(x - h)^2$. We have replaced x by $x - h$ in ax^2, and will therefore obtain a horizontal translation. The translation will be to the right if h is positive. The examples in the figure illustrate translations, and a summary is given below.

The graph of $f(x) = a(x - h)^2$:

a) opens upward if $a > 0$, downward if $a < 0$;

b) has $(h, 0)$ as a vertex;

c) has $x = h$ as a line of symmetry;

d) has a minimum 0 if $a > 0$, a maximum 0 if $a < 0$.

1. a) Graph $f(x) = 2x^2$.

  b) Does the graph open upward or does it open downward?

  c) What is the line of symmetry?

  d) What is the minimum value?

  e) What is the vertex?

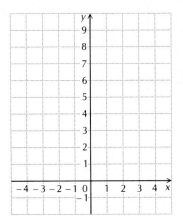

2. a) Graph $f(x) = -0.4x^2$.

  b) Does the graph open upward or does it open downward?

  c) What is the line of symmetry?

  d) What is the maximum value?

  e) What is the vertex?

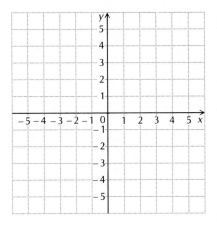

3. a) Graph $f(x) = 3x^2$.

b) Use the graph in (a) to graph

$$f(x) = 3(x - 2)^2.$$

c) What is the vertex of the graph in (b)?

d) What is the line of symmetry of the graph in (b)?

e) What is the minimum value?

f) Does the graph open upward or does it open downward?

g) Is the graph of $f(x) = 3(x - 2)^2$ a horizontal translation to the left or to the right?

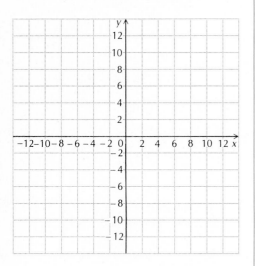

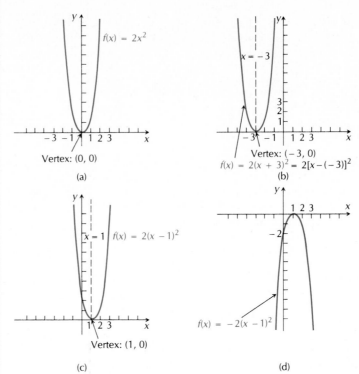

Vertex: (0, 0)

(a)

Vertex: (−3, 0)

$f(x) = 2(x + 3)^2 = 2[x - (-3)]^2$

(b)

$x = 1$ $f(x) = 2(x - 1)^2$

Vertex: (1, 0)

(c)

$f(x) = -2(x - 1)^2$

(d)

DO EXERCISES 3 AND 4. (NOTE THAT EXERCISE 4(g) IS ON THE FOLLOWING PAGE.)

Now consider $f(x) = a(x - h)^2 + k$, or $f(x) - k = a(x - h)^2$. We have replaced $f(x)$ by $f(x) - k$ in the equation $f(x) = a(x - h)^2$. Thus we have a translation. If k is positive, the translation is upward. If k is negative, the translation is downward. Consider these examples. Note that the vertex has been moved off the x-axis.

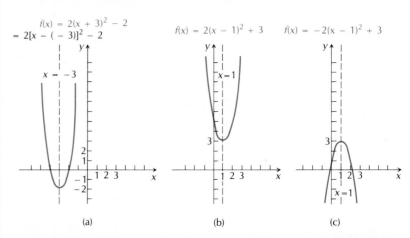

$f(x) = 2(x + 3)^2 - 2$
$= 2[x - (-3)]^2 - 2$

(a)

$f(x) = 2(x - 1)^2 + 3$

(b)

$f(x) = -2(x - 1)^2 + 3$

(c)

4. a) Graph $f(x) = -3x^2$.

b) Use the graph in (a) to graph

$$f(x) = -3(x + 2)^2$$
$$= -3[x - (-2)]^2.$$

c) What is the vertex of the graph in (b)?

d) What is the line of symmetry of the graph in (b)?

e) What is the maximum value?

f) Does the graph open upward or does it open downward?

> **The graph of $f(x) = a(x - h)^2 + k$:**
>
> a) opens upward if $a > 0$, downward if $a < 0$;
>
> b) has (h, k) as a vertex;
>
> c) has $x = h$ as the line of symmetry;
>
> d) has k as a minimum value (output) if $a > 0$, has k as a maximum value (output) if $a < 0$.

Thus without graphing, we can determine a lot of information about a function described by $f(x) = a(x - h)^2 + k$. The following table is an example.

Function	$f(x) = 3(x - \frac{1}{4})^2 - 2$	$\begin{aligned} g(x) &= -3(x + 5)^2 + 7 \\ &= -3[x - (-5)]^2 + 7 \end{aligned}$
a) What is the vertex?	$(\frac{1}{4}, -2)$	$(-5, 7)$
b) What is the line of symmetry?	$x = \frac{1}{4}$	$x = -5$
c) Is there a maximum? What is it?	No; graph extends upward; $3 > 0$.	Yes; 7; graph extends downward; $-3 < 0$.
d) Is there a minimum? What is it?	Yes; -2; graph extends upward; $3 > 0$.	No; graph extends downward; $-3 < 0$.

Note that the vertex (h, k) is used to find the maximum or minimum. The maximum or minimum is the number k, *not* the ordered pair (h, k).

DO EXERCISES 5–10.

Now let us consider a quadratic function $f(x) = ax^2 + bx + c$. Note that it is not in the form $f(x) = a(x - h)^2 + k$. We can put it into that form by *completing the square*.

EXAMPLE 1 For the function $f(x) = x^2 - 6x + 4$, complete the square.

We consider first the x^2- and x-terms:

$$f(x) = x^2 - 6x \qquad + 4$$

We construct a trinomial square. To do this we take half the coefficient of x and square it. That number is $(-6/2)^2$, or 9. We now add that number to complete the square. We also must subtract it to get an equivalent equation. We can think of this simply as adding $9 - 9$, which is 0.

$$\begin{aligned} f(x) &= x^2 - 6x + 9 - 9 + 4 \\ &= x^2 - 6x + 9 - 5 \end{aligned}$$

We now factor the trinomial square $x^2 - 6x + 9$ and we are finished:

$$f(x) = (x - 3)^2 - 5.$$

DO EXERCISE 11.

If the coefficient of x^2 is not 1, a preliminary step is needed.

EXAMPLE 2 For the function $f(x) = 2x^2 + 12x - 1$, complete the square.

Again we consider the x^2- and x-terms, but we begin by factoring out the x^2-coefficient, as follows:

$$f(x) = 2(x^2 + 6x) - 1.$$

g) Is the graph of $f(x) = -3(x + 2)^2$ a horizontal translation to the left or to the right?

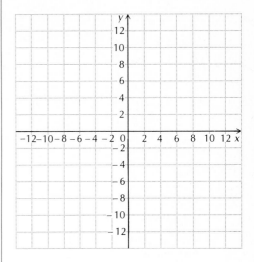

Answer the following questions in Exercises 5–10.

a) What is the vertex?

b) What is the line of symmetry?

c) Is there a maximum? What is it?

d) Is there a minimum? What is it?

5. Graph $f(x) = 3(x - 2)^2 + 4$.

6. Graph $f(x) = -3(x + 2)^2 - 1 = -3[x - (-2)]^2 - 1$.

Without graphing, answer the above questions for each function.

7. $f(x) = (x - 5)^2 + \pi$

8. $f(x) = -3(x - 5)^2$

9. $f(x) = 2\left(x + \frac{1}{4}\right)^2 - 6$

10. $f(x) = -\frac{1}{4}(x + 9)^2 + 3$

11. For the function

$$f(x) = x^2 - 4x + 7,$$

complete the square.

Complete the square.

12. $f(x) = 3x^2 + 24x + 10$

13. For the function

$$f(x) = 4x^2 - 12x - 5,$$

a) find the vertex and the line of symmetry;

b) determine whether there is a maximum or minimum value and find that value.

14. For the function

$$f(x) = -2x^2 - 10x + 9,$$

find the vertex and the maximum or minimum value.

Next, we proceed as before, *inside* the parentheses. We take half the coefficient of x and square it. That number is $(6/2)^2$, or 9. We add $9 - 9$, but do it *inside* the parentheses:

$$f(x) = 2(x^2 + 6x + 9 - 9) - 1.$$

We now have an extra, unwanted, term inside the parentheses so we rearrange things a bit:

$$f(x) = 2(x^2 + 6x + 9) - 2 \cdot 9 - 1.$$

We have used the distributive law and multiplication to get the $(-2 \cdot 9)$ outside. We now have

$$f(x) = 2(x^2 + 6x + 9) - 18 - 1$$
$$= 2(x + 3)^2 - 19.$$

To complete the square on a quadratic function $f(x) = ax^2 + bx + c$:

1. If $a \neq 1$, factor a out of the x^2- and x-terms.

2. Take half the resulting coefficient of x and square it to obtain Q. Add $Q - Q$ *inside* the parentheses.

3. Multiply to obtain $f(x) = a(x^2 + px + Q) - aQ + c$. Finish by factoring the polynomial and simplifying.

DO EXERCISE 12.

EXAMPLE 3 For the function $f(x) = 4x^2 + 12x + 7$, (a) find the vertex and the line of symmetry; and (b) determine whether there is a maximum or minimum value and find that value.

We complete the square:

$$f(x) = 4(x^2 + 3x) + 7$$
$$= 4(x^2 + 3x + \tfrac{9}{4} - \tfrac{9}{4}) + 7 \qquad (\tfrac{3}{2})^2 = \tfrac{9}{4}$$
$$= 4(x^2 + 3x + \tfrac{9}{4}) - 4 \cdot \tfrac{9}{4} + 7$$
$$= 4(x + \tfrac{3}{2})^2 - 2.$$

a) The vertex is $(-\tfrac{3}{2}, -2)$. The line of symmetry goes through this point, hence is the line $x = -\tfrac{3}{2}$.

b) Since the coefficient of x^2 is positive, the parabola opens upward. Thus we have a minimum value. The value is -2.

DO EXERCISE 13.

EXAMPLE 4 For the function $f(x) = -2x^2 + 10x - 7$, find the vertex and the maximum or minimum value.

We complete the square:

$$f(x) = -2(x^2 - 5x) - 7$$
$$= -2(x^2 - 5x + \tfrac{25}{4} - \tfrac{25}{4}) - 7 \qquad (\tfrac{-5}{2})^2 = \tfrac{25}{4}$$
$$= -2(x^2 - 5x + \tfrac{25}{4}) - (-2)\tfrac{25}{4} - 7 = -2(x - \tfrac{5}{2})^2 + \tfrac{11}{2}.$$

The vertex is $(\tfrac{5}{2}, \tfrac{11}{2})$. The maximum value is $\tfrac{11}{2}$, since the coefficient of x^2 is negative.

DO EXERCISE 14.

ii GRAPHING QUADRATIC FUNCTIONS

We know that the graph of any quadratic function is a parabola. If the coefficient of x^2 is positive, the parabola opens upward. Keeping this in mind, we can calculate and plot a few points and connect them. It also helps to know where the vertex and the line of symmetry are.

EXAMPLE 5 Graph $f(x) = -2x^2 + 10x - 7$.

We note that the coefficient of x^2 is negative. The parabola opens downward. We calculate several input–output pairs. (Note that $f(0)$ is always easy to find.) We then plot the ordered pairs and complete the graph.

x	$f(x)$
0	-7
-1	-19
1	1
2	5
3	5
4	1

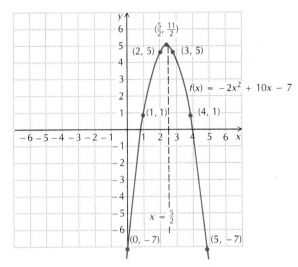

Sometimes, as in this case, the line of symmetry is apparent. (If it is not apparent, we could complete the square as in Example 4.) In this case it is the line $x = \frac{5}{2}$. We can sketch the line, and then:

a) Find the vertex. It is $(\frac{5}{2}, f(\frac{5}{2}))$, or $(\frac{5}{2}, \frac{11}{2})$.

b) Find other points to plot by reflecting across the line of symmetry.

For example, the point opposite $(0, -7)$ is $(5, -7)$.

To graph a quadratic function:

1. Note whether the coefficient of x^2 is positive or negative and thus determine whether the curve opens upward or downward.

2. Calculate several input–output pairs. (A calculator can be most useful.) Plot these ordered pairs and complete the graph.

3. Find the vertex and line of symmetry. Use completing the square if they are not apparent. Use reflection across the line of symmetry to find further points to plot.

DO EXERCISES 15 AND 16.

iii x-INTERCEPTS

The points at which a graph crosses the x-axis are called its x-intercepts. These are, of course, the points at which $f(x) = 0$. Thus the x-values at the intercepts* are the solutions of the equation $f(x) = 0$.

---

* Also called *zeros of the function*.

Graph.

15. $f(x) = x^2 - 6x + 4$

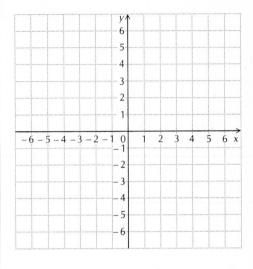

16. $f(x) = -4x^2 + 12x - 5$

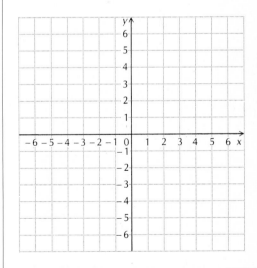

Find the *x*-intercepts.

17. $f(x) = x^2 - 2x - 5$

Find the *x*-intercepts if they exist.

18. $f(x) = x^2 - 2x - 3$

19. $f(x) = x^2 + 8x + 16$

20. $f(x) = -2x^2 - 4x - 3$

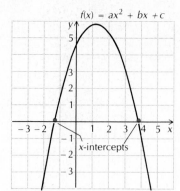

$f(x) = ax^2 + bx + c$

x-intercepts

In order to find the *x*-intercepts of a quadratic function $f(x) = ax^2 + bx + c$ we solve the equation $f(x) = 0$, or

$$ax^2 + bx + c = 0.$$

EXAMPLE 6   Find the *x*-intercepts of the graph of $f(x) = x^2 - 2x - 2$.

We solve the equation

$$x^2 - 2x - 2 = 0.$$

The equation is difficult to factor, so we use the quadratic formula and get $x = 1 \pm \sqrt{3}$. Thus the *x*-intercepts are $(1 - \sqrt{3}, 0)$ and $(1 + \sqrt{3}, 0)$. For plotting, we approximate, to get $(-0.7, 0)$ and $(2.7, 0)$. We sometimes refer to the *x*-coordinates as intercepts.

Note that the *x*-intercepts could also be used when graphing quadratic functions, but they are not essential.

DO EXERCISE 17.

The discriminant, $b^2 - 4ac$, tells us how many real-number solutions the equation $ax^2 + bx + c = 0$ has, so it also indicates how many intercepts there are. Compare.

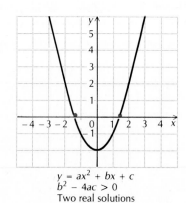

$y = ax^2 + bx + c$
$b^2 - 4ac > 0$
Two real solutions
Two *x*-intercepts

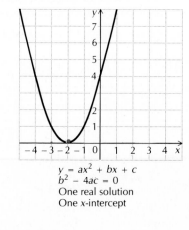

$y = ax^2 + bx + c$
$b^2 - 4ac = 0$
One real solution
One *x*-intercept

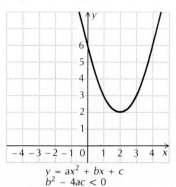

$y = ax^2 + bx + c$
$b^2 - 4ac < 0$
No real solutions
No *x*-intercepts

DO EXERCISES 18–20.

EXERCISE SET 4.3

i For each of the following functions, (a) find the vertex; (b) find the line of symmetry; and (c) determine whether there is a maximum or minimum function value and find that value.

1. $f(x) = x^2$

2. $f(x) = -5x^2$

3. $f(x) = -2(x - 9)^2$

4. $f(x) = 5(x - 7)^2$

5. $f(x) = 2(x - 1)^2 - 4$

6. $f(x) = -(x + 4)^2 - 3$

For each of the following functions, (a) find an equation of the type $f(x) = a(x - h)^2 + k$; (b) find the vertex; (c) determine whether there is a maximum or minimum function value and find that value.

7. $f(x) = -x^2 + 2x + 3$

8. $f(x) = -x^2 + 8x - 7$

9. $f(x) = x^2 + 3x$

10. $f(x) = x^2 - 9x$

11. $f(x) = -\frac{3}{4}x^2 + 6x$

12. $f(x) = \frac{3}{2}x^2 + 3x$

13. $f(x) = 3x^2 + x - 4$

14. $f(x) = -2x^2 + x - 1$

ii Graph.

15. $f(x) = -x^2 + 2x + 3$

16. $f(x) = x^2 - 3x - 4$

17. $f(x) = x^2 - 8x + 19$

18. $f(x) = -x^2 - 8x - 17$

19. $f(x) = -\frac{1}{2}x^2 - 3x + \frac{1}{2}$

20. $f(x) = 2x^2 - 4x - 2$

21. $f(x) = 3x^2 - 24x + 50$

22. $f(x) = -2x^2 + 2x + 1$

iii Find the x-intercepts.

23. $f(x) = -x^2 + 2x + 3$

24. $f(x) = x^2 - 3x - 4$

25. $f(x) = x^2 - 8x + 5$

26. $f(x) = -x^2 - 3x - 3$

27. $f(x) = -5x^2 + 6x - 5$

28. $f(x) = 2x^2 + x - 5$

Find an equation of the type $f(x) = a(x - h)^2 + k$ for each of the following.

29. $f(x) = ax^2 + bx + c$

30. $f(x) = 3x^2 + mx + m^2$

Graph.

31. $f(x) = |x^2 - 1|$ (*Hint:* Consider two cases, $x^2 - 1 \geq 0$ and $x^2 - 1 < 0$; or as another way graph $y = x^2 - 1$ and then reflect negative values across the x-axis.)

32. $f(x) = |3 - 2x - x^2|$

Find the maximum or minimum value for each of the following functions.

33. ▦ $f(x) = 2.31x^2 - 3.135x - 5.89$

34. ▦ $f(x) = -18.8x^2 + 7.92x + 6.18$

35. ▦ What is the minimum product of two numbers whose difference is 4.932? What are the numbers?

36. ▦ What is the maximum product of two numbers whose sum is 21.355? What are the numbers?

37. Find the dimensions and area of the largest rectangle that can be inscribed as shown in a right triangle ABC whose sides have lengths 9 cm, 12 cm, and 15 cm.

38. A farmer wants to build a rectangular fence near a river, and will use 120 ft of fencing. What is the area of the largest region that can be enclosed? Note that the side next to the river is not fenced.

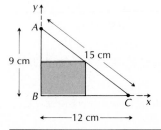

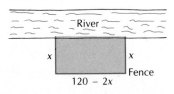

4.4

MATHEMATICAL MODELS

Mathematics is often constructed to fit certain situations. When this is done, we say that we have a *mathematical model*. One of the simplest examples is the natural numbers, that is, the numbers 1, 2, 3, 4, and so on. These numbers and operations on them were invented so that they would apply to situations in which counting is involved, in some way or other. By working within a mathematical model, we hope to be able to predict what will happen in an actual situation.

Recall our problem-solving strategy. We first *familiarize* ourselves with the problem situation. Then we *translate* to mathematical language. In effect, we are creating a mathematical model.

EXAMPLE 1   Use the mathematical model of the natural numbers to predict how much money a theater will collect if it charges $2 per admission and 147 people attend.

We *translate* the problem situation to the language of the mathematical model. In this case, the total number of dollars collected will be the product of the admission price and the number of admissions. We *carry out* some mathematical manipulation. In this case, we multiply, to get an answer of 294. On this basis we predict that when the owner counts the receipts, there will be $294.

In an example such as the preceding one, we expect our prediction to be exact. In most situations, mathematical models are not that good. Predictions will be only approximate.

EXAMPLE 2   Use the mathematical model of the natural numbers to predict how much mixture will result from mixing one liter of water and one liter of alcohol.

We solve this problem by adding 1 and 1, to get 2. On this basis, the answer is 2 liters.

In the latter example, our result is not exact. The actual result will be something less than 2 liters, because for some reason there is shrinkage when water and alcohol are mixed.

If a mathematical model does not give precise answers, we ordinarily try revising it so that it does, or we may even look for an entirely new model.

i LINEAR FUNCTIONS AS MATHEMATICAL MODELS

As a result of experiment or experience, we often acquire data that indicate that a function of some sort would be a good mathematical model.

How do we know when a linear function fits a situation?

EXAMPLE 3   Plot the following data and determine whether a linear equation gives an approximate fit.

Crickets are known to chirp faster when the temperature is higher. Here are some measurements made at different temperatures.

Temperature, °C	6	8	10	15	20
Number of chirps per min	11	29	47	75	107

We make a graph with a T- (temperature) axis and an N- (number per minute) axis and plot these data. We see that they do lie approximately on a straight line, so we can use a linear equation in this situation.

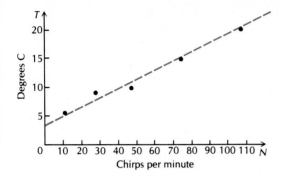

EXAMPLE 4   Wind friction, or *resistance*, increases with speed. Here are some measurements made in a wind tunnel. Plot the data and determine whether a linear equation will give an approximate fit.

Velocity, km/h	10	21	34	40	45	52
Force of resistance, kg	3	4.2	6.2	7.1	15.1	29.0

We make a graph with a V- (velocity) axis and an F- (force) axis and plot the data. They do not lie on a straight line, even approximately. Therefore we cannot use a linear equation in this situation.*

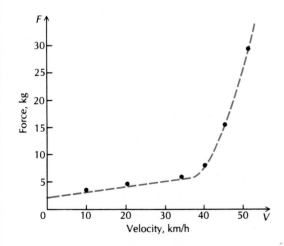

When we have a situation that a linear equation fits, we can find an equation and use it to solve problems and make predictions.

---
*The situation in Example 4 might fit an equation that is not linear. We shall consider such problems later.

1. A class of college students wanted to determine a function from which they could predict a final exam score S from a midterm test score d. To do this they checked with two students who took the class before. One student scored 70 on the midterm and 75 on the final. Another scored 85 on the midterm and 89 on the final.

   a) Assuming that S is a linear function of d, express S in terms of d.

   b) Use the equation obtained in (a) to predict a student's final exam score, given that an 81 was scored on the midterm.

EXAMPLE 5   When crickets chirp 40 times per minute, the temperature is 10°C. When they chirp 112 times per minute, the temperature is 20°C. We know that a linear equation fits this situation.

a)   Find a linear equation that fits the data.

b)   From your equation, find the temperature when crickets chirp 76 times per minute; 100 times per minute.

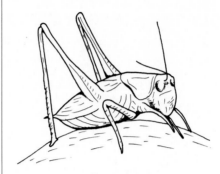

a)   We can use our problem-solving strategy.

   1. *Familiarize:* We know from the statement of the problem that a linear equation fits the situation. The wording of the problem gives us two ordered pairs: $(40, 10°)$ and $(112, 20°)$.

   2. *Translate:* To find the equation, we use the two known ordered pairs $(40, 10°)$ and $(112, 20°)$. We call these *data points*. We find the two-point equation, using these data points:

$$T - T_1 = \frac{T_2 - T_1}{N_2 - N_1}(N - N_1)$$

$$T - 10 = \frac{20 - 10}{112 - 40}(N - 40). \qquad \text{Substituting}$$

   3. *Carry out:* We simplify as follows:

$$T - 10 = \frac{10}{72}(N - 40)$$

$$T - 10 = \frac{5}{36}N - \frac{5}{36} \cdot 40$$

$$T = \frac{5}{36}N + \frac{40}{9}, \quad \text{or} \quad \frac{5N + 160}{36}. \qquad \text{Solving for } T$$

   4. *Check:* We can check by going over our work a second time.

b)   Using the formula $T = (5N + 160)/36$, we find T when $N = 76$:

$$T = \frac{5 \cdot 76 + 160}{36} = 15°.$$

When $N = 100$, we get

$$T = \frac{5 \cdot 100 + 160}{36} \approx 18.3°.$$

DO EXERCISE 1.

Equations give decent results only within certain limits. A negative number of chirps per minute would be meaningless. Also, imagine what would happen to a cricket at $-40°$ or at $100°$! When a function is used in an application, the domain is often restricted by physical factors.

EXAMPLE 6   (*Profit and loss analysis*). Boxowits, Inc., a computer manufacturing firm, is going to produce a new minicalculator. During the first year, the costs for setting up the new production line are $100,000. These are fixed costs such as rent, tools, etc., which must be incurred before any calculators are produced. The additional cost of producing a calculator is $20. This cost is directly related to production, such as material, wages, fuels, and so on. It is a variable cost, according to the number of calculators produced. If x calculators are produced, the variable cost is then $20x$ dollars. The *total* cost of producing x calculators is given by a function C:

$$\text{Total cost} = \text{Fixed cost plus Variable cost}$$
$$C(x) \quad = \quad 100,000 \quad + \quad 20x.$$

The firm determines that its revenue (money coming in) from the sale of the calculators is $45 per calculator, or $45x$ dollars, for x calculators. Thus we have a *revenue* function R:

$$R(x) = 45x.$$

The stockholders are most interested in the *profit* function P:

$$\text{Profit} = \text{Revenue} - \text{Total cost}$$
$$P(x) = \quad 45x \quad - (100,000 + 20x)$$
$$P(x) = \quad 25x \quad - 100,000.$$

Find those values of x for which (a) the company will break even; (b) the company will make a profit; and (c) the company will suffer a loss.

In this example, we have a model consisting of three linear functions. The function that will provide the answer to the question is the profit function P. When the profit is 0 the company breaks even. When it is positive the company makes money. When it is negative the company loses money. Let us first find the value(s) of x that make $P(x) = 0$:

$$0 = P(x) = 25x - 100,000$$
$$25x = 100,000$$
$$x = 4000.$$

The break-even point occurs when 4000 calculators are sold. When $x > 4000$ there is a profit. When $x < 4000$ there is a loss.

A graph of the profit function in the preceding example is of interest. In fact, graphs are often of considerable help with mathematical models.

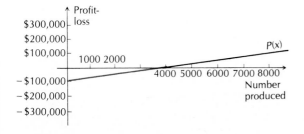

DO EXERCISE 2.

2. Rework Example 6, where $C(x) = 30x + 15,000$ and $R(x) = 90x$.

ii QUADRATIC FUNCTIONS AS MATHEMATICAL MODELS

EXAMPLE 7 (*A projectile problem*). When an object such as a bullet or a ball is shot or thrown upward with an initial velocity v_0, its height is given, approximately, by a quadratic function:

$$s = -4.9t^2 + v_0 t + h.$$

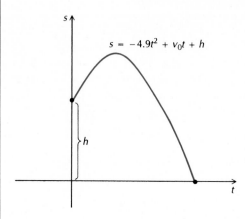

In this function, h is the starting height in meters, s is the actual height (also in meters), and t is the time from projection in seconds.

This model is constructed from theoretical principles, rather than experiment. It is based on the assumption that there is no air resistance, and that the force of gravity pulling the object earthward is constant. Neither of these conditions exists precisely, so this model (as is the case with most mathematical models) gives only approximate results.

A model rocket is fired upward. At the end of the burn it has an upward velocity of 49 m/sec and is 155 m high. Find (a) its maximum height and when it is attained; (b) when it reaches the ground.

We will start counting time at the end of the burn. Thus $v_0 = 49$ and $h = 155$. We graph the appropriate function, and begin by completing the square:

$$s = -4.9(t^2 - \tfrac{49}{4.9}t) + 155$$
$$= -4.9(t^2 - 10t + 25 - 25) + 155$$
$$= -4.9(t - 5)^2 + 4.9 \times 25 + 155$$
$$= -4.9(t - 5)^2 + 277.5.$$

The vertex of the graph is the point $(5, 277.5)$ and the graph is shown below. The maximum height reached is 277.5 m and it is attained 5 sec after the end of the burn.

To find when the rocket reaches the ground, we set $s = 0$ in our equation and solve for t:

$$-4.9(t - 5)^2 + 277.5 = 0$$
$$(t - 5)^2 = \tfrac{277.5}{4.9}$$
$$t - 5 = \sqrt{\tfrac{277.5}{4.9}} \approx 7.525$$
$$t \approx 12.525.$$

The rocket will reach the ground about 12.525 sec after the end of the burn.

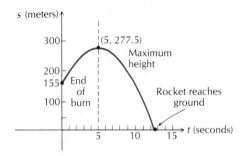

DO EXERCISE 3.

We will fit a quadratic function to a set of data in Chapter 5.

3. A ball is thrown upward from the top of a cliff 12 m high, at a velocity of 2.8 m/sec. Find:

a) its maximum height and when it is attained;

b) when it reaches the ground.

EXERCISE SET 4.4

1. (*Life expectancy of females in the United States*). In 1950 the life expectancy of females was 72 years. In 1970 it was 75 years. Let E represent the life expectancy and t the number of years since 1950. ($t = 0$ gives 1950 and $t = 10$ gives 1960.)

a) Assuming E is a linear function of t, express E in terms of t. [*Hint:* Data points are (0, 72) and (20, 75).]

b) Use the equation in part (a) to predict the life expectancy of females in 1980; in 1989.

2. (*Life expectancy of males in the United States*). In 1950 the life expectancy of males was 65 years. In 1970 it was 68 years. Let E represent life expectancy and t the number of years since 1950.

a) Assuming E is a linear function of t, express E in terms of t.

b) Use the equation in part (a) to predict the life expectancy of males in 1980; in 1989.

3. (*World record in the 100-meter dash*). In 1920 the world record for the 100-m dash was 10.43 sec. In 1970 it was 9.93 sec. Let R represent the record in the 100-m dash and t the number of years since 1920.

a) Assuming R is a linear function of t, express R in terms of t.

b) Use the function in part (a) to predict the record in 1988; in 1990.

c) According to this model, in what year will the record be 9.0 sec?

4. (*Natural gas demand*). In 1950 natural gas demand in the United States was 19 quadrillion BTU. In 1960 the demand was 21 quadrillion BTU. Let D represent the demand for natural gas t years after 1950.

a) Assuming D is a linear function of t, express D in terms of t.

b) Use the equation in part (a) to predict the natural gas demand in 1988; in 2000.

5. (*Profit and loss in the ski business*). A ski manufacturer is planning a new line of skis. For the first year the fixed costs for setting up the new production line are $22,500. Variable costs for producing each pair of skis are estimated to be $40. The sales department projects that 3000 pairs of skis can be sold the first year. The revenue from each pair is to be $85.

a) Formulate a function $C(x)$ for the total cost of producing x pairs of skis.

b) Formulate a function $R(x)$ for the total revenue from the sale of x pairs of skis.

c) Formulate a function $P(x)$ for the profit from the sale of x pairs of skis.

d) What profit or loss will the company realize if 3000 pairs are actually sold?

e) Find the break-even value of x.

f) Find the values of x that will result in a profit.

g) Find the values of x that will result in a loss.

6. (*Profit and loss*). A clothing firm has fixed costs of $10,000. To produce x units of a certain kind of suit it costs $20 per unit in addition to the fixed costs. Total revenue from the sale of x suits is $100 per suit.

a) Formulate a function $C(x)$ for the total cost of producing x suits.

b) Formulate a function $R(x)$ for the total revenue from the sale of x suits.

c) Formulate a function $P(x)$ for the profit from the sale of x suits.

d) Find the break-even value of x.

e) Find the values of x that will result in a loss.

f) Find the values of x that will result in a profit.

7. A rocket is fired upward. At the end of the burn it has an upward velocity of 147 m/sec and is 560 m high. Find (a) its maximum height and when it is attained; (b) when it reaches the ground.

9. The sum of the base and height of a triangle is 20. Find the dimensions for which the area is a maximum.

8. A rocket is fired upward. At the end of the burn it has an upward velocity of 245 m/sec and is 1240 m high. Find (a) its maximum height and when it is attained; (b) when it reaches the ground.

10. The sum of the base and height of a triangle is 14. Find the dimensions for which the area is a maximum.

11. (*Maximizing yield*). An orange grower finds that she gets an average yield of 40 bu per tree when she plants 20 trees on an acre of ground. Each time she adds a tree to an acre the yield per tree decreases by 1 bu, due to congestion. How many trees per acre should she plant for maximum yield?

13. (*Maximizing revenue*). When a theater charges $4 for admission it averages 500 people in attendance. For each 20¢ decrease in admission price the average number of people increases by 30. What should it charge to make the most money?

12. (*Maximizing revenue*). When a theater owner charges $2 for admission he averages 100 people attending. For each 10¢ increase in admission price the average number attending decreases by 1. What should he charge to make the most money?

14. (*Maximizing area*). A farmer wants to enclose two adjacent rectangular regions, as shown below, near a river, one for sheep and one for cattle. No fencing will be used next to the river, but 60 m of fencing will be used. What is the area of the largest region that can be enclosed?

OBJECTIVES

You should be able to:

i Find the union and intersection of sets like {1, 2, 3, 4} and {3, 4, 5}; and graph the union and intersection of sets like {x|x > 1} and {x|x ≤ 3}.

ii Solve conjunctions and disjunctions of inequalities.

4.5

SETS, SENTENCES, AND INEQUALITIES

i INTERSECTIONS AND UNIONS

The *intersection* of two sets consists of those elements common to the sets. Intersection is illustrated in the figure on the left. Note in the figure on the right that the intersection of two sets may be the empty set. The intersection of sets A and B is indicated as $A \cap B$.

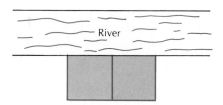

Graphs are set diagrams. They often show pictorially the solution set of an equation or inequality. We can find intersections of solution sets using graphs. In the following example, we find the intersection of the set of all x greater than 3 and the set of all x less than or equal to 5. These sets are indicated, and the symbolism is read, as follows:

$\{x \mid x > 3\}$ "the set of all x such that x is greater than 3,"

$\{x \mid x \leqslant 5\}$ "the set of all x such that x is less than or equal to 5."

EXAMPLE 1   Find and graph $\{x \mid x > 3\} \cap \{x \mid x \leqslant 5\}$.

We graph the two solution sets separately and then find the intersection.

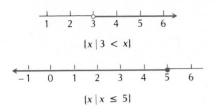

$\{x \mid 3 < x\}$

$\{x \mid x \le 5\}$

The open circle at 3 indicates that 3 is not in the solution set. The solid circle at 5 indicates that 5 is in the solution set. The intersection is as follows.

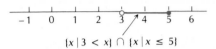

$\{x \mid 3 < x\} \cap \{x \mid x \le 5\}$

EXAMPLE 2 Graph $\{x \mid -3 \le x\} \cap \{x \mid x \le 0\}$.

Again, we graph the solution sets separately and then find the intersection.

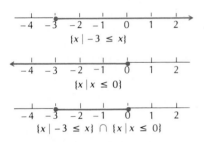

$\{x \mid -3 \le x\}$

$\{x \mid x \le 0\}$

$\{x \mid -3 \le x\} \cap \{x \mid x \le 0\}$

DO EXERCISES 1–6.

The *union* of two sets consists of the members that are in one or both of the sets. Union is illustrated in the following diagram. The union of sets A and B is indicated as $A \cup B$.

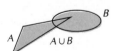

$A \cup B$

In the following examples, we find unions of solution sets. Note that we graph the individual sets separately and then combine them.

EXAMPLE 3 Graph $\{x \mid -1 \le x\} \cup \{x \mid x < 2\}$.

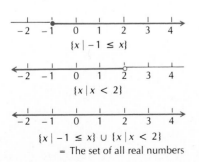

$\{x \mid -1 \le x\}$

$\{x \mid x < 2\}$

$\{x \mid -1 \le x\} \cup \{x \mid x < 2\}$
= The set of all real numbers

Find these intersections.

1. $\{-3, -4, 2, 3, 4\} \cap \{1, 4, -3, 8, 9, 11\}$

2. $\{2, b, c, d, e\} \cap \{1, 2, d, e, f, g\}$

Graph.

3. $\{x \mid 1 < x\} \cap \{x \mid x \le 3\}$

$-6\ -5\ -4\ -3\ -2\ -1\ \ 0\ \ 1\ \ 2\ \ 3\ \ 4\ \ 5\ \ 6$

4. $\{x \mid -4 < x\} \cap \{x \mid x < -1\}$

$-6\ -5\ -4\ -3\ -2\ -1\ \ 0\ \ 1\ \ 2\ \ 3\ \ 4\ \ 5\ \ 6$

5. $\{x \mid 0 \le x\} \cap \{x \mid x < \pi\}$

$-6\ -5\ -4\ -3\ -2\ -1\ \ 0\ \ 1\ \ 2\ \ 3\ \ 4\ \ 5\ \ 6$

6. $\{x \mid 1 \le x\} \cap \{x \mid x \le -2\}$

$-6\ -5\ -4\ -3\ -2\ -1\ \ 0\ \ 1\ \ 2\ \ 3\ \ 4\ \ 5\ \ 6$

Linear and Quadratic Functions and Inequalities

Find these unions.

7. $\{1, 2\} \cup \{2, 3, 4, 5\}$

8. $\{-3, -4, 2, 3, 4\} \cup \{1, 4, -3, 8, 9, 11\}$

9. $\{2, a, b, c\} \cup \{1, 2, c, d, e\}$

Graph.

10. $\{x \mid x < -3\} \cup \{x \mid x \geqslant 1\}$

11. $\{x \mid x \leqslant -3\} \cup \{x \mid x \geqslant 3\}$

12. $\{x \mid x > 0\} \cup \{x \mid x < 1\}$

13. $\{x \mid x > \frac{1}{2}\} \cup \{x \mid x = \frac{1}{2}\}$

Abbreviate each conjunction.

14. $4 < x \quad and \quad x < 8$

15. $-3 \leqslant x \quad and \quad x < 0$

16. $-5 \leqslant x \quad and \quad x \leqslant -2$

17. $-1 < x \quad and \quad x \leqslant -\frac{1}{4}$

Rewrite, using the word *and*.

18. $-\frac{1}{2} < x < 1$

19. $-\frac{17}{3} \leqslant x < -2$

20. $\frac{19}{4} \leqslant x \leqslant \frac{37}{6}$

EXAMPLE 4 Graph $\{x \mid x \leqslant -2\} \cup \{x \mid x > 1\}$.

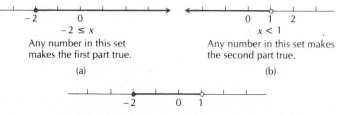

DO EXERCISES 7–13.

ⅱ COMPOUND SENTENCES

When two sentences are joined by the word *and*, a compound sentence is formed. Such a sentence is called a *conjunction*. (The word conjunction used in this way is a logical term; the meaning is not the same as in ordinary grammar.) A conjunction of two sentences is true when both parts are true. Thus the solution set is the intersection of the solution sets of the parts. Consider, for example, the conjunction

$$-2 \leqslant x \quad and \quad x < 1.$$

Any number that makes this sentence true must make both parts true. We can graph the sentence by graphing the parts and then finding their intersection.

Any number in this set makes the first part true.

(a)

Any number in this set makes the second part true.

(b)

Any number in this set (the intersection) makes both parts true ($-2 \leqslant x$ *and* $x < 1$ are both true).

(c)

We often abbreviate certain conjunctions of inequalities. In this case,

$$-2 \leqslant x \quad and \quad x < 1 \qquad \text{is abbreviated} \qquad -2 \leqslant x < 1.$$

The latter is read "-2 is less than or equal to x and x is less than 1," or "-2 is less than or equal to x is less than 1." Thus

$$\{x \mid -2 \leqslant x < 1\} = \{x \mid -2 \leqslant x \quad and \quad x < 1\}$$
$$= \{x \mid -2 \leqslant x\} \cap \{x \mid x < 1\}.$$

The word *and* corresponds to set *intersection*.

DO EXERCISES 14–20.

EXAMPLE 5 Solve and graph $-3 < 2x + 5 < 7$.

Method 1.
$\qquad -3 < 2x + 5 \quad and \quad 2x + 5 < 7 \qquad$ Rewriting using *and*
$\qquad\quad -8 < 2x \qquad and \qquad\quad 2x < 2 \qquad$ Adding -5
$\qquad\quad -4 < x \qquad and \qquad\quad\quad x < 1 \qquad$ Multiplying by $\frac{1}{2}$

Method 2. $\quad -3 < 2x + 5 < 7$

$\qquad\qquad -8 < 2x < 2$

$\qquad\qquad -4 < x < 1$

The solution set is

$$\{x \mid -4 < x\} \cap \{x \mid x < 1\}, \quad \text{or} \quad \{x \mid -4 < x < 1\}.$$

The graph is as follows.

EXAMPLE 6 Solve and graph:

$$-4 < \frac{5 - 3x}{2} \leqslant 5.$$

We have the following:

$\qquad -8 < 5 - 3x \leqslant 10 \qquad$ Multiplying by 2

$\qquad -13 < -3x \leqslant 5 \qquad$ Adding -5

$\qquad \frac{13}{3} > x \geqslant -\frac{5}{3} \qquad$ Multiplying by $-\frac{1}{3}$

$\qquad -\frac{5}{3} \leqslant x < \frac{13}{3} \qquad x \geqslant -\frac{5}{3}$ means $-\frac{5}{3} \leqslant x$, and $\frac{13}{3} > x$ means $x < \frac{13}{3}$

The solution set is $\left\{ x \mid -\frac{5}{3} \leqslant x < \frac{13}{3} \right\}$.

The graph is as follows.

DO EXERCISES 21–23.

When two sentences are joined by the word *or*, a compound sentence is formed. Such a sentence is called a *disjunction*. A disjunction of two sentences is true when either part is true. It is also true when both parts are true. The solution set of a disjunction is thus the union of the solution sets of the parts. Consider the disjunction

$$x < -2 \quad or \quad x > \tfrac{1}{4}.$$

Any number that makes either or both of the parts true makes the disjunction true.

We can graph the sentence by graphing the two parts and then finding their union. The word *or* corresponds to set *union*.

$x < -2$

Any number in this set makes the first part true.

(a)

$x > \frac{1}{4}$

Any number in this set makes the second part true.

(b)

$x < -2$ or $x > \frac{1}{4}$

$\{x \mid x < -2\} \cup \{x \mid x > \frac{1}{4}\}$

Any number in this set makes one or both of the parts true.

(c)

Note that $3x \leqslant 15$ is an abbreviation for $3x < 15$ *or* $3x = 15$.

Solve. Then graph.

21. $-4 < 3x - 2 \leqslant 10$

22. $-2 < \dfrac{3 - x}{4} < 2$

23. $\dfrac{2}{3} \leqslant 1 - 2x \leqslant \dfrac{5}{3}$

Solve. Then graph.

24. $x + 4 < -3$ *or* $x + 4 > 3$

$$-6\ -5\ -4\ -3\ -2\ -1\ \ 0\ \ 1\ \ 2\ \ 3\ \ 4\ \ 5\ \ 6$$

25. $2x - 3 \leqslant -5$ *or* $2x - 3 > 5$

$$-6\ -5\ -4\ -3\ -2\ -1\ \ 0\ \ 1\ \ 2\ \ 3\ \ 4\ \ 5\ \ 6$$

26. $4 - 3x \leqslant -1$ *or* $4 - 3x \geqslant 1$

$$-6\ -5\ -4\ -3\ -2\ -1\ \ 0\ \ 1\ \ 2\ \ 3\ \ 4\ \ 5\ \ 6$$

27. $\dfrac{4x + 5}{3} < -2$ *or* $\dfrac{4x + 5}{3} \geqslant 2$

$$-6\ -5\ -4\ -3\ -2\ -1\ \ 0\ \ 1\ \ 2\ \ 3\ \ 4\ \ 5\ \ 6$$

> **CAUTION!** There is no compact way to abbreviate disjunctions of inequalities, ordinarily. *Be careful about this!* For example, if you try to abbreviate "$-3 < x$ *or* $x < 4$" as $-3 < x < 4$ you will be *wrong*, because $-3 < x < 4$ is an abbreviation for the conjunction $-3 < x$ *and* $x < 4$.

EXAMPLE 7 Solve. Then graph.

$$2x - 5 < -7 \quad or \quad 2x - 5 > 7$$
$$2x < -2 \quad or \qquad 2x > 12 \qquad \text{Adding 5}$$
$$x < -1 \quad or \qquad x > 6 \qquad \text{Multiplying by } \tfrac{1}{2}$$

The solution set is $\{x \mid x < -1 \ or \ x > 6\}$, which is the union $\{x \mid x < -1\} \cup \{x \mid x > 6\}$. The graph is as follows.

$$-3\ \ -2\ \ -1\ \ \ 0\ \ \ 1\ \ \ 2\ \ \ 3\ \ \ 4\ \ \ 5\ \ \ 6\ \ \ 7\ \ \ 8$$

EXAMPLE 8 Solve. Then graph.

$$\frac{4 - 3x}{2} < -1 \qquad or \qquad \frac{4 - 3x}{2} \geqslant 1$$

$$4 - 3x < -2 \qquad or \quad 4 - 3x \geqslant 2 \qquad \text{Multiplying by 2}$$
$$4 < -2 + 3x \quad or \qquad 4 \geqslant 2 + 3x \qquad \text{Adding } 3x$$
$$6 < 3x \qquad or \qquad 2 \geqslant 3x$$
$$2 < x \qquad or \qquad \tfrac{2}{3} \geqslant x$$

The solution set is

$$\left\{ x \mid 2 < x \ or \ \tfrac{2}{3} \geqslant x \right\},$$

which is the union

$$\left\{ x \mid 2 < x \right\} \cup \left\{ x \mid \tfrac{2}{3} \geqslant x \right\}.$$

The graph is as follows.

$$-1\ \ \ 0\ \ \tfrac{2}{3}\,1\ \ \ 2\ \ \ 3\ \ \ 4$$

DO EXERCISES 24–27.

EXERCISE SET 4.5

i Find these unions or intersections.

1. $\{3, 4, 5, 8, 10\} \cap \{1, 2, 3, 4, 5, 6, 7\}$

2. $\{3, 4, 5, 8, 10\} \cup \{1, 2, 3, 4, 5, 6, 7\}$

3. $\{0, 2, 4, 6, 8\} \cup \{4, 6, 9\}$

4. $\{0, 2, 4, 6, 8\} \cap \{4, 6, 9\}$

5. $\{a, b, c\} \cap \{c, d\}$

6. $\{a, b, c\} \cup \{c, d\}$

Graph.

7. $\{x \mid 7 \leqslant x\} \cup \{x \mid x < 9\}$

8. $\{x \mid -\tfrac{1}{2} \leqslant x\} \cup \{x \mid x < \tfrac{1}{2}\}$

9. $\{x \mid -\tfrac{1}{2} \leqslant x\} \cap \{x \mid x < \tfrac{1}{2}\}$

10. $\{x \mid x > \tfrac{1}{4}\} \cap \{x \mid 1 \geqslant x\}$

11. $\{x \mid x < -\pi\} \cup \{x \mid x > \pi\}$

12. $\{x \mid -\pi \leqslant x\} \cap \{x \mid x < \pi\}$

13. $\{x \mid x < -7\} \cup \{x \mid x = -7\}$

14. $\{x \mid x > \tfrac{1}{2}\} \cup \{x \mid x = \tfrac{1}{2}\}$

15. $\{x \mid x \geqslant 5\} \cap \{x \mid x \leqslant -3\}$

16. $\{x \mid x \geqslant -3\} \cup \{x \mid x \leqslant 0\}$

ii Solve.

17. $-2 \leqslant x + 1 < 4$

18. $-3 < x + 2 \leqslant 5$

19. $5 \leqslant x - 3 \leqslant 7$

20. $-1 < x - 4 < 7$

21. $-3 \leqslant x + 4 \leqslant -3$

22. $-5 < x + 2 < -5$

23. $-2 < 2x + 1 < 5$

24. $-3 \leqslant 5x + 1 \leqslant 3$

25. $-4 \leqslant 6 - 2x < 4$

26. $-3 < 1 - 2x \leqslant 3$

27. $-5 < \frac{1}{2}(3x + 1) \leqslant 7$

28. $\frac{2}{3} \leqslant -\frac{4}{5}(x - 3) < 1$

29. $3x \leqslant -6 \quad or \quad x - 1 > 0$

30. $2x < 8 \quad or \quad x + 3 \geqslant 1$

31. $2x + 3 \leqslant -4 \quad or \quad 2x + 3 \geqslant 4$

32. $3x - 1 < -5 \quad or \quad 3x - 1 > 5$

33. $2x - 20 < -0.8 \quad or \quad 2x - 20 > 0.8$

34. $5x + 11 \leqslant -4 \quad or \quad 5x + 11 \geqslant 4$

35. $x + 14 \leqslant -\frac{1}{4} \quad or \quad x + 14 \geqslant \frac{1}{4}$

36. $x - 9 < -\frac{1}{2} \quad or \quad x - 9 > \frac{1}{2}$

Solve.

37. $x \leqslant 3x - 2 \leqslant 2 - x$

38. $2x \leqslant 5 - 7x < 7 + x$

39. $(x + 1)^2 > x(x - 3)$

40. $(x + 4)(x - 5) > (x + 1)(x - 7)$

41. $(x + 1)^2 \leqslant (x + 2)^2 \leqslant (x + 3)^2$

42. $(x - 1)(x + 1) < (x + 1)^2 \leqslant (x - 3)^2$

43. ▓ The length of a rectangle is 15.23 cm. What widths will give a perimeter greater than 40.23 cm and less than 137.8 cm?

44. The height of a triangle is 15 m. What lengths of the base will keep the area less than or equal to 305.4 m² (and, of course, positive)?

45. To get an A in a course, a student's average must be greater than or equal to 90%. It will of course be less than or equal to 100%. On the first three tests a student scored 83%, 87%, and 93%. What scores on the fourth test will produce an A? Is an A possible?

46. In Exercise 45, suppose the scores on the first three tests are 75%, 70%, and 83%. What scores on the fourth test will produce an A? Is an A possible?

Find the domain of each function.

47. $f(x) = \dfrac{\sqrt{x + 2}}{\sqrt{x - 2}}$

48. $f(x) = \dfrac{\sqrt{3 - x}}{\sqrt{x + 5}}$

4.6

EQUATIONS AND INEQUALITIES WITH ABSOLUTE VALUE

i Absolute value was defined in Section 1.1. An informal way of thinking of absolute value is that it is the distance from 0 of a number on a number line. For example, $|4|$ is 4 because 4 is 4 units from 0; $|-5|$ is 5 because -5 is 5 units from 0. This idea is helpful in solving equations and inequalities with absolute value.

EXAMPLE 1   Solve $|x| = 3$.

To solve we look for all numbers x whose distance from 0 is 3. There are two of them, so there are two solutions, 3 and -3. The graph is as follows.

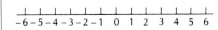

DO EXERCISES 1 AND 2.

OBJECTIVE

You should be able to:

i Solve and graph equations and inequalities with absolute value.

Solve and graph.

1. $|x| = 5$

2. $|x| = \dfrac{1}{4}$

Solve and graph.

3. $|x| < 5$

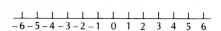

4. $|x| \leqslant \dfrac{1}{4}$

$$\underset{-6\,-5\,-4\,-3\,-2\,-1\ \ 0\ \ 1\ \ 2\ \ 3\ \ 4\ \ 5\ \ 6}{\bot\!\bot\!\bot\!\bot\!\bot\!\bot\!\bot\!\bot\!\bot\!\bot\!\bot\!\bot\!\bot}$$

Solve and graph.

5. $|x| \geqslant 5$

$$\underset{-6\,-5\,-4\,-3\,-2\,-1\ \ 0\ \ 1\ \ 2\ \ 3\ \ 4\ \ 5\ \ 6}{\bot\!\bot\!\bot\!\bot\!\bot\!\bot\!\bot\!\bot\!\bot\!\bot\!\bot\!\bot\!\bot}$$

6. $|x| > \dfrac{1}{4}$

$$\underset{-6\,-5\,-4\,-3\,-2\,-1\ \ 0\ \ 1\ \ 2\ \ 3\ \ 4\ \ 5\ \ 6}{\bot\!\bot\!\bot\!\bot\!\bot\!\bot\!\bot\!\bot\!\bot\!\bot\!\bot\!\bot\!\bot}$$

EXAMPLE 2   Solve $|x| < 3$.

This time we look for all numbers x whose distance from 0 is less than 3. These are the numbers between -3 and 3. The solution set and its graph are as follows.

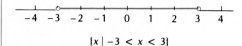

$$\{x \mid -3 < x < 3\}$$

DO EXERCISES 3 AND 4.

EXAMPLE 3   Solve $|x| \geqslant 3$.

This time we look for all numbers x whose distance from 0 is 3 or greater. The solution set and its graph are as follows.

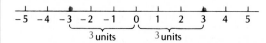

$$\{x \mid x \leq -3 \ or \ x \geq 3\}$$

DO EXERCISES 5 AND 6.

The results of the above examples can be generalized as follows.

> **For any $a > 0$:**
> i) $|x| = a$ is equivalent to $x = -a$ or $x = a$.
> ii) $|x| < a$ is equivalent to $-a < x < a$.
> iii) $|x| > a$ is equivalent to $x < -a$ or $x > a$.

Similar statements hold true for $|x| \leqslant a$ and $|x| \geqslant a$.

EXAMPLE 4   Solve $|x - 2| = 3$.

Note that this is a translation of $|x| = 3$, two units to the right. We first graph $|x| = 3$.

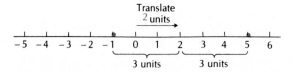

This consists of the numbers that are a distance of 3 from 0. We now translate.

This solution set consists of the numbers that are a distance of 3 from 2 (note that 2 is where 0 went in the translation). The solutions of $|x - 2| = 3$ are -1 and 5.

The results of this example can be generalized as follows.

> **For any a, $|x - a|$ is the distance from a to x.**

EXAMPLE 5

a) $|x - 5|$ is the distance from 5 to x.

b) $|x + 7|$ is the distance from -7 to x [because $x + 7 = x - (-7)$].

Here are some further examples of solving inequalities.

EXAMPLE 6 Solve $|x + 2| < 3$.

Method 1. We translate the graph of $|x| < 3$, to the left 2 units.

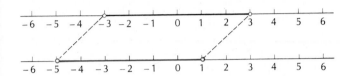

The solution set is $\{x \mid -5 < x < 1\}$.

Method 2. The solutions are those numbers x whose distance from -2 is less than 3. Thus to find the solutions graphically we locate -2. Then we locate those numbers that are less than 3 units to the left and less than 3 units to the right. Thus the solution set is $\{x \mid -5 < x < 1\}$.

Method 3. We use property (ii), replacing x by $x + 2$:

$$|x + 2| < 3$$
$$-3 < x + 2 < 3$$
$$-5 < x < 1.$$

The solution set is $\{x \mid -5 < x < 1\}$.

DO EXERCISES 7–9.

The sentences in the next examples are more complicated. Although we could continue using translations and graphs, this actually gets more difficult than if we use properties (i)–(iii).

EXAMPLE 7 Solve $|2x + 3| = 1$.

$$2x + 3 = -1 \quad or \quad 2x + 3 = 1 \qquad \text{Property (i)}$$
$$2x = -4 \quad or \quad 2x = -2 \qquad \text{Adding } -3$$
$$x = -2 \quad or \quad x = -1$$

The solution set is $\{-2, -1\}$.

EXAMPLE 8 Solve $|2x + 3| \leqslant 1$.

$$-1 \leqslant 2x + 3 \leqslant 1 \qquad \text{Property (ii)}$$
$$-4 \leqslant 2x \leqslant -2 \qquad \text{Adding } -3$$
$$-2 \leqslant x \leqslant -1$$

The solution set is $\{x \mid -2 \leqslant x \leqslant -1\}$.

Solve. Use all three methods.

7. $|x + 1| = 4$

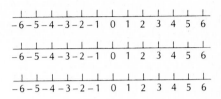

8. $|x + 7| < 2$

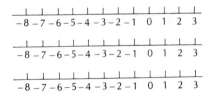

9. $|x - 3| > 2$

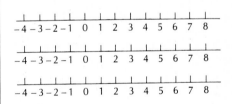

Solve.

10. $|3x - 4| = 7$

11. $\left|5x + \dfrac{1}{2}\right| < 1$

12. $|3x - 4| > 7$

EXAMPLE 9 Solve $|3 - 4x| > 2$.

$$3 - 4x < -2 \quad or \quad 3 - 4x > 2 \qquad \text{Property (iii)}$$
$$-4x < -5 \quad or \quad -4x > -1 \qquad \text{Adding } -3$$
$$x > \tfrac{5}{4} \quad or \qquad x < \tfrac{1}{4}$$

The solution set is $\{x \mid x > \tfrac{5}{4} \text{ or } x < \tfrac{1}{4}\}$.

DO EXERCISES 10–12.

EXERCISE SET 4.6

i Solve and graph.

1. $|x| = 7$ **2.** $|x| = \pi$ **3.** $|x| < 7$ **4.** $|x| \leqslant \pi$ **5.** $|x| \geqslant \pi$ **6.** $|x| > 7$

Solve. Use three methods.

7. $|x - 1| = 4$ **8.** $|x - 7| = 5$ **9.** $|x + 8| < 9$ **10.** $|x + 6| \leqslant 10$

11. $|x + 8| \geqslant 9$ **12.** $|x + 6| > 10$ **13.** $\left|x - \tfrac{1}{4}\right| < \tfrac{1}{2}$ **14.** $|x - 0.5| \leqslant 0.2$

Solve. Use any method.

15. $|3x| = 1$ **16.** $|5x| = 4$ **17.** $|3x + 2| = 1$ **18.** $|7x - 4| = 8$

19. $|3x| < 1$ **20.** $|5x| \leqslant 4$ **21.** $|2x + 3| \leqslant 9$ **22.** $|2x + 3| < 13$

23. $|x - 5| > 0.1$ **24.** $|x - 7| \geqslant 0.4$ **25.** $\left|x + \dfrac{2}{3}\right| \leqslant \dfrac{5}{3}$ **26.** $\left|x + \dfrac{3}{4}\right| < \dfrac{1}{4}$

27. $|6 - 4x| \leqslant 8$ **28.** $|5 - 2x| > 10$ **29.** $\left|\dfrac{2x + 1}{3}\right| > 5$ **30.** $\left|\dfrac{3x + 2}{4}\right| \leqslant 5$

31. $\left|\dfrac{13}{4} + 2x\right| > \dfrac{1}{4}$ **32.** $\left|\dfrac{5}{6} + 3x\right| < \dfrac{7}{6}$ **33.** $\left|\dfrac{3 - 4x}{2}\right| \leqslant \dfrac{3}{4}$ **34.** $\left|\dfrac{2x - 1}{3}\right| \geqslant \dfrac{5}{6}$

35. $|x| = -3$ **36.** $|x| < -3$ **37.** $|2x - 4| < -5$ **38.** $|3x + 5| < 0$

39. ▦ $|x - 2.0245| < 0.1011$ **40.** ▦ $|x + 17.217| > 5.0012$

41. ▦ $|3.0147x - 8.9912| \leqslant 6.0243$ **42.** ▦ $|-2.1437x + 7.8814| \geqslant 9.1132$

Solve. The definition of absolute value in Section 1.1 may be helpful for some of these exercises.

43. $|4x - 5| = x + 1$ **44.** $|2x + 3| = |x| + 8$ **45.** $\bigl||x| - 1\bigr| = 3$ **46.** $|x + 2| > x$

47. $|x + 2| \leqslant |x - 5|$ **48.** $|3x - 1| > 5x - 2$ **49.** $|x| + |x - 1| < 10$ **50.** $|x| - |x - 3| < 7$

51. $|x - 3| + |2x + 5| > 6$

52. Solve: $|x - 3| + |2x + 5| + |3x - 1| = 12$.

Prove the following for any real numbers a and b.

53. $-|a| \leqslant a \leqslant |a|$

54. (*The triangle inequality*). $|a + b| \leqslant |a| + |b|$

55. Show that if $|a| < \dfrac{e}{2}$ and $|b| < \dfrac{e}{2}$, then $|a + b| < e$.

56. a) Prove that $\left|x - \dfrac{a + b}{2}\right| < \dfrac{b - a}{2}$ is equivalent to $a < x < b$.

Use graphs or the result of part (a) to find an inequality with absolute value for each of the following.

     b) $-5 < x < 5$ c) $-6 < x < 6$ d) $-1 < x < 7$ e) $-5 < x < 1$

57. Use absolute value to prove that the number halfway between a and b is $\dfrac{a + b}{2}$.

4.7

QUADRATIC AND RATIONAL INEQUALITIES

▐i▐   QUADRATIC AND CERTAIN OTHER POLYNOMIAL INEQUALITIES

Quadratic Inequalities

Inequalities such as the following are called *quadratic inequalities*:

$$x^2 - 2x - 2 > 0, \qquad 3x^2 + 5x + 1 \leqslant 0.$$

In each case we have a polynomial of degree 2 on the left. One way of solving quadratic inequalities is by graphing.

EXAMPLE 1   Solve $x^2 + 2x - 3 > 0$.

Consider $f(x) = x^2 + 2x - 3$. Its graph opens upward. Function values will be positive to the left and right of the intercepts as shown. We find the intercepts by solving $f(x) = 0$:

$$x^2 + 2x - 3 = 0$$
$$(x + 3)(x - 1) = 0$$
$$x + 3 = 0 \quad \text{or} \quad x - 1 = 0$$
$$x = -3 \quad \text{or} \qquad x = 1.$$

Then the solution set of the inequality is

$$\{x \mid x < -3 \text{ or } x > 1\}.$$

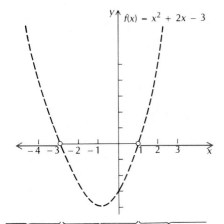

These inputs give positive outputs.   Inputs in this interval give negative or 0 outputs.   These inputs give positive outputs.

DO EXERCISE 1.

We can solve any quadratic inequality by considering its graph and finding intercepts as in Example 1. In some cases we will need to use the quadratic formula to find the intercepts.

EXAMPLE 2   Solve $x^2 - 2x \leqslant 2$.

Solve by graphing.

1. $x^2 - 3x - 10 < 0$

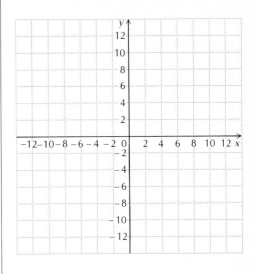

Solve by graphing.

2. $x^2 + 2x > 4$

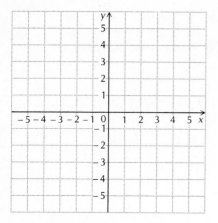

We first find standard form with 0 on one side:

$$x^2 - 2x - 2 \leqslant 0.$$

Consider $f(x) = x^2 - 2x - 2$. Its graph opens upward. Function values will be nonpositive between and including its intercepts as shown. We find the intercepts by solving $f(x) = 0$. This time we will need the quadratic formula. The intercepts are $1 + \sqrt{3}$ and $1 - \sqrt{3}$. The solution set of the inequality is

$$\{x \mid 1 - \sqrt{3} \leqslant x \leqslant 1 + \sqrt{3}\}.$$

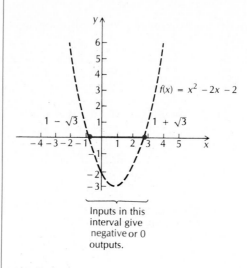

Inputs in this interval give negative or 0 outputs.

DO EXERCISE 2.

It should be pointed out that we need not actually draw graphs as in the preceding examples. Merely visualizing the graph will usually suffice.

Let us now consider another method of solving inequalities that works for polynomials of higher degree.

EXAMPLE 3 Solve $x^2 + 2x - 3 > 0$.

We factor the inequality, obtaining $(x + 3)(x - 1) > 0$. The solutions of $(x + 3)(x - 1) = 0$ are -3 and 1. They are not solutions of the inequality, but they divide the real-number line in a natural way, pictured as follows. The product $(x + 3)(x - 1)$ is positive or negative, for values other than -3 and 1, depending on the signs of the factors $x + 3$ and $x - 1$. We tabulate signs in these intervals.

Interval	$x + 3$	$x - 1$	Product: $(x + 3)(x - 1)$
$x < -3$	−	−	+
$-3 < x < 1$	+	−	−
$1 < x$	+	+	+

In order for the product $(x + 3)(x - 1)$ to be positive, both of the factors must be positive or both must be negative. We see from the table that the solution set of the inequality is $\{x|x > 1\} \cup \{x|x < -3\}$, or $\{x|x > 1$ or $x < -3\}$.

DO EXERCISES 3 AND 4.

Polynomial Inequalities of Higher Degree

We can extend the preceding method to polynomials with more than two factors.

EXAMPLE 4   Solve $4x(x + 1)(x - 1) < 0$.

The solutions of $4x(x + 1)(x - 1) = 0$ are -1, 0, and 1. They divide the real-number line in a natural way, pictured as follows. The product $4x(x + 1)(x - 1)$ is positive or negative depending on the signs of the factors $4x$, $x + 1$, and $x - 1$. We tabulate signs in these intervals.

Interval	4x	x + 1	x − 1	Product: 4x(x + 1)(x − 1)
$x < -1$	−	−	−	−
$-1 < x < 0$	−	+	−	+
$0 < x < 1$	+	+	−	−
$1 < x$	+	+	+	+

The product of three numbers is negative when it has an odd number of negative factors. We see from the table that the solution set is

$$\{x|x < -1\} \cup \{x|0 < x < 1\}, \quad \text{or} \quad \{x|x < -1 \text{ or } 0 < x < 1\}.$$

DO EXERCISE 5.

ii   RATIONAL INEQUALITIES

We can also use the preceding method when an inequality involves functions that are quotients of polynomials. Such functions are called *rational functions*.

EXAMPLE 5   Solve

$$\frac{x + 1}{x - 2} \geq 3.$$

We add -3, to get 0 on one side:

$$\frac{x + 1}{x - 2} - 3 \geq 0.$$

Next, we obtain a single fractional expression:

$$\frac{x + 1}{x - 2} - 3\left(\frac{x - 2}{x - 2}\right) = \frac{x + 1 - 3x + 6}{x - 2} \geq 0$$

$$\frac{-2x + 7}{x - 2} \geq 0.$$

Solve. Use the method of Example 3.

3. $x^2 - 3x > 4$

4. $x^2 - 3x < 4$

Solve.

5. $7x(x + 3)(x - 2) > 0$

Solve.

6. $\dfrac{x-3}{x+4} \geqslant 2$

7. $\dfrac{x}{x-5} \leqslant 2$

Let us consider the equality portion:

$$\frac{-2x+7}{x-2} = 0.$$

This has the solution $\frac{7}{2}$. Thus $\frac{7}{2}$ is in the solution set. Next, consider the inequality portion:

$$\frac{-2x+7}{x-2} > 0.$$

The solutions of $-2x+7 = 0$ and $x-2 = 0$ are 2 and $\frac{7}{2}$. They divide the real-number line in a natural way, pictured as follows. The quotient is positive or negative depending on the signs of $-2x+7$ and $x-2$. We tabulate signs in these intervals.

Interval	$-2x+7$	$x-2$	Quotient: $\dfrac{-2x+7}{x-2}$
$x < 2$	$+$	$-$	$-$
$2 < x < \dfrac{7}{2}$	$+$	$+$	$+$
$\dfrac{7}{2} < x$	$-$	$+$	$-$

We see from the table that the solution set of $(-2x+7)/(x-2) > 0$ is $\{x \mid 2 < x < \frac{7}{2}\}$. Thus the solution set of the inequality in question is $\{x \mid 2 < x \leqslant \frac{7}{2}\}$.

DO EXERCISES 6 AND 7.

EXERCISE SET 4.7

i Solve.

1. $x^2 - x - 2 < 0$

2. $x^2 - 4 > 0$

3. $x^2 \geqslant 1$

4. $x^2 - 2x < 5$

5. $x^2 - 2x + 1 \geqslant 0$

6. $x^2 + 6x + 9 < 0$

7. $x^2 < 6x - 4$

8. $5x - 2 > x^2$

9. $x^2 - 8x - 20 \leqslant 0$

10. $x^2 - 8x - 20 \geqslant 0$

11. $x^2 - 2x > 8$

12. $x^2 - 2x < 8$

13. $4x^2 + 7x < 15$

14. $4x^2 + 7x \geqslant 15$

15. $2x^2 + x > 5$

16. $2x^2 + x \leqslant 2$

17. $5x(x+1)(x-1) > 0$

18. $3x(x+2)(x-2) < 0$

19. $(x+3)(x+2)(x-1) < 0$

20. $(x-2)(x-3)(x+1) > 0$

ii Solve.

21. $\dfrac{1}{4-x} < 0$

22. $\dfrac{-4}{2x+5} > 0$

23. $\dfrac{3x+2}{x-3} > 0$

24. $\dfrac{5-2x}{4x+3} < 0$

25. $\dfrac{x+1}{2x-3} \geqslant 1$

26. $\dfrac{x-1}{x-2} \geqslant 3$

27. $\dfrac{x+1}{x+2} \leqslant 3$

28. $\dfrac{x+1}{2x-3} \leqslant 1$

29. $(x+1)(x-2) > (x+3)^2$

30. $(x-4)(x+3) \leqslant (x-1)^2$

31. $x^3 - x^2 > 0$

32. $x^3 - 4x > 0$

33. $x + \dfrac{4}{x} > 4$

34. $\dfrac{1}{x^2} \leqslant \dfrac{1}{x^3}$

35. $\dfrac{1}{x^3} \leqslant \dfrac{1}{x^2}$

36. $x + \dfrac{1}{x} > 2$

37. $\dfrac{2+x-x^2}{x^2+5x+6} < 0$

38. $\dfrac{4}{x^2} - 1 > 0$

Solve.

39. $\left| \dfrac{x+3}{x-4} \right| < 2$

40. $|x^2 - 5| = 5 - x^2$

41. $(7-x)^{-2} < 0$

42. $(1-x)^3 > 0$

43. $\left| 1 + \dfrac{1}{x} \right| < 3$

44. $(x+5)^{-2} > 0$

45. $\left| 2 - \dfrac{1}{x} \right| \leqslant 2 + \left| \dfrac{1}{x} \right|$

46. $\dfrac{(x-2)^2(x-3)^3(x+1)}{(x+2)(4-x)} \geqslant 0$

47. $|x^2 + 3x - 1| < 3$ (*Hint:* Solve $-3 < x^2 + 3x - 1 < 3$.)

48. $|1 + 5x - x^2| \geqslant 5$ (*Hint:* Solve the disjunction $1 + 5x - x^2 \leqslant -5$ or $1 + 5x - x^2 \geqslant 5$.)

49. The base of a triangle is 4 cm greater than the height. Find the possible heights h such that the area of the triangle will be greater than 10 cm^2.

50. The length of a rectangle is 3 m greater than the width. Find the possible widths w such that the area of the rectangle will be greater than 15 m^2.

51. A company has the following total cost and total revenue functions to use in producing and selling x units of a certain product (for reference, see p. 179):

$$R(x) = 50x - x^2, \qquad C(x) = 5x + 350.$$

a) Find the break-even values.
b) Find the values of x that produce a profit.
c) Find the values of x that result in a loss.

52. A company has the following total cost and total revenue functions to use in producing and selling x units of a certain product:

$$R(x) = 80x - x^2, \qquad C(x) = 10x + 600.$$

a) Find the break-even values.
b) Find the values of x that produce a profit.
c) Find the values of x that result in a loss.

53. Find the numbers k for which the quadratic equation $x^2 + kx + 1 = 0$ has (a) two real-number solutions; (b) no real-number solution.

54. Find the numbers k for which the quadratic equation $2x^2 - kx + 1 = 0$ has (a) two real-number solutions; (b) no real-number solution.

Find the domain of each function.

55. $f(x) = \sqrt{1 - x^2}$

56. $f(x) = \dfrac{1}{\sqrt{1 - x^2}}$

57. $g(x) = \sqrt{x^2 + 2x - 3}$

SUMMARY AND REVIEW: CHAPTER 4

The following contains a summary of what you should be able to do after completing this chapter. The review exercises are for practice. Answers are at the back of the book. If you miss an exercise, restudy the section indicated alongside the answer.

You should be able to:

Given a linear equation, graph it, and find its slope and y-intercept.

1. Graph $5y - 2x = 10$.

2. Find the slope and y-intercept of the line $-2x - y = 7$.

Find the slope, if it exists, of the line containing two points.

3. Find the slope of the line containing the points $(7, -2)$ and $(1, 4)$.

Given the slope and coordinates of one point of a line, find an equation of the line.

4. Find an equation of the line through $(-2, -1)$ with $m = 3$.

Given the coordinates of two points, find an equation of the line containing them, the distance between them, and the midpoint of the segment with these points as endpoints.

5. Find an equation of the line containing $(4, 1)$ and $(-2, -1)$.

6. Find the distance between $(3, 7)$ and $(-2, 4)$.

7. Find the midpoint of the segment with endpoints $(3, 7)$ and $(-2, 4)$.

Given an equation of a line and the coordinates of a point, find equations of lines parallel or perpendicular to that line and containing the given point.

Given the point $(1, -1)$ and the line $2x + 3y = 4$:

8. Find an equation of the line containing the given point and parallel to the given line.

9. Find an equation of the line containing the given point and perpendicular to the given line.

Given equations of two lines, tell whether they are parallel, perpendicular, or neither.

Determine whether the lines are parallel, perpendicular, or neither.

10. $3x - 2y = 8$,
$6x - 4y = 2$

11. $y - 2x = 4$,
$2y - 3x = -7$

12. $y = \frac{3}{2}x + 7$,
$y = -\frac{2}{3}x - 4$

Given a quadratic function, find the vertex of its graph, the line of symmetry, and its maximum or minimum value; and graph the function.

For the functions in Questions 13 and 14:

a) find an equation of the type $f(x) = a(x - h)^2 + k$;
b) find the vertex;
c) find the line of symmetry; and
d) determine whether the second coordinate of the vertex is a maximum or minimum, and find the maximum or minimum.

13. $f(x) = 3x^2 + 6x + 1$

14. $f(x) = -2x^2 - 3x + 6$

15. Graph $f(x) = 3x^2 + 6x + 1$.

Find the x-intercepts of a quadratic function.

16. Find the x-intercepts of $f(x) = -x^2 - x - 1$.

Find the union and intersection of two sets. Graph the union and intersection of two sets.

17. Find $\{4, 5, 8, 12, 13\} \cap \{3, 5, 7, 9, 11\}$.

18. Find $\{4, 5, 8, 12, 13\} \cup \{3, 5, 7, 9, 11\}$.

19. Graph $\{x | -2 < x\} \cap \{x | x \leq 3\}$ on a line.

20. Graph $\{x | x < -2\} \cup \{x | x > 2\}$ on a line.

Solve conjunctions and disjunctions of inequalities; equations and inequalities with absolute value; and quadratic and certain other polynomial inequalities and rational inequalities.

Solve.

21. $-3x + 2 \leq -4$ *and* $x - 1 \leq 3$

22. $|x - 6| < 5$

23. $|-3x - 2| > 2$

24. $|\frac{1}{2}x + 4| \leq 6$

25. $|2x + 5| = 9$

26. $x^2 - 9 < 0$

27. $2x^2 - 3x - 2 > 0$

28. $(1 - x)(x + 4)(x - 2) < 0$

29. $\dfrac{x - 2}{x + 3} < 4$

Given a situation that is described by a linear function, find that function from two sets of data and then use that function to make predictions. Given a situation described by a quadratic function, use that function to make predictions, including finding a maximum or minimum value of the function.

30. (*World records in the mile run*). It has been found experimentally that, within certain restrictions, world records in the mile run can be approximated by a linear function. In 1900 the record for the mile was 4:20 (4 minutes, 20 seconds). In 1975 it was 3:50. Let R represent the record in the mile run, in seconds, and t the number of years since 1900.

a) Find a linear function that fits the data.
b) Predict the record in the year 2000.

31. The sum of the length and width of a rectangle is 40. Find the dimensions for which the area is a maximum.

32. A company has the following total cost and total revenue functions to use in producing and selling x units of a product:

$$C(x) = 15x + 200, \qquad R(x) = 100 + x^2.$$

a) Find the break-even values.
b) Find the values of x that produce a profit.
c) Find the values of x that result in a loss.

Solve.

33. $4x \leqslant 5x + 2 < 4 - x$

34. $\left|1 - \dfrac{1}{x^2}\right| < 3$

35. $(x - 2)^{-3} < 0$

36. Under what conditions is the square of a linear function a quadratic function?

Find the domain of each function.

37. $f(x) = \sqrt{1 - |3x - 2|}$

38. $g(x) = \dfrac{1}{\sqrt{5 - |7x + 2|}}$

TEST: CHAPTER 4

1. Find the slope and y-intercept of the line $10 - 7y = 3x$.

2. Find an equation of the line through $(-2, 2)$ with $m = 5$.

3. Find an equation of the line containing $(2, -3)$ and $(5, 3)$.

4. Find the distance between $(4, -5)$ and $(6, -2)$.

5. Find the midpoint of the segment with endpoints $(4, -9)$ and $(7, 6)$.

6. Determine whether these lines are parallel, perpendicular, or neither.

$$2x - 3y = -9,$$
$$3x + 2y = 6$$

7. Find an equation of the line containing the given point and parallel to the given line: $(2, -4)$; $y - \frac{1}{3}x = 6$.

For the functions in Questions 8 and 9:

a) find an equation of the type $f(x) = a(x - h)^2 + k$;
b) find the vertex; and
c) determine whether there is a maximum or minimum function value and find that value.

8. $f(x) = 5x^2 - 10x + 3$

9. $f(x) = -4x^2 + 3x - 1$

10. Graph $f(x) = 5x^2 - 10x + 3$.

11. Find the x-intercepts of $f(x) = 3x^2 - x - 1$.

12. Find $\{3, 4, 7, 8, 11, 12\} \cup \{2, 4, 6, 8, 10\}$.

13. Graph $\{x \mid x \leqslant \frac{3}{2}\} \cap \{x \mid x > \frac{11}{2}\}$.

Solve.

14. $x - 3 > -8 \quad and \quad -2x + 3 \geqslant 9$

15. $|-4x + 3| \geqslant 9$

16. $|x - 3| < 2$

17. $|6x - 3| = 15$

18. $x^2 - 8x + 12 > 0$

19. $8x^2 + 10x - 3 < 0$

20. $\dfrac{x - 4}{2x + 3} < 1$

21. Find two numbers whose sum is -20 and whose product is a maximum.

22. To produce and sell x units of a certain product a company has the following total cost and total revenue functions:

$$R(x) = 65x - x^2, \qquad C(x) = 25x + 300.$$

a) Find the break-even values.
b) Find the values of x that produce a profit.
c) Find the values of x that result in a loss.

23. Solve $\left|2 + \dfrac{1}{x}\right| < 6$.

24. Find the domain of $f(x) = \sqrt{x^2 + 3x - 10}$.

How can chemists use systems of equations to mix certain kinds of solutions?

The hardest part of solving problems in algebra is almost always translating the problem situation to mathematical language. Once you get an equation, for example, the rest is easy. In this chapter we study *systems of equations* and how to solve them. One of the great advantages to using a system of equations is that many problem situations then become easier to translate to mathematical language.

A system of equations has solutions that are ordered pairs, ordered triples, and so on, depending on the number of variables. A solution of a system is an ordered *n*-tuple that makes *all* the equations of the system true (at the same time). Hence an older terminology refers to a system as *simultaneous equations.*

Systems of equations have extensive applications to many fields such as sociology, psychology, business, education, engineering, and science. Suppose a study is being made of whether stereo A is better than stereo B. The basis of the research done to make this determination is often the solution of a system of many equations in many variables.

Systems of inequalities are also useful in a rather new branch of mathematics called *linear programming.* In this chapter we give you a brief introduction to linear programming.

OBJECTIVES

You should be able to:

i   Determine whether an ordered pair is a solution of a system of equations.

ii   Solve a system of two linear equations in two variables by graphing.

iii   Solve a system of two linear equations in two variables by the substitution method.

iv   Solve a system of two linear equations in two variables by the addition method.

v   Solve applied problems by translating them to a system of two linear equations and solving the system.

1. Determine whether $(-3, 2)$ is a solution of the conjunction

$$2x - y = -8 \quad and \quad 3x + 4y = -1.$$

2. Determine whether $(0, \frac{1}{4})$ is a solution of the conjunction

$$5x + 12y = 13 \quad and \quad \sqrt{2}x + 9y = 10.$$

3. Solve graphically.

$$y - x = 1 \quad and \quad y + x = 3$$

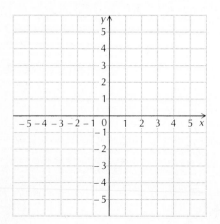

5.1

SYSTEMS OF EQUATIONS IN TWO VARIABLES

i   IDENTIFYING SOLUTIONS

Recall that a *conjunction* of sentences is formed by joining sentences with the word *and*. Here is an example:

$$x + y = 11 \quad and \quad 3x - y = 5.$$

A solution of a sentence with two variables such as $x + y = 11$ is an ordered pair. Some pairs in the solution set of $x + y = 11$ are

$$(5, 6), \quad (12, -1), \quad (4, 7), \quad (8, 3).$$

Some pairs in the solution set of $3x - y = 5$ are

$$(0, -5), \quad (4, 7), \quad (-2, -11), \quad (9, 22).$$

The solution set of the sentence

$$x + y = 11 \quad and \quad 3x - y = 5$$

consists of all pairs that make *both* sentences true. That is, it is the *intersection* of the solution sets. Note that $(4, 7)$ is a solution of the conjunction; in fact, it is the only solution.

DO EXERCISES 1 AND 2.

ii   SOLVING SYSTEMS OF EQUATIONS GRAPHICALLY

One way to find solutions is by trial and error. Another way is to graph the equations and look for points of intersection. For example, the following graph shows the solution sets of $x + y = 11$ and $3x - y = 5$. Their intersection is the single ordered pair $(4, 7)$.

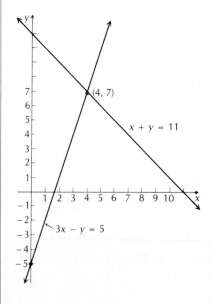

DO EXERCISE 3.

We often refer to a conjunction of equations as a *system* of equations. We usually omit the word *and,* and very often write one equation under the other.

We now consider two algebraic methods for solving systems of linear equations.

iii THE SUBSTITUTION METHOD

To use the substitution method we solve one equation for one of the variables. Then we substitute in the other equation and solve.

EXAMPLE 1 Solve the system

$$x + y = 11, \tag{1}$$
$$3x - y = 5. \tag{2}$$

First we solve Eq. (1) for y:*

$$y = 11 - x.$$

Then we substitute $11 - x$ for y in Eq. (2). This gives an equation in one variable, which we know how to solve from earlier work:

$$3x - (11 - x) = 5$$
$$x = 4.$$

Now substitute 4 for x in either Eq. (1) or (2) and solve for y. Let us use Eq. (1):

$$4 + y = 11$$
$$y = 7.$$

The solution is $(4, 7)$. Note the alphabetical listing: 4 for x, 7 for y.

Check:

$$\begin{array}{c|c} x + y = 11 \\ \hline 4 + 7 & 11 \\ 11 & \end{array} \qquad \begin{array}{c|c} 3x - y = 5 \\ \hline 3 \cdot 4 - 7 & 5 \\ 5 & \end{array}$$

DO EXERCISES 4 AND 5.

iv THE ADDITION METHOD

The *addition method* makes use of the addition and multiplication principles for solving equations. Consider the system

$$3x - 4y = -1,$$
$$-3x + 2y = 0.$$

We add the left-hand sides and obtain $-2y$, and add the right-hand sides† and obtain -1. Thus we obtain a new equation: $-2y = -1$. We can now solve for y and then substitute into one of the original equations to find x.

$$
\begin{aligned}
-2y &= -1 & -3x + 2(\tfrac{1}{2}) &= 0 & &\text{Substituting into the} \\
y &= \tfrac{1}{2} & -3x + 1 &= 0 & &\text{second equation} \\
& & -3x &= -1 & & \\
& & x &= \tfrac{1}{3} & &
\end{aligned}
$$

Solve, using the substitution method.

4. $2x + y = 6,$
  $3x + 4y = 4$

5. $8x - 3y = -31,$
  $2x + 6y = 26$

---

* We could just as well solve for x.
† When we do this we often say that we "added the two equations."

We now know that the solution of the original system is $(\frac{1}{3}, \frac{1}{2})$, because the solutions of this last system are obvious and we know that this system is equivalent to the original. We know that the last system is equivalent to the original, because the only computations we did consisted of using the addition principle, adding only expressions for which all replacements were sensible, and multiplying by a nonzero constant. Each produces results equivalent to the original equations. For this reason we do not need to check results, except to detect errors in computation.

EXAMPLE 2   Solve:

$$5x + 3y = 7,$$
$$3x - 5y = -23.$$

We first multiply the second equation by 5 to make the x-coefficient a multiple of 5:

$$5x + 3y = 7,$$
$$15x - 25y = -115.$$

Now we multiply the first equation by -3 and add it to the second equation. This gets rid of the x-term:

$$5x + 3y = 7,$$
$$- 34y = -136.$$

Next we solve the second equation for y. Then we substitute the result into the first equation to find x.

$$-34y = -136 \qquad 5x + 3(4) = 7$$
$$y = 4 \qquad\qquad 5x + 12 = 7$$
$$5x = -5$$
$$x = -1$$

The solution is $(-1, 4)$.

There are some preliminary things that might be done to make the steps of the addition method simpler. One is to first write the equations in the form $Ax + By = C$. Another is to interchange two equations before beginning. For example, if we have the system

$$5x + y = -2,$$
$$x + 7y = 3,$$

we may prefer to write the second equation first. This accomplishes two things. One is that the x-coefficient is 1 in the first equation. Having found y later, this will make solving for x easier. The other is that this makes the x-coefficient in the second equation a multiple of the first.

Something else that might be done is to multiply one or more equations by a power of 10 before beginning in order to eliminate decimal points. For example, if we have the system

$$-0.3x + 0.5y = 0.03,$$
$$0.01x - 0.4y = 1.2,$$

we can multiply both equations by 100 to clear of decimal points, obtaining

$$-30x + 50y = 3,$$
$$x - 40y = 120.$$

Still other procedures are possible. However, it is recommended that no others be used at this point, because we are establishing an *algorithm*, which could be done mechanically by a computer.

EXAMPLE 3 Solve:

$$5x + y = -2,$$
$$x + 7y = 3.$$

We first interchange the equations so that the *x*-coefficient of the second equation will be a multiple of the first:

$$x + 7y = 3,$$
$$5x + y = -2.$$

Next, we multiply the first equation by -5 and add the result to the second equation. This gets rid of the *x*-term in the second equation:

$$x + 7y = 3,$$
$$-34y = -17.$$

Now we solve the second equation for *y*. Then we substitute the result in the first equation to find *x*.

$$-34y = -17 \qquad x + 7(\tfrac{1}{2}) = 3$$
$$y = \tfrac{1}{2} \qquad x + \tfrac{7}{2} = 3$$
$$x = 3 - \tfrac{7}{2}, \quad \text{or} \quad -\tfrac{1}{2}$$

The solution is $(-\tfrac{1}{2}, \tfrac{1}{2})$.

DO EXERCISES 6–8.

■ v ■ PROBLEM SOLVING

As you have already noticed, in the four-step problem-solving process the most difficult and time-consuming part is the translation of a problem situation to mathematical language. Sometimes translating to an equation takes considerable thought and effort. This task becomes easier in many cases if we translate to more than one equation in more than one variable.

EXAMPLE 4 An airplane flies the 3000-mi distance from Los Angeles to New York, with a tail wind, in 5 hr. On the return trip, against the wind, it makes the trip in 6 hr. Find the speed of the plane and the speed of the wind.

1. *Familiarize:* We first make a drawing.

Solve.

6. $4x + 3y = -6,$
$-4x + 2y = 16$

7. $9x - 2y = -4,$
$3x + 4y = 1$

8. $0.2x + 0.3y = 0.1,$
$0.3x - 0.1y = 0.7$

9. The sum of two numbers is 10. The difference is 1. Find the numbers.

We let p represent the speed of the plane in still air and w represent the speed of the wind. Recall from Chapter 2 that distance, speed, and time are related by the definition of speed:

$$\text{Speed} = \frac{\text{Distance}}{\text{Time}}, \qquad r = \frac{d}{t}.$$

We often need to use other formulas that we can derive from this one, namely, $d = rt$ and $t = d/r$.

From the drawing we see that the distances are the same. When the plane flies east with the wind its speed is $p + w$. When it flies west against the wind its speed is $p - w$. We list the information in a table, the columns of the table coming from the formula $d = rt$.

	Distance	Speed	Time
East (with the wind)	3000	$p + w$	5
West (against the wind)	3000	$p - w$	6

10. The sum of two numbers is 1. The difference is 10. Find the numbers.

2. *Translate:* Using $d = rt$ in each row of the table, we get an equation. Thus we have a system of equations:

$$3000 = (p + w)5 = 5p + 5w,$$
$$3000 = (p - w)6 = 6p - 6w.$$

3. *Carry out:* We solve the system

$$5p + 5w = 3000,$$
$$6p - 6w = 3000.$$

We first multiply the second equation by 5 to make the p-coefficient a multiple of 5:

$$5p + 5w = 3000,$$
$$30p - 30w = 15{,}000.$$

Then we multiply the first equation by -6 and add it to the second equation. This gets rid of the p-term:

$$5p + 5w = 3000,$$
$$-60w = -3000.$$

11. It takes a boat 3 hr to travel 24 km upstream. It takes 2 hr to travel the 24 km downstream. Find the speed of the boat and the speed of the stream.

Now we solve the second equation for w. Then we substitute the result into the first equation to find p.

$$-60w = -3000 \qquad\qquad 5p + 5(50) = 3000$$
$$w = 50 \qquad\qquad\qquad 5p + 250 = 3000$$
$$5p = 2750$$
$$p = 550$$

4. *Check:* The solution of the system of equations is $(550, 50)$. That is, the speed of the plane is 550 mph and the speed of the wind is 50 mph. We leave it to the student to show that this does indeed check in the original problem.

DO EXERCISES 9–11.

EXAMPLE 5   Wine A is 5% alcohol and wine B is 15% alcohol. How many liters of each should be mixed to get a 10-L mixture that is 12% alcohol?

1. *Familiarize:*   We organize the information in a table.

	Amount of solution	Percent of alcohol	Amount of alcohol in solution
A	x liters	5%	5%x
B	y liters	15%	15%y
Mixture	10 liters	12%	0.12×10, or 1.2 liters

Note that we have used x for the number of liters of A and y for the number of liters of B. To get the amount of alcohol, we multiply by the percentages, of course.

2. *Translate:*   If we add x and y in the first column we get 10, and this gives us one equation:

$$x + y = 10.$$

If we add the amounts of alcohol in the third column we get 1.2, and this gives us another equation:

$$5\%x + 15\%y = 1.2.$$

After changing percents to decimals and clearing, we have this system:

$$x + y = 10,$$
$$5x + 15y = 120.$$

3. *Carry out:*   Then solve the system. We leave this to the student. The solution is $(3, 7)$. That is, 3 L of wine A and 7 L of wine B are possibilities for a solution to the original problem.

4. *Check:*   We add the amounts of wine: $3\,\text{L} + 7\,\text{L} = 10\,\text{L}$. Thus the amount of wine checks. Next we check the amount of alcohol:

$$5\%(3) + 15\%(7) = 0.15 + 1.05, \quad \text{or } 1.2\,\text{L}.$$

Thus the amount of alcohol checks. The solution of the problem is 3 L of wine A and 7 L of wine B.

DO EXERCISE 12.

12. Solution A is 25% acid. Solution B is 65% acid. How many liters of each should be mixed to get 8 L of a solution that is 40% acid?

EXERCISE SET 5.1

1. Determine whether $(\frac{1}{2}, 1)$ is a solution.

$$3x + y = \frac{5}{2},$$
$$2x - y = \frac{1}{4}$$

2. Determine whether $(-2, \frac{1}{4})$ is a solution.

$$x + 4y = -1,$$
$$2x + 8y = -2$$

ii Solve graphically.

3. $x + y = 2,$
  $3x + y = 0$

4. $x + y = 1,$
  $3x + y = 7$

iii Solve using the substitution method.

5. $x - 5y = 4,$
  $2x + y = 7$

6. $3x - y = 5,$
  $x + y = \frac{1}{2}$

iv Solve using the addition method.

7. $x - 3y = 2,$
  $6x + 5y = -34$

8. $x + 3y = 0,$
  $20x - 15y = 75$

9. $0.3x + 0.2y = -0.9,$
  $0.2x - 0.3y = -0.6$

10. $0.2x - 0.3y = 0.3,$
  $0.4x + 0.6y = -0.2$

11. $\frac{1}{5}x + \frac{1}{2}y = 6,$
  $\frac{3}{5}x - \frac{1}{2}y = 2$

12. $\frac{2}{3}x + \frac{3}{5}y = -17,$
  $\frac{1}{2}x - \frac{1}{3}y = -1$

13. $2a = 5 - 3b,$
  $4a = 11 - 7b$

14. $7(a - b) = 14,$
  $2a = b + 5$

v **Problem Solving**

15. Find two numbers whose sum is -10 and whose difference is 1.

16. Find two numbers whose sum is -1 and whose difference is 10.

17. A boat travels 46 km downstream in 2 hr. It travels 51 km upstream in 3 hr. Find the speed of the boat and the speed of the stream.

18. An airplane travels 3000 km with a tail wind in 3 hr. It travels 3000 km with a head wind in 4 hr. Find the speed of the plane and the speed of the wind.

19. Antifreeze A is 18% alcohol. Antifreeze B is 10% alcohol. How many liters of each should be mixed to get 20 L of a mixture that is 15% alcohol?

20. Beer A is 6% alcohol and beer B is 2% alcohol. How many liters of each should be mixed to get 50 L of a mixture that is 3.2% alcohol?

21. Two cars leave town traveling in opposite directions. One travels 80 km/h and the other 96 km/h. In how many hours will they be 528 km apart?

22. A train leaves a station and travels north at 75 km/h. Two hours later a second train leaves on a parallel track and travels north at 125 km/h. How far from the station will they meet?

23. Two planes travel toward each other from cities that are 780 km apart at rates of 190 and 200 km/h. They started at the same time. In how many hours will they meet?

24. Two motorcycles travel toward each other from Chicago and Indianapolis, which are about 350 km apart, at rates of 110 and 90 km/h. They started at the same time. In how many hours will they meet?

25. One week a business sold 40 scarves. White ones cost $4.95 and printed ones cost $7.95. In all, $282 worth of scarves were sold. How many of each kind were sold?

26. One day a store sold 30 sweatshirts. White ones cost $9.95 and yellow ones cost $10.50. In all, $310.60 worth of sweatshirts were sold. How many of each color were sold?

27. Paula is 12 years older than her brother Bob. Four years from now Bob will be $\frac{2}{3}$ as old as Paula. How old are they now?

28. Carlos is 8 years older than his sister Maria. Four years ago Maria was $\frac{2}{3}$ as old as Carlos. How old are they now?

29. The perimeter of a lot is 190 m. The width is one-fourth the length. Find the dimensions.

30. The perimeter of a rectangular field is 628 m. The width of the field is 6 m less than the length. Find the dimensions.

31. The perimeter of a rectangle is 384 m. The length is 82 m greater than the width. Find the length and width.

32. The perimeter of a rectangle is 86 cm. The length is 19 cm greater than the width. Find the area.

Solve.

33. $\dfrac{x + y}{4} - \dfrac{x - y}{3} = 1,$

  $\dfrac{x - y}{2} + \dfrac{x + y}{4} = -9$

34. $\dfrac{x + y}{2} - \dfrac{y - x}{3} = 0,$

  $\dfrac{x + y}{3} - \dfrac{x + y}{4} = 0$

Solve. Check by substituting.

35. 🮐 $2.35x - 3.18y = 4.82,$
$1.92x + 6.77y = -3.87$

36. 🮐 $0.0375x + 0.912y = -1.003,$
$463x - 801y = 946$

Problem Solving

37. Nancy jogs and walks to the university each day. She averages 4 km/h walking and 8 km/h jogging. The distance from home to the university is 6 km and she makes the trip in 1 hr. How far does she jog in a trip?

38. James and Joan are mathematics professors. They have a total of 46 years of teaching. Two years ago James had taught 2.5 times as many years as Joan. How long has each taught?

39. A limited edition of a book published by a historical society was offered for sale to its membership. The cost was one book for $12 or two books for $20. The society sold 880 books and the total amount of money taken in was $9840. How many members ordered two books?

40. The ten's digit of a two-digit positive integer is 2 more than three times the unit's digit. If the digits are interchanged, the new number is 13 less than half the given number. Find the given integer. (*Hint:* Let $x =$ the ten's place digit and $y =$ the unit's place digit; then $10x + y$ is the number.)

41. The numerator of a fraction is 12 more than the denominator. The sum of the numerator and the denominator is 5 more than three times the denominator. What is the reciprocal of the fraction?

42. The measure of one of two supplementary angles is 8° more than three times the other. Find the measure of the larger of the two angles.

43. A train leaves Union Station for Central Station, 216 km away, at 9 A.M. One hour later, a train leaves Central Station for Union Station. They meet at noon. If the second train had started at 9 A.M. and the first train at 10:30 A.M., they would still have met at noon. Find the speed of each train.

44. An automobile radiator contains 16 L of antifreeze and water. This mixture is 30% antifreeze. How much of this mixture should be drained and replaced with pure antifreeze so that there will be 50% antifreeze?

45. A stablehand agreed to work for one year. At the end of that time he was to receive $240 and one horse. After 7 months the boy quit the job, but still received the horse and $100. What is the value of the horse?

46. You are in line at a ticket window. There are two more people ahead of you in line than there are behind you. In the entire line there are three times as many people as there are behind you. How many people are ahead of you in line?

47. Phil and Phyllis are siblings. Phyllis has twice as many brothers as she has sisters. Phil has the same number of brothers as sisters. How many girls and how many boys are in the family?

48. An automobile gets 18 miles per gallon (mpg) in city driving and 24 mpg in highway driving. The car is driven 465 mi on a full tank of 23 gal of gasoline. How many miles were driven in the city and how many were driven on the highway?

Each of the following is a system of equations that is *not* linear. But each is *linear in form*, in that an appropriate substitution, say u for $1/x$ and v for $1/y$, yields a linear system. Solve for the new variable and then solve for the original variable.

49. $\dfrac{1}{x} - \dfrac{3}{y} = 2,$
$\dfrac{6}{x} + \dfrac{5}{y} = -34$

50. $2\sqrt[3]{x} + \sqrt{y} = 0,$
$5\sqrt[3]{x} + 2\sqrt{y} = -5$

51. $3|x| + 5|y| = 30,$
$5|x| + 3|y| = 34$

52. $15x^2 + 2y^3 = 6,$
$25x^2 - 2y^3 = -6$

5.2

SYSTEMS OF EQUATIONS IN THREE OR MORE VARIABLES

■ IDENTIFYING SOLUTIONS

A *solution* of a system of equations in three variables is an ordered triple that makes all three equations true.

EXAMPLE 1 Determine whether $(\frac{3}{2}, -4, 3)$ is a solution of the system

$$4x - 2y - 3z = 5,$$
$$-8x - y + z = -5,$$
$$2x + y + 2z = 5.$$

OBJECTIVES

You should be able to:

i Determine whether an ordered triple is a solution of a system of equations in three variables.

ii Solve systems of linear equations in three or more variables.

iii Solve applied problems by translating them to a system of three linear equations and solving the system.

iv Fit a quadratic function to data when three data points are given.

1. Consider the following system:

$$4x - y + z = 6,$$
$$2x + y + 2z = 3,$$
$$3x - 2y + z = 3.$$

a) Determine whether $(3, 0, \frac{1}{4})$ is a solution.

b) Determine whether $(2, 1, -1)$ is a solution.

We substitute $(\frac{3}{2}, -4, 3)$ into the equations, using alphabetical order.

$$\frac{4x - 2y - 3z = 5}{4 \cdot \frac{3}{2} - 2(-4) - 3 \cdot 3 \mid 5}$$
$$6 + 8 - 9$$
$$5$$

$$\frac{-8x - y + z = -5}{-8 \cdot \frac{3}{2} - (-4) + 3 \mid -5}$$
$$-12 + 4 + 3$$
$$-5$$

$$\frac{2x + y + 2z = 5}{2 \cdot \frac{3}{2} + (-4) + 2 \cdot 3 \mid 5}$$
$$3 - 4 + 6$$
$$5$$

The triple makes all three equations true, so it is a solution.

DO EXERCISE 1.

ii SOLVING SYSTEMS OF EQUATIONS IN THREE OR MORE VARIABLES

Graphical methods of solving linear equations in three variables are unsatisfactory, because a three-dimensional coordinate system is required. The substitution method becomes cumbersome for most systems of more than two equations. Therefore, we will use the addition method. It is essentially the same as for systems of two equations.

EXAMPLE 2 Solve:

$$2x - 4y + 6z = 22, \quad ①$$
$$4x + 2y - 3z = 4, \quad ②$$
$$3x + 3y - z = 4. \quad ③$$

These numbers indicate the equations in the first, second, and third positions, respectively.

We begin by multiplying ③ by 2, to make each x-coefficient a multiple of the first.* Then we have the following:

$$2x - 4y + 6z = 22,$$
$$4x + 2y - 3z = 4,$$
$$6x + 6y - 2z = 8.$$

Next, we multiply ① by -2 and add it to ②. We also multiply ① by -3 and add it to ③. Then we have the following:

$$2x - 4y + 6z = 22,$$
$$10y - 15z = -40,$$
$$18y - 20z = -58.$$

Now we multiply ③ by -5 to make the y-coefficient a multiple of the y-coefficient in ②:

$$2x - 4y + 6z = 22,$$
$$10y - 15z = -40,$$
$$-90y + 100z = 290.$$

Next, we multiply ② by 9 and add it to ③:

$$2x - 4y + 6z = 22,$$
$$10y - 15z = -40,$$
$$-35z = -70.$$

_____

* By proceeding in this manner, we avoid fractions. The method will work when fractions are allowed, but is more difficult that way.

Now we solve ③ for z:

$$2x - 4y + 6z = 22,$$
$$10y - 15z = -40,$$
$$z = 2.$$

Next we substitute 2 for z in ②, and solve for y:

$$10y - 15(2) = -40$$
$$10y - 30 = -40$$
$$10y = -10$$
$$y = -1.$$

Finally, we substitute -1 for y and 2 for z in ①:

$$2x - 4(-1) + 6(2) = 22$$
$$2x + 4 + 12 = 22$$
$$2x + 16 = 22$$
$$2x = 6$$
$$x = 3.$$

The solution is $(3, -1, 2)$. To be sure computational errors have not been made, one can check by substituting 3 for x, -1 for y, and 2 for z in all three equations. If all are true, then the triple is a solution.

DO EXERCISE 2.

iii PROBLEM SOLVING

Solving systems of three or more equations is important in many applications. Systems of equations arise very often in the use of statistics, for example, in such fields as the social sciences. They also come up in problems of business, science, and engineering.

EXAMPLE 3   In a triangle the largest angle is $70°$ greater than the smallest angle. The largest angle is twice as large as the remaining angle. Find the measure of each angle.

1. *Familiarize:* The first thing to do with a problem like this is to make a drawing, or a sketch.

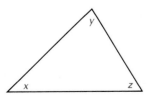

We don't know the size of any angle, so we have used x, y, and z for the measures of the angles. A geometric fact will be needed here, the fact that the measures of the angles of a triangle add up to $180°$.

2. *Translate:* The geometric fact about triangles gives us one equation:

$$x + y + z = 180.$$

There are two statements in the problem that we can translate almost directly.

2. Solve the system, using exactly the procedure used in the text.

$$x + 2y - z = 5,$$
$$2x - 4y + z = 0,$$
$$3x + 2y + 2z = 3$$

3. In a factory there are three machines A, B, and C. When all three are running, they produce 222 suitcases per day. If A and B work, but C does not, they produce 159 suitcases per day. If B and C work, but A does not, they produce 147 suitcases per day. What is the daily production of each machine?

$$\underbrace{\text{The largest angle}}_{z} \ \underbrace{\text{is}}_{\downarrow} \ \underbrace{70°}_{= \ 70} \ \underbrace{\text{greater than}}_{+} \ \underbrace{\text{the smallest angle.}}_{x}$$

$$\underbrace{\text{The largest angle}}_{z} \ \underbrace{\text{is}}_{\overset{\downarrow}{=}} \ \underbrace{\text{twice as large as the remaining angle.}}_{2y}$$

We now have a system of three equations.

$$\begin{aligned} x + y + z &= 180, & x + \ y + z &= 180, \\ x + 70 &= z, & \text{or} \quad x \quad\ \ - z &= -70, \\ 2y &= z; & 2y - z &= 0. \end{aligned}$$

3. *Carry out:* We solve the system. The details are left to you, but the solution is $(30, 50, 100)$.

4. *Check:* The sum of the numbers is 180, so that checks.

The largest angle measures $100°$ and the smallest measures $30°$. The largest angle is thus $70°$ greater than the smallest.

The remaining angle measures $50°$. The largest angle measures $100°$, so it is twice as large. We do have an answer to the problem.

The measures of the angles of the triangle are $30°$, $50°$, and $100°$.

DO EXERCISE 3.

iv MATHEMATICAL MODELS AND PROBLEM SOLVING

In a situation in which a quadratic function will serve as a mathematical model, we may wish to find an equation, or formula, for the function. Recall that for a linear model, we can find an equation if we know two data points. For a quadratic function we need three data points.

4. Find a quadratic function that fits the data points $(1, 0)$, $(-1, 4)$, and $(2, 1)$.

EXAMPLE 4 In a certain situation, it is believed that a quadratic function will be a good model. Find an equation of the function, given the data points $(1, -4)$, $(-1, -6)$, and $(2, -9)$.

We wish to find a quadratic function

$$f(x) = ax^2 + bx + c$$

containing the three given points, that is, a function for which the equation will be true when we substitute any of the ordered pairs of numbers into it. When we substitute, we get,

$$\begin{aligned} \text{for } (1, -4), \quad -4 &= a \cdot 1^2 \ + b \cdot 1 \ + c; \\ \text{for } (-1, -6), \quad -6 &= a(-1)^2 + b(-1) + c; \\ \text{for } (2, -9), \quad -9 &= a \cdot 2^2 \ + b \cdot 2 \ + c. \end{aligned}$$

We now have a system of equations in the three unknowns a, b, and c:

$$\begin{aligned} a + \ b + c &= -4, \\ a - \ b + c &= -6, \\ 4a + 2b + c &= -9. \end{aligned}$$

We solve this system of equations, obtaining $(-2, 1, -3)$. Thus the function we are looking for is

$$f(x) = -2x^2 + x - 3.$$

DO EXERCISE 4.

EXAMPLE 5 (*The cost of operating an automobile at various speeds*). It is found that the cost of operating an automobile as a function of speed is approximated by a quadratic function. Use the data given to find an equation of the function. Then use the equation to determine the cost of operating the automobile at 60 mph and at 80 mph.

Speed, in mph	Operating cost per mile, in cents
10	22
20	20
50	20

We use the three data points to obtain a, b, and c in the equation $f(x) = ax^2 + bx + c$:

$$22 = 100a + 10b + c,$$
$$20 = 400a + 20b + c, \quad \text{Substituting}$$
$$20 = 2500a + 50b + c.$$

We solve this system of equations, thus obtaining $(0.005, -0.35, 25)$. Thus

$$f(x) = 0.005x^2 - 0.35x + 25.$$

To find the cost of operating at 60 mph, we find $f(60)$:

$$f(60) = 0.005(60)^2 - 0.35(60) + 25 = 22¢.$$

We also find $f(80)$:

$$f(80) = 0.005(80)^2 - 0.35(80) + 25 = 29¢.$$

A graph of the cost function of Example 5 is as follows.

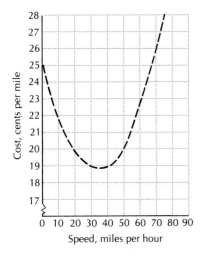

It should be noted that this cost function can give approximate results only within a certain interval. For example, $f(0) = 25$, meaning that it cost 25¢ per mile to stand still. This, of course, is absurd in the sense of mileage, although one does incur costs in owning a car whether one drives it or not.

DO EXERCISE 5.

5. The following table has values that will fit a quadratic function. Find the average number of accidents as a function of age. Then use your model to calculate the average number of accidents in which 16-year-olds are involved daily.

Age of driver	Average number of accidents per day
20	400
40	150
60	400

EXERCISE SET 5.2

i Consider the system

$$2x + 3y - 5z = 1,$$
$$6x - 6y + 10z = 3,$$
$$4x - 9y + 5z = 0.$$

1. Determine whether $(\frac{1}{2}, \frac{1}{3}, \frac{1}{5})$ is a solution of the system.

2. Determine whether $(-1, 1, 0)$ is a solution of the system.

ii Solve.

3. $x + y + z = 2,$
$6x - 4y + 5z = 31,$
$5x + 2y + 2z = 13$

4. $x + 6y + 3z = 4,$
$2x + y + 2z = 3,$
$3x - 2y + z = 0$

5. $x - y + 2z = -3,$
$x + 2y + 3z = 4,$
$2x + y + z = -3$

6. $x + y + z = 6,$
$2x - y - z = -3,$
$x - 2y + 3z = 6$

7. $4a + 9b = 8,$
$8a + 6c = -1,$
$6b + 6c = -1$

8. $3p + 2r = 11,$
$q - 7r = 4,$
$p - 6q = 1$

9. $x + y + z + w = 2,$
$2x + 2y + 4z + w = 1,$
$x - y - z - w = -6,$
$3x + y - z - w = -2$

10. $x - y + z + w = 0,$
$2x + 2y + z - w = 5,$
$3x + y - z - w = -4,$
$x + y - 3z - 2w = -7$

iii **Problem Solving**

11. The sum of three numbers is 26. Twice the first minus the second is 2 less than the third. The third is the second minus three times the first. Find the numbers.

12. The sum of three numbers is 5. The first number minus the second plus the third is 1. The first minus the third is 3 more than the second. Find the numbers.

13. In triangle ABC, the measure of angle B is three times the measure of angle A. The measure of angle C is 30° greater than the measure of angle A. Find the angle measures.

14. In triangle ABC, the measure of angle B is 2° more than three times the measure of angle A. The measure of angle C is 8° more than the measure of angle A. Find the angle measures.

15. Pat picked strawberries on three days. He picked a total of 87 quarts. On Tuesday he picked 15 quarts more than on Monday. On Wednesday he picked 3 quarts fewer than on Tuesday. How many quarts did he pick each day?

16. Gina sells magazines part time. On Thursday, Friday, and Saturday, she sold $66 worth. On Thursday she sold $3 more than on Friday. On Saturday she sold $6 more than on Thursday. How much did she take in each day?

17. Sawmills A, B, and C can produce 7400 board-feet of lumber per day. A and B together can produce 4700 board-feet, while B and C together can produce 5200 board-feet. How many board-feet can each mill produce by itself?

18. In a factory there are three polishing machines, A, B, and C. When all three of them are working, 5700 lenses can be polished in one week. When only A and B are working, 3400 lenses can be polished in one week. When only B and C are working, 4200 lenses can be polished in one week. How many lenses can be polished in a week by each machine?

19. Three welders, A, B, and C, can weld 37 linear feet per hour when working together. If A and B together can weld 22 linear feet per hour, and A and C together can weld 25 linear feet per hour, how many linear feet per hour can each weld alone?

20. When three pumps, A, B, and C, are running together, they can pump 3700 gallons per hour. When only A and B are running, 2200 gallons per hour can be pumped. When only A and C are running, 2400 gallons per hour can be pumped. What is the pumping capacity of each pump?

iv Solve.

21. (*Curve fitting*). Find numbers a, b, and c such that a quadratic function $ax^2 + bx + c$ fits the data points $(1, 4)$, $(-1, -2)$, and $(2, 13)$.

22. (*Curve fitting*). Find numbers a, b, and c such that a quadratic function $ax^2 + bx + c$ fits the data points $(1, 4)$, $(-1, 6)$, and $(-2, 16)$.

23. (*Predicting earnings*). A business earns $38 in the first week, $66 in the second week, and $86 in the third week. The manager graphs the points $(1, 38)$, $(2, 66)$, and $(3, 86)$ and finds that a quadratic function might fit the data.

a) Find a quadratic function that fits the data.
b) Using your model, predict the earnings for the fourth week.

24. (*Predicting earnings*). A business earns $1000 in its first month, $2000 in the second month, and $8000 in the third month. The manager plots the points $(1, 1000)$, $(2, 2000)$, and $(3, 8000)$ and finds that a quadratic function might fit the data.

a) Find a quadratic function that fits the data.
b) Using your model, predict the earnings for the fourth month.

Hint for Exercises 25 and 26: Let u represent $1/x$, v represent $1/y$, and w represent $1/z$. First solve for u, v, and w.

25. $\dfrac{2}{x} - \dfrac{1}{y} - \dfrac{3}{z} = -1,$

$\dfrac{2}{x} - \dfrac{1}{y} + \dfrac{1}{z} = -9,$

$\dfrac{1}{x} + \dfrac{2}{y} - \dfrac{4}{z} = 17$

26. $\dfrac{2}{x} + \dfrac{2}{y} - \dfrac{3}{z} = 3,$

$\dfrac{1}{x} - \dfrac{2}{y} - \dfrac{3}{z} = 9,$

$\dfrac{7}{x} - \dfrac{2}{y} + \dfrac{9}{z} = -39$

27. When A, B, and C are working together, they can do a job in 2 hr. When B and C work together, they can do the job in 4 hr. When A and B work together, they can do the job in $\frac{12}{5}$ hr. How long would it take each, working alone, to do the job?

28. Pipes A, B, and C are connected to the same tank. When all three pipes are running, they can fill the tank in 3 hr. When pipes A and C are running, they can fill the tank in 4 hr. When pipes A and B are running, they can fill the tank in 8 hr. How long would it take each, running alone, to fill the tank?

29. Find the sum of the angle measures at the tips of the star.

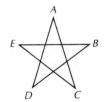

30. Find the year in which the first U.S. transcontinental railroad was completed. The following are some facts about the number. The sum of the digits in the year is 24. The one's digit is one more than the hundred's digit. Both the ten's and one's digits are multiples of three.

31. A theater had 100 people in attendance. The audience consisted of men, women, and children. The ticket prices were $10 for men, $3 for women, and 50¢ for children. The total amount of money taken in was $100. How many men, women, and children were in attendance? Does there seem to be some information missing? Do some more careful reasoning.

32. ▦ Solve:

$$3.12x + 2.14y - 0.988z = 3.79,$$
$$3.84x - 3.53y + 1.96z = 7.80,$$
$$4.63x - 1.08y + 0.011z = 6.34.$$

33. ▦ Solve:

$$1.01x - 0.905y + 2.12z = -2.54,$$
$$1.32x + 2.05y + 2.97z = 3.97,$$
$$2.21x + 1.35y + 0.001z = -3.15.$$

34. Art, Bob, Carl, Denny, and Fred are on the same bowling team. They are all being factual in the following statements regarding the last game they bowled one night.

Art: My score was a prime number. Fred finished third.

Bob: None of us bowled a score over 200.

Carl: Art beat me by exactly 23 pins. Denny's score was divisible by 10.

Denny: The sum of our five scores was exactly 885 pins. Bob's score was divisible by 8.

Fred: Art beat Bob by less than 10 pins. Denny beat Bob by exactly 14 pins.

Determine the score of each bowler in the game.

OBJECTIVE

You should be able to:

i In solving a system of linear equations in two variables, recognize whether the system is inconsistent or dependent, and describe the solutions, if they exist.

Solve.

1. $3x - 4y = 1,$
   $6x - 8y = 7$

Determine whether these systems are consistent or inconsistent.

2. $3x - y = 2,$
   $6x - 2y = 3$

3. $x + 4y = 2,$
   $2x - y = 1$

5.3

SPECIAL CASES

i INCONSISTENT AND DEPENDENT EQUATIONS

Inconsistent Equations

Definition

If a system of equations has a solution, we say that it is *consistent*. If a system does not have a solution, we say that it is *inconsistent*.

All of the systems in Sections 5.1 and 5.2 were consistent because they had solutions.

EXAMPLE 1 Solve:

$$x - 3y = 1,$$
$$-2x + 6y = 5.$$

We multiply the first equation by 2 and add. This gives us

$$x - 3y = 1,$$
$$0 = 7.$$

The second equation says that $0 \cdot x + 0 \cdot y = 7$. There are no numbers x and y for which this is true.

Whenever we obtain a statement such as $0 = 7$, which is obviously false, we will know that the system we are trying to solve has no solutions. It is inconsistent. The solution set is $\emptyset$.

The graphs of the above equations are as follows.

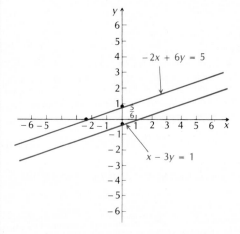

DO EXERCISES 1–3.

EXAMPLE 2 Solve:

$$x + 3y - 7z = 1, \qquad ①$$
$$2x - 4y + z = 3, \qquad ②$$
$$-2x + 4y - z = 10. \qquad ③$$

Each x-coefficient is a multiple of the first. Thus we can begin by multiplying ① by -2 and adding it to ②. We also multiply ① by 2 and add it to ③:

$$x + 3y - 7z = 1,$$
$$-10y + 15z = 1,$$
$$10y - 15z = 12.$$

Since the y-coefficient of ② is the additive inverse of the y-coefficient of ③, we proceed by simply adding ② to ③:

$$x + 3y - 7z = 1,$$
$$-10y + 15z = 1,$$
$$0 = 13.$$

Since in ③ we obtain the false equation $0 = 13$, the system is inconsistent; it has no solution. We need work no further.

DO EXERCISES 4–6.

Dependent Equations

Definition

If a system of n linear equations is equivalent to a system of fewer than n of them, we say that the system is *dependent*. If such is not the case, we call the system *independent*.

Consider the following system:

$$2x + 3y = 1,$$
$$4x + 6y = 2.$$

Note that if we multiply the first equation by 2, we get the second equation. Thus the two equations are equivalent. In other words, the conjunction of the *two* equations is equivalent to *one* of the equations. Thus the system is *dependent*.

Consider the following system:

$$x - 4y + 3z = 5,$$
$$x + y + z = 3,$$
$$x + y + z = 3.$$

Since the last two equations are identical, the system of *three* equations is equivalent to the following system of *two* equations:

$$x - 4y + 3z = 5,$$
$$x + y + z = 3.$$

The system is dependent.

In solving a system, how do we know it is dependent? If, at some stage, we find that two of the equations are identical, then we know that the system is dependent. If we obtain an obviously true statement, such as $0 = 0$, then we know that the system is dependent. We cannot know whether such a system is consistent or inconsistent without further analysis.

A dependent and consistent system may have an infinite number of solutions. In such a case for systems of two variables we can describe the solutions by expressing one variable in terms of the other.

4. Solve.

$$x + 2y - 4z = -3,$$
$$2x + y + z = 9,$$
$$-x - 2y + 4z = 2$$

Determine whether these systems are consistent or inconsistent.

5. $x + 2y + z = 1,$
$$3x + 3y + z = 2,$$
$$2x + y = 2$$

6. $x + z = 1,$
$$y + z = 1,$$
$$x + y = 1$$

Solve. Describe the solutions by expressing one variable in terms of the other. Give three of the ordered pairs in the solution set.

7. $-6x + 4y = 10,$
$\quad 3x - 2y = -5$

8. $2x + 5y = 1,$
$\quad 4x + 10y = 2$

EXAMPLE 3 Solve:

$$2x + 3y = 1,$$
$$4x + 6y = 2.$$

We multiply the first equation by -2 and add. This gives us

$$2x + 3y = 1,$$
$$0 + 0 = 0.$$

The last equation contributes nothing, so we consider only the first one. Let us solve it for x. We obtain

$$x = \frac{1}{2} - \frac{3}{2}y, \quad \text{or} \quad \frac{1 - 3y}{2}.$$

We can now describe the ordered pairs in the solution set, in terms of y only, as follows:

$$\left(\frac{1 - 3y}{2}, y \right).$$

Any value we choose for y then gives us a value for x, and thus an ordered pair in the solution set. Some of these solutions are

$$(-4, 3), \quad (-\tfrac{5}{2}, 2), \quad (\tfrac{7}{2}, -2).$$

DO EXERCISES 7 AND 8.

When a system of three or more equations is dependent and consistent, we can describe its solutions by expressing one or more of the variables in terms of the others.

EXAMPLE 4 Solve:

$$x + 2y + 3z = 4, \qquad \text{①}$$
$$2x - y + z = 3, \qquad \text{②}$$
$$3x + y + 4z = 7. \qquad \text{③}$$

Each x-coefficient is a multiple of the first. Thus we can begin solving by multiplying ① by -2 and adding it to ②. We also multiply ① by -3 and add it to ③:

$$x + 2y + 3z = 4,$$
$$-5y - 5z = -5,$$
$$-5y - 5z = -5.$$

Now ② and ③ are identical. We no longer have a system of three equations, but a system of two. Thus we know that the system is dependent. If we were to multiply ② by -1 and add it to ③, we would obtain $0 = 0$. We proceed by multiplying 2 by $-\frac{1}{5}$ since the coefficients have the common factor -5:

$$x + 2y + 3z = 4,$$
$$y + z = 1.$$

We will express two of the variables in terms of the other one. Let us choose z. Then we solve ② for y:

$$y = 1 - z.$$

We substitute this value of y in $①$, obtaining

$$x + 2(1 - z) + 3z = 4.$$

Solving for x, we get

$$x = 2 - z.$$

The solutions, then, are all of the form

$$(2 - z, 1 - z, z).$$

We obtain the solutions by choosing various values of z. If we let $z = 0$, we obtain the triple $(2, 1, 0)$. If we let $z = 2$, we obtain $(0, -1, 2)$, and so on.

DO EXERCISE 9.

Homogeneous Equations

When all the terms of a polynomial have the same degree, we say that the polynomial is *homogeneous*. Here are some examples:

$$3x^2 + 5y^2, \qquad 4x + 5y - 2z, \qquad 17x^3 - 4y^3 + 57z^3.$$

An equation formed by a homogeneous polynomial set equal to 0 is called a *homogeneous equation*. Let us now consider a system of homogeneous equations.

EXAMPLE 5   Solve:

$$\begin{aligned} 4x - 3y + z &= 0, & \quad ① \\ 2x - 3z &= 0, & \quad ② \\ -8x + 6y - 2z &= 0. & \quad ③ \end{aligned}$$

Any homogeneous system like this always has a solution (can never be inconsistent), because $(0, 0, 0)$ is a solution. This is called the *trivial solution*. There may or may not be other solutions. To find out, we proceed as in the case of nonhomogeneous equations.

First we interchange the first two equations so all the x-coefficients are multiples of the first:

$$\begin{aligned} 2x - 3z &= 0, & \quad ① \\ 4x - 3y + z &= 0, & \quad ② \\ -8x + 6y - 2z &= 0. & \quad ③ \end{aligned}$$

Now we multiply $①$ by -2 and add it to $②$. We also multiply $①$ by 4 and add it to $③$:

$$\begin{aligned} 2x - 3z &= 0, \\ -3y + 7z &= 0, \\ 6y - 14z &= 0. \end{aligned}$$

Next we multiply $②$ by 2 and add it to $③$:

$$\begin{aligned} 2x - 3z &= 0, \\ -3y + 7z &= 0. \\ 0 &= 0. \end{aligned}$$

9. Solve, giving the general form of the solutions. Then list three of the solutions.

$$\begin{aligned} 2x + y + z &= 9, \\ x + 2y - 4z &= -3, \\ x + y - z &= 2 \end{aligned}$$

Solve. If there is more than one solution, list three of them.

10. $x + y - z = 0,$
$x + 2y - 4z = 0,$
$2x + y + z = 0$

11. $x + y - z = 0,$
$x - y - z = 0,$
$x + y + 2z = 0$

Now we know that the system is dependent, hence it has an infinite set of solutions. Now we solve ② for y:

$$y = \tfrac{7}{3}z.$$

Typically, we would now substitute $\tfrac{7}{3}z$ for y in ①, but since the y-term is missing, we need only solve for x:

$$x = \tfrac{3}{2}z.$$

We can now describe the members of the solution set as follows:

$$(\tfrac{3}{2}z, \tfrac{7}{3}z, z).$$

Some of the ordered pairs in the solution set are

$$(\tfrac{3}{2}, \tfrac{7}{3}, 1), \qquad (3, \tfrac{14}{3}, 2), \qquad (-\tfrac{3}{2}, -\tfrac{7}{3}, -1).$$

DO EXERCISES 10 AND 11.

EXERCISE SET 5.3

i Solve. If a system has more than one solution, list three of them.

1. $9x - 3y = 15,$
$6x - 2y = 10$

2. $2s - 3t = 9,$
$4s - 6t = 9$

3. $5c + 2d = 24,$
$30c + 12d = 10$

4. $3x + 2y = 18,$
$9x + 6y = 5$

Solve.

5. $3x + 2y = 5,$
$4y = 10 - 6x$

6. $5x + 2 = 7y,$
$-14y + 4 = -10x$

7. $12y - 8x = 6,$
$4x + 3 = 6y$

8. $16x - 12y = 10,$
$6y + 5 = 8x$

Solve. If a system has more than one solution, list three of them.

9. $x + 2y - z = -8,$
$2x - y + z = 4,$
$8x + y + z = 2$

10. $x + 2y - z = 4,$
$4x - 3y + z = 8,$
$5x - y = 12$

11. $2x + y - 3z = 1,$
$x - 4y + z = 6,$
$4x - 16y + 4z = 24$

12. $4x + 12y + 16z = 4,$
$3x + 4y + 5z = 3,$
$x + 8y + 11z = 1$

13. $2x + y - 3z = 0,$
$x - 4y + z = 0,$
$4x - 16y + 4z = 0$

14. $4x + 12y + 16z = 0,$
$3x + 4y + 5z = 0,$
$x + 8y + 11z = 0$

15. $x + y - z = -3,$
$x + 2y + 2z = -1$

16. $x + y + 13z = 0,$
$x - y - 6z = 0$

17. $2x + y + z = 0,$
$x + y - z = 0,$
$x + 2y + 2z = 0$

18. $5x + 4y + z = 0,$
$10x + 8y - z = 0,$
$x - y - z = 0$

19. Classify each of the systems in the odd-numbered exercises 1–17 as consistent or inconsistent, dependent or independent.

20. Classify each of the systems in the even-numbered exercises 2–18 as consistent or inconsistent, dependent or independent.

Solve.

21. 📱 $4.026x - 1.448y = 18.32,$
$0.724y = -9.16 + 2.013x$

22. 📱 $0.0284y = 1.052 - 8.114x,$
$0.0142y + 4.057x = 0.526$

23. a) Solve:

$$x + y + z + w = 4,$$
$$x + y + z + w = 3,$$
$$x + y + z + w = 3.$$

b) Classify the system as consistent or inconsistent.
c) Classify the system as dependent or independent.

24. a) Solve:

$$-8x + 3y + 2z + w = 0,$$
$$5x + y - z - w = 0,$$
$$2x + 5y - w = 0,$$
$$3x - 4y - z = 0.$$

b) Classify the system as consistent or inconsistent.
c) Classify the system as dependent or independent.

Determine the constant k such that each system is dependent.

25. $6x - 9y = -3,$
$\quad -4x + 6y = k$

26. $8x - 16y = 20,$
$\quad 10x - 20y = k$

5.4

MATRICES AND SYSTEMS OF EQUATIONS

i SOLVING SYSTEMS USING MATRICES

In solving systems of equations, we perform computations with the constants. The variables play no important role in the process. We can simplify writing by omitting the variables. For example, the system

$$3x + 4y = 5,$$
$$x - 2y = 1$$

simplifies to

$$\begin{array}{rrr} 3 & 4 & 5 \\ 1 & -2 & 1 \end{array}$$

if we leave off the variables and omit the operation and equals signs.

In the above example we have written a rectangular array of numbers. Such an array is called a *matrix* (plural, *matrices*). We ordinarily write brackets around matrices, although this is not necessary when calculating with matrices. The following are matrices:

$$\begin{bmatrix} 4 & 1 & 3 & 5 \\ 1 & 0 & 1 & 2 \\ 6 & 3 & -2 & 0 \end{bmatrix}, \quad \begin{bmatrix} 6 & 2 & 1 & 4 & 7 \\ 1 & 2 & 1 & 3 & 1 \\ 4 & 0 & -2 & 0 & -3 \end{bmatrix}, \quad \begin{bmatrix} 1 & 2 \\ 145 & 0 \\ -7 & 9 \\ 8 & 1 \\ 0 & 0 \end{bmatrix}.$$

The *rows* of a matrix are horizontal, and the *columns* are vertical.

$$\begin{bmatrix} 5 & -2 & 2 \\ 1 & 0 & 1 \\ 0 & 1 & 2 \end{bmatrix} \begin{array}{l} \longleftarrow \text{ row 1} \\ \longleftarrow \text{ row 2} \\ \longleftarrow \text{ row 3} \end{array}$$

column 1 column 2 column 3

Let us now use matrices to solve systems of linear equations.

EXAMPLE 1 Solve:

$$2x - y + 4z = -3,$$
$$x \quad\quad - 4z = 5,$$
$$6x - y + 2z = 10.$$

OBJECTIVE

You should be able to:

i Solve systems of linear equations using matrices.

We first write a matrix, using only the constants. Note that where there are missing terms we must write 0's:

$$\begin{bmatrix} 2 & -1 & 4 & -3 \\ 1 & 0 & -4 & 5 \\ 6 & -1 & 2 & 10 \end{bmatrix}.$$

We do exactly the same calculations using the matrix that we would do if we wrote the entire equations. The first step, if possible, is to interchange the rows so that each number in the first column below the first number is a multiple of that number. We do this by interchanging rows 1 and 2:

$$\begin{bmatrix} 1 & 0 & -4 & 5 \\ 2 & -1 & 4 & -3 \\ 6 & -1 & 2 & 10 \end{bmatrix}.$$ This corresponds to interchanging equation ① with equation ②.

Next we multiply the first row by -2 and add it to the second row:

$$\begin{bmatrix} 1 & 0 & -4 & 5 \\ 0 & -1 & 12 & -13 \\ 6 & -1 & 2 & 10 \end{bmatrix}.$$ This corresponds to multiplying equation ① by -2 and adding it to equation ②.

Now we multiply the first row by -6 and add it to the third row:

$$\begin{bmatrix} 1 & 0 & -4 & 5 \\ 0 & -1 & 12 & -13 \\ 0 & -1 & 26 & -20 \end{bmatrix}.$$ This corresponds to multiplying equation ① by -6 and adding it to equation ③.

Next we multiply row 2 by -1 and add it to the third row:

$$\begin{bmatrix} 1 & 0 & -4 & 5 \\ 0 & -1 & 12 & -13 \\ 0 & 0 & 14 & -7 \end{bmatrix}.$$ This corresponds to multiplying equation ② by -1 and adding it to equation ③.

If we now put the variables back, we have

$$x \quad - \quad 4z = 5,$$
$$-y + 12z = -13,$$
$$14z = -7.$$

Now we proceed as before. We solve ③ for z and get $z = -\frac{1}{2}$. Next we substitute $-\frac{1}{2}$ for z in ② and solve for y: $-y + 12(-\frac{1}{2}) = -13$, so $y = 7$. Since there is no y-term in ① we need only substitute $-\frac{1}{2}$ for z in ① and solve for x: $x - 4(-\frac{1}{2}) = 5$, so $x = 3$. The solution is $(3, 7, -\frac{1}{2})$.

Note in the preceding that our goal was to get the matrix in the form

$$\begin{bmatrix} a & b & c & d \\ 0 & e & f & g \\ 0 & 0 & h & k \end{bmatrix},$$

where there are just 0's below the *main diagonal*, formed by a, e, and h. Then we put the variables back and complete the solution.

All the operations used in the preceding example correspond to operations with the equations and they produce equivalent systems of equations. We call the matrices *row-equivalent*, and the operations that produce them *row-equivalent operations*.

Theorem 1

Each of the following row-equivalent operations produces equivalent matrices:

a) **Interchanging any two rows of a matrix.**

b) **Multiplying each element of a row by the same nonzero number.**

c) **Multiplying each element of a row by a nonzero number and adding the result to another row.**

The best overall method for solving systems of equations is by row-equivalent matrices; even computers are programmed to use them.

DO EXERCISES 1 AND 2.

Solve, using matrices.

1. $5x - 2y = -3,$
$2x + 5y = -24$

2. $x - 2y + 3z = 4,$
$2x - y + z = -1,$
$4x + y + z = 1$

EXERCISE SET 5.4

i Solve, using matrices.

1. $4x + 2y = 11,$
$3x - y = 2$

2. $3x - 3y = 11,$
$9x - 2y = 5$

3. $x + 2y - 3z = 9,$
$2x - y + 2z = -8,$
$3x - y - 4z = 3$

4. $x - y + 2z = 0,$
$x - 2y + 3z = -1,$
$2x - 2y + z = -3$

5. $5x - 3y = -2,$
$4x + 2y = 5$

6. $3x + 4y = 7,$
$-5x + 2y = 10$

7. $4x - y - 3z = 1,$
$8x + y - z = 5,$
$2x + y + 2z = 5$

8. $3x + 2y + 2z = 3,$
$x + 2y - z = 5,$
$2x - 4y + z = 0$

9. $p + q + r = 1,$
$p + 2q + 3r = 4,$
$4p + 5q + 6r = 7$

10. $m + n + t = 9,$
$m - n - t = -15,$
$m + n + t = 3$

11. $2x + 2y - 2z - 2w = -10,$
$x + y + z + w = -5,$
$x - y + 4z + 3w = -2,$
$3x - 2y + 2z + w = -6$

12. $2x - 3y + z - w = -8,$
$x + y - z - w = -4,$
$x + y + z + w = 22,$
$x - y - z - w = -14$

---

13. A collection of 34 coins consists of dimes and nickels. The total value is $1.90. How many dimes and how many nickels are there?

14. A collection of 43 coins consists of dimes and quarters. The total value is $7.60. How many dimes and how many quarters are there?

15. A collection of 22 coins consists of nickels, dimes, and quarters. The total value is $2.90. There are 6 more nickels than dimes. How many of each type of coin are there?

16. A collection of 18 coins consists of nickels, dimes, and quarters. The total value is $2.55. There are 2 more quarters than there are dimes. How many of each type of coin are there?

17. A tobacco dealer has two kinds of tobacco. One is worth $4.05 per lb and the other is worth $2.70 per lb. The dealer wants to blend the two tobaccos to get a 15-lb mixture worth $3.15 per lb. How much of each kind of tobacco should be used?

18. A grocer mixes candy worth $0.80 per lb with nuts worth $0.70 per lb to get a 20-lb mixture worth $0.77 per lb. How many pounds of candy and how many pounds of nuts are used?

Recall the formula $I = prt$, for simple interest.

19. One year $8950 was received in interest from two investments. A certain amount was invested at $12\frac{1}{2}\%$ and $10,000 more than this at 13%. Find the amount of principal invested at each rate.

20. One year a person had some money invested at $13\frac{1}{2}\%$ and another amount at $13\frac{3}{4}\%$. The income from the investments was $3275. The income from the $13\frac{1}{2}\%$ investment was $575 less than that from the $13\frac{3}{4}\%$ investment. How much was invested at each rate?

Solve.

21. ▦ $4.83x + 9.06y = -39.42,$
$-1.35x + 6.67y = -33.99$

22. ▦ $3.11x - 2.04y = -24.39,$
$7.73x + 5.19y = -35.48$

23. ▦ $3.55x - 1.35y + 1.03z = 9.16,$
$-2.14x + 4.12y + 3.61z = -4.50,$
$5.48x - 2.44y - 5.86z = 0.813$

24. ▦ $4.12x - 1.35y - 18.2z = 601.3,$
$-3.41x + 68.9y + 38.7z = 1777,$
$0.955x - 0.813y - 6.53z = 160.2$

25. $\sqrt{2}x + \pi y = 3,$
$\pi x - \sqrt{2}y = 1$

26. $ax + by = c,$
$dx + ey = f$

OBJECTIVES

You should be able to:

i Evaluate determinants of 2×2 matrices.

ii Solve systems of two equations in two variables using Cramer's rule.

iii Evaluate determinants of 3×3 matrices.

iv Solve systems of three equations in three variables using Cramer's rule.

Evaluate.

1. $\begin{vmatrix} \sqrt{3} & -5 \\ -2 & -\sqrt{3} \end{vmatrix}$

2. $\begin{vmatrix} 1 & 2 \\ 3 & 4 \end{vmatrix}$

3. $\begin{vmatrix} -2 & -3 \\ 4 & x \end{vmatrix}$

5.5

DETERMINANTS AND CRAMER'S RULE

i DETERMINANTS OF TWO-BY-TWO MATRICES

A matrix of m rows and n columns is called an $m \times n$ matrix (read "m by n"). If a matrix has the same number of rows and columns, it is called a *square matrix*. With every square matrix is associated a number called its *determinant*, defined as follows for 2×2 matrices.*

Definition

The determinant of the matrix $\begin{bmatrix} a & c \\ b & d \end{bmatrix}$ is denoted $\begin{vmatrix} a & c \\ b & d \end{vmatrix}$ and is defined as follows:

$$\begin{vmatrix} a & c \\ b & d \end{vmatrix} = ad - bc.$$

EXAMPLE 1 Evaluate $\begin{vmatrix} \sqrt{2} & -3 \\ -4 & -\sqrt{2} \end{vmatrix}$.

$\begin{vmatrix} \sqrt{2} & -3 \\ -4 & -\sqrt{2} \end{vmatrix}$ The arrows indicate the products involved.

$= \sqrt{2}(-\sqrt{2}) - (-4)(-3) = -2 - 12 = -14$

DO EXERCISES 1–3.

ii CRAMER'S RULE

Determinants have many uses. One of these is in solving systems of non-homogeneous linear equations, where the number of variables is the same as the number of equations. Let us consider a system of two equations:

$$a_1 x + b_1 y = c_1,$$
$$a_2 x + b_2 y = c_2.$$

---

* The definition of the determinant of any square matrix is given in Chapter 6.

Using the methods of the preceding sections we can solve. We obtain

$$x = \frac{c_1 b_2 - c_2 b_1}{a_1 b_2 - a_2 b_1}, \qquad y = \frac{a_1 c_2 - a_2 c_1}{a_1 b_2 - a_2 b_1}.$$

The numerators and denominators of the expressions for x and y are determinants:

$$x = \frac{\begin{vmatrix} c_1 & b_1 \\ c_2 & b_2 \end{vmatrix}}{\begin{vmatrix} a_1 & b_1 \\ a_2 & b_2 \end{vmatrix}}, \qquad y = \frac{\begin{vmatrix} a_1 & c_1 \\ a_2 & c_2 \end{vmatrix}}{\begin{vmatrix} a_1 & b_1 \\ a_2 & b_2 \end{vmatrix}}.$$

The above equations make sense only if the denominator determinant is not 0. If the denominator *is* 0, then one of two things happens.

1. If the denominator is 0 and the other two determinants are also 0, then the system of equations is dependent.

2. If the denominator is 0 and at least one of the other determinants is not 0, then the system is inconsistent.

The equations with determinants above describe *Cramer's rule* for solving systems of equations. To use this rule, we compute the three determinants and compute x and y as shown above. Note that the denominator in both cases contains the coefficients of x and y, in the same position as in the original equations. For x the numerator is obtained by replacing the x-coefficients (the a's) by the c's. For y the numerator is obtained by replacing the y-coefficients (the b's) by the c's.

EXAMPLE 2 Solve, using Cramer's rule:

$$2x + 5y = 7,$$
$$5x - 2y = -3.$$

We have

$$x = \frac{\begin{vmatrix} 7 & 5 \\ -3 & -2 \end{vmatrix}}{\begin{vmatrix} 2 & 5 \\ 5 & -2 \end{vmatrix}}$$

$$= \frac{7(-2) - (-3)5}{2(-2) - 5 \cdot 5} = -\frac{1}{29},$$

$$y = \frac{\begin{vmatrix} 2 & 7 \\ 5 & -3 \end{vmatrix}}{\begin{vmatrix} 2 & 5 \\ 5 & -2 \end{vmatrix}}$$

$$= \frac{2(-3) - 5 \cdot 7}{-29} = \frac{41}{29}.$$

The solution is $\left(-\frac{1}{29}, \frac{41}{29}\right)$.

DO EXERCISES 4–6.

Solve, using Cramer's rule.

4. $2x - y = 5,$
   $x - 2y = 1$

5. $3x + 4y = -2,$
   $5x - 7y = 1$

6. $\sqrt{2}x - \pi y = 3,$
   $\pi x + \sqrt{2}y = 4$

Evaluate.

7. $\begin{vmatrix} 3 & 2 & 2 \\ -2 & 1 & 4 \\ 4 & -3 & 3 \end{vmatrix}$

8. $\begin{vmatrix} -5 & 0 & 0 \\ 4 & 2 & 0 \\ -3 & 5 & -6 \end{vmatrix}$

9. $\begin{vmatrix} x & 0 & x \\ 0 & x & 0 \\ 1 & 0 & x \end{vmatrix}$

iii Determinants of Three-by-Three Matrices

Definition

> The *determinant* of a three-by-three matrix is defined as follows:
>
> $$\begin{vmatrix} a_1 & b_1 & c_1 \\ a_2 & b_2 & c_2 \\ a_3 & b_3 & c_3 \end{vmatrix} = a_1 \cdot \begin{vmatrix} b_2 & c_2 \\ b_3 & c_3 \end{vmatrix} - a_2 \cdot \begin{vmatrix} b_1 & c_1 \\ b_3 & c_3 \end{vmatrix} + a_3 \cdot \begin{vmatrix} b_1 & c_1 \\ b_2 & c_2 \end{vmatrix}.$$

The two-by-two determinants on the right can be obtained from the three-by-three matrix by crossing out the row and column in which the a-coefficient occurs.

EXAMPLE 3 Evaluate.

$$\begin{vmatrix} -1 & 0 & 1 \\ -5 & 1 & -1 \\ 4 & 8 & 1 \end{vmatrix} = -1 \cdot \begin{vmatrix} 1 & -1 \\ 8 & 1 \end{vmatrix} - (-5) \cdot \begin{vmatrix} 0 & 1 \\ 8 & 1 \end{vmatrix} + 4 \cdot \begin{vmatrix} 0 & 1 \\ 1 & -1 \end{vmatrix}$$

$$= -1(1 + 8) + 5(-8) + 4(-1)$$

$$= -9 - 40 - 4 = -53$$

DO EXERCISES 7–9.

Let us consider three equations in three variables. Consider

$$a_1 x + b_1 y + c_1 z = d_1,$$
$$a_2 x + b_2 y + c_2 z = d_2,$$
$$a_3 x + b_3 y + c_3 z = d_3,$$

and the following determinants:

$$D = \begin{vmatrix} a_1 & b_1 & c_1 \\ a_2 & b_2 & c_2 \\ a_3 & b_3 & c_3 \end{vmatrix}, \qquad D_x = \begin{vmatrix} d_1 & b_1 & c_1 \\ d_2 & b_2 & c_2 \\ d_3 & b_3 & c_3 \end{vmatrix}.$$

$$D_y = \begin{vmatrix} a_1 & d_1 & c_1 \\ a_2 & d_2 & c_2 \\ a_3 & d_3 & c_3 \end{vmatrix}, \qquad D_z = \begin{vmatrix} a_1 & b_1 & d_1 \\ a_2 & b_2 & d_2 \\ a_3 & b_3 & d_3 \end{vmatrix}.$$

If we solve the system of equations, we obtain the following:

$$x = \frac{D_x}{D}, \qquad y = \frac{D_y}{D}, \qquad z = \frac{D_z}{D}.$$

Note that we obtain the determinant D_x in the numerator for x from D by replacing the x-coefficients by d_1, d_2, and d_3. A similar thing happens with D_y and D_z. We have thus extended *Cramer's rule* to solve systems of three equations in three variables. As before, when $D = 0$, Cramer's rule cannot be used. If $D = 0$, and D_x, D_y, and D_z are 0, the system is dependent. If $D = 0$ and one of D_x, D_y, or D_z is not zero, then the system is inconsistent.

iv CRAMER'S RULE

EXAMPLE 4 Solve, using Cramer's rule:

$$x - 3y + 7z = 13,$$
$$x + y + z = 1,$$
$$x - 2y + 3z = 4.$$

We have

$$D = \begin{vmatrix} 1 & -3 & 7 \\ 1 & 1 & 1 \\ 1 & -2 & 3 \end{vmatrix} = -10, \quad D_x = \begin{vmatrix} 13 & -3 & 7 \\ 1 & 1 & 1 \\ 4 & -2 & 3 \end{vmatrix} = 20,$$

$$D_y = \begin{vmatrix} 1 & 13 & 7 \\ 1 & 1 & 1 \\ 1 & 4 & 3 \end{vmatrix} = -6, \quad D_z = \begin{vmatrix} 1 & -3 & 13 \\ 1 & 1 & 1 \\ 1 & -2 & 4 \end{vmatrix} = -24.$$

Then

$$x = \frac{D_x}{D} = \frac{20}{-10} = -2,$$

$$y = \frac{D_y}{D} = \frac{-6}{-10} = \frac{3}{5},$$

$$z = \frac{D_z}{D} = \frac{-24}{-10} = \frac{12}{5}.$$

The solution is $(-2, \frac{3}{5}, \frac{12}{5})$. In practice, it is not necessary to evaluate D_z. When we have found values for x and y we can substitute them into one of the equations and find z.

DO EXERCISE 10.

10. Solve, using Cramer's rule.

$$x - 3y - 7z = 6,$$
$$2x + 3y + z = 9,$$
$$4x + y = 7$$

EXERCISE SET 5.5

i Evaluate.

1. $\begin{vmatrix} -2 & -\sqrt{5} \\ -\sqrt{5} & 3 \end{vmatrix}$ **2.** $\begin{vmatrix} \sqrt{5} & -3 \\ 4 & 2 \end{vmatrix}$ **3.** $\begin{vmatrix} x & 4 \\ x & x^2 \end{vmatrix}$ **4.** $\begin{vmatrix} y^2 & -2 \\ y & 3 \end{vmatrix}$

iii Evaluate.

5. $\begin{vmatrix} 3 & 1 & 2 \\ -2 & 3 & 1 \\ 3 & 4 & -6 \end{vmatrix}$ **6.** $\begin{vmatrix} 3 & -2 & 1 \\ 2 & 4 & 3 \\ -1 & 5 & 1 \end{vmatrix}$ **7.** $\begin{vmatrix} x & 0 & -1 \\ 2 & x & x^2 \\ -3 & x & 1 \end{vmatrix}$ **8.** $\begin{vmatrix} x & 1 & -1 \\ x^2 & x & x \\ 0 & x & 1 \end{vmatrix}$

ii Solve, using Cramer's rule.

9. $-2x + 4y = 3,$ **10.** $5x - 4y = -3,$ **11.** $\sqrt{3}x + \pi y = -5,$ **12.** $\pi x - \sqrt{5}y = 2,$
$\quad\ 3x - 7y = 1$ $\quad\ 7x + 2y = 6$ $\quad\ \pi x - \sqrt{3}y = 4$ $\quad\ \sqrt{5}x + \pi y = -3$

iv Solve, using Cramer's rule.

13. $3x + 2y - z = 4,$ **14.** $3x - y + 2z = 1,$ **15.** $6y + 6z = -1,$ **16.** $3x + 5y = 2,$
$\quad\ 3x - 2y + z = 5,$ $\quad\ x - y + 2z = 3,$ $\quad\ 8x + 6z = -1,$ $\quad\ 2x - 3z = 7,$
$\quad\ 4x - 5y - z = -1$ $\quad\ -2x + 3y + z = 1$ $\quad\ 4x + 9y = 8$ $\quad\ 4y + 2z = -1$

Solve.

17. $\begin{vmatrix} x & 5 \\ -4 & x \end{vmatrix} = 24$

18. $\begin{vmatrix} y & 2 \\ 3 & y \end{vmatrix} = y$

19. $\begin{vmatrix} x & -3 \\ -1 & x \end{vmatrix} \geqslant 0$

20. $\begin{vmatrix} y & -5 \\ -2 & y \end{vmatrix} < 0$

21. $\begin{vmatrix} x+3 & 4 \\ x-3 & 5 \end{vmatrix} = -7$

22. $\begin{vmatrix} m+2 & -3 \\ m+5 & -4 \end{vmatrix} = 3m - 5$

23. $\begin{vmatrix} 2 & x & 1 \\ 1 & 2 & -1 \\ 3 & 4 & -2 \end{vmatrix} = -6$

24. $\begin{vmatrix} x & 2 & x \\ 3 & -1 & 1 \\ 1 & -2 & 2 \end{vmatrix} = -10$

Rewrite each expression using determinants. Answers may vary.

25. $2L + 2W$

26. $\pi r + \pi h$

27. $a^2 + b^2$

28. $\frac{1}{2}h(a+b)$

29. $2\pi r^2 + 2\pi rh$

30. $x^2y^2 - Q^2$

OBJECTIVES

You should be able to:

i Graph linear inequalities in two variables in the plane.

ii Graph systems of linear inequalities and find vertices, if they exist.

1. Determine whether $(1, -4)$ is a solution of the inequality $4x - 5y < 12$.

5.6

SYSTEMS OF INEQUALITIES

i **GRAPHS OF INEQUALITIES IN TWO VARIABLES**

A *solution* of an inequality in two variables is an ordered pair of numbers that makes the inequality true when the coordinates are substituted in alphabetical order.

EXAMPLE 1 Determine whether $(-3, 2)$ is a solution of the inequality $5x - 4y \leqslant 13$.

We replace x by -3 and y by 2:

$$\begin{array}{c|c} 5x - 4y \leqslant 13 & \\ \hline 5(-3) - 4 \cdot 2 & 13 \\ -15 - 8 & \\ -23 & \end{array}$$

Since -23 is less than 13, this replacement makes the inequality true.

DO EXERCISE 1.

A *linear inequality* is an inequality that is equivalent to

$$ax + by < c \quad \text{or} \quad ax + by > c,$$

that is, there is a linear polynomial on one side and c on the other. The inequality signs can, of course, be $\leqslant$ or $\geqslant$. To graph linear inequalities, we shall use what we already know about graphing linear equations.

EXAMPLE 2 Graph $y < x + 2$.

For comparison, we first graph the line $y = x + 2$. We draw it dashed. Every solution of $y = x + 2$ is an ordered pair having a second coordinate that is 2 more than the first.

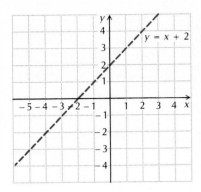

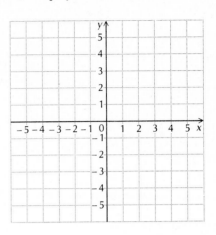

2. Graph $y > -2x$.

Now look at a vertical line and some ordered pairs on it. The one shown goes through $(-3, -1)$.

For any point here, the y-value is greater than two more than the x-value.

For any point here, the y-value is less than two more than the x-value.

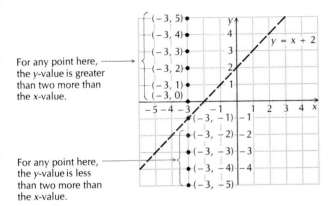

For all points below $y = x + 2$, the second coordinate is less than 2 more than the first. That is, $y < x + 2$. For all points above the line, $y > x + 2$. The same thing happens for any vertical line. Thus all points below the line $y = x + 2$ are solutions. We shade the half-plane below $y = x + 2$. The points on the line $y = x + 2$ are not included. That is why we draw the line dashed.

For any point here, $y > x + 2$.

For any point here, $y = x + 2$.

$y < x + 2$

For any point here, $y < x + 2$.

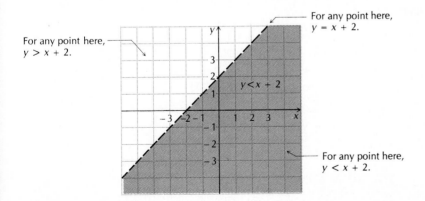

DO EXERCISE 2.

3. Graph $2x + y \geqslant 2$.

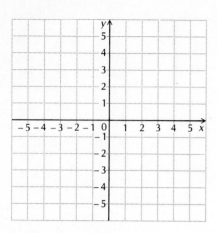

4. Graph $3x - y > -3$.

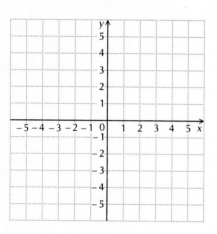

Graph.

5. $x \leqslant -1$

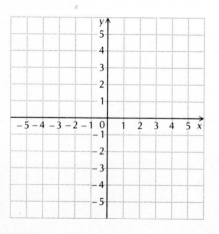

To graph a linear inequality, we first replace the inequality sign with an equals sign. Then we graph the resulting line. If the inequality symbol is $<$ or $>$, we draw the line dashed. If the inequality symbol is $\leqslant$ or $\geqslant$, we draw the line solid. The graph consists of a half-plane as in Example 2 and, if the line is solid, the line as well.

EXAMPLE 3 Graph $3y - 2x \geqslant 1$.

a) We first graph the line $3y - 2x = 1$ by any method. Since the inequality symbol is $\geqslant$, we draw the line solid. We find the intercepts. They are $(0, \frac{1}{3})$ and $(-\frac{1}{2}, 0)$. We plot them and then use them to draw the line. The points $(4, 3)$ and $(-2, -1)$ are also on the graph.

b) We determine which half-plane to shade by trying some point off the line. The point $(0, 0)$ is easy to check unless the line goes through it. Any point not on the line can be used as a check.

$$\begin{array}{c|c} 3y - 2x > 1 \\ \hline 3(0) - 2(0) & 1 \\ 0 & \end{array}$$

Since $0 \geqslant 1$ is false, the point is not on the graph. We shade the other half-plane. The graph consists of the shaded half-plane and the solid line as well.

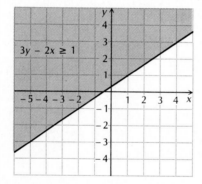

DO EXERCISES 3 AND 4.

EXAMPLE 4 Graph $x \leqslant 3$.

We first graph the equation $x = 3$, drawing it solid. Then we shade the appropriate half-plane.

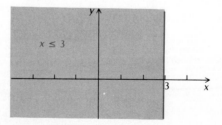

EXAMPLE 5 Graph $-1 < y \leqslant 2$.

This is a conjunction of two inequalities:

$$-1 < y \quad and \quad y \leqslant 2.$$

It will be true for any y that is both greater than -1 and less than or equal to 2. The graph is as shown here.

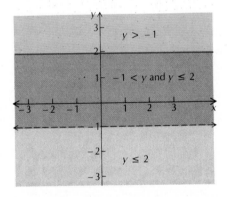

Since our inequality is a conjunction, the graph is the intersection of the graphs of the parts.

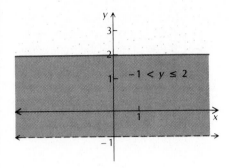

DO EXERCISES 5–8. (NOTE THAT EXERCISE 5 IS ON THE PRECEDING PAGE.)

ii SYSTEMS OF INEQUALITIES

To get a picture of the solution set of a system or *conjunction* of inequalities, we will graph the inequalities separately and find their intersection. A system of linear inequalities may have a graph that is a polygon and its interior. In certain applications it is important to find the vertices.

EXAMPLE 6 Graph this system of inequalities. Find the coordinates of any vertices formed.

$$2x + y \geqslant 2,$$
$$4x + 3y \leqslant 12,$$
$$\tfrac{1}{2} \leqslant x \leqslant 2,$$
$$y \geqslant 0$$

6. $y > -\tfrac{1}{2}$

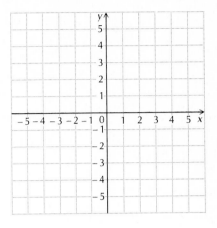

7. $-4 \leqslant x < 1$

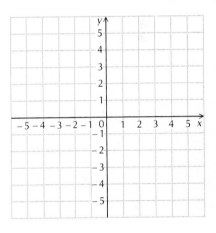

8. Graph $1 \leqslant y \leqslant 2\tfrac{1}{2}$.

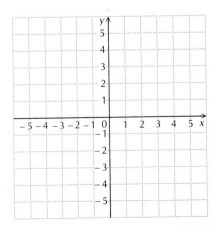

9. Graph. If a polygon is formed, find the vertices.

$$x + y \geq 1,$$
$$y - x \geq 2$$

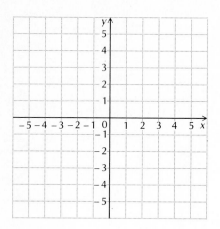

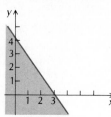

$2x + y \geq 2$

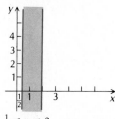

$4x + 3y \leq 12$

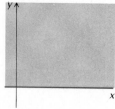

$\frac{1}{2} \leq x \leq 2$

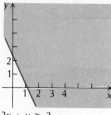

$y \geq 0$

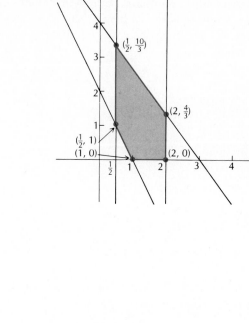

10. Graph. If a polygon is formed, find the vertices.

$$5x + 6y \leq 30,$$
$$0 \leq y \leq 3,$$
$$0 \leq x \leq 4$$

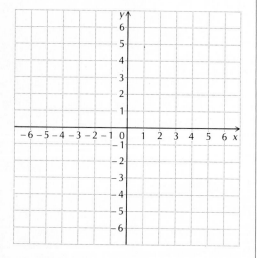

The separate graphs are shown on the left and the graph of the intersection, which is the graph of the system, is shown on the right.

We find the vertex $(\frac{1}{2}, 1)$ by solving the system

$$2x + y = 2,$$
$$x = \frac{1}{2}.$$

We find the vertex $(1, 0)$ by solving the system

$$2x + y = 2,$$
$$y = 0.$$

We find the vertex $(2, 0)$ from the system

$$x = 2$$
$$y = 0.$$

The vertices $(2, \frac{4}{3})$ and $(\frac{1}{2}, \frac{10}{3})$ were found by solving, respectively, the systems

$$x = 2, \qquad \qquad x = \frac{1}{2},$$
$$\text{and}$$
$$4x + 3y = 12; \qquad 4x + 3y = 12.$$

DO EXERCISES 9 AND 10.

EXERCISE SET 5.6

i Graph.

1. $y < x$	**2.** $y \geq x$	**3.** $y + x \geq 0$	**4.** $y + x < 0$
5. $3x - 2y < 6$	**6.** $2x - 5y > 10$	**7.** $2x + 3y \geq 6$	**8.** $x + 2y \leq 4$
9. $3x - 2 \leq 5x + y$	**10.** $2x - 6y \geq 8 + 4y$	**11.** $x < -4$	**12.** $y \geq 5$
13. $0 \leq x < 5\frac{1}{2}$	**14.** $-4 < y < -1$	**15.** $y \geq |x|$	**16.** $y < |x|$

ii Graph these systems. Find the coordinates of any vertices formed.

17. $x + y \leq 1,$
$x - y \leq 2$

18. $x + y \leq 3,$
$x - y \leq 4$

19. $y - 2x > 1,$
$y - 2x < 3$

20. $y + x > 0,$
$y + x < 2$

21. $2y - x \leq 2,$
$y - 3x \geq -1$

22. $x + 3y \geq 9,$
$3x - 2y \leq 5$

23. $y \leq 2x + 1,$
$y \geq -2x + 1,$
$x \leq 2$

24. $x - y \leq 2,$
$x + 2y \geq 8,$
$y \leq 4$

25. $x + 2y \leq 12,$
$2x + y \leq 12,$
$x \geq 0,$
$y \geq 0$

26. $8x + 5y \leq 40,$
$x + 2y \leq 8,$
$x \geq 0,$
$y \geq 0$

27. $3x + 4y \geq 12,$
$5x + 6y \leq 30,$
$1 \leq x \leq 3$

28. $y - 2x \geq 3,$
$y - 2x \leq 5,$
$6 \leq y \leq 8$

Graph.

29. $y \geq x^2 - 2,$
$y \leq 2 - x^2$

30. $x \geq 2y^2 - 1,$
$x < y^2$

31. $y < x + 1,$
$y \geq x^2$

32. $y \leq -x^2 + 5,$
$y > \frac{1}{2}x^2 - 1$

Graph.

33. $|x| + |y| \leq 1$

34. $|x + y| \leq 1$

35. $|x - y| > 0$

36. $|x| > |y|$

5.7

LINEAR PROGRAMMING

i Suppose you are taking a test in which different items are worth different numbers of points. Presumably those with higher point values take more time. How many items of each kind should you do, in order to make the best score? A kind of mathematics called *linear programming* (developed during World War II) may provide an answer. Let us consider an example.

EXAMPLE 1 You are taking a test in which items of type A are worth 10 points and items of type B are worth 15 points. It takes 3 minutes for each item of type A and 6 minutes for each item of type B. The total time allowed is 60 minutes and you are not allowed to answer more than 16 questions. Assuming that all of your answers are correct, how many items of each type should you answer to get the best score?

Let $x =$ the number of items of type A, and $y =$ the number of items of type B. The total score T is a function of the two variables x and y:

$$T = 10x + 15y.$$

This function has a domain that is a set of ordered pairs of numbers (x, y).

OBJECTIVE

You should be able to:

i Solve linear programming problems.

This domain is determined by the following conditions, called *constraints*:

Total number of questions allowed, not more than 16	$x + y \leqslant 16,$
Time, not more than 60 min	$3x + 6y \leqslant 60,$
Numbers of items answered will not be negative	$x \geqslant 0,$
	$y \geqslant 0.$

We now graph the domain of the function T. This is the graph of the system of inequalities above. We will determine the vertices, if any are formed.

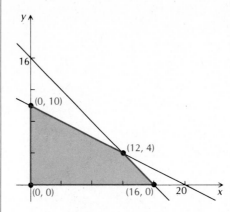

The domain consists of a *convex polygon** and its interior. Under this condition, our linear function does have a maximum value and a minimum value. Moreover, the maximum and minimum values occur at the vertices of the polygon. All we need do to find these values is to substitute the coordinates of the vertices in $T = 10x + 15y$.

Vertices (x, y)	Score $T = 10x + 15y$
(0, 0)	0
(16, 0)	160
(12, 4)	180
(0, 10)	150

From the table we see that the minimum value is 0 and the maximum is 180. To get this maximum you would answer 12 items of type A and 4 items of type B.

We now state the main theorem needed to perform linear programming.

Theorem 2

If a linear function $F = ax + by + c$ is defined on a domain described by a system of linear inequalities (constraints), then any maximum or minimum value of F will occur at a vertex. If the domain consists of a convex polygon and its interior, then both maximum and minimum values of F actually exist.

---

* A convex polygon, roughly speaking, is one that has no indentations, unlike this one:

EXAMPLE 2 Find the maximum and minimum values of $F = 9x + 40y$ subject to the constraints

$$y - x \geqslant 1,$$
$$y - x \leqslant 3,$$
$$2 \leqslant x \leqslant 5.$$

We graph the system of inequalities, determine the vertices, and find the function values for those ordered pairs.

Vertices (x, y)	Function $F = 9x + 40y$
(2, 3)	138
(2, 5)	218
(5, 6)	285
(5, 8)	365

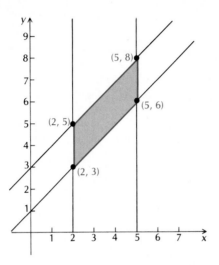

The maximum value of F is 365 when $x = 5$ and $y = 8$. The minimum value of F is 138 when $x = 2$ and $y = 3$.

DO EXERCISES 1 AND 2.

EXAMPLE 3 A company manufactures motorcycles and bicycles. To stay in business it must produce at least 10 motorcycles each month, but it does not have facilities to produce more than 60 motorcycles. It also does not have facilities to produce more than 120 bicycles. The total production of motorcycles and bicycles cannot exceed 160. The profit on a motorcycle is $134 and on a bicycle is $20. Find the number of each that should be manufactured to maximize profit.

Let $x =$ the number of motorcycles to be produced, and $y =$ the number of bicycles to be produced. The profit P is given by

$$P = \$134x + \$20y,$$

subject to the constraints

$$10 \leqslant x \leqslant 60,$$
$$0 \leqslant y \leqslant 120,$$
$$x + y \leqslant 160.$$

We graph the system of inequalities, determine the vertices, and find the function values for those ordered pairs.

1. Find the maximum and minimum values of

$$F = 34x + 6y$$

subject to

$$x + y \leqslant 6,$$
$$x + y \geqslant 1,$$
$$1 \leqslant x \leqslant 3.$$

2. Find the maximum and minimum values of

$$G = 3x - 5y + 27$$

subject to

$$x + 2y \leqslant 8,$$
$$0 \leqslant y \leqslant 3,$$
$$0 \leqslant x \leqslant 6.$$

3. A college snack bar cooks and sells hamburgers and hot dogs during the lunch hour. To stay in business it must sell at least 10 hamburgers but cannot cook more than 40. It must also sell at least 30 hot dogs but cannot cook more than 70. It cannot cook more than 90 sandwiches altogether. The profit on a hamburger is $0.33 and on a hot dog it is $0.21. How many of each kind of sandwich should they sell to make the maximum profit? What is the maximum profit?

Vertices (x, y)	Profit P = \$134x + \$20y
(10, 0)	$ 1,340
(60, 0)	$ 8,040
(60, 100)	$10,040
(40, 120)	$ 7,760
(10,120)	$ 3,740

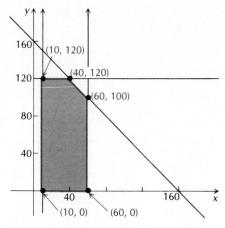

Thus the company will make a maximum profit of $10,040 by producing 60 motorcycles and 100 bicycles.

DO EXERCISE 3.

EXERCISE SET 5.7

[i] In Exercises 1–4, maximize and also minimize (find maximum and minimum values of the function, and the values of x and y where they occur).

1. $F = 4x + 28y$,
subject to
$5x + 3y \leqslant 34$,
$3x + 5y \leqslant 30$,
$x \geqslant 0$,
$y \geqslant 0$.

2. $G = 14x + 16y$,
subject to
$3x + 2y \leqslant 12$,
$7x + 5y \leqslant 29$,
$x \geqslant 0$,
$y \geqslant 0$.

3. $P = 16x - 2y + 40$,
subject to
$6x + 8y \leqslant 48$,
$0 \leqslant y \leqslant 4$,
$0 \leqslant x \leqslant 7$.

4. $Q = 24x - 3y + 52$,
subject to
$5x + 4y \leqslant 20$,
$0 \leqslant y \leqslant 4$,
$0 \leqslant x \leqslant 3$.

Solve.

5. You are about to take a test that contains questions of type A worth 4 points and questions of type B worth 7 points. You must do at least 5 questions of type A but time restricts doing more than 10. You must do at least 3 questions of type B but time restricts doing more than 10. In total, you can do no more than 18 questions. How many of each type of question must you do to maximize your score? What is this maximum score?

6. You are about to take a test that contains questions of type A worth 10 points and questions of type B worth 25 points. You must do at least 3 questions of type A but time restricts doing more than 12. You must do at least 4 questions of type B but time restricts doing more than 15. In total you can do no more than 20 questions. How many of each type of question must you do to maximize your score? What is this maximum score?

7. A man is planning to invest up to $22,000 in bank X or bank Y or both. He wants to invest at least $2000 but no more than $14,000 in bank X. Bank Y does not insure more than a $15,000 investment so he will invest no more than that in bank Y. The interest in bank X is 6% and in bank Y it is $6\frac{1}{2}\%$ and this will be simple interest for one year. How much should he invest in each bank to maximize his income? What is the maximum income?

8. A woman is planning to invest up to $40,000 in corporate or municipal bonds or both. The least she is allowed to invest in corporate bonds is $6000 and she does not want to invest more than $22,000 in corporate bonds. She also does not want to invest more than $30,000 in municipal bonds. The interest on corporate bonds is 8% and on municipal bonds it is $7\frac{1}{2}\%$. This is simple interest for one year. How much should she invest in each type of bond to maximize her income? What is the maximum income?

9. It takes a tailoring firm 2 hr of cutting and 4 hr of sewing to make a knit suit. To make a worsted suit it takes 4 hr of cutting and 2 hr of sewing. At most, 20 hr per day are available for cutting and, at most, 16 hr per day are available for sewing. The profit on a knit suit is $34 and on a worsted suit is $31. How many of each kind of suit should be made to maximize profit? What is the maximum profit?

10. A pipe tobacco company has 3000 lb of English tobacco, 2000 lb of Virginia tobacco, and 500 lb of Latakia tobacco. To make one batch of SMELLO tobacco it takes 12 lb of English tobacco and 4 lb of Latakia. To make one batch of ROPPO tobacco it takes 8 lb of English and 8 lb of Virginia tobacco. The profit is $10.56 per batch for SMELLO and $6.40 for ROPPO. How many batches of each kind of tobacco should be made to yield maximum profit? What is the maximum profit? (*Hint:* Organize the information in a table.)

11. An airline with two types of airplanes, P-1 and P-2, has contracted with a tour group to provide accommodations for a minimum of each of 2000 first-class, 1500 tourist, and 2400 economy-class passengers. Airplane P-1 costs $12,000 per mile to operate and can accommodate 40 first-class, 40 tourist, and 120 economy-class passengers, whereas airplane P-2 costs $10,000 per mile to operate and can accommodate 80 first-class, 30 tourist, and 40 economy-class passengers. How many of each type of airplane should be used to minimize the operating cost?

12. A new airplane P-3 becomes available, having an operating cost of $15,000 per mile and accommodating 40 first-class, 80 tourist, and 80 economy-class passengers. If airplane P-1 of Exercise 11 were replaced by airplane P-3, how many of P-2 and P-3 would be needed to minimize the operating cost?

SUMMARY AND REVIEW: CHAPTER 5

The following contains a summary of what you should be able to do after completing this chapter. The review exercises are for practice. Answers are at the back of the book. If you miss an exercise, restudy the section indicated alongside the answer.

You should be able to:

Solve a system of linear equations in two or more variables and classify it as consistent or inconsistent, dependent or independent.

Solve.

1. $5x - 3y = -4$
$3x - y = -4$

2. $2x + 3y = 2,$
$5x - y = -29$

3. $x + 5y = 12,$
$5x + 25y = 12$

4. $2x - 4y + 3z = -3,$
$-5x + 2y - z = 7,$
$3x + 2y - 2z = 4$

5. $x + 5y + 3z = 0,$
$3x - 2y + 4z = 0,$
$2x + 3y - z = 0$

6. $x - y = 5,$
$y - z = 6,$
$z - w = 7,$
$w + x = 8$

7. Classify each of the systems in Exercises 1–6 as consistent or inconsistent.

8. Classify each of the systems in Exercises 1–6 as dependent or independent.

Solve an applied problem by translating it to a system of linear equations and solving the system.

Solve.

9. The value of 75 coins, consisting of nickels and dimes, is $5.95. How many of each kind are there?

10. A family invested $5000, part at 10% and the remainder at 10.5%. The annual income from both investments is $517. What is the amount invested at each rate?

11. In triangle ABC, the measure of angle B is three times that of angle A. The measure of angle C is 20° more than that of angle A. Find the angle measures.

12. A student has a total of 225 on three tests. The sum of the scores on the first and second tests exceeds the third score by 61. The first score exceeds the second by 6. Find the three scores.

Solve a system of linear equations using matrices. If there is more than one solution, list several of them.

Solve, using matrices. If there is more than one solution, list three of them.

13. $x + 2y = 5,$
$2x - 5y = -8$

14. $3x + 4y + 2z = 3,$
$5x - 2y - 13z = 3,$
$4x + 3y - 3z = 6$

15. $3x + 5y + z = 0,$
$2x - 4y - 3z = 0,$
$x + 3y + z = 0$

16. $x + y + z + w = -2,$
$-2x + 3y + 2z - 3w = 10,$
$3x + 2y - z + 2w = -12,$
$4x - y + z + 2w = 1$

Fit a quadratic equation to data when three data points are given.

17. Find numbers a, b, and c such that the function $y = ax^2 + bx + c$ fits the data points $(0, 3)$, $(1, 0)$, and $(-1, 4)$. Then write the equation.

Evaluate determinants of 2 × 2 and 3 × 3 matrices.

Evaluate.

18. $\begin{vmatrix} 1 & -2 \\ 3 & 4 \end{vmatrix}$

19. $\begin{vmatrix} \sqrt{3} & -5 \\ -3 & -\sqrt{3} \end{vmatrix}$

20. $\begin{vmatrix} -2 & -3 \\ 4 & -x \end{vmatrix}$

21. $\begin{vmatrix} 2 & -1 & 1 \\ 1 & 2 & -1 \\ 3 & 4 & -3 \end{vmatrix}$

22. $\begin{vmatrix} -5.8 & 7.5 & 4.6 \\ 0 & 2.2 & 8.9 \\ 0 & 0 & 1.3 \end{vmatrix}$

23. $\begin{vmatrix} 3a & 3b & 3c \\ 5a & 5b & 5c \\ d & e & f \end{vmatrix}$

Solve systems of two equations in two variables and three equations in three variables using Cramer's rule.

Solve for (x, y) using Cramer's rule.

24. $5x - 2y = 19,$
$\quad\; 7x + 3y = 15$

25. $ax - by = a^2,$
$\quad\;\; bx + ay = ab$

26. Solve using Cramer's rule.

$$3x - 2y + z = 5,$$
$$4x - 5y - z = -1,$$
$$3x + 2y - z = 4$$

Graph a system of linear inequalities in two variables in the plane and find vertices, if they exist.

27. Graph this system. Find the coordinates of any vertices formed.

$$2x + \; y \geqslant 9,$$
$$4x + 3y \geqslant 23,$$
$$x + 3y \geqslant 8,$$
$$x \geqslant 0,$$
$$y \geqslant 0$$

Solve linear programming problems.

28. Maximize and minimize $T = 6x + 10y$ subject to

$$x + \; y \leqslant 10,$$
$$5x + 10y \leqslant 50,$$
$$x \geqslant 2,$$
$$y \geqslant 0.$$

29. You are about to take a test that contains questions of type A worth 7 points and questions of type B worth 12 points. The total number of questions worked must be at least 8. If you know that type A questions take 10 min and type B questions take 8 min and that the maximum time for the test is 80 min, how many of each type of question must you do to maximize your score? What is this maximum score?

30. One year a person invested a total of $40,000, part at 12%, part at 13%, and the rest at $14\frac{1}{2}$%. The total interest received on the investments was $5370. The interest received on the $14\frac{1}{2}$% investment was $1050 more than the interest received on the 13% investment. How much was invested at each rate?

Solve.

31. $\dfrac{2}{3x} + \dfrac{4}{5y} = 8,$

$\quad\; \dfrac{5}{4x} - \dfrac{3}{2y} = -6$

32. $\dfrac{3}{x} - \dfrac{4}{y} + \dfrac{1}{z} = -2,$

$\quad\; \dfrac{5}{x} + \dfrac{1}{y} - \dfrac{2}{z} = 1,$

$\quad\; \dfrac{7}{x} + \dfrac{3}{y} + \dfrac{2}{z} = 19$

Graph.

33. $|x| - |y| \leqslant 1$

34. $|xy| > 1$

TEST: CHAPTER 5

Solve.

1. $0.2x - 0.5y = -0.1,$
$0.01x + 0.1y = 0.12$

2. $\dfrac{7}{x} - \dfrac{3}{y} = -16,$

$\dfrac{2}{x} + \dfrac{5}{y} = 13$

3. A boat travels 36 km downstream in 2 hr. It travels 48 km upstream in 4 hr. Find the speed of the boat and the speed of the stream.

4. A chemist has one solution of acid and water that is 15% acid and a second that is 75% acid. Find how many gallons of each should be mixed together to get 20 gallons of a solution that is 39% acid.

5. A factory has three machines A, B, and C. With all three working, they make 500 toothbrushes per day. With A and B working, they make 284 toothbrushes per day. With A and C working, they make 260 toothbrushes per day. What is the daily production of each machine?

Solve using matrices. If there is more than one solution, list three of them.

6. $8x - 2y = 1,$
$12x + 8y = 7$

7. $12x - 6y - 19z = 0,$
$3x - 2y - 6z = 0,$
$6x + 2y + 3z = 0$

8. $2x - 7y + z = 11,$
$3x + 2y - 4z = 15,$
$5x - 4y + 6z = 7$

Classify as consistent or inconsistent, dependent or independent.

9. $-x + y = -6,$
$-x - y = 6$

10. $x + 4y - 2z = 1,$
$2x + 6y - z = 0,$
$-3x - 10y + 3z = -1$

11. Find numbers a, b, and c such that the function $y = ax^2 + bx + c$ fits the data points $(2, 1)$, $(0, 1)$, and $(-1, -5)$. Then write the equation.

Evaluate.

12. $\begin{vmatrix} -3 & 1 \\ -4 & \frac{2}{3} \end{vmatrix}$

13. $\begin{vmatrix} 1 & 1 & -2 \\ 0 & 2 & -6 \\ 4 & 0 & 3 \end{vmatrix}$

Solve using Cramer's rule.

14. $7x - 3y = 31,$
$4x + 2y = 14$

15. $x - 2y + 7z = 11,$
$2x + y - 3z = -5,$
$6x + z = 1$

16. Graph $4x - 3y \geqslant 12$.

17. Maximize and minimize $T = 20x + 60y$ subject to

$$x + 2y \leqslant 16,$$
$$2 \leqslant x \leqslant 5,$$
$$y \geqslant 0.$$

18. You are about to take a test that contains questions of type A worth 5 points and type B worth 12 points. You must complete the test in 72 minutes. Type A questions take 4 minutes; type B questions take 8 minutes. The total number of problems worked must not exceed 12. If you are told that you must work at least 2 questions of type B, how many of each type of question must you do in order to maximize your score? What is this maximum score?

19. Solve $\begin{vmatrix} \sqrt{x} & 3 \\ -2 & \sqrt{x} \end{vmatrix} = x^2.$

20. Graph $|2x + y| < 2$.

The windows of this building form an arrangement known as a *matrix*.

MATRICES AND DETERMINANTS

A *matrix* is a rectangular array, such as that formed by the rows and columns of a marching band, or the numbers in a table you might see in a newspaper.

If you studied Section 5.4, you already know how matrices can be used in solving systems of equations. This is a more efficient method of solution than either the substitution or addition methods.

Matrices can be studied as entities unto themselves and this is what we do in this chapter. This is an optional chapter, since you will not need it for the later chapters of the book. Since applications of matrices seem to spring up everywhere, however, a study of this chapter may help you in some of your future studies or in problem solving.

6

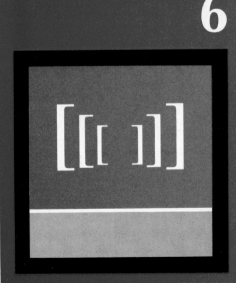

OBJECTIVES

You should be able to:

i Add, subtract, and multiply matrices when possible.

ii Write a matrix equation equivalent to a system of equations.

Find the dimensions of each matrix.

1. $\begin{bmatrix} -3 & 5 \\ 4 & \frac{1}{4} \\ -\pi & 0 \end{bmatrix}$

2. $\begin{bmatrix} -3 & 0 \\ 0 & 3 \end{bmatrix}$

3. $\begin{bmatrix} 1 & 2 & 3 \\ 0 & 1 & 8 \\ 0 & 0 & 1 \end{bmatrix}$

4. $[\pi \quad \sqrt{2}]$

5. $\begin{bmatrix} -5 \\ \pi \end{bmatrix}$

6. $[-3]$

7. Which of the above are square matrices?

8. Let

$$\mathbf{A} = \begin{bmatrix} 4 & -1 \\ 6 & -3 \end{bmatrix} \quad \text{and} \quad \mathbf{B} = \begin{bmatrix} -6 & -5 \\ 7 & 3 \end{bmatrix}.$$

a) Find $\mathbf{A} + \mathbf{B}$.

b) Find $\mathbf{B} + \mathbf{A}$.

9. Add.

$$\begin{bmatrix} -3 & -4 & -5 \\ 0 & 1 & -1 \end{bmatrix} + \begin{bmatrix} 4 & 5 & -5 \\ 2 & 3 & -2 \end{bmatrix}$$

6.1

OPERATIONS ON MATRICES*

DIMENSIONS OF A MATRIX

A matrix of m rows and n columns is called a matrix with *dimensions* $m \times n$ (read "m by n").

EXAMPLE 1 Find the dimensions of each matrix.

$$\begin{bmatrix} 2 & -3 & 4 \\ -1 & \frac{1}{2} & \pi \end{bmatrix}, \qquad \begin{bmatrix} -3 & 8 & 9 \\ \pi & -2 & 5 \\ -6 & 7 & 8 \end{bmatrix},$$

A 2 × 3 matrix   A 3 × 3 matrix

$$[-3 \quad 4], \qquad \begin{bmatrix} 10 \\ -7 \end{bmatrix}.$$

A 1 × 2 matrix   A 2 × 1 matrix

DO EXERCISES 1–7.

i OPERATIONS ON MATRICES

Matrix Addition

To add matrices, we add the corresponding members. For this to be possible, the matrices must have the same dimensions.

EXAMPLES

2. $\begin{bmatrix} -5 & 0 \\ 4 & \frac{1}{2} \end{bmatrix} + \begin{bmatrix} 6 & -3 \\ 2 & 3 \end{bmatrix} = \begin{bmatrix} -5 + 6 & 0 - 3 \\ 4 + 2 & \frac{1}{2} + 3 \end{bmatrix}$

$\qquad\qquad\qquad = \begin{bmatrix} 1 & -3 \\ 6 & 3\frac{1}{2} \end{bmatrix}$

3. $\begin{bmatrix} 1 & 3 & 2 \\ -1 & 5 & 4 \\ 6 & 0 & 1 \end{bmatrix} + \begin{bmatrix} -1 & -2 & 1 \\ 1 & -2 & 2 \\ -3 & 1 & 0 \end{bmatrix} = \begin{bmatrix} 0 & 1 & 3 \\ 0 & 3 & 6 \\ 3 & 1 & 1 \end{bmatrix}$

Addition of matrices is both commutative and associative.

DO EXERCISES 8 AND 9.

Zero Matrices

A matrix having zeros for all of its members is called a *zero matrix* and is often denoted by **O**. When a zero matrix is added to another matrix of the same dimensions, that same matrix is obtained. Thus a zero matrix is an *additive identity*.

---

* This chapter is optional.

EXAMPLE 4

$$\begin{bmatrix} 2 & -1 & 3 \\ 1 & 0 & -1 \end{bmatrix} + \begin{bmatrix} 0 & 0 & 0 \\ 0 & 0 & 0 \end{bmatrix} = \begin{bmatrix} 2 & -1 & 3 \\ 1 & 0 & -1 \end{bmatrix}$$

DO EXERCISE 10.

Inverses and Subtraction

To subtract matrices, we subtract the corresponding members. Of course, the matrices must have the same dimensions for this to be possible.

EXAMPLE 5

$$\begin{bmatrix} 1 & 2 \\ -2 & 0 \\ -3 & -1 \end{bmatrix} - \begin{bmatrix} 1 & -1 \\ 1 & 3 \\ 2 & 3 \end{bmatrix} = \begin{bmatrix} 0 & 3 \\ -3 & -3 \\ -5 & -4 \end{bmatrix}$$

DO EXERCISES 11 AND 12.

The additive inverse of a matrix can be obtained by replacing each member by its additive inverse. Of course, when two matrices that are inverses of each other are added, a zero matrix is obtained.

EXAMPLE 6

$$\begin{bmatrix} 1 & 0 & 2 \\ 3 & -1 & 5 \end{bmatrix} + \begin{bmatrix} -1 & 0 & -2 \\ -3 & 1 & -5 \end{bmatrix} = \begin{bmatrix} 0 & 0 & 0 \\ 0 & 0 & 0 \end{bmatrix}$$

$$\mathbf{A} \quad + \quad (-\mathbf{A}) \quad = \quad \mathbf{O}.$$

DO EXERCISES 13 AND 14.

With numbers, we can subtract by adding an inverse. This is also true of matrices. If we denote matrices by **A** and **B** and an additive inverse by $-\mathbf{B}$, this fact can be stated as follows:

$$\mathbf{A} - \mathbf{B} = \mathbf{A} + (-\mathbf{B}).$$

EXAMPLE 7

$$\begin{bmatrix} 3 & -1 \\ -2 & 4 \end{bmatrix} - \begin{bmatrix} 2 & 1 \\ 3 & -2 \end{bmatrix} = \begin{bmatrix} 1 & -2 \\ -5 & 6 \end{bmatrix}$$

$$\mathbf{A} \quad - \quad \mathbf{B}$$

$$\begin{bmatrix} 3 & -1 \\ -2 & 4 \end{bmatrix} + \begin{bmatrix} -2 & -1 \\ -3 & 2 \end{bmatrix} = \begin{bmatrix} 1 & -2 \\ -5 & 6 \end{bmatrix}$$

$$\mathbf{A} \quad + \quad (-\mathbf{B})$$

DO EXERCISE 15.

Multiplying Matrices and Numbers

We define a product of a matrix and a number.

10. Let

$$\mathbf{A} = \begin{bmatrix} 4 & -3 \\ 5 & 8 \end{bmatrix} \quad \text{and} \quad \mathbf{O} = \begin{bmatrix} 0 & 0 \\ 0 & 0 \end{bmatrix}.$$

a) Find $\mathbf{A} + \mathbf{O}$.

b) Find $\mathbf{O} + \mathbf{A}$.

Subtract.

11. $\begin{bmatrix} 1 & 3 & -2 \\ 4 & 0 & 5 \end{bmatrix} - \begin{bmatrix} 2 & -1 & 5 \\ 6 & 4 & -3 \end{bmatrix}$

12. $\begin{bmatrix} 1 & 2 \\ 4 & 1 \\ -5 & 4 \end{bmatrix} - \begin{bmatrix} 7 & -4 \\ 3 & 5 \\ 2 & -1 \end{bmatrix}$

13. Find the additive inverse.

$$\begin{bmatrix} 2 & -1 & 5 \\ 6 & 4 & -3 \end{bmatrix}$$

14. Add.

$$\begin{bmatrix} 2 & -1 & 5 \\ 6 & 4 & -3 \end{bmatrix} + \begin{bmatrix} -2 & 1 & -5 \\ -6 & -4 & 3 \end{bmatrix}$$

15. Add. Compare with Exercise 11.

$$\begin{bmatrix} 1 & 3 & -2 \\ 4 & 0 & 5 \end{bmatrix} + \begin{bmatrix} -2 & 1 & -5 \\ -6 & -4 & 3 \end{bmatrix}$$

Compute these products.

16. $5\begin{bmatrix} 1 & -2 & x \\ 4 & y & 1 \\ 0 & -5 & x^2 \end{bmatrix}$

Definition

The product of a number k and a matrix A is the matrix, denoted kA, obtained by multiplying each number in A by the number k.

EXAMPLES   Let

$$\mathbf{A} = \begin{bmatrix} -3 & 0 \\ 4 & 5 \end{bmatrix}.$$

Then:

8. $3\mathbf{A} = 3\begin{bmatrix} -3 & 0 \\ 4 & 5 \end{bmatrix} = \begin{bmatrix} -9 & 0 \\ 12 & 15 \end{bmatrix}$

9. $(-1)\mathbf{A} = -1\begin{bmatrix} -3 & 0 \\ 4 & 5 \end{bmatrix} = \begin{bmatrix} 3 & 0 \\ -4 & -5 \end{bmatrix}.$

DO EXERCISES 16 AND 17.

Products of Matrices

We do not multiply two matrices by multiplying their corresponding members. The motivation for defining matrix products comes from systems of equations.

Let us begin by considering one equation,

$$3x + 2y - 2z = 4.$$

17. $t\begin{bmatrix} 1 & -1 & 4 & x \\ y & 3 & -2 & y \\ 1 & 4 & -5 & y \end{bmatrix}$

We will write the coefficients on the left side in a 1×3 matrix (a *row matrix*) and the variables in a 3×1 matrix (a *column matrix*). The 4 on the right is written in a 1×1 matrix:

$$\begin{bmatrix} 3 & 2 & -2 \end{bmatrix} \begin{bmatrix} x \\ y \\ z \end{bmatrix} = \begin{bmatrix} 4 \end{bmatrix}.$$

We can return to our original equation by multiplying the members of the row matrix by those of the column matrix, and adding:

$$\begin{bmatrix} 3 & 2 & -2 \end{bmatrix} \begin{bmatrix} x \\ y \\ z \end{bmatrix} = \begin{bmatrix} 3x + 2y - 2z \end{bmatrix}.$$

We define multiplication accordingly. In this special case, we have a *row matrix* **A** and a *column matrix* **B**. Their product **AB** is a 1×1 matrix, having the single member 4 (also called $3x + 2y - 2z$).

18. Multiply.

$$\begin{bmatrix} 4 & -2 & 3 \end{bmatrix} \begin{bmatrix} 2 \\ 3 \\ -5 \end{bmatrix}$$

EXAMPLE 10   Find the product of these matrices.

$$\begin{bmatrix} 3 & 2 & -1 \end{bmatrix} \begin{bmatrix} 1 \\ -2 \\ 3 \end{bmatrix} = \begin{bmatrix} 3 \cdot 1 + 2(-2) + (-1) \cdot 3 \end{bmatrix} = \begin{bmatrix} -4 \end{bmatrix}$$

DO EXERCISE 18.

Let us continue by considering a system of equations:

$$\begin{aligned} 3x + 2y - 2z &= 4, \\ 2x - y + 5z &= 3, \\ -x + y + 4z &= 7. \end{aligned}$$

Consider the following matrices:

$$\begin{bmatrix} 3 & 2 & -2 \\ 2 & -1 & 5 \\ -1 & 1 & 4 \end{bmatrix} \begin{bmatrix} x \\ y \\ z \end{bmatrix} \begin{bmatrix} 4 \\ 3 \\ 7 \end{bmatrix}.$$
$$\quad\quad \mathbf{A} \quad\quad\quad \mathbf{X} \quad \mathbf{B}$$

If we multiply the first row of **A** by the (only) column of **X**, as we did above, we get $3x + 2y - 2z$. If we multiply the second row of **A** by the column in **X**, in the same way, we get the following:

$$[2 \quad -1 \quad 5] \begin{bmatrix} x \\ y \\ z \end{bmatrix} = 2x - y + 5z.$$

Note that the first members are multiplied, the second members are multiplied, the third members are multiplied, and the results are added, to get the single number $2x - y + 5z$. What do we get when we multiply the third row of **A** by the column in **X**?

$$[-1 \quad 1 \quad 4] \begin{bmatrix} x \\ y \\ z \end{bmatrix} = -x + y + 4z$$

We define the product **AX** to be the column matrix

$$\begin{bmatrix} 3x + 2y - 2z \\ 2x - y + 5z \\ -x + y + 4z \end{bmatrix}.$$

Now consider this matrix equation:

$$\begin{bmatrix} 3x + 2y - 2z \\ 2x - y + 5z \\ -x + y + 4z \end{bmatrix} = \begin{bmatrix} 4 \\ 3 \\ 7 \end{bmatrix}.$$

Equality for matrices is the same as for numbers, that is, a sentence such as $a = b$ says that a and b are two names for the same thing. Thus if the above matrix equation is true, the "two" matrices are really the same one. This means that $3x + 2y - 2z$ is 4, $2x - y + 5z$ is 3, and $-x + y + 4z$ is 7, or that

$$3x + 2y - 2z = 4,$$
$$2x - y + 5z = 3,$$
and
$$-x + y + 4z = 7.$$

Thus the matrix equation **AX = B** is equivalent to the original system of equations.

EXAMPLE 11   Multiply.

$$\begin{bmatrix} 3 & 1 & -1 \\ 1 & 2 & 2 \\ -1 & 0 & 5 \\ 4 & 1 & 2 \end{bmatrix} \begin{bmatrix} 1 \\ 2 \\ 1 \end{bmatrix} = \begin{bmatrix} 3 \cdot 1 + 1 \cdot 2 - 1 \cdot 1 \\ 1 \cdot 1 + 2 \cdot 2 + 2 \cdot 1 \\ -1 \cdot 1 + 0 \cdot 2 + 5 \cdot 1 \\ 4 \cdot 1 + 1 \cdot 2 + 2 \cdot 1 \end{bmatrix} = \begin{bmatrix} 4 \\ 7 \\ 4 \\ 8 \end{bmatrix}$$

DO EXERCISE 19.

19. Multiply.

$$\begin{bmatrix} 1 & 4 & 2 \\ -1 & 6 & 3 \\ 3 & 2 & -1 \\ 5 & 0 & 2 \end{bmatrix} \begin{bmatrix} 2 \\ 1 \\ 3 \end{bmatrix}$$

20. Multiply.

$$\begin{bmatrix} 4 & 1 & 2 \\ -3 & 2 & 3 \\ 2 & 0 & 5 \\ 3 & 1 & 4 \end{bmatrix} \begin{bmatrix} 1 & 4 \\ 2 & 0 \\ -3 & 5 \end{bmatrix}$$

21. Find **AB** and **BA** if possible.

$$\mathbf{A} = \begin{bmatrix} -2 & 4 & 0 \\ -3 & 0 & -8 \end{bmatrix},$$

$$\mathbf{B} = \begin{bmatrix} -1 & -2 & -3 \\ 0 & 1 & 0 \\ 4 & 5 & 2 \end{bmatrix}$$

22. Multiply.

$$\begin{bmatrix} 4 & 1 & 0 & 2 \end{bmatrix} \begin{bmatrix} 1 & 0 & 1 \\ 2 & -1 & 0 \\ 3 & 5 & 1 \\ 1 & 3 & 0 \end{bmatrix}$$

23. Find **AB** and **BA** and compare.

$$\mathbf{A} = \begin{bmatrix} -8 & 3 \\ -4 & 4 \end{bmatrix},$$

$$\mathbf{B} = \begin{bmatrix} 1 & -4 \\ 2 & 0 \end{bmatrix}$$

24. Find **AI** and **IA**. Comment.

$$\mathbf{A} = \begin{bmatrix} 3 & 2 \\ -1 & 5 \end{bmatrix},$$

$$\mathbf{I} = \begin{bmatrix} 1 & 0 \\ 0 & 1 \end{bmatrix}$$

In all the examples discussed so far, the second matrix had only one column. If the second matrix has more than one column, we treat it in the same way when multiplying that we treated the single column. The product matrix will have as many columns as the second matrix.

EXAMPLE 12 Multiply (compare with Example 11).

$$\begin{bmatrix} 3 & 1 & -1 \\ 1 & 2 & 2 \\ -1 & 0 & 5 \\ 4 & 1 & 2 \end{bmatrix} \begin{bmatrix} 1 & 0 \\ 2 & 1 \\ 1 & 3 \end{bmatrix} = \begin{bmatrix} 4 & 3 \cdot 0 + 1 \cdot 1 + (-1)3 \\ 7 & 1 \cdot 0 + 2 \cdot 1 + 2 \cdot 3 \\ 4 & -1 \cdot 0 + 0 \cdot 1 + 5 \cdot 3 \\ 8 & 4 \cdot 0 + 1 \cdot 1 + 2 \cdot 3 \end{bmatrix} = \begin{bmatrix} 4 & -2 \\ 7 & 8 \\ 4 & 15 \\ 8 & 7 \end{bmatrix}$$

A   **B**

Same as in Example 11

The rows of **A** multiplied by the second column of **B**

DO EXERCISE 20.

EXAMPLE 13 Multiply.

$$\begin{bmatrix} 3 & 1 & -1 \\ 2 & 0 & 3 \end{bmatrix} \begin{bmatrix} 1 & 4 & 6 \\ 3 & -1 & 9 \\ 2 & 5 & 1 \end{bmatrix}$$

$$= \begin{bmatrix} 3 \cdot 1 + 1 \cdot 3 - 1 \cdot 2 & 3 \cdot 4 + 1 \cdot (-1) - 1 \cdot 5 & 3 \cdot 6 + 1 \cdot 9 - 1 \cdot 1 \\ 2 \cdot 1 + 0 \cdot 3 + 3 \cdot 2 & 2 \cdot 4 + 0 \cdot (-1) + 3 \cdot 5 & 2 \cdot 6 + 0 \cdot 9 + 3 \cdot 1 \end{bmatrix}$$

$$= \begin{bmatrix} 4 & 6 & 26 \\ 8 & 23 & 15 \end{bmatrix}$$

If matrix A has *n* columns and matrix B has *n* rows, then we can compute the product AB, regardless of the other dimensions. The product will have as many rows as A and as many columns as B.

CAUTION! Given any two matrices **A** and **B**, you may or may not be able to add, subtract, or multiply them. **A + B** and **A − B** exist only when the dimensions are the same. **AB** exists only when the number of columns in **A** is the same as the number of rows in **B**.

For example, consider the matrices

$$\mathbf{A} = \begin{bmatrix} 3 & 1 & -1 \\ 2 & 0 & 3 \end{bmatrix} \quad \text{and} \quad \mathbf{B} = \begin{bmatrix} 1 & 4 & 6 \\ 3 & -1 & 9 \\ 2 & 5 & 1 \end{bmatrix}.$$

A + B and **A − B** do not exist because the dimensions of **A** and **B** are *not* the same. **AB** does exist because the number of columns in **A**, 3, is the same as the number of rows in **B**, 3. **AB** is given in Example 13. But **BA** does *not* exist because the number of columns in **B**, 3, is not the same as the number of rows in **A**, 2. In this context it can also be pointed out that matrix multiplication is not commutative, **AB** exists, and **BA** does not, so **AB** ≠ **BA**.

DO EXERCISES 21–24.

ii   EQUIVALENT MATRIX EQUATIONS

For later purposes it is important to be able to write a matrix equation equivalent to a system of equations.

EXAMPLE 14   Write a matrix equation equivalent to this system of equations:

$$
\begin{aligned}
4x + 2y - z &= 3, \\
9x + z &= 5, \\
4x + 5y - 2z &= 1, \\
x + y + z &= 0.
\end{aligned}
$$

We write the coefficients on the left in a matrix. We write the product of that matrix by the column matrix containing the variables, and set the result equal to the column matrix containing the constants on the right:

$$
\begin{bmatrix} 4 & 2 & -1 \\ 9 & 0 & 1 \\ 4 & 5 & -2 \\ 1 & 1 & 1 \end{bmatrix}
\begin{bmatrix} x \\ y \\ z \end{bmatrix}
=
\begin{bmatrix} 3 \\ 5 \\ 1 \\ 0 \end{bmatrix}
$$

DO EXERCISE 25.

A SUMMARY OF PROPERTIES OF SQUARE MATRICES

We now list a summary of some of the properties of square matrices of the same dimensions whose elements are real numbers. We make the restriction of square matrices so all additions and multiplications are possible. Some of the proofs will be considered in the exercise set. Note that not all the field properties hold.

Theorem 1

For any square matrices A, B, and C of the same dimensions the following hold:

Commutativity. $A + B = B + A.$

Associativity. $A + (B + C) = (A + B) + C,\quad A(BC) = (AB)C.$

Identity.   There exists a unique matrix O, such that

$$A + O = O + A = A.$$

Inverses.   There exists a unique matrix $-A$, such that

$$A + (-A) = -A + A = O.$$

Distributivity. $A(B + C) = AB + AC.$
For any square matrices A and B, of the same dimensions, and any real numbers k and m,

$$k(A + B) = kA + kB;$$

$$(k + m)A = kA + mA,$$

and

$$(km)A = k(mA).$$

25. Write a matrix equation equivalent to this system of equations.

$$
\begin{aligned}
3x + 4y - 2z &= 5, \\
2x - 2y + 5z &= 3, \\
6x + 7y - z &= 0
\end{aligned}
$$

CAUTION! Note that, even with these restrictions, matrix multiplication is still *not* commutative. For example, let

$$\mathbf{A} = \begin{bmatrix} 1 & 0 \\ 2 & 0 \end{bmatrix} \quad \text{and} \quad \mathbf{B} = \begin{bmatrix} 3 & 4 \\ 0 & 0 \end{bmatrix}.$$

Then

$$\mathbf{AB} = \begin{bmatrix} 3 & 4 \\ 6 & 8 \end{bmatrix} \quad \text{and} \quad \mathbf{BA} = \begin{bmatrix} 11 & 0 \\ 0 & 0 \end{bmatrix},$$

so $\mathbf{AB} \neq \mathbf{BA}$.

EXERCISE SET 6.1

i For Exercises 1–16, let

$$\mathbf{A} = \begin{bmatrix} 1 & 2 \\ 4 & 3 \end{bmatrix}, \quad \mathbf{B} = \begin{bmatrix} -3 & 5 \\ 2 & -1 \end{bmatrix}, \quad \mathbf{C} = \begin{bmatrix} 1 & -1 \\ -1 & 1 \end{bmatrix}, \quad \mathbf{D} = \begin{bmatrix} 1 & 1 \\ 1 & 1 \end{bmatrix},$$

$$\mathbf{E} = \begin{bmatrix} 1 & 3 \\ 2 & 6 \end{bmatrix}, \quad \mathbf{F} = \begin{bmatrix} 3 & 3 \\ -1 & -1 \end{bmatrix}, \quad \mathbf{O} = \begin{bmatrix} 0 & 0 \\ 0 & 0 \end{bmatrix}, \quad \text{and} \quad \mathbf{I} = \begin{bmatrix} 1 & 0 \\ 0 & 1 \end{bmatrix}.$$

Find.

1. $\mathbf{A} + \mathbf{B}$

2. $\mathbf{B} + \mathbf{A}$

3. $\mathbf{E} + \mathbf{O}$

4. $2\mathbf{A}$

5. $3\mathbf{F}$

6. $(-1)\mathbf{D}$

7. $3\mathbf{F} + 2\mathbf{A}$

8. $\mathbf{A} - \mathbf{B}$

9. $\mathbf{B} - \mathbf{A}$

10. $\mathbf{AB}$

11. $\mathbf{BA}$

12. $\mathbf{OF}$

13. $\mathbf{CD}$

14. $\mathbf{EF}$

15. $\mathbf{AI}$

16. $\mathbf{IA}$

In Exercises 17–20, let

$$\mathbf{A} = \begin{bmatrix} 1 & 0 & -2 \\ 0 & -1 & 3 \\ 3 & 2 & 4 \end{bmatrix}, \quad \mathbf{B} = \begin{bmatrix} -1 & -2 & 5 \\ 1 & 0 & -1 \\ 2 & -3 & 1 \end{bmatrix}, \quad \mathbf{C} = \begin{bmatrix} -2 & 9 & 6 \\ -3 & 3 & 4 \\ 2 & -2 & 1 \end{bmatrix}, \quad \text{and} \quad \mathbf{I} = \begin{bmatrix} 1 & 0 & 0 \\ 0 & 1 & 0 \\ 0 & 0 & 1 \end{bmatrix}.$$

Find.

17. $\mathbf{AB}$

18. $\mathbf{BA}$

19. $\mathbf{CI}$

20. $\mathbf{IC}$

Multiply.

21. $[-3 \quad 2] \begin{bmatrix} 4 \\ -2 \end{bmatrix}$

22. $[-2 \quad 0 \quad 4] \begin{bmatrix} 8 \\ -6 \\ \frac{1}{2} \end{bmatrix}$

23. $[-5 \quad 1 \quad 2] \begin{bmatrix} 1 & 3 \\ -1 & 0 \\ 4 & -2 \end{bmatrix}$

24. $\begin{bmatrix} -3 & 2 \\ 0 & 1 \\ -4 & 5 \end{bmatrix} \begin{bmatrix} 4 \\ 2 \end{bmatrix}$

ii Write a matrix equation equivalent to each of the following systems of equations.

25. $3x - 2y + 4z = 17,$
$\quad 2x + y - 5z = 13$

26. $3x + 2y + 5z = 9,$
$\quad 4x - 3y + 2z = 10$

27. $x - y + 2z - 4w = 12,$
$\quad 2x - y - z + w = 0,$
$\quad x + 4y - 3z - w = 1,$
$\quad 3x + 5y - 7z + 2w = 9,$

28. $2x + 4y - 5z + 12w = 2,$
$\quad 4x - y + 12z - w = 5,$
$\quad -x + 4y \qquad + 2w = 13,$
$\quad 2x + 10y + z \qquad = 5$

Compute.

29. $\blacksquare$ $\begin{bmatrix} 3.61 & -2.14 & 16.7 \\ -4.33 & 7.03 & 12.9 \\ 5.82 & -6.95 & 2.34 \end{bmatrix} \begin{bmatrix} 3.05 & 0.402 & -1.34 \\ 1.84 & -1.13 & 0.024 \\ -2.83 & 2.04 & 8.81 \end{bmatrix}$

30. $\blacksquare$ $\begin{bmatrix} -1.23 & 4.51 & -17.4 \\ 61.2 & -8.81 & 0.123 \\ 14.14 & 6.92 & -14.4 \end{bmatrix} \begin{bmatrix} 4.24 & 16.1 & 41.3 \\ 0.146 & -6.06 & -18.9 \\ -8.43 & 1.12 & 0.0245 \end{bmatrix}$

For Exercises 31 and 32, let

$$\mathbf{A} = \begin{bmatrix} -1 & 0 \\ 2 & 1 \end{bmatrix}, \quad \text{and} \quad \mathbf{B} = \begin{bmatrix} 1 & -1 \\ 0 & 2 \end{bmatrix}.$$

31. Show that $(\mathbf{A} + \mathbf{B})(\mathbf{A} - \mathbf{B}) \neq \mathbf{A}^2 - \mathbf{B}^2$, where

$$\mathbf{A}^2 = \mathbf{A}\mathbf{A} \quad \text{and} \quad \mathbf{B}^2 = \mathbf{B}\mathbf{B}.$$

32. Show that $(\mathbf{A} + \mathbf{B})(\mathbf{A} + \mathbf{B}) \neq \mathbf{A}^2 + 2\mathbf{A}\mathbf{B} + \mathbf{B}^2$.

Let $\mathbf{A} = \begin{bmatrix} a_{11} & a_{12} \\ a_{21} & a_{22} \end{bmatrix}$, $\mathbf{B} = \begin{bmatrix} b_{11} & b_{12} \\ b_{21} & b_{22} \end{bmatrix}$, $\mathbf{C} = \begin{bmatrix} c_{11} & c_{12} \\ c_{21} & c_{22} \end{bmatrix}$, $\mathbf{I} = \begin{bmatrix} 1 & 0 \\ 0 & 1 \end{bmatrix}$. Prove the following.

33. $\mathbf{A} + \mathbf{B} = \mathbf{B} + \mathbf{A}$

34. $\mathbf{A} + (\mathbf{B} + \mathbf{C}) = (\mathbf{A} + \mathbf{B}) + \mathbf{C}$

35. $k(\mathbf{A} + \mathbf{B}) = k\mathbf{A} + k\mathbf{B}$

36. $(k + m)\mathbf{A} = k\mathbf{A} + m\mathbf{A}$

37. $\mathbf{A}(\mathbf{B}\mathbf{C}) = (\mathbf{A}\mathbf{B})\mathbf{C}$

38. $\mathbf{A}\mathbf{I} = \mathbf{I}\mathbf{A} = \mathbf{A}$

6.2

DETERMINANTS OF HIGHER ORDER

i EVALUATING DETERMINANTS OF HIGHER ORDER

Further Notation

We shall define the determinant function for square matrices of any dimension. To do this we need some new notation. The members, or elements, of a matrix will now be denoted by lower-case letters with two subscripts, as follows:

$$\mathbf{A} = \begin{bmatrix} a_{11} & a_{12} & a_{13} \\ a_{21} & a_{22} & a_{23} \\ a_{31} & a_{32} & a_{33} \end{bmatrix}.$$

The element in the ith row and jth column is denoted a_{ij}. We may also name the above matrix $\mathbf{A}$ as

$$[a_{ij}].$$

EXAMPLE 1 Consider

$$[a_{ij}] = \begin{bmatrix} -8 & 0 & 6 \\ 4 & -6 & 7 \\ -1 & -3 & 5 \end{bmatrix}.$$

OBJECTIVE

You should be able to:

i Find a specified element, a_{ij}, of a matrix, its minor, and its cofactor, and be able to expand its determinant about any row or column.

1. For the matrix of Example 1, find a_{11}, a_{13}, a_{22}, a_{31}, and a_{32}.

Find a_{12}, a_{23}, and a_{33}.

$\qquad a_{12} = 0$ (This is the intersection of the first row and second column.)

$\qquad a_{23} = 7$ (This is the intersection of the second row and third column.)

$\qquad a_{33} = 5$ (This is the intersection of the third row and third column.)

DO EXERCISE 1.

Minors

We shall restrict our attention to square matrices.

Definition

> In a matrix $[a_{ij}]$, the *minor* M_{ij} of an element a_{ij} is the determinant of the matrix found by deleting the *i*th row and *j*th column.

Note that a minor is a certain determinant, hence is a number.

EXAMPLE 2 In the matrix given in Example 1, find M_{11} and M_{23}.

To find M_{11} we delete the first row and the first column:

$$\begin{bmatrix} -8 & 0 & 6 \\ 4 & -6 & 7 \\ -1 & -3 & 5 \end{bmatrix}.$$

2. For the matrix of Example 1, find the minors M_{22}, M_{31}, and M_{13}.

We calculate the determinant of the matrix formed by the remaining elements:

$$M_{11} = \begin{vmatrix} -6 & 7 \\ -3 & 5 \end{vmatrix} = (-6) \cdot 5 - (-3) \cdot 7 = -30 - (-21)$$

$$= -30 + 21 = -9.$$

To find M_{23} we delete the second row and the third column:

$$\begin{bmatrix} -8 & 0 & 6 \\ 4 & -6 & 7 \\ -1 & -3 & 5 \end{bmatrix}.$$

We calculate the determinant of the matrix formed by the remaining elements:

$$M_{23} = \begin{vmatrix} -8 & 0 \\ -1 & -3 \end{vmatrix} = -8(-3) - (-1)0 = 24.$$

DO EXERCISE 2.

Definition

> In a matrix $[a_{ij}]$, the *cofactor* of an element a_{ij} is denoted A_{ij} and is given by
>
> $$A_{ij} = (-1)^{i+j} M_{ij},$$
>
> where M_{ij} is the minor of a_{ij}. In other words, to find the cofactor of an element, find its minor and multiply it by $(-1)^{i+j}$.

Note that $(-1)^{i+j}$ is 1 if $i + j$ is even and is -1 if $i + j$ is odd. Thus in calculating a cofactor, find the minor. Then add the number of the row and the number of the column. The sum is $i + j$. If this sum is odd, change the sign of the minor. If this sum is even, leave the minor as is.* Note also that the cofactor of an element is a number.

EXAMPLE 3 In the matrix given in Example 1, find A_{11} and A_{23}.

In Example 2 we found that $M_{11} = -9$. In A_{11} the sum of the subscripts is even, so

$$A_{11} = -9.$$

In Example 2 we found that $M_{23} = 24$. In A_{23} the sum of the subscripts is odd, so

$$A_{23} = -24.$$

DO EXERCISE 3.

3. For the matrix of Example 1, find the cofactors A_{22}, A_{32}, and A_{13}.

Evaluating Determinants Using Cofactors

Consider the matrix **A** given by

$$\mathbf{A} = \begin{bmatrix} a_{11} & a_{12} & a_{13} \\ a_{21} & a_{22} & a_{23} \\ a_{31} & a_{32} & a_{33} \end{bmatrix}.$$

The determinant of the matrix, denoted $|\mathbf{A}|$, can be found as follows:

$$|\mathbf{A}| = a_{11}\mathbf{A}_{11} + a_{21}\mathbf{A}_{21} + a_{31}\mathbf{A}_{31}.$$

That is, multiply each element of the first column by its cofactor and add:

$$|\mathbf{A}| = a_{11} \cdot \begin{vmatrix} a_{22} & a_{23} \\ a_{32} & a_{33} \end{vmatrix} - a_{21} \cdot \begin{vmatrix} a_{12} & a_{13} \\ a_{32} & a_{33} \end{vmatrix} + a_{31} \cdot \begin{vmatrix} a_{12} & a_{13} \\ a_{22} & a_{23} \end{vmatrix}.$$

The last line is equivalent to the definition on p. 224. It can be shown that $|\mathbf{A}|$ can be found by picking *any* row or column, multiplying each element by its cofactor, and adding. This is called *expanding* about a row or column. We just expanded about the first column. We now define the determinant function for square matrices of any dimension.

Definition

For any square matrix A of dimensions $n \times n$ $(n > 1)$, we define the *determinant* of A, denoted $|A|$, as follows. Pick any row or column. Multiply each element in that row or column by its cofactor and add the results. The determinant of a 1×1 matrix is simply the element of the matrix.

---

* $(-1)^{i+j}$ can also be found by counting through the matrix horizontally and/or vertically, starting with a_{11} and $(+)$, saying $+, -, +, -$, and so on, until you come to a_{ij}.

Start here $(+)$

$$\begin{bmatrix} a_{11}^+ \rightarrow {}^- \rightarrow {}^+ \rightarrow \downarrow^- \\ \quad\quad\quad\quad \downarrow^+ \\ \quad\quad\quad\quad \downarrow^- \\ \quad\quad\quad a_{ij}^+ \end{bmatrix}$$ The path does not matter.

4. Consider the matrix of Example 1. Find $|\mathbf{A}|$ by expanding about the second column.

The value of a determinant will be the same no matter how it is evaluated.

EXAMPLE 4 Consider the matrix $\mathbf{A}$ of Example 1. Evaluate $|\mathbf{A}|$ by expanding about the third row.

$$|\mathbf{A}| = (-1)A_{31} + (-3)A_{32} + 5A_{33}$$

$$= (-1)(-1)^{3+1} \cdot \begin{vmatrix} 0 & 6 \\ -6 & 7 \end{vmatrix} + (-3)(-1)^{3+2} \cdot \begin{vmatrix} -8 & 6 \\ 4 & 7 \end{vmatrix}$$

$$+ 5(-1)^{3+3} \cdot \begin{vmatrix} -8 & 0 \\ 4 & -6 \end{vmatrix}$$

$$= (-1) \cdot 1 \cdot [0 \cdot 7 - (-6)6] + (-3)(-1)[-8 \cdot 7 - 4 \cdot 6]$$

$$+ 5 \cdot 1 \cdot [-8(-6) - 4 \cdot 0]$$

$$= -[36] + 3[-80] + 5[48]$$

$$= -36 - 240 + 240$$

$$= -36$$

5. Consider the matrix of Example 1. Evaluate $|\mathbf{A}|$ by expanding about the third column.

The value of this determinant is -36 no matter how we evaluate it. That is, if we use the second column, we still get -36.

DO EXERCISES 4 AND 5.

EXERCISE SET 6.2

i Use the following matrix for Exercises 1–10.

$$\mathbf{A} = \begin{bmatrix} 7 & -4 & -6 \\ 2 & 0 & -3 \\ 1 & 2 & -5 \end{bmatrix}$$

1. Find a_{11}, a_{32}, and a_{22}.

2. Find a_{13}, a_{31}, and a_{23}.

3. Find M_{11}, M_{32}, and M_{22}.

4. Find M_{13}, M_{31}, and M_{23}.

5. Find A_{11}, A_{32}, and A_{22}.

6. Find A_{13}, A_{31}, and A_{23}.

7. Evaluate $|\mathbf{A}|$ by expanding about the second row.

8. Evaluate $|\mathbf{A}|$ by expanding about the second column.

9. Evaluate $|\mathbf{A}|$ by expanding about the third column.

10. Evaluate $|\mathbf{A}|$ by expanding about the first row.

Use the following matrix for Exercises 11–16.

$$\mathbf{A} = \begin{bmatrix} 1 & 0 & 0 & -2 \\ 4 & 1 & 0 & 0 \\ 5 & 6 & 7 & 8 \\ -2 & -3 & -1 & 0 \end{bmatrix}$$

11. Find M_{41} and M_{33}.

12. Find M_{12} and M_{44}.

13. Find A_{24} and A_{43}.

14. Find A_{22} and A_{34}.

15. Evaluate $|\mathbf{A}|$ by expanding about the first row.

16. Evaluate $|\mathbf{A}|$ by expanding about the third column.

Evaluate.

17. $\begin{vmatrix} 5 & -4 & 2 & -2 \\ 3 & -3 & -4 & 7 \\ -2 & 3 & 2 & 4 \\ -8 & 9 & 5 & -5 \end{vmatrix}$

18. $\begin{vmatrix} x & p & q & r \\ 0 & y & s & t \\ 0 & 0 & z & u \\ 0 & 0 & 0 & w \end{vmatrix}$

19. If a line contains the points (x_1, y_1) and (x_2, y_2), an equation of the line can be written as follows:

$$\begin{vmatrix} x & y & 1 \\ x_1 & y_1 & 1 \\ x_2 & y_2 & 1 \end{vmatrix} = 0.$$

Prove this.

20. Show that three points (x_1, y_1), (x_2, y_2), and (x_3, y_3) are collinear (on the same straight line) if and only if

$$\begin{vmatrix} x_1 & y_1 & 1 \\ x_2 & y_2 & 1 \\ x_3 & y_3 & 1 \end{vmatrix} = 0.$$

6.3

PROPERTIES OF DETERMINANTS

i EVALUATION OF CERTAIN DETERMINANTS

We can simplify the evaluation of certain determinants using the following properties.

Theorem 2

If a row (or column) of a matrix A has all elements 0, then $|A| = 0$.

    Proof. Just evaluate by expanding about the row (or column) that has all 0's.

EXAMPLES   Evaluate.

1. $\begin{vmatrix} 0 & 6 \\ 0 & 7 \end{vmatrix} = 0$

2. $\begin{vmatrix} 4 & 5 & -7 \\ 0 & 0 & 0 \\ -3 & 9 & 6 \end{vmatrix} = 0$

DO EXERCISES 1 AND 2.

Theorem 3

If two rows (or columns) of a matrix A are interchanged to obtain a new matrix B, then $|A| = -|B|$.

    Proof. Pick one of the rows (or columns) to be interchanged and evaluate $|A|$ by expanding about that row. Expand about the same row to evaluate $|B|$. For that row each $(-1)^{i+j}$ has changed signs, so $|A| = -|B|$.

EXAMPLES

3. $\begin{vmatrix} 6 & 7 & 8 \\ 4 & 1 & 2 \\ 2 & 9 & 0 \end{vmatrix} = -1 \cdot \begin{vmatrix} 6 & 8 & 7 \\ 4 & 2 & 1 \\ 2 & 0 & 9 \end{vmatrix}$

4. $\begin{vmatrix} -6 & 8 \\ 4 & -3 \end{vmatrix} = -1 \cdot \begin{vmatrix} 4 & -3 \\ -6 & 8 \end{vmatrix}$

DO EXERCISES 3 AND 4.

Theorem 4

If two rows (or columns) of a matrix A are the same, then $|A| = 0$.

    Proof. Interchanging the rows (or columns) that are the same does not change **A**. Thus by Theorem 3, $|A| = -|A|$. This is possible only when $|A| = 0$.

5. Evaluate.

$$\begin{vmatrix} -1 & 3 & -7 \\ -1 & 2 & -6 \\ -1 & 3 & -7 \end{vmatrix}$$

Solve for x.

6. $\begin{vmatrix} -16 & 32 \\ -5 & -3 \end{vmatrix} = x \cdot \begin{vmatrix} 4 & -8 \\ -5 & -3 \end{vmatrix}$

7. $\begin{vmatrix} -3 & 12 & 2 \\ 5 & -6 & 3 \\ 0 & 18 & 5 \end{vmatrix} = x \cdot \begin{vmatrix} -3 & 2 & 2 \\ 5 & -1 & 3 \\ 0 & 3 & 5 \end{vmatrix}$

8. Without expanding, evaluate $|\mathbf{A}|$.

$$\mathbf{A} = \begin{bmatrix} 1 & 0 & -7 \\ -8 & 6 & -4 \\ 24 & -18 & 12 \end{bmatrix}$$

EXAMPLES Evaluate.

5. $\begin{vmatrix} 6 & 7 & 8 \\ -2 & 6 & 5 \\ -2 & 6 & 5 \end{vmatrix} = 0$

6. $\begin{vmatrix} -5 & 4 & -5 \\ 3 & 7 & 3 \\ 0 & 12 & 0 \end{vmatrix} = 0$

DO EXERCISE 5.

Theorem 5

If all the elements of a row (or column) of a matrix A are multiplied by k, $|A|$ is multiplied by k. Or, if all the elements of a row (or column) of A have a common factor, we can factor it out of the determinant of A.

EXAMPLES

7. $\begin{vmatrix} 2 & 4 & 6 \\ -2 & 5 & 9 \\ 4 & -1 & -3 \end{vmatrix} = 3 \cdot \begin{vmatrix} 2 & 4 & 2 \\ -2 & 5 & 3 \\ 4 & -1 & -1 \end{vmatrix}$

8. $\begin{vmatrix} 10 & 25 \\ -4 & -7 \end{vmatrix} = 5 \cdot \begin{vmatrix} 2 & 5 \\ -4 & -7 \end{vmatrix}$

DO EXERCISES 6 AND 7.

Proof of Theorem 5. Evaluate the determinants by expanding about the row (or column) in question. The cofactors are the same and the k can be factored out. Consider the case of a 3×3 matrix and the second column. Let

$$\mathbf{A} = \begin{bmatrix} a_{11} & a_{12} & a_{13} \\ a_{21} & a_{22} & a_{23} \\ a_{31} & a_{32} & a_{33} \end{bmatrix} \quad \text{and} \quad \mathbf{B} = \begin{bmatrix} a_{11} & ka_{12} & a_{13} \\ a_{21} & ka_{22} & a_{23} \\ a_{31} & ka_{32} & a_{33} \end{bmatrix}.$$

Then

$$\begin{aligned} |\mathbf{B}| &= ka_{12}A_{12} + ka_{22}A_{22} + ka_{32}A_{32} \\ &= k(a_{12}A_{12} + a_{22}A_{22} + a_{32}A_{32}) \\ &= k|\mathbf{A}|. \end{aligned}$$

EXAMPLE 9 Without expanding, find $|\mathbf{A}|$.

$$\mathbf{A} = \begin{bmatrix} -6 & 3 & 8 \\ 15 & -9 & -20 \\ -9 & -1 & 12 \end{bmatrix}$$

$$|\mathbf{A}| = (-3) \cdot \begin{vmatrix} 2 & 3 & 8 \\ -5 & -9 & -20 \\ 3 & -1 & 12 \end{vmatrix} \qquad \text{Factoring } -3 \text{ out of the first column}$$

$$= (-3)(4) \cdot \begin{vmatrix} 2 & 3 & 2 \\ -5 & -9 & -5 \\ 3 & -1 & 3 \end{vmatrix} \qquad \text{Factoring 4 out of the third column}$$

$$= 0 \qquad \text{By Theorem 4}$$

DO EXERCISE 8.

Theorem 6

If each element in a row (or column) is multiplied by a number *k*, and the products are added to the corresponding elements of another row (or column), we do not change the value of the determinant.

EXAMPLE 10   Find a determinant having the same value as the one on the left by adding three times the second column to the first column.

$$\begin{vmatrix} 0 & 1 & 2 \\ 4 & 5 & 6 \\ 7 & 8 & 9 \end{vmatrix} = \begin{vmatrix} 0 + 3(1) & 1 & 2 \\ 4 + 3(5) & 5 & 6 \\ 7 + 3(8) & 8 & 9 \end{vmatrix} = \begin{vmatrix} 3 & 1 & 2 \\ 19 & 5 & 6 \\ 31 & 8 & 9 \end{vmatrix}$$

EXAMPLE 11   Find a determinant having the same value as the one on the left by adding two times the third row to the first row.

$$\begin{vmatrix} 0 & 1 & 2 \\ 4 & 5 & 6 \\ 7 & 8 & 9 \end{vmatrix} = \begin{vmatrix} 0 + 2(7) & 1 + 2(8) & 2 + 2(9) \\ 4 & 5 & 6 \\ 7 & 8 & 9 \end{vmatrix} = \begin{vmatrix} 14 & 17 & 20 \\ 4 & 5 & 6 \\ 7 & 8 & 9 \end{vmatrix}$$

Proof of Theorem 6.   We prove the theorem for the case of a 3×3 matrix where *k* times the first column has been added to the third column. Let

$$\mathbf{A} = \begin{bmatrix} a_{11} & a_{12} & a_{13} \\ a_{21} & a_{22} & a_{23} \\ a_{31} & a_{32} & a_{33} \end{bmatrix} \quad \text{and} \quad \mathbf{B} = \begin{bmatrix} a_{11} & a_{12} & ka_{11} + a_{13} \\ a_{21} & a_{22} & ka_{21} + a_{23} \\ a_{31} & a_{32} & ka_{31} + a_{33} \end{bmatrix}.$$

To show that $|\mathbf{A}| = |\mathbf{B}|$, we evaluate $|\mathbf{B}|$ by expanding about the third column:

$$|\mathbf{B}| = (ka_{11} + a_{13})A_{13} + (ka_{21} + a_{23})A_{23} + (ka_{31} + a_{33})A_{33}$$

$$= k(a_{11}A_{13} + a_{21}A_{23} + a_{31}A_{33}) + (a_{13}A_{13} + a_{23}A_{23} + a_{33}A_{33})$$

$$= k(a_{11}A_{13} + a_{21}A_{23} + a_{31}A_{33}) + |\mathbf{A}|$$

$$= k \cdot \begin{vmatrix} a_{11} & a_{12} & a_{11} \\ a_{21} & a_{22} & a_{21} \\ a_{31} & a_{32} & a_{31} \end{vmatrix} + |\mathbf{A}|$$

$$= k(0) + |\mathbf{A}| \qquad \text{By Theorem 4}$$

$$= |\mathbf{A}|.$$

DO EXERCISE 9.

We can use the properties of determinants to simplify their evaluation. We try to find another determinant where in some row or column one element is 1 and the rest are 0.

EXAMPLE 12   Evaluate by first simplifying to a determinant where in one row or column one element is 1 and the rest are 0.

$$\begin{vmatrix} 6 & 2 & 3 \\ 6 & -1 & 5 \\ -2 & 3 & 1 \end{vmatrix}$$

We will try to get two 0's and a 1 in the third row. It already has a 1; that is why we picked the third row. We first factor a 2 out of column 1:

$$2 \cdot \begin{vmatrix} 3 & 2 & 3 \\ 3 & -1 & 5 \\ -1 & 3 & 1 \end{vmatrix}. \qquad \text{Theorem 5}$$

9. Find a determinant equal to this one by adding twice the first row to the second row.

$$\begin{vmatrix} -2 & 3 & 4 \\ 1 & 4 & -3 \\ 0 & 9 & 7 \end{vmatrix}$$

Evaluate by first simplifying to a determinant where in one row or column one element is 1 and the rest are 0.

10. $\begin{vmatrix} 3 & -1 & 1 \\ 2 & 2 & -4 \\ 2 & 4 & 1 \end{vmatrix}$

11. $\begin{vmatrix} 5 & -4 & 2 & -2 \\ 3 & -3 & -4 & 7 \\ -2 & 3 & 2 & 4 \\ -8 & 9 & 5 & -5 \end{vmatrix}$

12. Factor.

$$\begin{vmatrix} a^2 & b^2 & c^2 \\ a & b & c \\ 1 & 1 & 1 \end{vmatrix}$$

Now multiply each element in column 3 by -3 and add the corresponding elements to column 2:

$$2 \cdot \begin{vmatrix} 3 & -7 & 3 \\ 3 & -16 & 5 \\ -1 & 0 & 1 \end{vmatrix}. \qquad \text{Theorem 6}$$

Now add the elements in column 3 to the corresponding elements in column 1 (Theorem 6):

$$2 \cdot \begin{vmatrix} 6 & -7 & 3 \\ 8 & -16 & 5 \\ 0 & 0 & 1 \end{vmatrix}.$$

Finally, evaluate the determinant by expanding about the last row:

$$2 \cdot \left(0 - 0 + 1 \cdot \begin{vmatrix} 6 & -7 \\ 8 & -16 \end{vmatrix}\right) = 2 \cdot [6(-16) - 8(-7)] = -80.$$

DO EXERCISES 10 AND 11.

11 FACTORING CERTAIN DETERMINANTS

EXAMPLE 13 Factor

$$\begin{vmatrix} 1 & x & x^2 \\ 1 & y & y^2 \\ 1 & z & z^2 \end{vmatrix}.$$

$$\begin{vmatrix} 1 & x & x^2 \\ 1 & y & y^2 \\ 1 & z & z^2 \end{vmatrix} = \begin{vmatrix} 0 & x-y & x^2-y^2 \\ 1 & y & y^2 \\ 0 & z-y & z^2-y^2 \end{vmatrix} \qquad \begin{array}{l}\text{By Theorem 6, adding } -1 \text{ times the second} \\ \text{row to the first row; and } -1 \text{ times the} \\ \text{second row to the third row}\end{array}$$

$$= (x-y)(z-y) \cdot \begin{vmatrix} 0 & 1 & x+y \\ 1 & y & y^2 \\ 0 & 1 & z+y \end{vmatrix} \qquad \begin{array}{l}\text{By Theorem 5, factoring } x-y \\ \text{out of the first row and } z-y \\ \text{out of the third row}\end{array}$$

$$= (x-y)(z-y) \cdot \begin{vmatrix} 0 & 0 & x-z \\ 1 & y & y^2 \\ 0 & 1 & z+y \end{vmatrix} \qquad \begin{array}{l}\text{By Theorem 6, adding } -1 \text{ times} \\ \text{the third row to the first row}\end{array}$$

$$= (x-y)(z-y)(x-z) \cdot \begin{vmatrix} 0 & 0 & 1 \\ 1 & y & y^2 \\ 0 & 1 & z+y \end{vmatrix} \qquad \begin{array}{l}\text{By Theorem 5, factoring} \\ x-z \text{ out of the first row}\end{array}$$

$$= (x-y)(z-y)(x-z) \qquad \begin{array}{l}\text{Expanding the determinant about the} \\ \text{first row, we get 1.}\end{array}$$

DO EXERCISE 12.

EXERCISE SET 6.3

[i] Evaluate by first simplifying to a determinant where in one row or column one element is 1 and the rest are 0.

1. $\begin{vmatrix} -4 & 5 \\ 6 & 10 \end{vmatrix}$

2. $\begin{vmatrix} 3 & -9 \\ -2 & 4 \end{vmatrix}$

3. $\begin{vmatrix} 2 & 1 & 1 \\ 2 & -3 & -1 \\ -4 & 5 & 2 \end{vmatrix}$

4. $\begin{vmatrix} 1 & 2 & 4 \\ 2 & 3 & 5 \\ 3 & 1 & 6 \end{vmatrix}$

5. $\begin{vmatrix} 11 & -15 & 20 \\ 16 & 24 & -8 \\ 6 & 9 & 15 \end{vmatrix}$

6. $\begin{vmatrix} 4 & -24 & 15 \\ -3 & 18 & -6 \\ 5 & -4 & 3 \end{vmatrix}$

7. $\begin{vmatrix} -3 & 0 & 2 & 6 \\ 2 & 4 & 0 & -1 \\ -1 & 0 & -5 & 2 \\ 0 & -1 & -2 & -3 \end{vmatrix}$

8. $\begin{vmatrix} -2 & 1 & 0 & 5 \\ 3 & 0 & -4 & -2 \\ 4 & -6 & -8 & -1 \\ 8 & 0 & -2 & -3 \end{vmatrix}$

Evaluate each determinant without expanding.

9. $\begin{vmatrix} x & y & z \\ 0 & 0 & 0 \\ p & q & r \end{vmatrix}$

10. $\begin{vmatrix} 5 & 5 & 5 \\ 3 & 3 & 3 \\ 2 & -7 & 8 \end{vmatrix}$

11. $\begin{vmatrix} 2a & t & -7a \\ 2b & u & -7b \\ 2c & v & -7c \end{vmatrix}$

12. $\begin{vmatrix} a & -1 & 4a \\ b & 2 & 4b \\ x & -3 & 4x \end{vmatrix}$

ii Factor.

13. $\begin{vmatrix} x^2 & x & 1 \\ y^2 & y & 1 \\ z^2 & z & 1 \end{vmatrix}$

14. $\begin{vmatrix} 1 & 1 & 1 \\ a & b & c \\ a^2 & b^2 & c^2 \end{vmatrix}$

15. $\begin{vmatrix} x & x^2 & x^3 \\ y & y^2 & y^3 \\ z & z^2 & z^3 \end{vmatrix}$

16. $\begin{vmatrix} 1 & 1 & 1 \\ a & b & c \\ a^3 & b^3 & c^3 \end{vmatrix}$

17. Consider a triangle with vertices (x_1, y_1), (x_2, y_2), and (x_3, y_3). The area of this triangle is the absolute value of

$$\frac{1}{2} \cdot \begin{vmatrix} x_1 & y_1 & 1 \\ x_2 & y_2 & 1 \\ x_3 & y_3 & 1 \end{vmatrix}.$$

Prove this. (*Hint:* Look at this drawing. The area of triangle *ABC* is the area of trapezoid *ABDE* plus the area of trapezoid *AEFC* minus the area of trapezoid *BDFC*.)

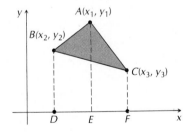

18. Prove that the lines $a_1 x + b_1 y = c_1$ and $a_2 x + b_2 y = c_2$ are parallel when

$$\begin{vmatrix} a_1 & b_1 \\ a_2 & b_2 \end{vmatrix} = 0$$

and either

$$\begin{vmatrix} c_1 & b_1 \\ c_2 & b_2 \end{vmatrix} \neq 0 \quad \text{or} \quad \begin{vmatrix} a_1 & c_1 \\ a_2 & c_2 \end{vmatrix} \neq 0.$$

6.4

INVERSES OF MATRICES

We use the symbol **I** to represent matrices of the type

$$\begin{bmatrix} 1 & 0 \\ 0 & 1 \end{bmatrix}, \quad \begin{bmatrix} 1 & 0 & 0 \\ 0 & 1 & 0 \\ 0 & 0 & 1 \end{bmatrix}.$$

Note that these are square matrices with 1's extending from the upper left to the lower right (along the main diagonal), and 0's elsewhere.

OBJECTIVES

You should be able to:

i Calculate the inverse of a square matrix, if it exists, using the cofactor method.

ii Calculate the inverse of a square matrix, if it exists, using the Gauss–Jordan reduction method.

iii Use matrix inverses to solve systems of *n* equations in *n* variables.

1. Let

$$\mathbf{A} = \begin{bmatrix} a & b \\ c & d \end{bmatrix}$$

and

$$\mathbf{I} = \begin{bmatrix} 1 & 0 \\ 0 & 1 \end{bmatrix}.$$

a) Find $\mathbf{AI}$.

b) Find $\mathbf{IA}$.

The following property can be shown easily for the 2×2 and 3×3 cases (see Exercise 33 in Exercise Set 6.4).

DO EXERCISE 1.

Theorem 7

For any $n \times n$ matrices A and I, AI = IA = A (I is a multiplicative identity).

We usually say simply that $\mathbf{I}$ is an *identity matrix*.

Suppose a matrix $\mathbf{A}$ has a *multiplicative inverse*, or simply *inverse*, $\mathbf{A}^{-1}$. Then $\mathbf{A}^{-1}$ is a matrix for which

$$\mathbf{A} \cdot \mathbf{A}^{-1} = \mathbf{A}^{-1} \cdot \mathbf{A} = \mathbf{I}.$$

For example, for the matrix

$$\mathbf{A} = \begin{bmatrix} 5 & 3 \\ 3 & 2 \end{bmatrix},$$

we have

$$\mathbf{A}^{-1} = \begin{bmatrix} 2 & -3 \\ -3 & 5 \end{bmatrix}.$$

We can check that $\mathbf{A} \cdot \mathbf{A}^{-1} = \mathbf{I}$ as follows:

$$\mathbf{A} \cdot \mathbf{A}^{-1} = \begin{bmatrix} 5 & 3 \\ 3 & 2 \end{bmatrix}\begin{bmatrix} 2 & -3 \\ -3 & 5 \end{bmatrix} = \begin{bmatrix} 1 & 0 \\ 0 & 1 \end{bmatrix} = \mathbf{I}.$$

We leave it to the student to verify that $\mathbf{A}^{-1} \cdot \mathbf{A} = \mathbf{I}$.

DO EXERCISE 2.

CALCULATING MATRIX INVERSES

In this section we consider two ways of calculating the inverse of a square matrix, if it exists. We shall see later that such inverses exist only when the determinant of the matrix is nonzero.

2. Let

$$\mathbf{A} = \begin{bmatrix} 3 & 1 & 0 \\ 1 & -1 & 2 \\ 1 & 1 & 1 \end{bmatrix}$$

and

$$\mathbf{A}^{-1} = \frac{1}{8}\begin{bmatrix} 3 & 1 & -2 \\ -1 & -3 & 6 \\ -2 & 2 & 4 \end{bmatrix}.$$

a) Find $\mathbf{AA}^{-1}$.

b) Find $\mathbf{A}^{-1}\mathbf{A}$.

c) Compare $\mathbf{AA}^{-1}$ and $\mathbf{A}^{-1}\mathbf{A}$.

▪ The Cofactor Method

Before describing this method we need some additional notation.

Definition

The *transpose* of a matrix A, denoted A^t, is found by interchanging the rows and columns of A.

EXAMPLE 1 Find $\mathbf{A}^t$, $\mathbf{B}^t$, $\mathbf{C}^t$, and $\mathbf{D}^t$.

$$\mathbf{A} = \begin{bmatrix} 2 & 4 & 6 \\ 9 & 8 & -2 \\ 0 & -1 & 4 \end{bmatrix}, \quad \mathbf{B} = \begin{bmatrix} -3 & 0 & 4 \\ 7 & 1 & 6 \end{bmatrix}, \quad \mathbf{C} = \begin{bmatrix} -4 \\ 3 \\ 2 \end{bmatrix}, \quad \mathbf{D} = [-1\ 2\ 3\ 0].$$

The transposes are as follows:

$$\mathbf{A}^t = \begin{bmatrix} 2 & 9 & 0 \\ 4 & 8 & -1 \\ 6 & -2 & 4 \end{bmatrix},$$

The transpose of a square matrix can be found by "reflecting" across the main diagonal.

$$\mathbf{B}^t = \begin{bmatrix} -3 & 7 \\ 0 & 1 \\ 4 & 6 \end{bmatrix}, \qquad \mathbf{C}^t = \begin{bmatrix} -4 & 3 & 2 \end{bmatrix}, \qquad \mathbf{D}^t = \begin{bmatrix} -1 \\ 2 \\ 3 \\ 0 \end{bmatrix}.$$

DO EXERCISE 3.

The following is a procedure for calculating the inverse of a square matrix.

The cofactor method

To find the inverse A^{-1} of a square matrix A,

a) Find the cofactor of each element;

b) Replace each element by its cofactor;

c) Find the transpose of the matrix found in (b);

d) Multiply the matrix in (c) by $1/|A|$. The result is A^{-1}.

EXAMPLE 2   Find $\mathbf{A}^{-1}$.

$$\mathbf{A} = \begin{bmatrix} 3 & 5 \\ 1 & -2 \end{bmatrix}$$

a)   Find the cofactor of each element:

$A_{11} = (-1)^{1+1}(-2)$
$\qquad = -2,$
$A_{12} = (-1)^{1+2}(1) = -1,$
$A_{21} = (-1)^{2+1}(5) = -5,$
$A_{22} = (-1)^{2+2}(3) = 3.$

The determinant of a 1×1 matrix is just the number. For example, $|[-2]| = -2$. Don't confuse determinant notation with absolute-value notation.

b)   Replace each element by its cofactor.

$$\mathbf{A} = \begin{bmatrix} 3 & 5 \\ 1 & -2 \end{bmatrix} \qquad \begin{bmatrix} -2 & -1 \\ -5 & 3 \end{bmatrix}$$

c)   Find the transpose of the matrix found in (b):

The transpose of $\begin{bmatrix} -2 & -1 \\ -5 & 3 \end{bmatrix}$ is $\begin{bmatrix} -2 & -5 \\ -1 & 3 \end{bmatrix}$.

d)   Multiply by $\dfrac{1}{|A|}$:

$$|\mathbf{A}| = 3(-2) - 1(5) = -11,$$

$$\mathbf{A}^{-1} = \frac{1}{-11} \cdot \begin{bmatrix} -2 & -5 \\ -1 & 3 \end{bmatrix} = \begin{bmatrix} \frac{2}{11} & \frac{5}{11} \\ \frac{1}{11} & -\frac{3}{11} \end{bmatrix}.$$

DO EXERCISES 4 AND 5.

3. Find $\mathbf{A}^t$, $\mathbf{B}^t$, $\mathbf{C}^t$, and $\mathbf{D}^t$.

$$\mathbf{A} = \begin{bmatrix} -8 & 1 & -2 \\ -4 & 0 & -1 \\ 6 & 7 & 8 \end{bmatrix},$$

$$\mathbf{B} = \begin{bmatrix} -7 & 9 & 10 & 4 \end{bmatrix},$$

$$\mathbf{C} = \begin{bmatrix} -20 \\ 11 \end{bmatrix},$$

$$\mathbf{D} = \begin{bmatrix} -4 & 5 \\ 1 & 0 \\ 0 & 1 \end{bmatrix}$$

Find $\mathbf{A}^{-1}$. Use the cofactor method.

4. $\mathbf{A} = \begin{bmatrix} 3 & 5 \\ 2 & 4 \end{bmatrix}$

5. $\mathbf{A} = \begin{bmatrix} 1 & 0 \\ 2 & -1 \end{bmatrix}$

6. Find A^{-1}. Use the cofactor method.

$$A = \begin{bmatrix} 3 & 1 & 0 \\ 1 & -1 & 2 \\ 1 & 1 & 1 \end{bmatrix}$$

EXAMPLE 3 Find A^{-1}.

$$A = \begin{bmatrix} 2 & -1 & 1 \\ 1 & -2 & 3 \\ 4 & 1 & 2 \end{bmatrix}$$

a) Find the cofactor of each element.

$$A_{11} = (-1)^{1+1} \cdot \begin{vmatrix} -2 & 3 \\ 1 & 2 \end{vmatrix} = -7, \qquad A_{12} = (-1)^{1+2} \cdot \begin{vmatrix} 1 & 3 \\ 4 & 2 \end{vmatrix} = 10,$$

$$A_{13} = (-1)^{1+3} \cdot \begin{vmatrix} 1 & -2 \\ 4 & 1 \end{vmatrix} = 9, \qquad A_{21} = (-1)^{2+1} \cdot \begin{vmatrix} -1 & 1 \\ 1 & 2 \end{vmatrix} = 3,$$

$$A_{22} = (-1)^{2+2} \cdot \begin{vmatrix} 2 & 1 \\ 4 & 2 \end{vmatrix} = 0, \qquad A_{23} = (-1)^{2+3} \cdot \begin{vmatrix} 2 & -1 \\ 4 & 1 \end{vmatrix} = -6,$$

$$A_{31} = (-1)^{3+1} \cdot \begin{vmatrix} -1 & 1 \\ -2 & 3 \end{vmatrix} = -1, \qquad A_{32} = (-1)^{3+2} \cdot \begin{vmatrix} 2 & 1 \\ 1 & 3 \end{vmatrix} = -5,$$

$$A_{33} = (-1)^{3+3} \cdot \begin{vmatrix} 2 & -1 \\ 1 & -2 \end{vmatrix} = -3.$$

b) Replace each element by its cofactor:

$$A = \begin{bmatrix} 2 & -1 & 1 \\ 1 & -2 & 3 \\ 4 & 1 & 2 \end{bmatrix} \longrightarrow \begin{bmatrix} -7 & 10 & 9 \\ 3 & 0 & -6 \\ -1 & -5 & -3 \end{bmatrix}.$$

c) Find the transpose of the matrix found in (b).

The transpose of $\begin{bmatrix} -7 & 10 & 9 \\ 3 & 0 & -6 \\ -1 & -5 & -3 \end{bmatrix}$ is $\begin{bmatrix} -7 & 3 & -1 \\ 10 & 0 & -5 \\ 9 & -6 & -3 \end{bmatrix}.$

d) Multiply by $\dfrac{1}{|A|}$:

$$|A| = a_{11}A_{11} + a_{21}A_{21} + a_{31}A_{31} \qquad \text{Expanding about the first column to evaluate } |A|$$
$$= 2(-7) + 1(3) + 4(-1)$$
$$= -15,$$

$$A^{-1} = \frac{1}{-15} \begin{bmatrix} -7 & 3 & -1 \\ 10 & 0 & -5 \\ 9 & -6 & -3 \end{bmatrix}$$

$$= \begin{bmatrix} \frac{7}{15} & -\frac{1}{5} & \frac{1}{15} \\ -\frac{2}{3} & 0 & \frac{1}{3} \\ -\frac{3}{5} & \frac{2}{5} & \frac{1}{5} \end{bmatrix}.$$

If $|A|$ is 0, then $1/|A|$ is not defined and A^{-1} does not exist.

DO EXERCISE 6.

ii The Gauss–Jordan Reduction Method

We now consider a different method. We again find the inverse of the matrix of Example 3:

$$A = \begin{bmatrix} 2 & -1 & 1 \\ 1 & -2 & 3 \\ 4 & 1 & 2 \end{bmatrix}.$$

First we form a new (*augmented*) matrix consisting, on the left, of the matrix **A** and, on the right, of the corresponding identity matrix **I**:

$$\begin{bmatrix} 2 & -1 & 1 \\ 1 & -2 & 3 \\ 4 & 1 & 2 \end{bmatrix} \begin{bmatrix} 1 & 0 & 0 \\ 0 & 1 & 0 \\ 0 & 0 & 1 \end{bmatrix}.$$

The matrix **A** ————⟋   ⟍——— The identity matrix **I**

We now proceed in a manner very much like that described in Section 5.4. We attempt to transform **A** to an identity matrix, but whatever operations we perform, we do on the entire augmented matrix. When we finish we will get a matrix like the following:

$$\begin{bmatrix} 1 & 0 & 0 & a & b & c \\ 0 & 1 & 0 & d & e & f \\ 0 & 0 & 1 & g & h & i \end{bmatrix}.$$

The matrix on the right,

$$\begin{bmatrix} a & b & c \\ d & e & f \\ g & h & i \end{bmatrix},$$

will be $\mathbf{A}^{-1}$.

EXAMPLE 4   Find $\mathbf{A}^{-1}$.

$$\mathbf{A} = \begin{bmatrix} 2 & -1 & 1 \\ 1 & -2 & 3 \\ 4 & 1 & 2 \end{bmatrix}$$

a) We find the augmented matrix consisting of **A** and **I**:

$$\begin{bmatrix} 2 & -1 & 1 & 1 & 0 & 0 \\ 1 & -2 & 3 & 0 & 1 & 0 \\ 4 & 1 & 2 & 0 & 0 & 1 \end{bmatrix}.$$

b) We interchange the first and second rows so that the elements of the first column are multiples of the top number on the main diagonal:

$$\begin{bmatrix} 1 & -2 & 3 & 0 & 1 & 0 \\ 2 & -1 & 1 & 1 & 0 & 0 \\ 4 & 1 & 2 & 0 & 0 & 1 \end{bmatrix}.$$

c) Next we obtain 0's in the rest of the first column. We multiply the first row by -2 and add it to the second row. Then we multiply the first row by -4 and add it to the third row:

$$\begin{bmatrix} 1 & -2 & 3 & 0 & 1 & 0 \\ 0 & 3 & -5 & 1 & -2 & 0 \\ 0 & 9 & -10 & 0 & -4 & 1 \end{bmatrix}.$$

d) Next we move down the main diagonal to the number 3. We note that the number below it, 9, is a multiple of 3. We multiply the second row by -3 and add it to the third:

$$\begin{bmatrix} 1 & -2 & 3 & 0 & 1 & 0 \\ 0 & 3 & -5 & 1 & -2 & 0 \\ 0 & 0 & 5 & -3 & 2 & 1 \end{bmatrix}.$$

Find $\mathbf{A}^{-1}$. Use the Gauss–Jordan reduction method.

7. A $= \begin{bmatrix} 1 & 0 & 1 \\ 2 & 1 & 0 \\ 1 & -1 & 1 \end{bmatrix}$

8. A $= \begin{bmatrix} 3 & 5 \\ 2 & 4 \end{bmatrix}$

e) Now we move down the main diagonal to the number 5. We check to see if each number above 5 in the third column is a multiple of 5. Since this is not the case, we multiply the first row by -5:

$$\begin{bmatrix} -5 & 10 & -15 & 0 & -5 & 0 \\ 0 & 3 & -5 & 1 & -2 & 0 \\ 0 & 0 & 5 & -3 & 2 & 1 \end{bmatrix}.$$

f) Now we work back up. We add the third row to the second. We also multiply the third row by 3 and add it to the first:

$$\begin{bmatrix} -5 & 10 & 0 & -9 & 1 & 3 \\ 0 & 3 & 0 & -2 & 0 & 1 \\ 0 & 0 & 5 & -3 & 2 & 1 \end{bmatrix}.$$

g) We move back to the number 3 on the main diagonal. We multiply the first row by -3, so the element on the top of the second column is a multiple of 3:

$$\begin{bmatrix} 15 & -30 & 0 & 27 & -3 & -9 \\ 0 & 3 & 0 & -2 & 0 & 1 \\ 0 & 0 & 5 & -3 & 2 & 1 \end{bmatrix}.$$

h) We multiply the second row by 10 and add it to the first:

$$\begin{bmatrix} 15 & 0 & 0 & 7 & -3 & 1 \\ 0 & 3 & 0 & -2 & 0 & 1 \\ 0 & 0 & 5 & -3 & 2 & 1 \end{bmatrix}.$$

i) Finally, we get all 1's on the main diagonal. We multiply the first row by $\frac{1}{15}$, the second by $\frac{1}{3}$, and the third by $\frac{1}{5}$:

$$\begin{bmatrix} 1 & 0 & 0 & \frac{7}{15} & -\frac{1}{5} & \frac{1}{15} \\ 0 & 1 & 0 & -\frac{2}{3} & 0 & \frac{1}{3} \\ 0 & 0 & 1 & -\frac{3}{5} & \frac{2}{5} & \frac{1}{5} \end{bmatrix}.$$

We now have the matrix $\mathbf{I}$ on the left. Thus

$$\mathbf{A}^{-1} = \begin{bmatrix} \frac{7}{15} & -\frac{1}{5} & \frac{1}{15} \\ -\frac{2}{3} & 0 & \frac{1}{3} \\ -\frac{3}{5} & \frac{2}{5} & \frac{1}{5} \end{bmatrix}.$$

With either method the student can check by doing the multiplication $\mathbf{A}^{-1}\mathbf{A}$ or $\mathbf{A}\mathbf{A}^{-1}$. If we cannot obtain the identity matrix on the left using the Gauss–Jordan reduction method, as would be the case when a system has no solution or infinitely many solutions, then $\mathbf{A}^{-1}$ does not exist.

DO EXERCISES 7 AND 8.

iii SOLVING SYSTEMS USING INVERSES

We can use matrix inverses to solve certain kinds of systems. Consider the system

$$3x_1 + 5x_2 = -1,$$
$$x_1 - 2x_2 = 4.$$

We write a matrix equation equivalent to this system:

$$\begin{bmatrix} 3 & 5 \\ 1 & -2 \end{bmatrix} \cdot \begin{bmatrix} x_1 \\ x_2 \end{bmatrix} = \begin{bmatrix} -1 \\ 4 \end{bmatrix}.$$

Now we let

$$\begin{bmatrix} 3 & 5 \\ 1 & -2 \end{bmatrix} = \mathbf{A}, \quad \begin{bmatrix} x_1 \\ x_2 \end{bmatrix} = \mathbf{X}, \quad \text{and} \quad \begin{bmatrix} -1 \\ 4 \end{bmatrix} = \mathbf{B}.$$

Then we have

$$\mathbf{A} \cdot \mathbf{X} = \mathbf{B}.$$

To solve this equation, we first find $\mathbf{A}^{-1}$. We found in Example 2 that

$$\mathbf{A}^{-1} = -\frac{1}{11} \begin{bmatrix} -2 & -5 \\ -1 & 3 \end{bmatrix}.$$

We solve the matrix equation $\mathbf{A} \cdot \mathbf{X} = \mathbf{B}$:

$\mathbf{A}^{-1} \cdot \mathbf{A} \cdot \mathbf{X} = \mathbf{A}^{-1} \cdot \mathbf{B}$ Multiplying by $\mathbf{A}^{-1}$ on the left on each side

$\mathbf{I} \cdot \mathbf{X} = \mathbf{A}^{-1} \cdot \mathbf{B}$ Since $\mathbf{A}^{-1} \cdot \mathbf{A} = \mathbf{I}$

$\mathbf{X} = \mathbf{A}^{-1} \cdot \mathbf{B}$ Since $\mathbf{I}$ is an identity

Now we have, substituting,

$$\mathbf{X} = \mathbf{A}^{-1} \cdot \mathbf{B}$$

$$\begin{bmatrix} x_1 \\ x_2 \end{bmatrix} = -\frac{1}{11} \begin{bmatrix} -2 & -5 \\ -1 & 3 \end{bmatrix} \cdot \begin{bmatrix} -1 \\ 4 \end{bmatrix}$$

$$= -\frac{1}{11} \begin{bmatrix} -18 \\ 13 \end{bmatrix}$$

$$= \begin{bmatrix} \frac{18}{11} \\ -\frac{13}{11} \end{bmatrix}.$$

The solution of the system of equations is therefore $x_1 = \frac{18}{11}$ and $x_2 = -\frac{13}{11}$.

DO EXERCISE 9.

9. Consider the system

$$4x_1 - 2x_2 = -1,$$
$$x_1 + 5x_2 = 1.$$

a) Write a matrix equation equivalent to the system.

b) Find the coefficient matrix $\mathbf{A}$.

c) Find $\mathbf{A}^{-1}$.

d) Use the inverse of the coefficient matrix to solve the system.

EXERCISE SET 6.4

i Find $\mathbf{A}^{-1}$, if it exists. Use the cofactor method. Check your answers by calculating $\mathbf{AA}^{-1}$ and $\mathbf{A}^{-1}\mathbf{A}$.

1. $\mathbf{A} = \begin{bmatrix} 3 & 2 \\ 5 & 3 \end{bmatrix}$

2. $\mathbf{A} = \begin{bmatrix} 3 & 5 \\ 1 & 2 \end{bmatrix}$

3. $\mathbf{A} = \begin{bmatrix} 11 & 3 \\ 7 & 2 \end{bmatrix}$

4. $\mathbf{A} = \begin{bmatrix} 8 & 5 \\ 5 & 3 \end{bmatrix}$

5. $\mathbf{A} = \begin{bmatrix} 4 & -3 \\ 1 & 2 \end{bmatrix}$

6. $\mathbf{A} = \begin{bmatrix} 0 & -1 \\ 1 & 0 \end{bmatrix}$

7. $\mathbf{A} = \begin{bmatrix} 3 & 1 & 0 \\ 1 & 1 & 1 \\ 1 & -1 & 2 \end{bmatrix}$

8. $\mathbf{A} = \begin{bmatrix} 1 & 0 & 1 \\ 2 & 1 & 0 \\ 1 & -1 & 1 \end{bmatrix}$

9. $\mathbf{A} = \begin{bmatrix} 1 & -1 & 2 \\ 0 & 1 & 3 \\ 2 & 1 & -2 \end{bmatrix}$

10. $\mathbf{A} = \begin{bmatrix} 1 & -1 & 2 \\ 0 & 1 & 2 \\ 1 & -3 & -4 \end{bmatrix}$

11. $\mathbf{A} = \begin{bmatrix} 1 & -4 & 8 \\ 1 & -3 & 2 \\ 2 & -7 & 10 \end{bmatrix}$

12. $\mathbf{A} = \begin{bmatrix} -2 & 5 & 3 \\ 4 & -1 & 3 \\ 7 & -2 & 5 \end{bmatrix}$

13. $\mathbf{A} = \begin{bmatrix} 1 & 2 & 3 & 4 \\ 0 & 1 & 3 & -5 \\ 0 & 0 & 1 & -2 \\ 0 & 0 & 0 & -1 \end{bmatrix}$

14. $\mathbf{A} = \begin{bmatrix} -2 & -3 & 4 & 1 \\ 0 & 1 & 1 & 0 \\ 0 & 4 & -6 & 1 \\ -2 & -2 & 5 & 1 \end{bmatrix}$

ii **15–28.** Find the inverse of each matrix in Exercises 1–14. Use the Gauss–Jordan reduction method.

iii For Exercises 29–32, write a matrix equation equivalent to the system. Find the inverse of the coefficient matrix. Use the inverse of the coefficient matrix to solve each system. Show your work.

29. $7x - 2y = -3,$
$\quad 9x + 3y = 4$

30. $5x_1 + 3x_2 = -2,$
$\quad 4x_1 - x_2 = 1$

31. $x_1 \qquad + x_3 = 1,$
$\quad 2x_1 + x_2 \qquad = 3,$
$\quad x_1 - x_2 + x_3 = 4$

32. $x + 2y + 3z = -1,$
$\quad 2x - 3y + 4z = 2,$
$\quad -3x + 5y - 6z = 4$

33. Let

$$A = \begin{bmatrix} a & b & c \\ d & e & f \\ g & h & i \end{bmatrix} \quad \text{and} \quad I = \begin{bmatrix} 1 & 0 & 0 \\ 0 & 1 & 0 \\ 0 & 0 & 1 \end{bmatrix}.$$

Show that $AI = IA = A$.

In each of the following, state the conditions under which A^{-1} exists. Then find a formula for A^{-1}.

34. $A = [x]$

35. $A = \begin{bmatrix} x & 0 \\ 0 & y \end{bmatrix}$

36. $A = \begin{bmatrix} 0 & 0 & x \\ 0 & y & 0 \\ z & 0 & 0 \end{bmatrix}$

37. $A = \begin{bmatrix} x & 1 & 1 & 1 \\ 0 & y & 0 & 0 \\ 0 & 0 & z & 0 \\ 0 & 0 & 0 & w \end{bmatrix}$

38. Consider

$$a_{11}x + a_{12}y = c_1,$$
$$a_{21}x + a_{22}y = c_2.$$

Use the cofactor method and the equivalent matrix equation $AX = C$ to prove Cramer's rule.

SUMMARY AND REVIEW: CHAPTER 6

The following contains a summary of what you should be able to do after completing this chapter. The review exercises are for practice. Answers are at the back of the book. If you miss an exercise, restudy the section indicated alongside the answer.

You should be able to:

Add, subtract, and multiply matrices when possible, and find the transpose of a matrix.

Let

$$A = \begin{bmatrix} 1 & -1 & 0 \\ 2 & 3 & -2 \\ -2 & 0 & 1 \end{bmatrix}, \quad B = \begin{bmatrix} -1 & 0 & 6 \\ 1 & -2 & 0 \\ 0 & 1 & -3 \end{bmatrix}, \quad \text{and} \quad C = \begin{bmatrix} -2 & 0 \\ 1 & 3 \end{bmatrix}.$$

Find each of the following, if possible.

1. $A + B$

2. $-3A$

3. $-A$

4. AB

5. $B + C$

6. $A - B$

7. A^t

8. B^t

Calculate the inverse of a square matrix, if it exists, using either the cofactor method or the Gauss–Jordan reduction method.

Find A^{-1} if it exists.

9. $A = \begin{bmatrix} -2 & 0 \\ 1 & 3 \end{bmatrix}$

10. $A = \begin{bmatrix} 0 & 0 & 3 \\ 0 & -2 & 0 \\ 4 & 0 & 0 \end{bmatrix}$

11. $A = \begin{bmatrix} 1 & 0 & 0 & 0 \\ 0 & 4 & -5 & 0 \\ 0 & 2 & 2 & 0 \\ 0 & 0 & 0 & 1 \end{bmatrix}$

Write a matrix equation equivalent to a system of equations.

12. Write a matrix equation equivalent to this system of equations.

$$3x - 2y + 4z = 13,$$
$$x + 5y - 3z = 7,$$
$$2x - 3y + 7z = -8$$

Find a specified element, a_{ij}, of a square matrix, its minor, and its cofactor, and be able to expand its determinant about any row or column.

Evaluate.

13. $\begin{vmatrix} -4 & \sqrt{3} \\ \sqrt{3} & 7 \end{vmatrix}$

14. $\begin{vmatrix} 1 & -1 & 2 \\ -1 & 2 & 0 \\ -1 & 3 & 1 \end{vmatrix}$

15. $\begin{vmatrix} 0 & a & b \\ -a & 0 & c \\ -b & -c & 0 \end{vmatrix}$

16. $\begin{vmatrix} 4 & -7 & 6 & 7 \\ 0 & -3 & 9 & -8 \\ 0 & 0 & -2 & 6 \\ 0 & 0 & 0 & 5 \end{vmatrix}$

Use properties of determinants to simplify their evaluation and to factor certain determinants.

17. Without expanding, show that

$$\begin{vmatrix} 5a & 5b & 5c \\ 3a & 3b & 3c \\ d & e & f \end{vmatrix} = 0.$$

Factor.

18. $\begin{vmatrix} 1 & a & bc \\ 1 & b & ac \\ 1 & c & ab \end{vmatrix}$

19. $\begin{vmatrix} 1 & x^2 & x^3 \\ 1 & y^2 & y^3 \\ 1 & z^2 & z^3 \end{vmatrix}$

20. $\begin{vmatrix} 1 & a & a^2 & a^3 \\ 1 & b & b^2 & b^3 \\ 1 & c & c^2 & c^3 \\ 1 & d & d^2 & d^3 \end{vmatrix}$

Use matrix inverses to solve systems of n equations in n variables.

21. Write a matrix equation equivalent to this system. Find the inverse of the coefficient matrix. Use the inverse of the coefficient matrix to solve each system. Show your work.

$$2x + 3y = 2,$$
$$5x - y = -29$$

22. On the basis of Exercise 16, conjecture and prove a theorem regarding determinants.

TEST: CHAPTER 6

Let

$$A = \begin{bmatrix} -1 & 3 \\ 0 & 4 \end{bmatrix}, \quad B = \begin{bmatrix} -2 & 1 & 0 \\ 3 & 2 & 5 \end{bmatrix}, \quad C = \begin{bmatrix} 4 & 0 \\ 1 & -1 \\ 2 & -3 \end{bmatrix}, \quad D = \begin{bmatrix} 0 & 2 & -1 \\ 3 & -1 & 1 \\ 0 & 4 & 3 \end{bmatrix}, \quad E = \begin{bmatrix} 5 & -1 & 3 \\ 0 & 4 & 2 \\ 1 & 0 & -6 \end{bmatrix},$$

$$F = [2 \quad -1 \quad -3], \quad G = \begin{bmatrix} -2 & -5 \\ 6 & -3 \end{bmatrix}, \quad O = \begin{bmatrix} 0 & 0 \\ 0 & 0 \end{bmatrix}, \quad I = \begin{bmatrix} 1 & 0 \\ 0 & 1 \end{bmatrix}.$$

Find each of the following, if possible.

1. GC

2. O + A

3. A + I

4. D + E

5. A − G

6. BC

7. B + C

8. −F

9. 2D − E

10. D^t

Find $\mathbf{A}^{-1}$, if it exists.

11. $\mathbf{A} = \begin{bmatrix} -3 & 1 \\ 2 & 0 \end{bmatrix}$

12. $\mathbf{A} = \begin{bmatrix} 3 & 4 & 3 \\ 1 & 0 & 1 \\ -2 & -5 & -2 \end{bmatrix}$

13. $\mathbf{A} = \begin{bmatrix} 2 & -1 & 0 \\ 3 & 0 & 1 \\ -2 & 4 & 0 \end{bmatrix}$

14. Let

$$\mathbf{A} = \begin{bmatrix} 3 & 1 & -2 \\ 2 & 0 & -1 \\ -5 & 0 & 4 \end{bmatrix}.$$

Find a_{12}, M_{12}, and A_{12}.

Evaluate.

15. $\begin{vmatrix} 1 & -1 & 3 \\ -2 & 4 & 2 \\ 1 & 2 & 1 \end{vmatrix}$

16. $\begin{vmatrix} -2 & 7 & -3 \\ 1 & 1 & 1 \\ 4 & -14 & 6 \end{vmatrix}$

17. $\begin{vmatrix} -2 & 3 & 4 & 6 \\ 1 & 0 & 0 & 0 \\ -1 & 1 & 2 & -3 \\ 0 & 5 & 1 & 3 \end{vmatrix}$

18. $\begin{vmatrix} 1 & -1 & -1 \\ 2 & -6 & 4 \\ -3 & 0 & 2 \end{vmatrix}$

19. Factor:

$$\begin{vmatrix} c^3 & b^3 & a^3 \\ c & b & a \\ 1 & 1 & 1 \end{vmatrix}.$$

20. Write a matrix equation equivalent to this system of equations and use the inverse of the coefficient matrix to solve the system. Show all your work.

$$2x - 3y = -9,$$
$$x + 4y = 1$$

21. Find k such that

$$\begin{vmatrix} 2 & 0 & 1 & -1 \\ 0 & 3 & -1 & 0 \\ 4 & 2 & 0 & k \\ 0 & 1 & 0 & -2 \end{vmatrix} = 18.$$

22. State the conditions under which $\mathbf{A}^{-1}$ exists and find a formula for $\mathbf{A}^{-1}$.

$$\mathbf{A} = \begin{bmatrix} x & 0 & z \\ 0 & y & 0 \\ 1 & 0 & 0 \end{bmatrix}$$

EXPONENTIAL AND LOGARITHMIC FUNCTIONS

Exponential functions are used in the radiocarbon dating of old bones.

In this chapter we shall consider two kinds of functions, closely related. The first kind, called *exponential functions*, are functions defined by using variables for the exponents. Such functions have many applications—one, for example, is to problems of population growth.

Imagine putting a function machine into reverse. If it works, it will represent what we call the *inverse* of the original function. Functions that are inverses of each other are thus closely related. The inverses of the exponential functions, called *logarithmic functions*, or *logarithm* functions, are also important in many applications. We shall also study logarithm functions and their properties in this chapter.

7

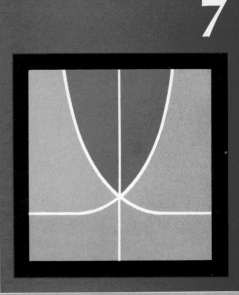

OBJECTIVES

You should be able to:

i Given an equation defining a relation, write an equation of the inverse relation.

ii Given a graph of a relation, sketch a graph of its inverse; or given an equation defining a relation, graph it and then graph its inverse.

iii Given an equation defining a relation, determine whether the graph is symmetric with respect to the line $y = x$.

iv Given the graph of a relation or an equation for a relation, determine whether its inverse is a function.

v Given a function defined by a simple formula, find a formula for its inverse.

vi For a function f whose inverse is also a function, quickly find $f^{-1}(f(x))$ and $f(f^{-1}(x))$ for any number x in the domains of the functions.

1. Write an equation of the inverse of each relation.

   a) $y = 3x + 2$ b) $y = x$
   c) $x^2 + 3y^2 = 4$ d) $y = 5x^2 + 2$
   e) $y^2 = 4x - 5$ f) $xy = 5$

2. a) Below, graph the relation containing (4, 1), (4, 2), (3, 2), (2, 3), (2, 4), and (1, 4).

   b) Interchange the members in each ordered pair to obtain the inverse.

   c) Graph the inverse and compare the graphs.

(1, 4)	(2, 4)	(3, 4)	(4, 4)
(1, 3)	(2, 3)	(3, 3)	(4, 3)
(1, 2)	(2, 2)	(3, 2)	(4, 2)
(1, 1)	(2, 1)	(3, 1)	(4, 1)

7.1

INVERSES OF RELATIONS

i INVERSES AND EQUATIONS

The *inverse* of a relation is the relation obtained by interchanging first and second members in each ordered pair. A relation is shown here, in color; its inverse is shaded.

(1, 4)	(2, 4)	(3, 4)	(4, 4)
(1, 3)	(2, 3)	(3, 3)	(4, 3)
(1, 2)	(2, 2)	(3, 2)	(4, 2)
(1, 1)	(2, 1)	(3, 1)	(4, 1)

If a relation is defined by an equation, an equation of the inverse can be found by interchanging the variables. The solutions of the second equation will be the same as those of the first, except that first and second members will be interchanged in each ordered pair.

> When a relation is defined by an equation, interchanging x and y produces an equation of the *inverse* relation.

EXAMPLE 1 Find an equation of the inverse of $y = x^2 - 5$.

We interchange x and y, and obtain $x = y^2 - 5$. This is an equation of the inverse relation.

DO EXERCISE 1.

ii GRAPHS OF INVERSE RELATIONS

Interchanging x and y in an equation, or interchanging first and second coordinates in each ordered pair of a relation, has the effect of interchanging the x-axis and the y-axis.

DO EXERCISES 2 AND 3 (EXERCISE 3 IS ON THE FOLLOWING PAGE).

Interchanging the x-axis and the y-axis has the effect of reflecting across the diagonal line whose equation is $y = x$, as shown below. Thus the graphs of a relation and its inverse are always reflections of each other across the line $y = x$. (This assumes, of course, that the same scale is used on both axes.)

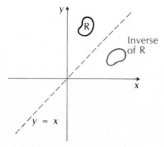

EXAMPLES 2–4 In each case a relation is shown in black. Graph the inverse.

The graph of each inverse is shown in color.

2. 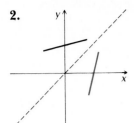   **3.** 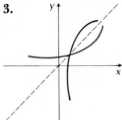   **4.**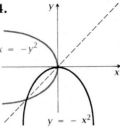

DO EXERCISE 4.

3. a) On a piece of thin paper, draw coordinate axes.

b) Draw the line $y = x$.

c) Draw a relation as shown here.

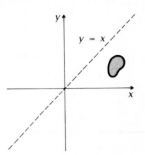

d) Flip the paper over (to interchange the x-axis and y-axis). Look through the paper. How do the graphs of the relation and its inverse compare?

4. Graph the inverse of each relation by reflecting across the line $y = x$.

a)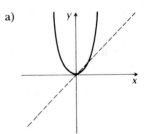

b)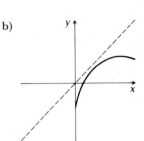

c) Graph $y = 4 - x^2$. Then by reflecting across the line $y = x$, graph its inverse.

iii Symmetry with Respect to the Line $y = x$

It can happen that a relation is its own inverse; that is, when x and y are interchanged or the relation is reflected across the line $y = x$, there is no change. Such a relation is symmetric with respect to the line $y = x$. The following are two examples of relations symmetric with respect to the line $y = x$.

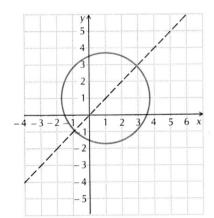 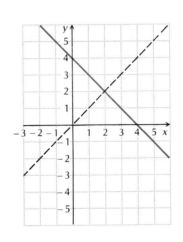

None of the relations shown in black in Examples 2–4 are symmetric with respect to the line $y = x$. If a relation is symmetric with respect to that line, that relation is its own inverse.

> For a relation defined by an equation: If interchanging x and y produces an equivalent equation, the relation is its own inverse, and the graph of the equation is symmetric with respect to the line $y = x$.

EXAMPLE 5   Test $3x + 3y = 5$ for symmetry with respect to the line $y = x$.

a) Interchange x and y. This amounts to replacing each occurrence of x by a y and each y by an x.

$$3x + 3y = 5$$
$$3y + 3x = 5$$

b) Is the resulting equation equivalent to the original?

The commutative law of addition guarantees that the resulting equation is equivalent to the original. Thus the graph is symmetric with respect to the line $y = x$.

Test the following for symmetry with respect to the line $y = x$.

5. $y = -x$

6. $x + y = 4$

7. $xy = 3$

8. $y = |x|$

9. $3x^2 + 3y^2 = 4$

10. $|x| = |y|$

11. $y = x^3$

12. $x - y = 4$

13. Which of the following have inverses that are functions?

a)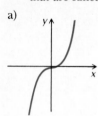

b)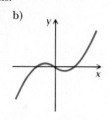

c)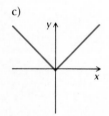

d)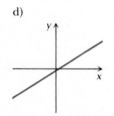

14. Graph the relation $y = x^2 - 1$ and determine whether it is a function. Then graph the inverse and determine whether it is a function.

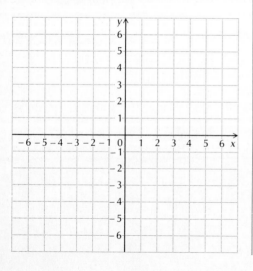

EXAMPLE 6 Test $y = x^2$ for symmetry with respect to the line $y = x$.

a) Interchange x and y.

$$y = x^2$$
$$x = y^2$$

b) Is the resulting equation equivalent to the original?

Note that $(2, 4)$ is a solution of the original equation $y = x^2$, but it is not a solution of the resulting equation $x = y^2$.

$y = x^2$	
4	2^2
	4

$x = y^2$	
2	4^2
	16

Thus the equations are not equivalent, so the graph of $y = x^2$ is *not* symmetric with respect to the line $y = x$.

DO EXERCISES 5–12.

iv INVERSES OF FUNCTIONS

Every function has an inverse, but that inverse may not be a function, as the following examples show.

EXAMPLE 7 The relation g given by

$$g = \{(5, 3), (3, 1), (-7, 2), (2, 3)\}$$

is a function. Find the inverse of g and determine whether it is a function.

The inverse of g is found by interchanging first and second members in each ordered pair, and is $\{(3, 5), (1, 3), (2, -7), (3, 2)\}$. It is *not* a function because the pairs $(3, 5)$ and $(3, 2)$ have the same first coordinates but different second coordinates.

EXAMPLE 8 Graph the relation $y = x^2$ and determine whether it is a function. Then graph its inverse and determine whether it is a function.

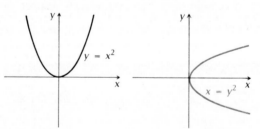

A function. It passes the vertical-line test.

Its inverse: not a function. It fails the vertical-line test.

It is possible to tell without graphing an inverse whether the inverse is a function. Simply apply a *horizontal-line test*. If a horizontal line can intersect a graph at more than one point, then the inverse is not a function. A horizontal line could intersect the black graph of Example 8 at more than one point, hence the inverse is not a function.

DO EXERCISES 13 AND 14.

v Finding Formulas for Inverses of Functions

All functions have inverses, but in only some cases is the inverse also a function. If the inverse of a function f is also a function, we denote it by f^{-1} (read "f inverse"). [*Caution:* This is *not* exponential notation!] Recall that we obtain the inverse of any relation by reversing the coordinates in each ordered pair. In terms of mapping, let us see what this means. A function *f maps* the set of first coordinates (the domain) D to the set of second coordinates (the range) R. The inverse mapping f^{-1} maps the range of f onto the domain of f. Each y in R is mapped onto just one x in D, provided f^{-1} is a function. Note that the domain of f is the range of f^{-1} and the range of f is the domain of f^{-1}.

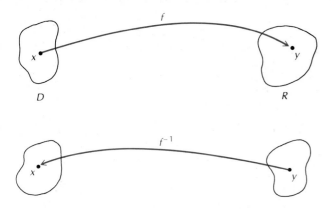

Let us consider inverses of functions in terms of function machines. Suppose that the function f programmed into a machine has an inverse that is also a function. Suppose then that the function machine has a reverse switch. When the switch is thrown, the machine is then programmed to do the inverse mapping f^{-1}. Inputs then enter at the opposite end and the entire process is reversed.

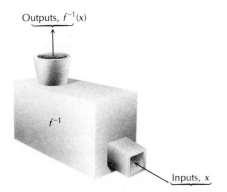

Outputs, $f^{-1}(x)$

f^{-1}

Inputs, x

When a function is defined by a formula, we can sometimes find a formula for its inverse by interchanging x and y.

EXAMPLE 9   Given $f(x) = x + 1$, find a formula for $f^{-1}(x)$.

a)   Let us think of this as $y = x + 1$.

b)   To find the inverse, we interchange x and y: $x = y + 1$.

15. Let $g(x) = 3x - 2$. Find a formula for $g^{-1}(x)$.

c)   Now we solve for y: $y = x - 1$.

d)   Thus $f^{-1}(x) = x - 1$.

Note in Example 9 that f maps any x onto $x + 1$ (this function adds 1 to each number of the domain). Its inverse, f^{-1}, maps any number x onto $x - 1$ (this inverse function subtracts 1 from each member of its domain). Thus the function and its inverse do opposite things.

EXAMPLE 10   Let $g(x) = 2x + 3$. Find a formula for $g^{-1}(x)$.

a)   Let us think of this as $y = 2x + 3$.

b)   To find the inverse, we interchange x and y: $x = 2y + 3$.

c)   Now we solve for y: $y = (x - 3)/2$.

d)   Thus $g^{-1}(x) = (x - 3)/2$.

Note in Example 10 that g maps any x onto $2x + 3$ (this function doubles each input and then adds 3). Its inverse, g^{-1}, maps any input onto $(x - 3)/2$ (this inverse function subtracts 3 from each input and then divides it by 2). Thus the function and its inverse do opposite things.

EXAMPLE 11   Consider $f(x) = x^2$. Find a formula for $f^{-1}(x)$.

Let us think of this as $y = x^2$. To find the inverse, we interchange x and y: $x = y^2$. Now we solve for y: $y = \pm\sqrt{x}$. Note that for each positive x we get two y's. For example, the pairs $(4, 2)$ and $(4, -2)$ belong to the relation. Look at the following graphs. Note also that the function fails the horizontal-line test, so the inverse fails the vertical-line test.

16. Let $f(x) = x^2 - 1$, and the domain of f is the set of all nonnegative real numbers. Find a formula for $f^{-1}(x)$.

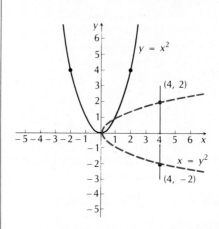

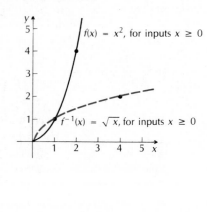

If we restrict the domain of $f(x) = x^2$ to nonnegative numbers, then its inverse is a function, $f^{-1}(x) = \sqrt{x}$.

DO EXERCISES 15 AND 16.

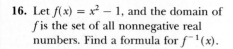 **Simplifying $f^{-1}(f(x))$ and $f(f^{-1}(x))$**

Suppose that the inverse of a function f is also a function. Then f and f^{-1} do opposite things. For an input x, if we find $f(x)$ and then find f^{-1} for the result, we will be back at x, that is,

$$f^{-1}(f(x)) = x.$$

Similarly, if we find $f^{-1}(x)$ and then find f for the result, we will be back where we started, at x, that is,

$$f(f^{-1}(x)) = x.$$

For the above statements to be true, the function values must all lie within the domains of the relevant functions. Following is a precise statement.

Theorem 1

For any function f whose inverse is also a function,

$$f^{-1}(f(a)) = a$$

and

$$f(f^{-1}(a)) = a,$$

where a is any number for which the function values are defined.

Proof. Suppose a is in the domain of f. Then $f(a) = b$, for some b in the range of f. Then the ordered pair (a, b) is in f, and by definition of f^{-1}, (b, a) is in f^{-1}. It follows that $f^{-1}(b) = a$. Then substituting $f(a)$ for b, we get $f^{-1}(f(a)) = a$. A similar proof shows that $f(f^{-1}(a)) = a$, for any a in the domain of f^{-1}.

EXAMPLE 12 For the function g of Example 10, find $g(4)$. Then find $g^{-1}(g(4))$.

$$g(x) = 2x + 3, \quad \text{so } g(4) = 11.$$

Now

$$g^{-1}(x) = \frac{x - 3}{2},$$

so

$$g^{-1}(11) = \frac{11 - 3}{2} = 4.$$

Thus

$$g^{-1}(g(4)) = 4.$$

DO EXERCISE 17.

EXAMPLE 13 For the function g of Example 12, find $g^{-1}(g(283))$. Find also $g(g^{-1}(-12{,}045))$.

We note that every real number is in the domain of both g and g^{-1}. Thus we may immediately write the answers, without calculating, using Theorem 1:

$$g^{-1}(g(283)) = 283,$$
$$g(g^{-1}(-12{,}045)) = -12{,}045.$$

DO EXERCISE 18.

17. Let $g(x) = 3x - 2$ as in Exercise 15. Find $g^{-1}(g(5))$. Find $g(g^{-1}(5))$ as in Example 12.

18. Let $g(x) = 3x - 2$ as in Exercise 17. For any number a, find $g^{-1}(g(a))$. Find $g(g^{-1}(a))$.

EXERCISE SET 7.1

i Write an equation of the inverse relation.

1. $y = 4x - 5$

2. $y = 3x + 5$

3. $x^2 - 3y^2 = 3$

4. $2x^2 + 5y^2 = 4$

5. $y = 3x^2 + 2$

6. $y = 5x^2 - 4$

7. $xy = 7$

8. $xy = -5$

ii

9. Graph $y = x^2 + 1$. Then by reflection across the line $y = x$, graph its inverse.

10. Graph $y = x^2 - 3$. Then by reflection across the line $y = x$, graph its inverse.

11. Graph $y = |x|$. Then by reflection across the line $y = x$, graph its inverse.

12. Graph $x = |y|$. Then by reflection across the line $y = x$, graph its inverse.

iii Test for symmetry with respect to the line $y = x$.

13. $3x + 2y = 4$

14. $5x - 2y = 7$

15. $4x + 4y = 3$

16. $5x + 5y = -1$

17. $xy = 10$

18. $xy = 12$

19. $3x = \dfrac{4}{y}$

20. $4y = \dfrac{5}{x}$

21. $y = |2x|$

22. $3x = |2y|$

23. $4x^2 + 4y^2 = 3$

24. $3x^2 + 3y^2 = 5$

iv

25. Which of the following have inverses that are functions?

a)
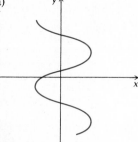

b)
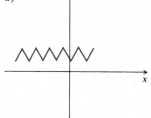

c)
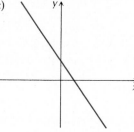

d)
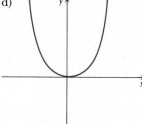

26. Which of the following have inverses that are functions?

a)
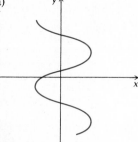

b)
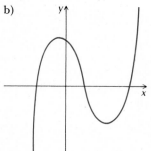

c)
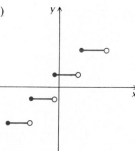

d)
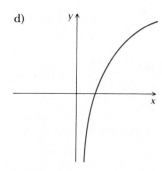

v

27. Let $f(x) = 2x + 5$. Find a formula for $f^{-1}(x)$.

28. Let $g(x) = 3x - 1$. Find a formula for $g^{-1}(x)$.

29. Let $f(x) = \sqrt{x + 1}$. Find a formula for $f^{-1}(x)$.

30. Let $g(x) = \sqrt{x - 1}$. Find a formula for $g^{-1}(x)$.

vi

31. Let $f(x) = 35x - 173$. Find $f^{-1}(f(3))$. Find $f(f^{-1}(-125))$.

32. Let $g(x) = \dfrac{-173x + 15}{3}$. Find $g^{-1}(g(5))$. Find $g(g^{-1}(-12))$.

33. Let $f(x) = x^3 + 2$. Find $f^{-1}(f(12{,}053))$. Find $f(f^{-1}(-17{,}243))$.

34. Let $g(x) = x^3 - 486$. Find $g^{-1}(g(489))$. Find $g(g^{-1}(-17{,}422))$.

35. Carefully graph $y = x^2$, using a large scale. Then use the graph to approximate $\sqrt{3.1}$.

36. Carefully graph $y = x^3$, using a large scale. Then use the graph to approximate $\sqrt[3]{-5.2}$.

37. Graph this equation and its inverse. Then test for symmetry with respect to the x-axis, the y-axis, the origin, and the line $y = x$.

$$y = \frac{1}{x^2}$$

38. Ice melts at $0°$ Celsius or $32°$ Fahrenheit. Water boils at $100°$ Celsius or $212°$ Fahrenheit.

   a) What linear transformation (see Section 3.5) converts Celsius temperature to Fahrenheit?

   b) Find the inverse of your answer to (a). Is it a function? What kind of conversion does it accomplish?

   c) At what temperature are the Celsius and Fahrenheit scales the same?

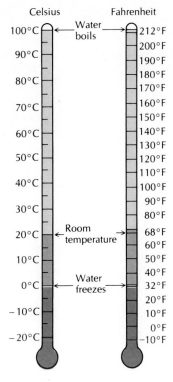

Graph each equation and its inverse. Then test for symmetry with respect to the x-axis, the y-axis, the origin, and the line $y = x$.

39. $|x| - |y| = 1$

40. $y = \dfrac{|x|}{x}$

7.2

EXPONENTIAL AND LOGARITHMIC FUNCTIONS

EXPONENTS AND FUNCTIONS

In this section, we define exponential functions and their inverses, logarithmic functions. We begin by considering irrational numbers as exponents.

Irrational Exponents

We have defined exponential notation for cases in which the exponent is a rational number. For example, $x^{2.34}$, or $x^{234/100}$, means to take the 100th root of x and raise the result to the 234th power. We now consider irrational exponents, such as π or $\sqrt{2}$.

Let us consider 2^{π}. We know that π has an unending decimal representation.

$$3.1415926535\ldots.$$

Now consider this sequence of numbers:

$$3, \quad 3.1, \quad 3.14, \quad 3.141, \quad 3.1415, \quad 3.14159, \ldots.$$

Each of these numbers is an approximation to π—the more decimal places the better the approximation. Let us use these (rational) numbers to form

OBJECTIVES

You should be able to:

i Graph exponential and logarithmic functions.

ii Given an exponential equation of the type $a^b = c$, write an equivalent logarithmic equation $\log_a c = b$, and vice versa.

iii Solve equations like $\log_3 9 = x$, $\log_x 9 = 2$, and $\log_3 x = 2$.

iv Simplify expressions like $a^{\log_a x}$ and $\log_a a^x$.

1. Use the graph of $y = 2^x$.

  a) Is this function increasing or decreasing?

  b) What is the domain?

  c) What is the range?

  d) What is the y-intercept?

  e) Use the graph to approximate $2^{\sqrt{3}}$ ($\sqrt{3} \approx 1.732$).

2. Graph $y = 4^x$, and compare it with the graph of $y = 2^x$.

  a) Is the function increasing or decreasing?

  b) What is the domain of $y = 4^x$?

  c) What is the range?

  d) What is the y-intercept?

  e) Use the graph to approximate $4^{\sqrt{2}}$ ($\sqrt{2} \approx 1.414$).

  f) Which function increases faster, 2^x or 4^x?

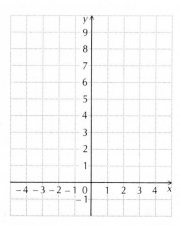

a sequence as follows:

$$2^3, \quad 2^{3.1}, \quad 2^{3.14}, \quad 2^{3.141}, \quad 2^{3.1415}, \quad 2^{3.14159}, \ldots .$$

Each of the numbers in this sequence is already defined, the exponent being rational. The numbers in this sequence get closer and closer to some real number. We define that number to be 2^π.

We can define exponential notation for any irrational exponent in a similar way. Thus any exponential expression a^x, $a > 0$, now has meaning, whether the exponent is rational or irrational. The usual laws of exponents still hold, in case exponents are irrational. We will not prove that fact here, however.

Exponential Functions

Exponential functions are defined using exponential notation, as follows.

Definition

The function $f(x) = a^x$, where a is some positive constant different from 1, is called the *exponential function, base a.*

It is important to keep in mind that for an exponential function as just defined, the variable is the exponent. The function $f(x) = 2^x$ is an exponential function, but the function $g(x) = x^2$ is not.

▐ⁱ GRAPHS OF EXPONENTIAL AND LOGARITHMIC FUNCTIONS

Exponential Functions

EXAMPLE 1 Graph $y = 2^x$. Use the graph to approximate 2^π.

We find some solutions, plot them, and then draw the graph.

x	0	1	2	3	-1	-2	-3
y	1	2	4	8	$\frac{1}{2}$	$\frac{1}{4}$	$\frac{1}{8}$

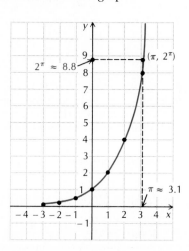

▦ (Note that as x increases, the function values increase. Check this on your calculator. As x decreases, the function values decrease toward 0. Check this on your calculator.)

To approximate 2^π, we locate π on the x-axis, at about 3.1. Then we find the corresponding function value. It is about 8.8.

DO EXERCISES 1 AND 2.

Let us now look at some other exponential functions. We will make comparisons, using transformations.

EXAMPLE 2 Graph $y = 4^x$.

We could plot points and connect them, but let us be more clever. We note that $4^x = (2^2)^x = 2^{2x}$. Thus the function we wish to graph is

$$y = 2^{2x}.$$

Compare this with $y = 2^x$, graphed on the preceding page. The graph of $y = 2^{2x}$ is a compression, in the x-direction toward the y-axis, of the graph of $y = 2^x$. Knowing this allows us to graph $y = 2^{2x}$ at once. Each point on the graph of 2^x is moved half the distance to the y-axis.

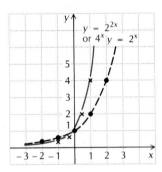

DO EXERCISES 3–5.

EXAMPLE 3 Graph $y = (\tfrac{1}{2})^x$.

We could plot points and connect them, but again let us be more clever. We note that $(\tfrac{1}{2})^x = 1/2^x = 2^{-x}$. Thus the function we wish to graph is

$$y = 2^{-x}.$$

Compare this with the graph of $y = 2^x$ in Example 1. The graph of $y = 2^{-x}$ is a reflection, across the y-axis, of the graph of $y = 2^x$. Knowing this allows us to graph $y = 2^{-x}$ at once.

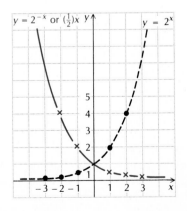

DO EXERCISE 6.

3. Graph $y = 8^x$. Use graph paper. [*Hint:* $8^x = (2^3)^x = 2^{3x}$, hence this is a horizontal compression of the graph of $y = 2^x$.]

4. Graph $y = 1^x$. Use graph paper.

5. Graph $y = (\tfrac{1}{2})^x$. Use graph paper.

6. Graph $y = (\tfrac{1}{3})^x$. [*Hint:* $(\tfrac{1}{3})^x = 3^{-x}$.]

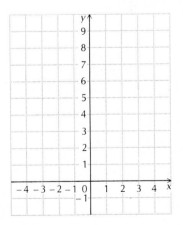

The preceding examples and exercises illustrate exponential functions of various bases. If $a = 1$, then $f(x) = a^x = 1^x = 1$ and the graph is a horizontal line. This is why we exclude 1 as a base for an exponential function. We summarize.

Theorem 2

> 1. When $a > 1$, the function $f(x) = a^x$ is an increasing function. The greater the value of a, the faster the function increases.
>
> 2. When $0 < a < 1$, the function $f(x) = a^x$ is a decreasing function. The greater the value of a, the more slowly the function decreases.

It should be noted that for any value of a, the y-intercept is $(0, 1)$.

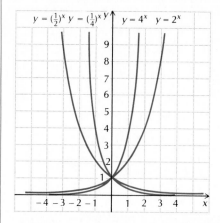

As an application of exponential functions we consider the compound interest formula, $A = P(1 + i)^t$. Suppose principal P of $1000 is invested at an interest rate of 13%, compounded annually. Then the amount A in the account after time t, in years, is given by the exponential function $A = \$1000(1.13)^t$.

Logarithmic Functions

The inverse of an exponential function, for $a > 0$ and $a \neq 1$, is a function. It is called a *logarithmic*, or *logarithm*, *function*. Thus one way to describe a logarithm function is to interchange variables in $y = a^x$:

$$x = a^y.$$

The most useful and interesting logarithmic functions are those for which $a > 1$. The graph of such a function is a reflection of $y = a^x$ across the line $y = x$, as shown below. Note that the domain of a logarithmic function is the set of all positive real numbers. Negative numbers and 0 *do not have logarithms*!

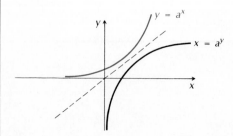

For logarithmic functions we use the notation $\log_a (x)$ or $\log_a x$.* That is, we use the symbol $\log_a x$ to denote the second coordinates of a function $x = a^y$. In other words, a logarithmic function can be described as $y = \log_a x$.

Definition

The following are equivalent:

1. $x = a^y$; and
2. $y = \log_a x$ (read "the log, base a, of x").

Thus $\log_a x$ is the exponent in the equation $x = a^y$, so the logarithm, base a, of a number x is the power to which a is raised to get x.

EXAMPLE 4   Graph $y = \log_3 x$. Determine the domain and range.

The equation $y = \log_3 x$ is equivalent to $x = 3^y$. The graph is a reflection of $y = 3^x$ across the line $y = x$. We make a table of values for $y = 3^x$ and then interchange x and y.

For $y = 3^x$:

x	0	1	2	-1	-2
y	1	3	9	$\frac{1}{3}$	$\frac{1}{9}$

For $y = \log_3 x$ (or $x = 3^y$):

x	1	3	9	$\frac{1}{3}$	$\frac{1}{9}$
y	0	1	2	-1	-2

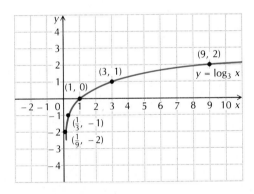

The graph of $y = \log_a x$, for any a, has the x-intercept $(1, 0)$. The domain is the set of all positive numbers. The range is the set of all real numbers.

DO EXERCISES 7 AND 8.

ii CONVERTING EXPONENTIAL AND LOGARITHMIC EQUATIONS

It is important to be able to convert from an exponential equation to a logarithmic equation.

EXAMPLES   Convert to logarithmic equations.

5. $8 = 2^x \rightarrow x = \log_2 8$ It helps, in such conversions, to remember that *the logarithm is the exponent.*

---

*The parentheses in $\log_a (x)$ are like those in $f(x)$. In the case of logarithmic functions we usually omit the parentheses.

7. Graph $y = \log_2 x$. What is the domain of this function? What is the range?

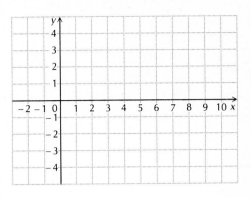

8. Graph $y = \log_4 x$. What is the domain of this function? What is the range?

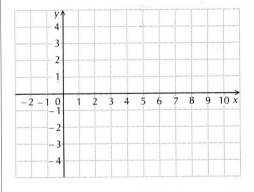

Write equivalent logarithmic equations.

9. $6^0 = 1$

10. $10^{-3} = 0.001$

11. $16^{1/4} = 2$

12. $\left(\frac{6}{5}\right)^{-2} = \frac{25}{36}$

Write equivalent exponential equations.

13. $\log_2 32 = 5$

14. $\log_{10} 1000 = 3$

15. $\log_{10} 0.01 = -2$

16. $\log_{\sqrt{5}} 5 = 2$

Solve.

17. $\log_{10} x = 4$

18. $\log_x 81 = 4$

19. $\log_2 16 = x$

20. $\log_3 3 = x$

21. $\log_5 \frac{1}{25} = x$

6. $y^{-1} = 4 \rightarrow -1 = \log_y 4$

7. $a^b = c \rightarrow b = \log_a c$

DO EXERCISES 9–12.

It is also important to be able to convert from a logarithmic equation to an exponential equation.

EXAMPLES   Convert to exponential equations.

8. $y = \log_3 5 \rightarrow 3^y = 5$ *Again, it helps to remember that the logarithm is the exponent.*

9. $-2 = \log_a 7 \rightarrow a^{-2} = 7$

10. $a = \log_b d \rightarrow b^a = d$

DO EXERCISES 13–16.

iii **SOLVING LOGARITHMIC EQUATIONS**

Certain equations containing logarithmic notation can be solved by first converting to exponential notation.

EXAMPLE 11   Solve $\log_2 x = -3$.

Since $\log_2 x = -3$ is equivalent to $2^{-3} = x$, $x = \frac{1}{8}$.

EXAMPLE 12   Solve $\log_{27} 3 = x$.

Since $\log_{27} 3 = x$ is equivalent to $27^x = 3$, and $27^{1/3} = 3$, we have $x = \frac{1}{3}$.

EXAMPLE 13   Solve $\log_x 4 = \frac{1}{2}$.

Since $\log_x 4 = \frac{1}{2}$ is equivalent to $x^{1/2} = 4$, and $(x^{1/2})^2 = 4^2$, we have $x = 16$.

DO EXERCISES 17–21.

iv **SIMPLIFYING THE EXPRESSIONS $a^{\log_a x}$ AND $\log_a a^x$**

The exponential and logarithmic functions are inverses of each other. Let us recall an important fact about functions and their inverses (Section 7.1). If the domains are suitable, then for any x,

$$f(f^{-1}(x)) = x$$

and

$$f^{-1}(f(x)) = x.$$

We apply this fact to exponential and logarithmic functions. Suppose f is the exponential function, base a:

$$f(x) = a^x.$$

Then f^{-1} is the logarithmic function, base a:

$$f^{-1}(x) = \log_a x.$$

Now let us find $f(f^{-1}(x))$:

$$f(f^{-1}(x)) = a^{f^{-1}(x)} = a^{\log_a x} = x.$$

Thus for any suitable base a, $a^{\log_a x} = x$ for any positive number x (negative numbers and 0 do not have logarithms).

Next, let us find $f^{-1}(f(x))$:

$$f^{-1}(f(x)) = \log_a f(x) = \log_a a^x = x.$$

Thus for any suitable base a, $\log_a a^x = x$ for any number x whatever.

These facts are important in simplification and should be learned well.

Theorem 3

For any number a, suitable as a logarithm base,

1. $a^{\log_a x} = x$, for any positive number x; and
2. $\log_a a^x = x$, for any number x.

EXAMPLES Simplify.

14. $2^{\log_2 5} = 5$
15. $10^{\log_{10} t} = t$
16. $\log_e e^{-3} = -3$
17. $\log_{10} 10^{5.6} = 5.6$

DO EXERCISES 22–27.

Simplify.

22. $4^{\log_4 3}$

23. $7^{\log_7 \pi}$

24. $b^{\log_b 42}$

25. $\log_5 5^{37}$

26. $\log_e e^M$

27. $\log_{10} 10^{3.2}$

EXERCISE SET 7.2

i Graph.

1. a) $y = 4^x$ b) $y = (\frac{1}{4})^x$ c) $y = \log_4 x$
2. a) $y = 3^x$ b) $y = (\frac{2}{3})^x$ c) $y = \log_3 x$
3. $y = 5^x$ 4. $y = 2^x$ 5. $y = \log_2 x$ 6. $y = \log_{10} x$

ii Write equivalent exponential equations.

7. $\log_2 32 = 5$ 8. $\log_{10} 1000 = 3$ 9. $\log_{10} 0.01 = -2$
10. $\log_{\sqrt{5}} 5 = 2$ 11. $\log_6 6 = 1$ 12. $\log_b M = N$

Write equivalent logarithmic equations.

13. $6^0 = 1$ 14. $10^{-3} = 0.001$ 15. $(\frac{6}{5})^{-2} = \frac{25}{36}$ 16. $5^4 = 625$
17. $5^{-2} = \frac{1}{25}$ 18. $8^{1/3} = 2$ 19. $e^{0.08} = 1.0833$ 20. $10^{0.4771} = 3$

iii Solve.

21. $\log_{10} x = 4$ 22. $\log_3 x = 2$ 23. $\log_x \frac{1}{32} = 5$ 24. $\log_8 x = \frac{1}{3}$
25. $\log_2 16 = x$ 26. $\log_3 3 = x$ 27. $\log_4 2 = x$ 28. $\log_2 64 = x$
29. $\log_{10} 10^2 = x$ 30. $\log_3 3^4 = x$ 31. $\log_\pi \pi = x$ 32. $\log_a a = x$
33. $\log_{10} 0.001 = x$ 34. $\log_{10} 1000 = x$

iv Simplify.

35. $3^{\log_3 4x}$ 36. $5^{\log_5 (4x-5)}$ 37. $\log_Q Q^{\sqrt{5}}$ 38. $\log_e e^{|x-4|}$

Graph.

39. $y = \log_2 (x + 3)$ 40. $y = \log_3 (x - 2)$ 41. $y = 2^x - 1$ 42. $y = 2^{x-3}$
43. $f(x) = 2^{|x|}$ 44. $f(x) = \log_3 |x|$ 45. $f(x) = 2^x + 2^{-x}$ 46. $f(x) = 2^{-(x-1)}$

What is the domain of each function?

47. $f(x) = 3^x$
48. $f(x) = \log_{10} x$
49. $f(x) = \log_a x^2$
50. $f(x) = \log_4 x^3$
51. $f(x) = \log_{10} (3x - 4)$
52. $f(x) = \log_5 |x|$
53. $f(x) = \log_6 (x^2 - 9)$

Solve using graphing.

54. $2^x > 1$
55. $3^x \leqslant 1$
56. $\log_2 x < 0$
57. $\log_2 x \geqslant 4$
58. $\log_2 (x - 3) \leqslant 5$
59. $2^{x+3} > 1$

60. ▦ Approximate each of the following to six decimal places.

$$3^3, \quad 3^{3.1}, \quad 3^{3.14}, \quad 3^{3.141}, \quad 3^{3.1415}, \quad 3^{3.14159}$$

61. ▦ Which is larger, 5^π or π^5?
62. ▦ Which is larger, $\sqrt{8^3}$ or $8^{\sqrt{3}}$?
63. ▦ Graph $y = (0.745)^x$.

Graph.

64. $y = 2^{-x^2}$
65. $y = 3^{-(x+1)^2}$
66. $y = |2^{x^2} - 8|$
67. Solve $3^{3x} = 1$.

OBJECTIVE

You should be able to:

[i] Use properties of logarithms to express sums and differences as a single logarithm, and vice versa, and to find certain logarithms given values of others.

7.3

PROPERTIES OF LOGARITHM FUNCTIONS

[i] Let us now establish some basic properties of logarithmic functions, properties that are fundamental to the use of the functions.

Theorem 4

For any positive numbers x and y,

$$\log_a (x \cdot y) = \log_a x + \log_a y,$$

where a is any positive number different from 1.

Theorem 4 says that the logarithm of a *product* is the *sum* of the logarithms of the factors. Note that the base a must remain constant. The logarithm of a sum is *not* the sum of the logarithms of the summands.

Proof of Theorem 4. Since a is positive and different from 1, it can serve as a logarithm base. Since x and y are assumed positive, they are in the domain of the function $y = \log_a x$. Now let $b = \log_a x$ and $c = \log_a y$.

Writing equivalent exponential equations, we have

$$x = a^b \quad \text{and} \quad y = a^c.$$

Next we multiply, to obtain

$$xy = a^b a^c = a^{b+c}.$$

Now writing an equivalent logarithmic equation, we obtain

$$\log_a (xy) = b + c,$$

or

$$\log_a (xy) = \log_a x + \log_a y,$$

which was to be shown.

EXAMPLE 1 Express as a sum of logarithms and simplify.

$$\log_2 (4 \cdot 16) = \log_2 4 + \log_2 16 = 2 + 4 = 6$$

DO EXERCISES 1–4.

Theorem 5

> For any positive number x and any number p,
>
> $$\log_a x^p = p \cdot \log_a x,$$
>
> where a is any logarithm base.

Theorem 5 says that the logarithm of a power of a number is the exponent times the logarithm of the number.

Proof of Theorem 5. Let $b = \log_a x$. Then, writing an equivalent exponential equation, we have $x = a^b$. Next we raise both sides of the latter equation to the pth power. This gives us

$$x^p = (a^b)^p, \quad \text{or} \quad a^{bp}.$$

Now we can write an equivalent logarithmic equation,

$$\log_a x^p = \log_a a^{bp} = bp.$$

But $b = \log_a x$, so we have

$$\log_a x^p = p \cdot \log_a x,$$

which was to be shown.

EXAMPLES Express as products.

2. $\log_b 9^{-5} = -5 \cdot \log_b 9$
3. $\log_a \sqrt[4]{5} = \log_a 5^{1/4} = \frac{1}{4} \log_a 5$

DO EXERCISES 5 AND 6.

Theorem 6

> For any positive numbers x and y,
>
> $$\log_a \frac{x}{y} = \log_a x - \log_a y,$$
>
> where a is any logarithm base.

Theorem 6 says that the logarithm of a quotient is the difference of the logarithms (that is, the logarithm of the dividend minus the logarithm of the divisor).

Proof of Theorem 6. Since $x/y = x \cdot y^{-1}$,

$$\log_a \frac{x}{y} = \log_a (xy^{-1}).$$

By Theorem 4,

$$\log_a (xy^{-1}) = \log_a x + \log_a y^{-1},$$

and by Theorem 5,

$$\log_a y^{-1} = -1 \cdot \log_a y,$$

Express as a sum of logarithms.

1. $\log_a MN$

2. $\log_5 (25 \cdot 5)$

Express as a single logarithm.

3. $\log_3 7 + \log_3 5$

4. $\log_a C + \log_a A + \log_a B + \log_a I + \log_a N$

Express as a product.

5. $\log_7 4^5$

6. $\log_a \sqrt{5}$

7. Express as a difference.

a) $\log_a \dfrac{M}{N}$

b) $\log_c \dfrac{1}{4}$

8. Express as sums and differences of logarithms without exponential notation or radicals.

$$\log_{10} \dfrac{4\pi}{\sqrt{23}}$$

9. Express in terms of logarithms of x, y, and z.

$$\log_a \sqrt{\dfrac{z^3}{xy}}$$

10. Express as a single logarithm.

$$5 \log_a x - \log_a y + \dfrac{1}{4} \log_a z$$

11. Given that $\log_a 2 = 0.301$ and $\log_a 3 = 0.477$, find:

a) $\log_a 9$;

b) $\log_a \sqrt{2}$;

c) $\log_a \sqrt[3]{2}$;

d) $\log_a \dfrac{3}{2}$;

e) $\dfrac{\log_a 3}{\log_a 2}$.

so we have

$$\log_a \frac{x}{y} = \log_a x - \log_a y,$$

which was to be shown.

EXAMPLE 4 Express as sums and differences of logarithms without exponential notation or radicals.

$$\log_a \frac{\sqrt{17}}{5\pi} = \log_a \sqrt{17} - \log_a 5\pi$$

$$= \log_a 17^{1/2} - (\log_a 5 + \log_a \pi)$$

$$= \frac{1}{2} \log_a 17 - \log_a 5 - \log_a \pi$$

EXAMPLE 5 Express in terms of logarithms of x, y, and z.

$$\log_a \sqrt[4]{\frac{xy}{z^3}} = \log_a \left(\frac{xy}{z^3}\right)^{1/4}$$

$$= \frac{1}{4} \cdot \log_a \frac{xy}{z^3}$$

$$= \frac{1}{4} \left[\log_a xy - \log_a z^3\right]$$

$$= \frac{1}{4} \left[\log_a x + \log_a y - 3 \log_a z\right]$$

$$= \frac{1}{4} \log_a x + \frac{1}{4} \log_a y - \frac{3}{4} \log_a z$$

EXAMPLE 6 Express as a single logarithm.

$$\frac{1}{2} \log_a x - 7 \log_a y + \log_a z = \log_a \sqrt{x} - \log_a y^7 + \log_a z$$

$$= \log_a \frac{\sqrt{x}}{y^7} + \log_a z$$

$$= \log_a \frac{z\sqrt{x}}{y^7}$$

DO EXERCISES 7–10.

EXAMPLE 7 Given that $\log_a 2 = 0.301$ and $\log_a 3 = 0.477$, find each of the following.

a) $\log_a 6 = \log_a 2 \cdot 3$
$= \log_a 2 + \log_a 3$
$= 0.301 + 0.477 = 0.778$

b) $\log_a \sqrt{3} = \log_a 3^{1/2} = \frac{1}{2} \cdot \log_a 3 = \frac{1}{2} \cdot 0.477 = 0.2385$

c) $\log_a \frac{2}{3} = \log_a 2 - \log_a 3 = 0.301 - 0.477 = -0.176$

d) $\log_a 5$ *No way to find, using Theorems 4–6* ($\log_a 5 \neq \log_a 2 + \log_a 3$)

e) $\dfrac{\log_a 2}{\log_a 3} = \dfrac{0.301}{0.477} = 0.63$ Note that we could not use Theorems 4–6; we simply divided.

DO EXERCISE 11.

For any base a, $\log_a a = 1$. This is easily seen by writing an equivalent exponential equation, $a^1 = a$. Similarly, for any base a, $\log_a 1 = 0$. These facts are important and should be remembered.

Theorem 7

For any base a,

$$\log_a a = 1 \quad \text{and} \quad \log_a 1 = 0.$$

DO EXERCISES 12–15.

Simplify.

12. $\log_\pi \pi$

13. $\log_9 1$

14. $\log_e 1$

15. $\log_{1/4} \dfrac{1}{4}$

EXERCISE SET 7.3

i Express in terms of logarithms of x, y, and z.

1. $\log_a x^2 y^3 z$

2. $\log_a 5xy^4 z^3$

3. $\log_b \dfrac{xy^2}{z^3}$

4. $\log_c \sqrt[3]{\dfrac{x^4}{y^3 z^2}}$

Express as a single logarithm and simplify if possible.

5. $\frac{2}{3}\log_a 64 - \frac{1}{2}\log_a 16$

6. $\frac{1}{2}\log_a x + 3\log_a y - 2\log_a x$

7. $\log_a 2x + 3(\log_a x - \log_a y)$

8. $\log_a x^2 - 2\log_a \sqrt{x}$

9. $\log_a \dfrac{a}{\sqrt{x}} - \log_a \sqrt{ax}$

10. $\log_a (x^2 - 4) - \log_a (x - 2)$

11. $\log_a (x^3 + y^3) - \log_a (x + y)$

12. $\log_a (x - y) + \log_a (x^2 + xy + y^2)$

Express as a sum and/or difference of logarithms.

13. $\log_a \sqrt{1 - x^2}$

14. $\log_a \dfrac{x + t}{\sqrt{x^2 - t^2}}$

Given $\log_{10} 2 = 0.301$, $\log_{10} 3 = 0.477$, and $\log_{10} 10 = 1$, find each of the following.

15. $\log_{10} 4$

16. $\log_{10} 5$ (*Hint:* $5 = \frac{10}{2}$)

17. $\log_{10} 50$ (*Hint:* $50 = \frac{100}{2}$)

18. $\log_{10} 12$

19. $\log_{10} 60$

20. $\log_{10} \frac{1}{3}$

21. $\log_{10}\sqrt{\frac{2}{3}}$

22. $\log_{10} \sqrt[5]{12}$

23. $\log_{10} 90$

24. $\log_{10} \frac{9}{8}$

25. $\log_{10} \frac{9}{10}$

26. $\log_{10} \frac{1}{4}$

Which of the following are false?

27. $\dfrac{\log_a M}{\log_a N} = \log_a M - \log_a N$

28. $\dfrac{\log_a M}{\log_a N} = \log_a \dfrac{M}{N}$

29. $\dfrac{\log_a M}{c} = \log_a M^{1/c}$

30. $\log_N (M \cdot N)^x = x\log_N M + x$

31. $\log_a 2x = 2\log_a x$

32. $\log_a 2x = \log_a 2 + \log_a x$

33. $\log_a (M + N) = \log_a M + \log_a N$

34. $\log_a x^3 = 3\log_a x$

Solve.

35. $\log_\pi \pi^{2x+3} = 4$

36. $3^{\log_3 (8x-4)} = 5$

37. $4^{2\log_4 x} = 7$

38. $8^{2\log_8 x + \log_8 x} = 27$

39. $(x + 3) \cdot \log_a a^x = x$

40. $\log_a x^2 = 2\log_a x$

41. $\log_a 5x = \log_a 5 + \log_a x$

42. $\log_b \dfrac{5}{x + 2} = \log_b 5 - \log_b (x + 2)$

43. If $\log_a x = 2$, what is $\log_a \left(\dfrac{1}{x}\right)$?

44. If $\log_a x = 2$, what is $\log_{1/a} x$?

Prove the following for any base a and any positive number x.

45. $\log_a \left(\dfrac{1}{x}\right) = -\log_a x$

46. $\log_a \left(\dfrac{1}{x}\right) = \log_{1/a} x$

47. Show that $\log_a \left(\dfrac{x + \sqrt{x^2 - 5}}{5}\right) = -\log_a (x - \sqrt{x^2 - 5})$.

48. Graph and compare: $y = \log_2 |x|$ and $y = |\log_2 x|$.

OBJECTIVES

You should be able to:

i Find logarithms and antilogarithms, base 10, using a calculator.

ii Find logarithms and antilogarithms, base e, using a calculator.

iii Find logarithms and antilogarithms, base 10, using a table.

iv Find logarithms and antilogarithms, base e, using a base-ten table.

1. Find log 3,531,700.

2. Find log 0.0004968.

7.4

LOGARITHM FUNCTION VALUES; TABLES AND CALCULATORS

Any positive number different from 1 can be used as a base for logarithmic functions. However, some numbers are easier to use than others, and there are logarithm bases that fit into certain applications more naturally than others. Base-ten logarithms are called *common logarithms*. They are useful because they are of the same base as the decimal numeration system.

Before calculators became so widely available, common logarithms were extensively used in calculations. In fact, that is why logarithms were invented. Another logarithm base that is used a great deal today is, strangely enough, an irrational number, which we name e. Logarithms to the base e are called *natural logarithms*.

i COMMON LOGARITHMS ON A CALCULATOR

The abbreviation *log* is used for the logarithm function base ten, or common logarithms. Thus the symbol log 23 means $\log_{10} 23$. On scientific calculators the key for the common logarithm function is marked $\boxed{\text{log}}$. To find the common logarithm of a number, enter that number and then press the $\boxed{\text{log}}$ key.

EXAMPLE 1 Find log 475,000.

We enter 475,000 and then press the $\boxed{\text{log}}$ key. We find that log 475,000 ≈ 5.6767 (we have rounded the answer).

EXAMPLE 2 Find log 0.00372.

We enter 0.00372 and then press the $\boxed{\text{log}}$ key. We find that log 0.00372 ≈ -2.4295 (we have again rounded the answer).

DO EXERCISES 1 AND 2.

Antilogarithms

The inverse of a logarithmic function is, of course, an exponential function. The inverse of finding a logarithm is also called finding an *antilogarithm*. To find an antilogarithm, we *exponentiate*:

$$\text{antilog}_{10} x = 10^x.$$

Generally, there is no key on a calculator marked "antilog." It is up to you to know that to find the inverse, or antilogarithm, you must use the $\boxed{10^x}$ key, if there is one. If there is no such key, then you must raise 10 to the · x power using an exponential key.

EXAMPLE 3   Find the antilogarithm, base 10, of 3.2546.

a)   (*Using the 10^x key*): We enter 3.2546 and then press the $\boxed{10^x}$ key. We find that

$$\text{antilog } 3.2546 = 10^{3.2546} \approx 1797.21.$$

b)   (*Using an exponential key*): We enter 3.2546 and also 10. Then, pressing keys in the order appropriate for your particular calculator (you must read the instructions), you will find that

$$\text{antilog } 3.2546 = 10^{3.2546} \approx 1797.21.$$

DO EXERCISES 3 AND 4.

ii   NATURAL LOGARITHMS ON A CALCULATOR

Natural logarithms have an irrational number, named e, for the base. Since the number e is irrational, its decimal representation does not terminate or repeat. Here is an approximation of e:

$$e \approx 2.71828 \ldots .$$

The abbreviation *ln* is generally used with natural logarithms. Thus the symbol

$$\ln 23$$

denotes the logarithm, base e, of 23.

On scientific calculators the key for the natural logarithm function is marked $\boxed{\ln}$.

EXAMPLE 4   Find ln 3568.

We enter 3568 and then press the $\boxed{\ln}$ key. We find that $\ln 3568 \approx 8.1798$ (we have rounded the answer).

EXAMPLE 5   Find ln 0.0007659.

We enter 0.0007659 and then press the $\boxed{\ln}$ key. We find that ln 0.000765 ≈ -7.1745. (Again, we have rounded the answer.)

DO EXERCISES 5 AND 6.

Antilogarithms

To find the antilogarithm, base e, we use the $\boxed{e^x}$ key, if there is one.

EXAMPLE 6   Find the antilogarithm, base e, of 2.4837.

a)   (*Using the e^x key*): We enter 2.4837 and then press the $\boxed{e^x}$ key. We find that

$$\text{antilog}_e \, 2.4837 = e^{2.4837} \approx 11.9855.$$

3. Find the antilogarithm, base 10, of 4.0165; that is, find $10^{4.0165}$.

4. Find the antilogarithm, base 10, of -2.4141; that is, find $10^{-2.4141}$.

5. Find ln 67,415.

6. Find ln 0.002338.

7. Find the antilogarithm, base e, of 3.5974; that is, find $e^{3.5974}$.

b) (*Using an exponential key*): We enter 2.4837 and also an approximate value of e, say, 2.71828. Then, pressing keys in the order appropriate for your calculator (you must read the instructions), you will find that

$$\text{antilog}_e\, 2.4837 = e^{2.4837} \approx 11.9855.$$

DO EXERCISES 7 AND 8.

iii TABLES OF COMMON LOGARITHMS (OPTIONAL)

Table 2 at the back of the book is a table of common (base-10) logarithms, for numbers from 1 to 10. Part of that table is shown below.

x	0	1	2	3	4	5	6	7	8	9
5.0	0.6990	0.6998	0.7007	0.7016	0.7024	0.7033	0.7042	0.7050	0.7059	0.7067
5.1	0.7076	0.7084	0.7093	0.7101	0.7110	0.7118	0.7126	0.7135	0.7143	0.7152
5.2	0.7160	0.7168	0.7177	0.7185	0.7193	0.7202	0.7210	0.7218	0.7226	0.7235
5.3	0.7243	0.7251	0.7259	0.7267	0.7275	0.7284	0.7292	0.7300	0.7308	0.7316
5.4	0.7324	0.7332	0.7340	0.7348	0.7356	0.7364	0.7372	0.7380	0.7388	0.7396

8. Find the antilogarithm, base e, of -2.9768; that is, find $e^{-2.9768}$.

The values in the table are the same as those stored in a calculator, except that the table values have been rounded to four decimal places.

At this point a question will surely arise. Since calculators are rather easily available, why bother learning to use a table? The answer is that if all you are interested in is obtaining values to solve problems when a calculator is available, then there is little reason to fuss with a table. The educational value in learning to use a table lies essentially in two areas:

1. So that you can use a table if necessary, when a calculator is inoperative or unavailable.

2. So that you have some facility in working with tables in general. There are many functions, including functions that you might devise for yourself, through experiment in business or science, whose values are not stored in a calculator. For such functions, tables are important, and your ability to use tables is therefore also important.

Now let us see how to use the table of common logarithms.

9. Find log 5.38.

EXAMPLE 7   Find log 5.24.

At the left of the table, we find the row headed 5.2. Then we move across to the column headed (at the top of the table) 4. We find log 5.24 (the color entry in the table):

$$\log 5.24 \approx 0.7193.$$

10. Find antilog 0.7093.

EXAMPLE 8   Find antilog 0.7193.

We reverse the procedure of Example 7. We first look for 0.7193 in the body of the table (the color entry). We find the heading 5.2 at the left and the heading 4 at the top. Thus

$$\text{antilog}\, 0.7193 = 10^{0.7193} \approx 5.24.$$

DO EXERCISES 9 AND 10.

Numbers Not Between 1 and 10

Table 2 contains logarithms of numbers from 1 to 10 only. How do we find logarithms of other numbers? The answer is that we write scientific notation for the number and then use the properties of the logarithmic function.

First, we should note that for powers of ten, the common logarithms are integers. For example:

log 10 = 1, because $10 = 10^1$.   The log is the exponent.

log 100 = 2, because $100 = 10^2$.

log 1000 = 3, because $1000 = 10^3$.

log 0.1 = −1, because $0.1 = 10^{-1}$.

log 0.001 = −3, because $0.001 = 10^{-3}$.

EXAMPLE 9   Find log 5430.

We first write scientific notation for the number:

$$5430 = 5.43 \times 10^3.$$

By Theorem 4,

$$\log (5.43 \times 10^3) = \log 5.43 + \log 10^3$$

$$= \log 5.43 + 3 \quad \text{The log of } 10^3 \text{ is 3.}$$

$$= 0.7348 + 3 \quad \text{Finding log 5.43 from the table}$$

$$= 3.7348. \quad \text{Adding}$$

EXAMPLE 10   Find log 0.000507.

We first write scientific notation for the number:

$$0.000507 = 5.07 \times 10^{-4}.$$

By Theorem 4,

$$\log (5.07 \times 10^{-4}) = \log 5.07 + \log 10^{-4}$$

$$= \log 5.07 + (-4) \quad \text{The log of } 10^{-4} \text{ is } -4.$$

$$= 0.7050 + (-4) \quad \text{Finding log 5.07 from the table}$$

$$= -3.2950. \quad \text{Doing the addition}$$

DO EXERCISES 11 AND 12.

Antilogarithms

To find antilogarithms, we reverse the procedure of Example 9 or Example 10.

EXAMPLE 11   Find antilog 4.7308.

By the definition of antilog, we have

antilog 4.7308 = $10^{4.7308}$

$$= 10^{4 + 0.7308} \quad \text{Separating the exponent into an integer and a number between 0 and 1}$$

$$= 10^{0.7308} \times 10^4. \quad \text{Using a property of exponents } (a^x \cdot a^y = a^{x+y})$$

Now, using the table, we find $10^{0.7308}$. It is 5.38. Therefore, antilog 4.7308 = 5.38×10^4, or 53,800.

11. Find log 6380.

12. Find log 0.0000284.

13. Find antilog 3.5717; that is, find $10^{3.5717}$.

EXAMPLE 12 Find antilog (-1.2874).

By the definition of antilog, we have

$$\text{antilog}\,(-1.2874) = 10^{-1.2874}.$$

We need to separate -1.2874 into two parts, one of which is an integer and one of which is between 0 and 1. We do that by adding and subtracting 2 (or any integer greater than 1):

$$2 - 1.2874 - 2.$$

Doing the subtraction $2 - 1.2874$, we get

$$0.7126 - 2.$$

Now we have

$$\begin{aligned}
\text{antilog}\,(-1.2874) &= \text{antilog}\,(0.7126 - 2) \\
&= 10^{(0.7126-2)} \\
&= 10^{0.7126} \times 10^{-2}. \quad \text{Using a property of exponents}
\end{aligned}$$

Using the table, we find $10^{0.7126}$. It is 5.16. Therefore,

$$\begin{aligned}
\text{antilog}\,(-1.2874) &= \text{antilog}\,(0.7126 - 2) \\
&= 5.16 \times 10^{-2}, \quad \text{or } 0.0516.
\end{aligned}$$

DO EXERCISES 13 AND 14.

14. Find antilog (-2.8416); that is, find $10^{-2.8416}$.

iv NATURAL LOGARITHMS AND TABLES (OPTIONAL)

There are tables of natural logarithms, and they can be used in the same fashion as indicated in the preceding examples for common logarithms. The difficulty arises because powers of 10 do not have integers as their natural logarithms. Rather than dealing with natural logarithms of powers of ten, it is usually easier to find a natural logarithm by first finding its base-ten logarithm. We then apply a conversion formula, which is stated in the following theorem.

Theorem 8

For any positive number M,

$$\ln M = \frac{\log M}{\log e}.$$

Proof. Let $x = \ln M$ or $\log_e M$. Then $e^x = M$, so that

$$\log_{10} M = \log_{10} e^x \quad \text{Substituting } e^x \text{ for } M$$
$$\log_{10} M = x \log_{10} e. \quad \text{By Theorem 5}$$

We now solve for x:

$$x = \frac{\log_{10} M}{\log_{10} e}$$

$$x = \frac{\log M}{\log e}. \quad \text{Simplifying notation}$$

Now recall what x represents and substitute, to obtain

$$\ln M = \frac{\log M}{\log e}.$$

This is the desired formula.

To apply the formula of Theorem 8 it is easier to compute $1/\log e$, which is about 2.3026. Then to find the natural logarithm of a number, simply find its common logarithm and then multiply by 2.3026.

EXAMPLE 13 Find ln 5480.

We first find log 5480, or 5.48×10^3:

$$\log 5840 = 3.7388.$$

We then multiply by 2.3026, to obtain 8.6090. Therefore, $\ln 5480 \approx 8.6090$.

DO EXERCISE 15.

Antilogarithms

The natural antilogarithm of a number x is of course e^x. If we take the common logarithm, we obtain $\log e^x$. If we then take the base-10 antilogarithm, we have

$$10^{\log e^x}, \quad \text{which is also } e^x.$$

Also, by Theorem 5,

$$10^{\log e^x} = 10^{x \log e}$$

All of this tells us that to find the natural antilogarithm e^x, of a number x, we can multiply x by $\log e$ and then take the base-10 antilogarithm.

> **To find the base-e antilogarithm**   To find e^x (the antilogarithm of x), multiply by $\log e$ (which is about 0.4343) and then find the base-10 antilogarithm.

EXAMPLE 14 Find antilog$_e$ 8.5867.

We first multiply by $\log e$, which is about 0.4343:

$$8.5867 \times 0.4343 \approx 3.7292.$$

Now we find the base-10 antilog of 3.7292. It is

$$5.36 \times 10^3, \quad \text{or } 5360.$$

Thus antilog$_e$ $8.5867 = e^{8.5867} \approx 5360$.

DO EXERCISE 16.

INTERPOLATION

Suppose that a number falls between two entries in a table. How, then, can we find its logarithm or antilogarithm? We can approximate an answer using a technique called *interpolation*. That technique applies to tables in general, not just to logarithm tables. There is a treatment of the topic of interpolation in the Appendix.

15. Find ln 38,600.

16. Find antilog$_e$ 6.3818.

EXERCISE SET 7.4

| i | Find the following common logarithms using a calculator.

1. log 3.921

2. log 8.494

3. log 37,590

4. log 560,300

5. log 5149

6. log 3204

7. log 0.1414

8. log 0.3257

9. log 0.0005123

10. log 0.009808

11. log 0.00001234

12. log 0.01234

13. log (5.621×10^5)

14. log (6.123×10^7)

15. log (8.042×10^{38})

16. log (1.453×10^{29})

17. log (4.625×10^{-12})

18. log (5.032×10^{-20})

19. log (-4.923)

20. log (-7.891)

Find the following base-10 antilogarithms using a calculator.

21. antilog 4.6524

22. antilog 3.0214

23. $10^{5.9231}$

24. $10^{6.4132}$

25. antilog 0.003215

26. antilog 0.0004128

27. antilog (-3.0143)

28. antilog (-4.9212)

29. $10^{-5.9231}$

30. $10^{-6.0342}$

31. antilog (8.1111×10^{-4})

32. antilog (9.008×10^{-3})

33. $10^{(3.8146 \times 10^{-3})}$

34. $10^{(5.097 \times 10^{-2})}$

35. antilog (2.8307×10^{-8})

36. antilog (1.9984×10^{-6})

37. $10^{(4.0060 \times 10^{-7})}$

38. $10^{(1.1111 \times 10^{-5})}$

ii Find the following natural logarithms using a calculator.

39. ln 3285

40. ln 92,540

41. ln 0.1248

42. ln 0.05435

43. ln 0.0005127

44. ln 0.00003828

45. ln (-8.762)

46. ln (-12.873)

47. ln (8.041×10^{28})

48. ln (2.031×10^{12})

49. ln (5.043×10^{-14})

50. ln (3.051×10^{-15})

Find the following base-*e* antilogarithms using a calculator.

51. antilog$_e$ 3.4935

52. antilog$_e$ 7.2851

53. $e^{1.0312}$

54. $e^{6.7152}$

55. antilog$_e$ 31.891

56. antilog$_e$ 71.432

57. $e^{17.814}$

58. $e^{125.6}$

59. antilog$_e$ (-17.123)

60. antilog$_e$ (-3.821)

61. $e^{-12.832}$

62. $e^{-5.043}$

iii Find the following common logarithms using Table 2.

63. log 6.29

64. log 8.94

65. log 1.03

66. log 2.59

67. log 1340

68. log 25,300

69. log 82,400

70. log 8,450,000

71. log 925,000

72. log 2,030,000

73. log 0.0241

74. log 0.000387

75. log 0.00814

76. log 0.0000504

77. log (-8.24)

78. log (-8740)

Find the following base-10 antilogarithms using Table 2.

79. antilog 0.8791

80. antilog 0.5821

81. $10^{0.5999}$

82. $10^{0.7868}$

83. antilog 2.9355

84. antilog 3.7218

85. $10^{4.7832}$

86. $10^{3.1173}$

87. antilog $(0.8555 - 3)$

88. antilog $(0.5551 - 5)$

89. antilog $(3.9795 - 5)$

90. antilog $(2.8848 - 7)$

91. antilog (-2.1445)

92. antilog (-4.4449)

93. antilog (-1.0205)

94. antilog (-4.1152)

iv Find the following natural logarithms using Table 2.

95. ln 3820

96. ln 485

97. ln 7240

98. ln 92,400

99. ln 0.417

100. ln 0.0512

101. ln 0.00815

102. ln 0.0000109

103. ln (-6.49)

104. ln (-0.184)

Find the following base-*e* antilogarithms using Table 2. Choose the nearest values in the table, not attempting any interpolation.

105. antilog$_e$ 0.8791

106. antilog$_e$ 0.5821

107. $e^{0.5999}$

108. $e^{0.7868}$

109. antilog$_e$ (2.9555)

110. antilog$_e$ 3.7218

111. $e^{4.7832}$

112. $e^{3.1173}$

113. Using function values obtained on a calculator, plot points and draw a precise graph of $y = 10^x$.

114. Using the values obtained in Exercise 113, plot points and draw a precise graph of $y = \log x$.

115. Using function values obtained on a calculator, plot points and draw a precise graph of $y = e^x$.

116. Using the values obtained in Exercise 115, plot points and draw a precise graph of $y = \ln x$.

117. Prove the following theorem, which is a generalization of Theorem 8:

$$\text{For any logarithm bases } a \text{ and } b, \text{ and any positive number } M, \log_b M = \frac{\log_a M}{\log_a b}.$$

7.5

EXPONENTIAL AND LOGARITHMIC EQUATIONS

i EXPONENTIAL EQUATIONS

An equation with variables in exponents, such as $3^{2x-1} = 4$, is called an *exponential equation*. We can solve such equations by taking logarithms on both sides and then using Theorem 5.

The use of a calculator is recommended for the rest of this chapter.

EXAMPLE 1   Solve $3^x = 8$.

$\log 3^x = \log 8$ Taking the log on both sides
                     Remember $\log m = \log_{10} m$.

$x \log 3 = \log 8$ Using Theorem 5

$x = \dfrac{\log 8}{\log 3}$ Solving for x

$x \approx \dfrac{0.9031}{0.4771} \approx 1.8929$ We look up the logs, or find them on a calculator, and divide.

DO EXERCISE 1.

EXAMPLE 2   Solve $2^{3x-5} = 16$.

$\log 2^{3x-5} = \log 16$ Taking the log on both sides

$(3x - 5) \log 2 = \log 16$ Using Theorem 5

$3x - 5 = \dfrac{\log 16}{\log 2}$

$x = \dfrac{\dfrac{\log 16}{\log 2} + 5}{3} \approx \dfrac{\dfrac{1.2041}{0.3010} + 5}{3}$ Solving for x and evaluating logarithms

$x \approx 3.0001$ Calculating

The answer is approximate because the logarithms are approximate.

DO EXERCISE 2.

EXAMPLE 3   Solve $\dfrac{e^x + e^{-x}}{2} = t$, for x.

Note that we are to solve for x. However, we have more than one term with x in the exponent. To get a single expression with x in the exponent, we do the following:

$e^x + e^{-x} = 2t$ Multiplying by 2

$e^x + \dfrac{1}{e^x} = 2t$ Rewriting to eliminate the minus sign in an exponent

$e^{2x} + 1 = 2t \cdot e^x$ Multiplying on both sides by e^x

$(e^x)^2 - 2t \cdot e^x + 1 = 0.$

1. Solve $2^x = 7$.

2. Solve $4^{2x-3} = 64$, using Table 2.

3. Solve $\dfrac{e^x - e^{-x}}{2} = t$, for x.

This equation is reducible to quadratic, with $u = e^x$. Using the quadratic formula, we obtain

$$e^x = \frac{2t \pm \sqrt{4t^2 - 4}}{2} = t \pm \sqrt{t^2 - 1}.$$

We can now take the natural logarithm on both sides:

$$\ln e^x = x = \ln\left(t \pm \sqrt{t^2 - 1}\right).$$

DO EXERCISE 3.

LOGARITHMIC EQUATIONS

Equations that contain logarithmic expressions are called *logarithmic equations*. We solved some such equations in Section 7.2. We did so by converting to an equivalent exponential equation. For example, to solve $\log_2 x = -3$, we convert to $x = 2^{-3}$ and find that $x = \frac{1}{8}$.

To solve logarithmic equations, we first try to obtain a single logarithmic expression on one side of the equation and then write an equivalent exponential equation.

Solve.

4. $\log_5 x = 3$

EXAMPLE 4 Solve $\log_3 (5x + 7) = 2$.

We already have a single logarithmic expression, so we write an equivalent exponential equation:

$$
\begin{aligned}
5x + 7 &= 3^2 &&\text{Writing an equivalent exponential equation} \\
5x + 7 &= 9 \\
5x &= 2 \\
x &= \tfrac{2}{5}.
\end{aligned}
$$

Check:

$$
\begin{array}{c|c}
\multicolumn{2}{c}{\log_3 (5x + 7) = 2} \\
\hline
\log_3 (5 \cdot \tfrac{2}{5} + 7) & 2 \\
\log_3 (2 + 7) & \\
\log_3 9 & \\
2 &
\end{array}
$$

5. $\log_4 (8x - 6) = 3$

DO EXERCISES 4 AND 5.

EXAMPLE 5 Solve $\log x + \log (x - 3) = 1$.

Here we must first obtain a single logarithmic equation:

$$
\begin{aligned}
\log x + \log (x - 3) &= 1 \\
\log x(x - 3) &= 1 &&\text{Using Theorem 4 to obtain} \\
\log_{10} x(x - 3) &= 1 &&\text{a single logarithm} \\
x(x - 3) &= 10^1 &&\text{Converting to an equivalent} \\
x^2 - 3x &= 10 &&\text{exponential equation} \\
x^2 - 3x - 10 &= 0 \\
(x + 2)(x - 5) &= 0 &&\text{Factoring and using the principle of zero products} \\
x = -2 \ \ &\text{or} \ \ x = 5.
\end{aligned}
$$

Possible solutions to logarithmic equations must be checked because domains of logarithmic functions consist only of positive numbers.

Check:

$$\frac{\log x + \log (x - 3) = 1}{\log (-2) + \log (-2 - 3) \mid 1}$$

$$\begin{array}{r|l} \log x + \log (x - 3) = 1 \\ \hline \log 5 + \log (5 - 3) & 1 \\ \log 5 + \log 2 \\ \log 10 \\ 1 \end{array}$$

The number -2 is not a solution because negative numbers do not have logarithms. The solution is 5.

DO EXERCISE 6.

6. Solve $\log x + \log (x + 3) = 1$.

ii APPLICATIONS

Exponential and logarithmic functions and equations have many applications. We shall consider a few of them.

EXAMPLE 6 (*Compound interest*). The amount A that principal P will be worth after t years at interest rate i, compounded annually, is given by the formula $A = P(1 + r)^t$. Suppose \$4000 principal is invested at 12% interest and is eventually worth \$10,706. For how many years was it invested?

Since it is time that we want to find, we will solve the formula for t:

$$A = P(1 + r)^t$$
$$\log A = \log P(1 + r)^t \qquad \text{Taking the logarithm on both sides}$$
$$\log A = \log P + \log (1 + r)^t \qquad \text{Using Theorem 4}$$
$$\log A = \log P + t \log (1 + r) \qquad \text{Using Theorem 5}$$
$$\log A - \log P = t \log (1 + r) \qquad \text{Subtracting } \log P$$
$$\frac{\log A - \log P}{\log (1 + r)} = t. \qquad \text{Dividing by } \log (1 + r)$$

We now substitute into the formula as we have it solved for t:

$$\frac{\log 10{,}706 - \log 4000}{\log (1 + 0.12)} = t \qquad \text{Substituting}$$

$$\frac{4.0296 - 3.6021}{0.0492} = t \qquad \text{Finding the logarithms}$$

$$8.69 \approx t. \qquad \text{Calculating}$$

The money was invested for about 8.69 years.

DO EXERCISE 7.

7. \$5000 was invested at 14%, compounded annually, and it was eventually worth \$18,540. For how long was it invested?

EXAMPLE 7 (*Population growth*). The exponential equation

$$P = P_0 e^{kt}$$

under certain conditions allows us to calculate a population P, the number of people (or other living organism) at time t, where P_0 is the population when $t = 0$ and k is a positive constant that depends on the situation.

8. The approximate population of Dallas was 680,000 in 1960. In 1969 it was 815,000. Find k in the growth formula and estimate the population in 1990.

The population of the United States was 208 million in 1970. It was 225 million in 1980. Use these data to predict what the population will be in the year 2000.

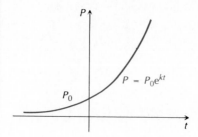

Let us start the clock in 1970. Then in 1980, t will be 10 and in 2000, t will be 30. We'll solve for k and then substitute, to find k:

$$P = P_0 e^{kt}$$

$\ln P = \ln P_0 e^{kt}$ Taking the natural logarithm on both sides

$\ln P = \ln P_0 + \ln e^{kt}$ Using Theorem 4

$\ln P = \ln P_0 + kt$ Simplifying ($\ln e^{kt} = kt$)

$\ln P - \ln P_0 = kt$ Subtracting $\ln P_0$

$\dfrac{\ln P - \ln P_0}{t} = k.$ Dividing by t

In 1980, we have $t = 10$ and $P = 225$ million. Of course, $P_0 = 208$ million, the population in 1970. We substitute and calculate:

$$\frac{\ln 225 - \ln 208}{10} = k$$ We are omitting the "millions."

$$\frac{5.4161 - 5.3375}{10} = k$$ Finding the logarithms

$$0.008 = k.$$ Calculating

The population equation in this situation is now

$$P = P_0 e^{0.008t}$$
$$P = 208 e^{0.008t}$$

In the year 2000, t will be 30, so we have for the population in 2000

$$P = 208 e^{0.008(30)}$$
$$= 208 e^{0.24}.$$

We find that $e^{0.24} \approx 1.2712$. Hence

$$P \approx 208 \times 1.2712 \approx 264.$$

The population should be about 264 million in the year 2000.

DO EXERCISE 8.

EXAMPLE 8 (*Radioactive decay*). In a radioactive substance such as radium, some of the atoms are always ceasing to become radioactive. Thus the amount of a radioactive substance decreases. This is called *radioactive decay*. A model for radioactive decay is as follows:

$$N = N_0 e^{-kt},$$

where N_0 is the amount of a radioactive substance at time 0, N is the amount at time t, and k is a positive constant depending on the situation. Strontium 90 has a *half-life* of 25 years. This means that half of a sample of the substance will cease to become radioactive in 25 yr. Find k in the formula and then use the formula to find how much of a 36-gram sample will remain after 100 years.

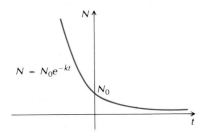

When $t = 25$ (half-life), N will be half of N_0, so we have

$$\tfrac{1}{2}N_0 = N_0 e^{-25k} \quad \text{or} \quad \tfrac{1}{2} = e^{-25k}.$$

We take the natural log on both sides:

$$\ln \tfrac{1}{2} = \ln e^{-25k} = -25k.$$

Thus

$$k = -\frac{\ln 0.5}{25} \approx 0.0277.$$

Now to find the amount remaining after 100 years, we use the formula

$$N = 36 e^{-0.0277 \cdot 100}$$
$$= 36 e^{-2.77}.$$

From a calculator or Table 2 we find that $e^{-2.8} = 0.0608$, and thus

$$N \approx 2.2 \text{ grams}.$$

DO EXERCISE 9.

EXAMPLE 9   (*Atmospheric pressure*). Under standard conditions of temperature, the atmospheric pressure at height h is given by

$$P = P_0 e^{-kh},$$

where P is the pressure, P_0 is the pressure where $h = 0$, and k is a positive constant. Standard sea level pressure is 1013 millibars. Suppose that the pressure at 18,000 ft is half that at sea level. Then find k and find the pressure at 1000 ft.

At 18,000 ft P is half of P_0, so we have

$$\frac{P_0}{2} = P_0 e^{-18,000k}.$$

Taking natural logarithms on both sides, we obtain

$$\ln \tfrac{1}{2} = -18,000k, \quad \text{or} \quad -\ln 2 = -18,000k.$$

9. Radioactive bismuth has a half-life of 5 days. A scientist buys 224 grams of it. How much of it will remain radioactive in 30 days?

10. Calculate the atmospheric pressure at 10,000 ft.

Then

$$k = \frac{\ln 2}{18,000} = 3.85 \times 10^{-5}.$$

To find the pressure at 1000 ft we use the fact that $P_0 = 1013$ and we let $h = 1000$:

$$P = 1013e^{-3.85 \times 10^{-5} \times 1000} = 1013e^{-0.0385}.$$

Rounding the exponent to -0.04 and using a calculator or Table 2, we calculate the pressure to be 973 millibars at 1000 ft.

DO EXERCISE 10.

EXAMPLE 10 (*Loudness of sound*). The sensation of loudness of sound is not proportional to the energy intensity, but rather is a logarithmic function. *Loudness*, in Bels (after Alexander Graham Bell), of a sound of intensity I is defined to be

11. Find the loudness, in decibels, of the sound in a library that is 2510 times as intense as the minimum intensity I_0.

$$L = \log \frac{I}{I_0},$$

where I_0 is the minimum intensity detectable by the human ear (such as the tick of a watch at 20 ft under quiet conditions). When a sound is 10 times as intense as another, its loudness is 1 Bel greater. If a sound is 100 times as intense as another, it is louder by 2 Bels, and so on. The Bel is a large unit, so a subunit, a *decibel*, is generally used. For L in decibels, the formula is as follows:

$$L = 10 \log \frac{I}{I_0}.$$

a) Find the loudness, in decibels, of the sound in a radio studio, for which the intensity I is 199 times I_0.

We substitute into the formula and calculate, using Table 2:

$$L = 10 \log \frac{199 \cdot I_0}{I_0} = 10 \log 199$$
$$= 10(2.2989)$$
$$= 23 \text{ decibels.}$$

12. Find the loudness, in decibels, of conversational speech, having an intensity that is 10^6 times as intense as the minimum, I_0.

b) Find the loudness of the sound of a heavy truck, for which the intensity is 10^9 times I_0.

$$L = 10 \log \frac{10^9 \cdot I_0}{I_0} = 10 \log 10^9$$
$$= 10 \cdot 9$$
$$= 90 \text{ decibels.}$$

DO EXERCISES 11 AND 12.

EXAMPLE 11 (*Earthquake magnitude*). The magnitude R (on the Richter scale) of an earthquake of intensity I is defined as follows:

$$R = \log \frac{I}{I_0},$$

where I_0 is a minimum intensity used for comparison. The Mexico City earthquake of 1978 had an intensity $10^{7.85}$ times I_0. What is its magnitude on the Richter scale?

We substitute into the formula:

$$R = \log \frac{10^{7.85} \cdot I_0}{I_0}$$

$$= \log 10^{7.85}$$

$$= 7.85.$$

DO EXERCISE 13.

EXAMPLE 12 (*Forgetting*). Here is a mathematical model from psychology. A group of people take a test and make an average score of S. After a time t they take an equivalent form of the same test. At that time the average score is $S(t)$. According to this model, $S(t)$ is given by the following function:

$$S(t) = A - B \log (t + 1),$$

where t is in months and the constants A and B are determined by experiment in various kinds of learning situations. The model is appropriate only over the interval $[0, 10^{A/B} - 1]$. Students in a zoology class took a final exam, and took equivalent forms of the test at monthly intervals thereafter. The average scores were found to be given by the function

$$S(t) = 78 - 15 \log (t + 1).$$

What was the average score (a) when they took the test originally? (b) after 4 months?

We substitute into the equation defining the function:

a) $S(0) = 78 - 15 \log (0 + 1)$
     $= 78 - 15 \log 1$
     $= 78 - 0 = 78.$ Original score

b) $S(4) = 78 - 15 \log (4 + 1)$
     $= 78 - 15 \log 5$
     $= 78 - 15 \cdot 0.6990$
     $= 78 - 10.49 = 67.51.$ Score after 4 months

DO EXERCISE 14.

ALTERNATIVE METHODS OF SOLVING (OPTIONAL)

Following are some other methods of solving exponential equations.

EXAMPLE 13   Solve $3^x = 8$.

Instead of taking the common logarithm on both sides we take the logarithm, base 3, on both sides. Then we have

$$\log_3 3^x = \log_3 8$$
$$x = \log_3 8.$$ Theorem 3

The only problem with this method compared to that of Example 1 is that as yet we do not know how to estimate $\log_3 8$. We shall consider this in a later section.

13. An earthquake has an intensity that is $10^{7.8}$ times I_0. What is its magnitude on the Richter scale?

14. Students in an accounting class took a final exam and then equivalent forms of the same test at monthly intervals. The average score $S(t)$, after t months, was found to be given by

$$S(t) = 68 - 20 \log (t + 1).$$

a) What was the initial average score?

b) What was the average score after 4 months?

c) What was the average score after 24 months?

15. Solve $2^x = 7$. Use the method of Example 13.

16. Solve $4^{2x-3} = 64$. Use the method of Example 14.

EXAMPLE 14   Solve $2^{3x-5} = 16$.

Note that $16 = 2^4$. Then we have

$$2^{3x-5} = 2^4.$$

Since the base is the same, 2, on both sides, the exponents must be the same. Thus

$$3x - 5 = 4$$
$$3x = 9$$
$$x = 3.$$

DO EXERCISES 15 AND 16.

EXERCISE SET 7.5

i   Solve.

1. $2^x = 32$

2. $3^{x-1} + 3 = 30$

3. $4^{2x} = 8^{3x-4}$

4. $3^{x^2+4x} = \frac{1}{27}$

5. $3^{5x} \cdot 9^{x^2} = 27$

6. $4^x = 7$

7. $2^x = 3^{x-1}$

8. $3^{x+2} = 5^{x-1}$

9. $\log x + \log (x - 9) = 1$

10. $\log x - \log (x + 3) = -1$

11. $\log (x + 9) - \log x = 1$

12. $\log (2x + 1) - \log (x - 2) = 1$

13. $\log_4 (x + 3) + \log_4 (x - 3) = 2$

14. $\log_8 (x + 1) - \log_8 x = \log_8 4$

15. $\log x^2 = (\log x)^2$

16. $(\log_3 x)^2 - \log_3 x^2 = 3$

17. $\log_3 (\log_4 x) = 0$

18. $\log (\log x) = 2$

19. $e^t = 100$

20. $e^t = 1000$

21. $e^x = 60$

22. $e^k = 90$

23. $e^{-t} = 0.1$

24. $e^{-t} = 0.01$

25. $e^{-0.02k} = 0.06$

26. $e^{0.07t} = 2$

Solve for x.

27. $\dfrac{e^x - e^{-x}}{t} = 5$

28. $e^x + e^{-x} = 5$

29. $\dfrac{e^x + e^{-x}}{e^x - e^{-x}} = t$

30. $\dfrac{5^x - 5^{-x}}{5^x + 5^{-x}} = t$

ii

31. (*Doubling time*). How many years will it take an investment of $1000 to double itself when interest is compounded annually at 6%?

32. (*Tripling time*). How many years will it take an investment of $1000 to triple itself when interest is compounded annually at 5%?

33. (*Population growth*). The population of Dallas was 680,000 in 1960. In 1969 it was 815,000. Find k in the growth formula and estimate the population in 1990.

34. (*Population growth*). The population of Kansas City was 475,000 in 1960. In 1970 it was 507,000. Find k in the growth formula and estimate the population in 2000.

35. (*Radioactive decay*). The half-life of polonium is 3 min. After 30 min, how much of a 410-g sample will remain radioactive?

36. (*Radioactive decay*). The half-life of a lead isotope is 22 yr. After 66 yr, how much of a 1000-g sample will remain radioactive?

37. (*Radioactive decay*). A certain radioactive substance decays from 66,560 g to 6.5 g in 16 days. What is its half-life?

38. Ten grams of uranium will decay to 2.5 grams in 496,000 years. What is its half-life?

39. (*Radiocarbon dating*). Carbon-14, an isotope of carbon, has a half-life of 5750 years. Organic objects contain carbon-14 as well as nonradioactive carbon, in known proportions. When a living organism dies, it takes in no more carbon. The carbon-14 decays, thus changing the proportions of the kinds of carbon in the organism. By determining the amount of carbon-14 it is possible to determine how long the organism has been dead, hence how old it is.

a) How old is an animal bone that has lost 30% of its carbon-14?

b) A mummy discovered in the pyramid Khufu in Egypt had lost 46% of its carbon-14. Determine its age.

40. (*Radiocarbon dating*).

a) How old is an animal bone that has lost 20% of its carbon-14?

b) The Statue of Zeus at Olympia in Greece is one of the seven wonders of the world. It is made of gold and ivory. The ivory was found to have lost 35% of its carbon-14. Determine the age of the statue.

41. (*Atmospheric pressure*). What is the pressure at the top of Mt. Shasta in California, 14,162 ft high?

42. (*Atmospheric pressure*). Blood will boil when atmospheric pressure goes below about 62 millibars. At what altitude, in an unpressurized vehicle, will a pilot's blood boil?

43. Students in an industrial mathematics course take a final exam and are then retested at monthly intervals. The forgetting function is given by the equation $S(t) = 82 - 18 \log (t + 1)$.

    a) What was the average score originally on the final exam?
    b) What was the average score after 5 months had elapsed?

44. Students graduating from a cosmetology curriculum take a final exam and are then retested at monthly intervals. The forgetting function is given by the equation $S(t) = 75 - 20 \log (t + 1)$.

    a) What was the average score originally on the final exam?
    b) What was the average score after 6 months had elapsed?

45. Refer to Exercise 43. How much time will elapse before the average score has decreased to 64?

46. Refer to Exercise 44. How much time will elapse before the average score has decreased to 61?

47. Find the loudness of the sound of an automobile, having an intensity 3,100,000 times I_0.

48. Find the loudness of the sound of a dishwasher, having an intensity 2,500,000 times I_0.

49. Find the loudness of the threshold of sound pain, for which the intensity is 10^{14} times I_0.

50. Find the loudness of a jet aircraft, having an intensity 10^{12} times I_0.

51. The Los Angeles earthquake of 1971 had an intensity $10^{6.7}$ times I_0. What was its magnitude on the Richter scale?

52. The San Francisco earthquake of 1906 had an intensity $10^{8.25}$ times I_0. What was its magnitude on the Richter scale?

53. An earthquake has a magnitude of 5 on the Richter scale. What is its intensity?

54. An earthquake has a magnitude of 7 on the Richter scale. What is its intensity?

55. (*Consumer price index*). The *consumer price index* compares the costs of goods and services over various years. The base year is 1967. The same goods and services that cost \$100 (P_0) in 1967 cost \$184.50 in 1977. Assuming the exponential model:

    a) Find the value k and write the equation.
    b) Estimate what the same goods and services will cost in 1987.
    c) When did the same goods and services cost double that of 1967?

56. (*Cost of a double-dip ice cream cone*). In 1970 the cost of a double-dip ice cream cone was 52¢. In 1978 it was 66¢. Assuming the exponential model:

    a) Find the value k and write the equation.
    b) Estimate the cost of a cone in 1986.
    c) When will the cost of a cone be twice that of 1978?

57. In chemistry, pH is defined as follows:

$$pH = -\log [H^+],$$

where $[H^+]$ is the hydrogen ion concentration in moles per liter. For example, the hydrogen ion concentration in milk is 4×10^{-7} moles per liter, so

$$pH = -\log (4 \times 10^{-7}) = -[\log 4 + (-7)] \approx 6.4.$$

For tomatoes, $[H^+]$ is about 6.3×10^{-5}. Find the pH.

58. For eggs, $[H^+]$ is about 1.6×10^{-8}. Find the pH.

Solve.

59. $\log \sqrt{x} = \sqrt{\log x}$

60. $\log_5 \sqrt{x^2 + 1} = 1$

61. $(\log_a x)^{-1} = \log_a x^{-1}$

62. $|\log_5 x| = 2$

63. $\log_3 |x| = 2$

64. $\dfrac{(e^{3x+1})^2}{e^4} = e^{10x}$

65. $\dfrac{\sqrt{(e^{2x} \cdot e^{-5x})^{-4}}}{e^x \div e^{-x}} = e^7$

66. $\log x^{\log x} = 4$

67. Solve $y = ax^n$ for n. Use $\log_x$.

68. Solve $y = ke^{at}$ for t. Use $\log_e$.

Solve for t.

69. $P = P_0 e^{kt}$

70. $P = P_0 e^{-kt}$

71. $T = T_0 + (T_1 - T_0) 10^{-kt}$

72. Solve for n. Use $\log_V$.

73. Solve for Q.

74. Solve for y.

$$PV^n = c$$

$$\log_a Q = \tfrac{1}{3} \log_a y + b$$

$$\log_a y = 2x + \log_a x$$

Solve for x.

75. $x^{\log x} = \dfrac{x^3}{100}$

76. $x^{\log x} = 100x$

77. ▦ Given $f(x) = (1 + x)^{1/x}$, find $f(1)$, $f(0.5)$, $f(0.2)$, $f(0.1)$, $f(0.01)$, and $f(0.001)$ to six decimal places. This sequence of numbers approaches the number e.

78. ▦ Given $f(t) = t^{1/(t-1)}$, find $f(0.5)$, $f(0.9)$, $f(0.99)$, $f(0.999)$, and $f(0.9999)$ to six decimal places. This sequence of numbers approaches the number e.

Solve.

79. $|\log_5 x| + 3 \log_5 |x| = 4$ **80.** $|\log_a x| = \log_a |x|$

81. $(0.5)^x < \frac{4}{5}$ **82.** $8x^{0.3} - 8x^{-0.3} = 63$

83. Solve the system of equations.

$$5^{x+y} = 100,$$
$$3^{2x-y} = 1000$$

84. Given that

$$\log_2 [\log_3 (\log_4 x)] =$$
$$\log_3 [\log_2 (\log_4 y)] =$$
$$\log_4 [\log_3 (\log_2 z)] = 0,$$

find $x + y + z$.

85. If $2 \log_3 (x - 2y) = \log_3 x + \log_3 y$, find $\frac{x}{y}$.

86. Find the ordered pair (x, y) for which $4^{\log_{16} 27} = 2^x 3^y$.

SUMMARY AND REVIEW: CHAPTER 7

The following contains a summary of what you should be able to do after completing this chapter. The review exercises are for practice. Answers are at the back of the book. If you miss an exercise, restudy the section indicated alongside the answer.

You should be able to:

Given an equation defining a relation, write an equation of the inverse relation.

Write an equation of the inverse.

1. $y = 3x^2 + 2x - 1$

2. $y = \sqrt{x + 2}$

Given an equation defining a relation, determine whether the graph is symmetric with respect to the line $y = x$.

3. Which of the following relations have graphs that are symmetric with respect to the line $y = x$?

a) $y = 7$ b) $x^2 + y^2 = 4$ c) $x^3 = y^3 - y$ d) $y^2 = x^2 + 3$
e) $x + y = 3$ f) $x = 3$ g) $y = x^2$ h) $y = x^3$

Given the graph of a relation or an equation for a relation, determine whether its inverse is a function.

4. Which of the following have inverses that are functions?

a)
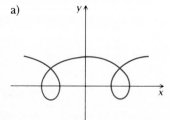

b)
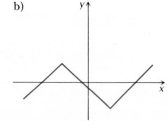

c)
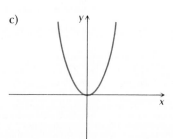

d)
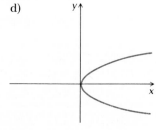

Given a function defined by a simple formula, find a formula for its inverse.

Find a formula for $f^{-1}(x)$.

5. $f(x) = \dfrac{\sqrt{x}}{2} + 2$

6. $f(x) = x^2 + 2$

For a function whose inverse is also a function, find $f^{-1}(f(x))$ and $f(f^{-1}(x))$ for any number x in the domains of the functions.

7. Find $f(f^{-1}(a))$.

$$f(x) = x^3 + 2$$

8. Find $h^{-1}(h(t))$.

$$h(x) = x^{17} + x^{65}$$

Graph exponential and logarithmic functions.

Graph.

9. $y = \log_2 (x - 1)$

10. $y = (\frac{1}{2})^x$

Given an exponential equation of the type $a^b = c$, write an equivalent logarithmic equation $\log_a c = b$, and vice versa.

11. Write an exponential equation equivalent to $\log_8 \frac{1}{4} = -\frac{2}{3}$.

12. Write a logarithmic equation equivalent to $7^{2.3} = x$.

Use the properties of logarithms to express sums and differences as a single logarithm, and vice versa, and to find certain logarithms given values of others.

13. Write an equivalent expression containing a single logarithm:

$$\frac{1}{2} \log_b a + \frac{3}{2} \log_b c - 4 \log_b d$$

14. Express in terms of logarithms of M and N:

$$\log \sqrt[3]{M^2/N}.$$

Given that $\log_a 2 = 0.301$, $\log_a 3 = 0.477$, and $\log_a 7 = 0.845$, find each of the following.

15. $\log_a 18$

16. $\log_a \frac{7}{2}$

17. $\log_a \frac{1}{4}$

18. $\log_a \sqrt{3}$

Simplify expressions like $a^{\log_a x}$ and $\log_a a^x$.

Simplify.

19. $\log_{12} 12^{x^2+1}$

20. $\log_8 8^{\sqrt{9}}$

Solve exponential and logarithmic equations and applied problems involving exponential and logarithmic equations.

Solve.

21. $\log_x 64 = 3$

22. $\log_{16} 4 = x$

23. $\log_5 125 = x$

24. $3^{1-x} = 9^{2x}$

25. $e^x = 80$

26. $\log x^2 = \log x$

27. $\log (x^2 - 1) - \log (x - 1) = 1$

28. $\log 2 + 2 \log x = \log (5x + 3)$

29. $\log_2 (x - 1) + \log_2 (x + 1) = 3$

30. How many years will it take an investment of $1000 to double if interest is compounded annually at 13%?

31. What is the loudness, in decibels, of a sound whose intensity is 1000 times I_0?

32. The half-life of a radioactive substance is 15 days. How much of a 25-gram sample will remain radioactive after 30 days?

Find logarithms and antilogarithms, base 10 and base e, using a calculator.

Find the following logarithms, base 10 and base e, using a calculator.

33. $\log 0.00216$

34. $\log 1,342,000$

35. $\log (2.037 \times 10^{-5})$

36. $\ln 87,380$

37. $\ln 0.00002776$

38. $\ln (4.369 \times 10^{10})$

Find the following antilogarithms, base 10 and base e, using a calculator.

39. antilog 3.0287

40. $10^{-5.4632}$

41. antilog 0.0008162

42. $\text{antilog}_e 6.5692$

43. $e^{11.637}$

44. $\text{antilog}_e (-4.089)$

Solve.

45. $|\log_4 x| = 3$

46. $\log x = \ln x$

Graph.

47. $y = |\log_3 x|$

48. $y = |e^x - 4|$

Find the domain.

49. $f(x) = \dfrac{17}{\sqrt{5 \ln x - 6}}$

50. $f(x) = \dfrac{8}{e^{4x} - 10}$

TEST: CHAPTER 7

1. Write an equation of the inverse of the relation $y = |x|$.

2. Which of the following relations have graphs that are symmetric with respect to the line $y = x$?

a) $2y - \dfrac{5}{x} = 0$

b) $x = -3$

c) $y = 2 - x$

d) $x^3 - x = y^3$

e) $x^2 - 1 = y^2$

f) $3y = |x|$

3. Which of the following have inverses that are functions?

a)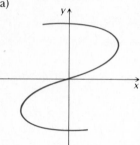

b)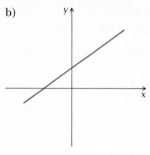

c)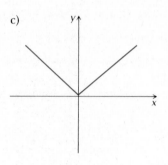

d)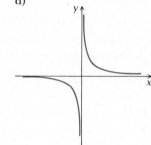

4. Find a formula for $f^{-1}(x)$:

$$f(x) = \sqrt{x - 6}.$$

5. Find $h(h^{-1}(3))$:

$$h(x) = \frac{-21x + 3}{20}.$$

6. Graph $y = \log_3 x$.

7. Write an exponential equation equivalent to $\log_{\sqrt{3}} 9 = 4$.

8. Write a logarithmic equation equivalent to $x^5 = 0.03125$.

9. Simplify $6^{\log_6 3x}$.

10. Write an equivalent expression containing a single logarithm:

$$3 \log_c x - 4 \log_c y + \tfrac{1}{2} \log_c z.$$

11. Express in terms of logarithms of x and r: $\log \sqrt[4]{wr^3}$.

Given that $\log_a 2 = 0.301$, $\log_a 5 = 0.699$, and $\log_a 6 = 0.778$, find each of the following.

12. $\log_a 3$

13. $\log_a 50$

14. $\log_a \sqrt[3]{5}$

15. Solve for x: $\log_b b^{2x^2} = x$.

Solve.

16. $\log_4 x = 2$

17. $4^{2x-1} - 3 = 61$

18. $\log_2 x + \log_2 (x - 2) = 3$

19. $e^{-x} = 0.2$

Find using a calculator.

20. $\log 0.005243$

21. $\ln 3.86$

22. $10^{2.0755}$

23. antilog$_e$ 2.037

24. $\log 87{,}200$

25. antilog (-2.003)

26. $\ln 0.00000029$

27. $\log (-14.68)$

28. How many years will it take an investment of \$5000 to double if interest is compounded annually at 12%?

29. What is the loudness, in decibels, of a sound whose intensity is 50,000 times I_0?

30. The population of a city was 80,000 in 1970 and 100,000 in 1980. Estimate the population in 2000.

31. True or false:

$$\log_a (3x^5) = 15 \log_a x.$$

32. Find the domain:

$$f(x) = \log_3 (\ln x).$$

Complex numbers find important application to problems of fluid flow. Predictions made mathematically are checked out in a wind tunnel.

It is a shame that any numbers were ever named *imaginary*, because the name seems to imply that such numbers are not good for anything practical. Curiously, the negative numbers were very nearly given the name "imaginary," and we know that they are useful in many practical applications.

In this chapter, we study the numbers that today bear the name *imaginary*. Those numbers are square roots of negative numbers. We will then allow addition of real numbers and imaginary numbers, thereby forming a system of numbers known as *complex* numbers.

The complex numbers have a great many applications to practical situations—for example, in electricity and engineering. Those applications are too involved to include in this book, but we want you to know that they exist.

8

OBJECTIVES

You should be able to:

i   Express imaginary numbers (square roots of negative numbers) in terms of *i*, and simplify.

ii   Add, subtract, and multiply complex numbers, expressing the answer as *a* + *bi*. Also factor sums of squares.

iii   Determine whether a complex number is a solution of an equation.

iv   Use the fact that for equality of complex numbers, the real parts must be the same and the imaginary parts must be the same.

Express in terms of *i*.

1. $\sqrt{-6}$

2. $-\sqrt{-10}$

3. $\sqrt{-4}$

4. $-\sqrt{-25}$

5. Simplify $\sqrt{-5}\sqrt{-2}$

Simplify.

6. $\dfrac{\sqrt{-22}}{\sqrt{-2}}$

7. $\dfrac{\sqrt{-21}}{\sqrt{3}}$

8. $\sqrt{-16} + \sqrt{-9}$

9. $\sqrt{-25} - \sqrt{-4}$

10. $\sqrt{-17} + \sqrt{-9}$

8.1

IMAGINARY AND COMPLEX NUMBERS

i   IMAGINARY NUMBERS

Negative numbers do not have square roots in the system of real numbers. Certain equations have no solutions. A new kind of number, called *imaginary*, was invented so that negative numbers would have square roots and certain equations would have solutions. These numbers were devised, starting with an imaginary unit named *i*, with the agreement that $i^2 = -1$ or $i = \sqrt{-1}$. All other imaginary numbers can be expressed as a product of *i* and a real number.

EXAMPLE 1   Express $\sqrt{-5}$ and $-\sqrt{-7}$ in terms of *i*.
$$\sqrt{-5} = \sqrt{-1 \cdot 5} = \sqrt{-1}\sqrt{5} = i\sqrt{5},$$
$$-\sqrt{-7} = -\sqrt{-1 \cdot 7} = -\sqrt{-1}\sqrt{7} = -i\sqrt{7}$$

It is also to be understood that the imaginary unit obeys the familiar laws of real numbers, such as the commutative and associative laws.

EXAMPLE 2   Simplify $\sqrt{-3}\sqrt{-7}$.

Important. We first express the two imaginary numbers in terms of *i*:
$$\sqrt{-3}\sqrt{-7} = i\sqrt{3} \cdot i\sqrt{7}.$$

Now, rearranging and combining, we have
$$i^2\sqrt{3}\sqrt{7} = -1 \cdot \sqrt{21}$$
$$= -\sqrt{21}.$$

Had we not expressed imaginary numbers in terms of *i* at the outset, we would have obtained $\sqrt{21}$ instead of $-\sqrt{21}$.

Important: All imaginary numbers must be expressed in terms of *i* before simplifying.

DO EXERCISES 1–5.

EXAMPLE 3   Simplify $-\sqrt{20}/\sqrt{-5}$.
$$\frac{-\sqrt{20}}{\sqrt{-5}} = \frac{-\sqrt{20}}{i\sqrt{5}} \cdot \frac{i}{i} = \frac{-i\sqrt{20}}{i^2\sqrt{5}}$$
$$= \frac{-i}{-1}\sqrt{\frac{20}{5}}$$
$$= 2i$$

EXAMPLE 4   Simplify $\sqrt{-9} + \sqrt{-25}$.
$$\sqrt{-9} + \sqrt{-25} = i\sqrt{9} + i\sqrt{25}$$
$$= 3i + 5i = (3+5)i$$
$$= 8i$$

DO EXERCISES 6–10.

Powers of i

Let us look at the powers of i:

$$i^2 = -1, \qquad\qquad i^5 = i^4 \cdot i = 1 \cdot i = i,$$
$$i^3 = i^2 \cdot i = -1 \cdot i = -i, \qquad i^6 = i^5 \cdot i = i \cdot i = -1,$$
$$i^4 = i^2 \cdot i^2 = -1(-1) = 1, \qquad i^7 = i^6 \cdot i = -1 \cdot i = -i.$$

The first four powers of i are all different, but thereafter there is a repeating pattern, in cycles of four. Note that $i^4 = 1$ and that all powers of i^4, such as i^8, i^{16}, and so on, are 1. To find a higher power we express it in terms of the nearest power of 4 less than the given one.

EXAMPLE 5   Simplify i^{17} and i^{23}.

$$i^{17} = i^{16} \cdot i$$

Since i^{16} is a power of i^4 and $i^4 = 1$, we know that $i^{16} = 1$ so

$$i^{17} = 1 \cdot i = i,$$
$$i^{23} = i^{20} \cdot i^3.$$

Since i^{20} is a power of i^4, $i^{20} = 1$, so $i^{23} = 1 \cdot i^3 = 1 \cdot (-i) = -i$.

DO EXERCISES 11–13.

ⅱ COMPLEX NUMBERS

The equation $x^2 + 1 = 0$ has no solution in real numbers, but it has the imaginary solutions i and $-i$. There are still rather simple-looking equations that do not have either real or imaginary solutions. For example, $x^2 - 2x + 2 = 0$ does not. If we allow sums of real and imaginary numbers, however, this equation and many others have solutions, as we shall show.

In order that more equations will have solutions, we invent a new system of numbers called the *system of complex numbers*.* A complex number is a sum of a real number and an imaginary number.

Definition

The set of complex numbers consists of all numbers $a + bi$, where a and b are real numbers.

For the complex number $a + bi$ we say that the *real part* is a and the *imaginary part* is bi.

We also agree that the familiar properties of real numbers (the *field* properties) hold also for complex numbers[†]. We list them.

Simplify.

11. i^{25}

12. i^{18}

13. i^{31}

---

*One may wonder why we do not invent a system of imaginary numbers. That would not make sense because products of imaginary numbers are not necessarily imaginary. Consider, for example, $i^2 = -1$.

[†]In a more rigorous treatment, addition and multiplication are defined for complex numbers. It is then proved that the field properties hold.

Simplify.

14. $(8 - i) + (4 + 2i)$

15. $(9 + 2i) - (4 + 3i)$

16. $(2 + 4i)(3 + i)$

17. $(4 + 5i) + (4 - 5i)$

18. $2i(4 + 3i)$

19. $(5 + 6i) - (5 + 3i)$

Factor.

20. $x^2 + 4$

21. $9 + y^2$

Commutativity. Addition and multiplication are commutative.

Associativity. Addition and multiplication are associative.

Distributivity. Multiplication is distributive over addition and also over subtraction.

Identities. The additive identity is $0 + 0i$, or 0. The multiplicative identity is $1 + 0i$, or 1.

Additive inverses. Every complex number $a + bi$ has the additive inverse $-a - bi$.

Multiplicative inverses. Every nonzero complex number has a multiplicative inverse, or reciprocal.

The complex-number system is an extension of the real-number system. Any number $a + bi$, where a and b are real numbers, is a complex number. The number b can be 0, in which case we have $a + 0i$, which simplifies to the real number a. Thus the complex numbers include all of the real numbers. They also include all of the imaginary numbers, because any imaginary number bi is equal to $0 + bi$.

CALCULATIONS

Since in the system of complex numbers the field properties hold, calculations are much the same as for real numbers. The primary difference is that one must remember that $i^2 = -1$.

EXAMPLE 6 Simplify $(8 + 6i) + (3 + 2i)$.

$$(8 + 6i) + (3 + 2i) = 8 + 3 + 6i + 2i$$
$$= 11 + 8i$$

EXAMPLE 7 Simplify $(1 + 2i)(1 + 3i)$.

$$(1 + 2i)(1 + 3i) = 1 + 6i^2 + 2i + 3i$$
$$= 1 - 6 + 2i + 3i \qquad i^2 = -1$$
$$= -5 + 5i$$

EXAMPLE 8 Simplify $(3 + 2i) - (5 - 2i)$.

$$(3 + 2i) - (5 - 2i) = (3 + 2i) - 5 + 2i$$
$$= 3 - 5 + 2i + 2i$$
$$= -2 + 4i$$

DO EXERCISES 14–19.

In the field of real numbers, a sum of two squares cannot be factored. In the system of complex numbers, a sum of squares is always factorable.

EXAMPLE 9 Factor $x^2 + y^2$.

$$(x + yi)(x - yi)$$

A check by multiplying will show that this is correct.

DO EXERCISES 20 AND 21.

iii SOLUTIONS OF EQUATIONS

In the system of complex numbers, a great many equations have solutions. In fact, any equation $P(x) = 0$, where $P(x)$ is a nonconstant polynomial, has a solution.*

EXAMPLE 10   Determine whether $1 + i$ is a solution of $x^2 - 2x + 2 = 0$.

$$\frac{x^2 - 2x + 2 = 0}{\begin{array}{c|c} (1 + i)^2 - 2(1 + i) + 2 & 0 \\ 1 + 2i + i^2 - 2 - 2i + 2 & \\ 1 + 2i - 1 - 2 - 2i + 2 & \\ 0 & \end{array}}$$ Substituting

The number $1 + i$ is a solution.

DO EXERCISE 22.

iv EQUALITY FOR COMPLEX NUMBERS

An equation $a + bi = c + di$ will be true if and only if the real parts are the same and the imaginary parts are the same. In other words, we have the following:

$$a + bi = c + di \quad \textbf{if and only if} \quad a = c \textbf{ and } b = d.$$

EXAMPLE 11   Suppose that $3x + yi = 5x + 1 + 2i$. Find x and y.

We equate the real parts: $3x = 5x + 1$. Solving this equation, we obtain $x = -\frac{1}{2}$. We equate the imaginary parts: $yi = 2i$. Thus $y = 2$.

DO EXERCISE 23.

22. Determine whether $1 - i$ is a solution of $x^2 - 2x + 2 = 0$.

23. Given that $3x + 1 + (y + 2)i = 2x + 2yi$, find x and y.

EXERCISE SET 8.1

i In Exercises 1–6, express in terms of i.

1. $\sqrt{-15}$ **2.** $\sqrt{-17}$ **3.** $\sqrt{-16}$
4. $\sqrt{-25}$ **5.** $-\sqrt{-12}$ **6.** $-\sqrt{-20}$

In Exercises 7–20, simplify. Leave answers in terms of i in Exercises 7–10.

7. $\sqrt{-16} + \sqrt{-25}$ **8.** $\sqrt{-36} - \sqrt{-4}$ **9.** $\sqrt{-7} - \sqrt{-10}$ **10.** $\sqrt{-5} + \sqrt{-7}$
11. $\sqrt{-5}\sqrt{-11}$ **12.** $\sqrt{-7}\sqrt{-8}$ **13.** $-\sqrt{-4}\sqrt{-5}$ **14.** $-\sqrt{-9}\sqrt{-7}$
15. $\dfrac{-\sqrt{5}}{\sqrt{-2}}$ **16.** $\dfrac{\sqrt{-7}}{-\sqrt{5}}$ **17.** $\dfrac{\sqrt{-9}}{-\sqrt{4}}$ **18.** $\dfrac{-\sqrt{25}}{\sqrt{-16}}$
19. $\dfrac{-\sqrt{-36}}{\sqrt{-9}}$ **20.** $\dfrac{\sqrt{-25}}{-\sqrt{-16}}$

ii In Exercises 21–38, simplify.

21. $(2 + 3i) + (4 + 2i)$ **22.** $(5 - 2i) + (6 + 3i)$ **23.** $(4 + 3i) + (4 - 3i)$
24. $(2 + 3i) + (-2 - 3i)$ **25.** $(8 + 11i) - (6 + 7i)$ **26.** $(9 - 5i) - (4 + 2i)$

---

* See Chapter 9 for more about this.

27. $2i - (4 + 3i)$

28. $3i - (5 + 2i)$

29. $(1 + 2i)(1 + 3i)$

30. $(1 + 4i)(1 - 3i)$

31. $(1 + 2i)(1 - 3i)$

32. $(2 + 3i)(2 - 3i)$

33. $3i(4 + 2i)$

34. $5i(3 - 4i)$

35. $(2 + 3i)^2$

36. $(3 - 2i)^2$

37. i^{13}

38. i^{72}

Factor.

39. $4x^2 + 25y^2$

40. $16a^2 + 49b^2$

iii

41. Determine whether $1 + 2i$ is a solution of $x^2 - 2x + 5 = 0$.

42. Determine whether $1 - 2i$ is a solution of $x^2 - 2x + 5 = 0$.

iv In Exercises 43 and 44, solve for x and y.

43. $4x + 7i = -6 + yi$

44. $-4 + (x + y)i = 2x - 5y + 5i$

45. A function f is defined as follows: $f(z) = z^2 - 4z + i$. Find $f(3 + i)$.

46. A function g is defined as follows: $g(z) = 2z^2 + z - 2i$. Find $g(2 - i)$.

47. Show that the general rule for radicals, in real numbers, $\sqrt{a \cdot b} = \sqrt{a} \cdot \sqrt{b}$, does not hold for complex numbers.

48. Show that the general rule for radicals, in real numbers, $\sqrt{\dfrac{a}{b}} = \dfrac{\sqrt{a}}{\sqrt{b}}$, does not hold for complex numbers.

49. Multiply.

$$\begin{bmatrix} i & 0 \\ 0 & 3i \end{bmatrix} \begin{bmatrix} i & 2i \\ 1 - i & 4 - i \end{bmatrix}$$

50. Solve $z^2 = -2i$. (*Hint:* Let $z = a + bi$.)

51. Solve $\dfrac{z^2}{4} = i$. (*Hint:* Let $z = a + bi$.)

OBJECTIVES

You should be able to:

i   Find the conjugate of a complex number and divide complex numbers.

ii   Find the reciprocal of a complex number and express it in the form $a + bi$.

iii   Given a polynomial in z, find a polynomial in $\bar{z}$ that is its conjugate.

Find the conjugate of each number.

1. $7 + 2i$

2. $6 - 4i$

3. $-5i$

4. $3i$

5. -3

6. 8

8.2

CONJUGATES AND DIVISION

i   CONJUGATES

We shall define the conjugate of a complex number as follows.

Definition

The conjugate of a complex number $a + bi$ is $a - bi$, and the conjugate of $a - bi$ is $a + bi$.

We illustrate:

The conjugate of $3 + 4i$ is $3 - 4i$.

The conjugate of $5 - 7i$ is $5 + 7i$.

The conjugate of $5i$ is $-5i$.

The conjugate of 6 is 6.

DO EXERCISES 1–6.

DIVISION

Fractional notation is useful for division of complex numbers. We also use the notion of conjugates.

EXAMPLE 1 Divide $4 + 5i$ by $1 + 4i$.

We shall write fractional notation and then multiply by 1.

$$\frac{4 + 5i}{1 + 4i} = \frac{4 + 5i}{1 + 4i} \cdot \frac{1 - 4i}{1 - 4i}$$ Note that $1 - 4i$ is the conjugate of the divisor.

$$= \frac{(4 + 5i)(1 - 4i)}{1^2 - 4^2 i^2}$$

$$= \frac{24 - 11i}{1 + 16} = \frac{24 - 11i}{17}$$ $i^2 = -1$

$$= \frac{24}{17} - \frac{11}{17} i$$

The procedure of the last example allows us always to find a quotient of two numbers and express it in the form $a + bi$. This is true because the product of a number and its conjugate is always a real number (we will prove this later), giving us a real-number denominator.

DO EXERCISES 7 AND 8.

ii RECIPROCALS

We can find the reciprocal, or multiplicative inverse, of a complex number by division. The reciprocal of a complex number z is $1/z$.

EXAMPLE 2 Find the reciprocal of $2 - 3i$ and express it in the form $a + bi$.

a) The reciprocal of $2 - 3i$ is $1/(2 - 3i)$.

b) We can express it in the form $a + bi$ as follows:

$$\frac{1}{2 - 3i} = \frac{1}{2 - 3i} \cdot \frac{2 + 3i}{2 + 3i} = \frac{2 + 3i}{2^2 - 3^2 i^2}$$

$$= \frac{2 + 3i}{4 + 9}$$

$$= \frac{2}{13} + \frac{3}{13} i.$$

DO EXERCISE 9.

iii PROPERTIES OF CONJUGATES

We can use a single letter for a complex number. For example, we could shorten $a + bi$ to z. To denote the conjugate of a number, we use a bar. The conjugate of z is $\bar{z}$. Or, the conjugate of $a + bi$ is $\overline{a + bi}$. Of course, by definition of conjugates, $\overline{a + bi} = a - bi$ and $\overline{a - bi} = a + bi$. We have already noted that the product of a number and its conjugate is always a real number. Let us state this formally and prove it.

Divide.

7. $\dfrac{1 + 3i}{3 + 2i}$

8. $\dfrac{2 + i}{3 - 2i}$

9. Find the reciprocal of $3 + 4i$ and express it in the form $a + bi$.

10. Compare $\overline{(3 + 2i) + (4 - 5i)}$ and $\overline{(3 + 2i)} + \overline{(4 - 5i)}$.

Theorem 1

For any complex number z, $z \cdot \bar{z}$ is a real number.

Proof. Let $z = a + bi$. Then

$$z \cdot \bar{z} = (a + bi)(a - bi)$$
$$= a^2 - b^2 i^2$$
$$= a^2 + b^2.$$

Since a and b are real numbers, so is $a^2 + b^2$. Thus $z \cdot \bar{z}$ is real.

The sum of a number and its conjugate is also always real. We state this as the second property.

Theorem 2

For any complex number z, $z + \bar{z}$ is a real number.

Proof. Let $z = a + bi$. Then

$$z + \bar{z} = (a + bi) + (a - bi)$$
$$= 2a.$$

Since a is a real number, $2a$ is real. Thus $z + \bar{z}$ is real.

Now we consider the conjugate of a sum and compare it with the sum of the conjugates.

EXAMPLE 3 Compare $\overline{(2 + 4i) + (5 + i)}$ and $\overline{(2 + 4i)} + \overline{(5 + i)}$.

a) $\overline{(2 + 4i) + (5 + i)} = \overline{7 + 5i}$ Adding the complex numbers
$$= 7 - 5i \qquad \text{Taking the conjugate}$$

b) $\overline{(2 + 4i)} + \overline{(5 + i)} = (2 - 4i) + (5 - i)$ Taking conjugates
$$= 7 - 5i \qquad \text{Adding}$$

DO EXERCISE 10.

Taking the conjugate of a sum gives the same result as adding the conjugates. Let us state and prove this.

Theorem 3

For any complex numbers z and w, $\overline{z + w} = \bar{z} + \bar{w}$.

Proof. Let $z = a + bi$ and $w = c + di$. Then

$$\overline{z + w} = \overline{(a + bi) + (c + di)}$$
$$= \overline{(a + c) + (b + d)i},$$

by adding. We now take the conjugate and obtain $(a + c) - (b + d)i$. Now

$$\bar{z} + \bar{w} = \overline{(a + bi)} + \overline{(c + di)}$$
$$= (a - bi) + (c - di),$$

taking the conjugates. We will now add to obtain $(a + c) - (b + d)i$, the same result as before. Thus $\overline{z + w} = \bar{z} + \bar{w}$.

Let us next consider the conjugate of a product.

EXAMPLE 4 Compare $\overline{(3 + 2i)(4 - 5i)}$ and $\overline{(3 + 2i)} \cdot \overline{(4 - 5i)}$.

a) $\overline{(3 + 2i)(4 - 5i)} = \overline{22 - 7i}$ Multiplying
$= 22 + 7i$ Taking the conjugate

b) $\overline{(3 + 2i)} \cdot \overline{(4 - 5i)} = (3 - 2i)(4 + 5i)$ Taking conjugates
$= 22 + 7i$ Multiplying

DO EXERCISE 11.

The conjugate of a product is the product of the conjugates. This is our next result.

Theorem 4

For any complex numbers z and w, $\overline{z \cdot w} = \bar{z} \cdot \bar{w}$.

Proof. Let $z = a + bi$ and $w = c + di$. Then

$$\overline{z \cdot w} = \overline{(a + bi)(c + di)}$$
$$= \overline{(ac - bd) + (bc + ad)i}. \quad \text{Multiplying}$$

Taking the conjugate, we obtain $(ac - bd) - (bc + ad)i$. Now

$$\bar{z} \cdot \bar{w} = \overline{(a + bi)} \cdot \overline{(c + di)} = (a - bi)(c - di),$$

taking conjugates. By multiplication we obtain $(ac - bd) - (bc + ad)i$, the same result as before. Thus $\overline{z \cdot w} = \bar{z} \cdot \bar{w}$.

Let us now consider conjugates of powers, using the preceding result.

EXAMPLE 5 Show that for any complex number z, $\overline{z^2} = \bar{z}^2$.

$$\overline{z^2} = \overline{z \cdot z} \quad \text{By the definition of exponents}$$
$$= \bar{z} \cdot \bar{z} \quad \text{By Theorem 4}$$
$$= \bar{z}^2 \quad \text{By the definition of exponents}$$

DO EXERCISE 12.

We now state our next result.

Theorem 5

For any complex number z, $\overline{z^n} = \bar{z}^n$. In other words, the conjugate of a power is the power of the conjugate. We understand that the exponent is a natural number.

The conjugate of a real number $a + 0i$ is $a - 0i$, and both are equal to a. Thus a real number is its own conjugate. We state this as our next result.

Theorem 6

If z is a real number, then $\bar{z} = z$.

11. Compare $\overline{(2 + 5i)(1 + 3i)}$ and $\overline{(2 + 5i)} \cdot \overline{(1 + 3i)}$.

12. Show that for any complex number z, $\overline{z^3} = \bar{z}^3$.

Find a polynomial in $\bar{z}$ that is the conjugate of each of the following.

13. $5z^3 + 4z^2 - 2z + 1$

14. $7z^5 - 3z^3 + 8z^2 + z$

CONJUGATES OF POLYNOMIALS

Given a polynomial in z, where z is a variable for a complex number, we can find its conjugate in terms of $\bar{z}$.

EXAMPLE 6 Find a polynomial in $\bar{z}$ that is the conjugate of $3z^2 + 2z - 1$.

We write an expression for the conjugate and then use the properties of conjugates.

$$\overline{3z^2 + 2z - 1} = \overline{3z^2} + \overline{2z} - \overline{1} \qquad \text{By Theorem 3}$$
$$= \overline{3}\,\overline{z^2} + \overline{2} \cdot \overline{z} - \overline{1} \qquad \text{By Theorem 4}$$
$$= 3\overline{z^2} + 2\overline{z} - 1 \qquad \begin{array}{l}\text{The conjugate of a real number}\\\text{is the number itself.}\end{array}$$
$$= 3\bar{z}^2 + 2\bar{z} - 1 \qquad \text{By Theorem 5}$$

DO EXERCISES 13 AND 14.

EXERCISE SET 8.2

i In Exercises 1–16, simplify.

1. $\dfrac{4 + 3i}{1 - i}$

2. $\dfrac{2 - 3i}{5 - 4i}$

3. $\dfrac{\sqrt{2} + i}{\sqrt{2} - i}$

4. $\dfrac{\sqrt{3} + i}{\sqrt{3} - i}$

5. $\dfrac{3 + 2i}{i}$

6. $\dfrac{2 + 3i}{i}$

7. $\dfrac{i}{2 + i}$

8. $\dfrac{3}{5 - 11i}$

9. $\dfrac{1 - i}{(1 + i)^2}$

10. $\dfrac{1 + i}{(1 - i)^2}$

11. $\dfrac{3 - 4i}{(2 + i)(3 - 2i)}$

12. $\dfrac{(4 - i)(5 + i)}{(6 - 5i)(7 - 2i)}$

13. $\dfrac{1 + i}{1 - i} \cdot \dfrac{2 - i}{1 - i}$

14. $\dfrac{1 - i}{1 + i} \cdot \dfrac{2 + i}{1 + i}$

15. $\dfrac{3 + 2i}{1 - i} + \dfrac{6 + 2i}{1 - i}$

16. $\dfrac{4 - 2i}{1 + i} + \dfrac{2 - 5i}{1 + i}$

ii In Exercises 17–24, find the reciprocal and express it in the form $a + bi$.

17. $4 + 3i$

18. $4 - 3i$

19. $5 - 2i$

20. $2 + 5i$

21. i

22. $-i$

23. $-4i$

24. $5i$

iii In Exercises 25–28, find a polynomial in $\bar{z}$ that is the conjugate.

25. $3z^5 - 4z^2 + 3z - 5$

26. $7z^4 + 5z^3 - 12z$

27. $4z^7 - 3z^5 + 4z$

28. $5z^{10} - 7z^8 + 13z^2 - 4$

29. Solve $z + 6\bar{z} = 7$.

30. Solve $5z - 4\bar{z} = 7 + 8i$.

31. Let $z = a + bi$. Find $\frac{1}{2}(z + \bar{z})$.

32. Let $z = a + bi$. Find $\frac{1}{2}i(\bar{z} - z)$.

33. Find $f(3 + i)$, where $f(z) = \dfrac{\bar{z}}{z - 1}$.

34. Show that for any complex number z, $z + \bar{z}$ is real.

35. Let $z = a + bi$. Find a general expression for $1/z$.

36. Let $z = a + bi$ and $w = c + di$. Find a general expression for w/z.

37. State and prove a theorem about the conjugate of a polynomial.

8.3

EQUATIONS AND COMPLEX NUMBERS

i LINEAR EQUATIONS

Linear equations with complex coefficients are solved in the same way as equations with real-number coefficients. The steps used in solving depend on the field properties in each case.

EXAMPLE 1   Solve $3ix + 4 - 5i = (1 + i)x + 2i$.

$$3ix - (1 + i)x = 2i - (4 - 5i) \qquad \text{Adding } -(1+i)x \text{ and } -(4-5i)$$

$$(-1 + 2i)x = -4 + 7i \qquad \text{Simplifying}$$

$$x = \frac{-4 + 7i}{-1 + 2i} \qquad \text{Dividing}$$

$$x = \frac{-4 + 7i}{-1 + 2i} \cdot \frac{-1 - 2i}{-1 - 2i} \qquad \text{Simplifying}$$

$$x = \frac{18 + i}{5} = \frac{18}{5} + \frac{1}{5}i$$

DO EXERCISE 1.

ii QUADRATIC EQUATIONS

In Chapter 2 we derived the quadratic formula for equations with real-number coefficients. A glance at that derivation shows that all the steps can also be done with complex-number coefficients with one possible exception. We need to know that the principle of zero products holds for complex numbers. This is the case, as will be shown later. Thus the quadratic formula holds for equations with complex coefficients.

Theorem 7

Any equation $ax^2 + bx + c = 0$, where $a \neq 0$ and a, b, and c are complex numbers, has solutions

$$\frac{-b \pm \sqrt{b^2 - 4ac}}{2a}.$$

EXAMPLE 2   Solve $x^2 + (1 + i)x - 2i = 0$.

We note that $a = 1$, $b = 1 + i$, and $c = -2i$. Thus

$$x = \frac{-(1 + i) \pm \sqrt{(1 + i)^2 - 4 \cdot 1 \cdot (-2i)}}{2 \cdot 1} \qquad \text{Substituting in the formula}$$

$$= \frac{-1 - i \pm \sqrt{10i}}{2}. \qquad \text{Simplifying}$$

DO EXERCISE 2.

In Example 2 we have not attempted to evaluate $\sqrt{10i}$. That will be done later (Example 6). Let us say, however, that $10i$ has two square roots that are additive inverses of each other. The same is true of every nonzero

OBJECTIVES

You should be able to:

i Solve linear equations with complex-number coefficients.

ii Solve quadratic equations with complex coefficients, using the quadratic formula, but not attempting to take square roots that occur in the formula.

iii Solve quadratic equations with real-number coefficients using the quadratic formula.

iv Find an equation having specified complex numbers as solutions.

v Find square roots of complex numbers by solving an equation $z^2 = a + bi$.

1. Solve.

$$3 - 4i + 2ix = 3i - (1 - i)x$$

2. Solve.

$$2x^2 + (1 - i)x + 2i = 0$$

3. Solve.

$$5x^2 + 6x + 5 = 0$$

4. Find an equation having i and $1 + i$ as solutions.

5. Find an equation having 2, i, and $-i$ as solutions.

complex number. There is no problem of negative numbers not having square roots, because in the system of complex numbers there are no positive or negative numbers. Whenever we speak of positive or negative numbers, we will be referring to the system of real numbers.

iii Quadratic Equations with Real Coefficients

Since any real number a is $a + 0i$, we can consider real numbers to be special kinds of complex numbers. If a quadratic equation has real coefficients, the quadratic formula always gives solutions in the system of complex numbers. Recall that in using the formula, we take the square root of $b^2 - 4ac$. If this *discriminant* is negative, the solutions will have nonzero imaginary parts.

EXAMPLE 3   Solve $x^2 + 2x + 5 = 0$.

We note that $a = 1$, $b = 2$, and $c = 5$. Thus

$$x = \frac{-2 \pm \sqrt{2^2 - 4 \cdot 1 \cdot 5}}{2 \cdot 1} = \frac{-2 \pm \sqrt{-16}}{2}$$

$$= \frac{-2 \pm 4i}{2}.$$

The solutions are $-1 + 2i$ and $-1 - 2i$.

It is easy to see from Example 3 that when there are nonzero imaginary parts, an equation has two solutions that are conjugates of each other.

DO EXERCISE 3.

iv WRITING EQUATIONS WITH SPECIFIED SOLUTIONS

The principle of zero products for real numbers states that a product is 0 if and only if at least one of the factors is 0. This principle also holds for complex numbers, as we will show in Section 8.4. Since the principle holds for complex numbers, we can write equations having specified solutions.

EXAMPLE 4   Find an equation having the numbers 1, i, and $-i$ as solutions.

The factors we use will be $x - 1$, $x - i$, and $x + i$. Next we set the product of these factors equal to 0:

$$(x - 1)(x - i)(x + i) = 0$$

Now we multiply and simplify:

$$(x - 1)(x^2 - i^2) = 0$$
$$(x - 1)(x^2 + 1) = 0$$
$$x^3 - x^2 + x - 1 = 0.$$

EXAMPLE 5   Find an equation having -1, i, and $1 + i$ as solutions.

$$(x + 1)(x - i)[x - (1 + i)] = 0$$
$$(x^2 + x - ix - i)(x - 1 - i) = 0$$
$$x^3 - 2ix^2 - ix - 2x - 1 + i = 0$$

DO EXERCISES 4 AND 5.

v   FINDING SQUARE ROOTS

In Example 2 we used the quadratic formula, but did not evaluate $\sqrt{10i}$. We wish to find complex numbers z for which $z^2 = 10i$. We do so in the following example.

EXAMPLE 6   Solve $z^2 = 10i$.

Let $z = x + yi$. Then we have $(x + yi)^2 = 10i$, or $x^2 + 2xyi + y^2i^2 = 10i$, or $x^2 - y^2 + 2xyi = 0 + 10i$. Since the real parts must be the same and the imaginary parts the same, $x^2 - y^2 = 0$ and $2xy = 10$. Here we have two equations in real numbers. We can find a solution by solving the second equation for y and substituting in the first. Since

$$y = \frac{5}{x},$$

then

$$x^2 - \left(\frac{5}{x}\right)^2 = 0$$

$$x^2 - \frac{25}{x^2} = 0 \quad \text{or} \quad x^4 - 25 = 0.$$

We now have an equation reducible to quadratic:

$$(x^2 + 5)(x^2 - 5) = 0 \qquad \text{Factoring}$$

$$x^2 + 5 = 0 \quad \text{or} \quad x^2 - 5 = 0. \qquad \text{Principle of zero products}$$

The real solutions are $\sqrt{5}$ and $-\sqrt{5}$. Now, going back to $2xy = 10$, we see that if $x = \sqrt{5}$, then $y = \sqrt{5}$, and if $x = -\sqrt{5}$, then $y = -\sqrt{5}$. Thus the solutions of our equation (and the square roots of $10i$) are $\sqrt{5} + \sqrt{5}i$ and $-\sqrt{5} - \sqrt{5}i$.

DO EXERCISE 6.

6. Solve $z^2 = 3 - 4i$.

EXERCISE SET 8.3

i   Solve.

1. $(3 + i)x + i = 5i$

2. $(2 + i)x - i = 5 + i$

3. $2ix + 5 - 4i = (2 + 3i)x - 2i$

4. $5ix + 3 + 2i = (3 - 2i)x + 3i$

5. $(1 + 2i)x + 3 - 2i = 4 - 5i + 3ix$

6. $(1 - 2i)x + 2 - 3i = 5 - 4i + 2x$

7. $(5 + i)x + 1 - 3i = (2 - 3i)x + 2 - i$

8. $(5 - i)x + 2 - 3i = (3 - 2i)x + 3 - i$

ii   In Exercises 9–14, solve using the quadratic formula. You need not evaluate square roots.

9. $x^2 + (1 - i)x + i = 0$

10. $x^2 + (1 + i)x - i = 0$

11. $2x^2 + ix + 1 = 0$

12. $2x^2 - ix + 1 = 0$

13. $3x^2 + (1 + 2i)x + 1 - i = 0$

14. $3x^2 + (1 - 2i)x + 1 + i = 0$

iii   Solve.

15. $x^2 - 2x + 5 = 0$

16. $x^2 - 4x + 5 = 0$

17. $x^2 - 4x + 13 = 0$

18. $x^2 - 6x + 13 = 0$

19. $x^2 + 3x + 4 = 0$

20. $3x^2 + x + 2 = 0$

iv   Find an equation having the specified solutions.

21. $2i, -2i$

22. $3i, -3i$

23. $1 + i, 1 - i$

24. $2 + i, 2 - i$

25. $2 + 3i, 2 - 3i$ **26.** $4 + 3i, 4 - 3i$ **27.** $3, i$ **28.** $5, i$

29. $1, 3i, -3i$ **30.** $1, 2i, -2i$ **31.** $2, 1 + i, i$ **32.** $3, 1 - i, i$

33. $i, 2i, -i$ **34.** $i, -2i, -i$

v Solve.

35. $z^2 = 4i$ **36.** $z^2 = -4i$ **37.** $z^2 = 3 + 4i$ **38.** $z^2 = 5 - 12i$

Solve. (*Hint:* Factor first.)

39. $x^3 - 8 = 0$ **40.** $x^3 + 8 = 0$

Solve.

41. $2x + 3y = 7 - 7i,$
    $3x - 2y = 4 + 9i$

42. $3x + 4y = 11 + 9i,$
    $2x - y = -5i$

Solve for x and y.

43. $\log x + 2^y i = 2 + 16i$ **44.** $e^{2x} + ie^y = 1 + i$

OBJECTIVES

You should be able to:

i Graph a complex number $a + bi$ and graph the sum of two numbers.

ii Given rectangular, or binomial, notation for a complex number, find polar notation.

iii Given polar notation for a complex number, find binomial notation.

iv Use polar notation to multiply and divide complex numbers.

8.4

GRAPHICAL REPRESENTATION AND POLAR NOTATION

i GRAPHS

The real numbers can be graphed on a line. Complex numbers are graphed on a plane. We graph a complex number $a + bi$ in the same way that we graph an ordered pair of real numbers (a, b). In place of an x-axis we have a *real axis*, and in place of a y-axis we have an *imaginary axis*.

EXAMPLES   Graph the following.

1. $3 + 2i$

2. $-4 + 5i$

3. $-5 - 4i$

See the figure.

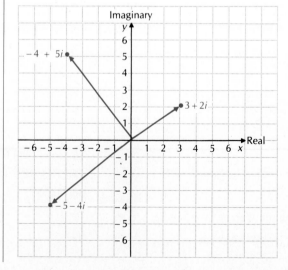

Horizontal distances correspond to the real part of a number. Vertical distances correspond to the imaginary part. Graphs of the numbers are shown above as vectors. The horizontal component of the vector for $a + bi$ is a and the vertical component is b. Complex numbers are sometimes used in the study of vectors.

DO EXERCISE 1.

Adding complex numbers is like adding vectors using components. For example, to add $3 + 2i$ and $5 + 4i$ we add the real parts and the imaginary parts to obtain $8 + 6i$. Graphically, then, the sum of two complex numbers looks like a vector sum. It is the diagonal of a parallelogram.

EXAMPLE 4   Show graphically $2 + 2i$ and $3 - i$. Show also their sum.

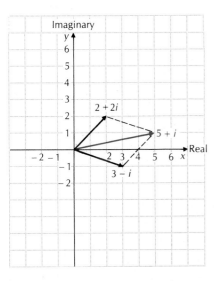

DO EXERCISE 2.

ii   POLAR NOTATION FOR COMPLEX NUMBERS

We can locate endpoints of arrows as shown in the preceding examples or by giving the length of the arrow and the angle that it makes with the positive half of the real axis. We now develop *polar notation* using this idea. From the figure below, we see that the length of the vector is $\sqrt{a^2 + b^2}$. Note that this quantity is a real number. It is called the absolute value of $a + bi$.

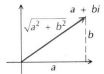

Definition

The *absolute value* of a complex number $a + bi$ is denoted $|a + bi|$ and is defined to be $\sqrt{a^2 + b^2}$.

1. Graph these complex numbers.

a) $5 - 3i$

b) $-3 + 4i$

c) $-5 - 2i$

d) $5 + 5i$

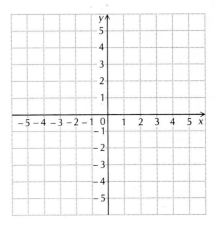

2. Graph each pair of complex numbers and graph their sum.

a) $2 + 3i, \quad 1 - 5i$

b) $-5 + 2i, \quad -1 - 4i$

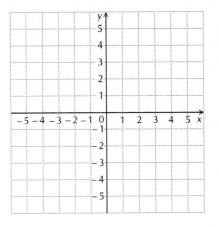

3. Find the absolute value.

   a) $|4 - 3i|$

   b) $|-12 - 5i|$

EXAMPLE 5 Find $|3 + 4i|$.

$$|3 + 4i| = \sqrt{3^2 + 4^2} = \sqrt{9 + 16} = 5$$

DO EXERCISE 3.

[That which follows is *only* for those who have studied trigonometry.]

Now let us consider any complex number $a + bi$. Suppose that its absolute value is r. Let us also suppose that the angle that the vector makes with the real axis is θ. As this diagram shows, we have

$$a = r \cos \theta \quad \text{and} \quad b = r \sin \theta.$$

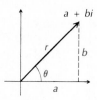

4. Write rectangular notation for

$$\sqrt{2} \, (\cos 315° + i \sin 315°).$$

Thus

$$a + bi = r \cos \theta + ir \sin \theta = r(\cos \theta + i \sin \theta).$$

This is polar notation for $a + bi$. The angle θ is called the *argument*.

Polar notation for complex numbers is also called *trigonometric notation*, and is often abbreviated $r \, cis \, \theta$.

ii, iii CHANGE OF NOTATION

To change from polar notation to *binomial*, or *rectangular*, notation $a + bi$, we use components, noting that $a = r \cos \theta$ and $b = r \sin \theta$.

EXAMPLE 6 Write binomial notation for $2(\cos 120° + i \sin 120°)$.

$$a = 2 \cos 120° = -1,$$
$$b = 2 \sin 120° = \sqrt{3}$$

Thus $2(\cos 120° + i \sin 120°) = -1 + i\sqrt{3}$.

5. Write binomial notation for

$$2 \, cis \left(-\frac{\pi}{6}\right).$$

EXAMPLE 7 Write binomial notation for $\sqrt{8} \, cis \, \dfrac{7\pi}{4}$.

$$a = \sqrt{8} \, \cos \frac{7\pi}{4} = \sqrt{8} \cdot \frac{1}{\sqrt{2}} = 2,$$

$$b = \sqrt{8} \, \sin \frac{7\pi}{4} = \sqrt{8} \cdot \frac{-1}{\sqrt{2}} = -2$$

Thus $\sqrt{8} \, cis \, \dfrac{7\pi}{4} = 2 - 2i$.

DO EXERCISES 4 AND 5.

To change from binomial notation to polar notation, we remember that $r = \sqrt{a^2 + b^2}$ and θ is an angle for which $\sin \theta = b/r$ and $\cos \theta = a/r$.

EXAMPLE 8   Find polar notation for $1 + i$.

We note that $a = 1$ and $b = 1$. Then

$$r = \sqrt{1^2 + 1^2} = \sqrt{2},$$

$$\sin \theta = \frac{1}{\sqrt{2}} \quad \text{and} \quad \cos \theta = \frac{1}{\sqrt{2}}.$$

Thus $\theta = \pi/4$, or $45°$, and we have

$$1 + i = \sqrt{2} \text{ cis } \frac{\pi}{4} \quad \text{or} \quad 1 + i = \sqrt{2} \text{ cis } 45°.$$

EXAMPLE 9   Find polar notation for $\sqrt{3} - i$.

$$r = \sqrt{(\sqrt{3})^2 + (-1)^2} = 2,$$

$$\sin \theta = -\frac{1}{2} \quad \text{and} \quad \cos \theta = \frac{\sqrt{3}}{2}.$$

Thus $\theta = 11\pi/6$, or $330°$, and we have

$$\sqrt{3} - i = 2 \text{ cis } \frac{11\pi}{6} = 2 \text{ cis } 330°.$$

In changing to polar notation, note that there are many angles satisfying the given conditions. We ordinarily choose the smallest positive angle.

DO EXERCISES 6 AND 7.

iv   MULTIPLICATION AND POLAR NOTATION

Multiplication of complex numbers is somewhat easier to do with polar notation than with rectangular notation. We simply multiply the absolute values and add the arguments. Let us state this more formally and prove it.

Theorem 8

For any complex numbers r_1 cis θ_1 and r_2 cis θ_2,

$$(r_1 \text{ cis } \theta_1)(r_2 \text{ cis } \theta_2) = r_1 \cdot r_2 \text{ cis } (\theta_1 + \theta_2).$$

Proof.   Let us first multiply $a_1 + b_1 i$ by $a_2 + b_2 i$:

$$(a_1 + b_1 i)(a_2 + b_2 i) = (a_1 a_2 - b_1 b_2) + (a_2 b_1 + a_1 b_2)i.$$

Recall that

$$a_1 = r_1 \cos \theta_1, \qquad b_1 = r_1 \sin \theta_1,$$

and

$$a_2 = r_2 \cos \theta_2, \qquad b_2 = r_2 \sin \theta_2.$$

We substitute these in the product above, to obtain

$$r_1(\cos \theta_1 + i \sin \theta_1) \cdot r_2(\cos \theta_2 + i \sin \theta_2)$$
$$= (r_1 r_2 \cos \theta_1 \cos \theta_2 - r_1 r_2 \sin \theta_1 \sin \theta_2)$$
$$+ (r_1 r_2 \sin \theta_1 \cos \theta_2 + r_1 r_2 \cos \theta_1 \sin \theta_2)i.$$

This simplifies to

$$r_1 r_2(\cos \theta_1 \cos \theta_2 - \sin \theta_1 \sin \theta_2) + r_1 r_2(\sin \theta_1 \cos \theta_2 + \cos \theta_1 \sin \theta_2)i.$$

Write polar notation.

6. $1 - i$

7. $-3\sqrt{2} - 3\sqrt{2}\, i$

Now, using identities for sums of angles, we simplify, obtaining

$$r_1 r_2 \cos (\theta_1 + \theta_2) + r_1 r_2 \sin (\theta_1 + \theta_2) i,$$

or

$$r_1 r_2 \operatorname{cis} (\theta_1 + \theta_2),$$

which was to be shown.

To divide complex numbers we do the reverse of the above. We state that fact, but will omit the proof.

Theorem 9

> For any complex numbers $r_1 \operatorname{cis} \theta_1$ and $r_2 \operatorname{cis} \theta_2$ $(r_2 \neq 0)$,
>
> $$\frac{r_1 \operatorname{cis} \theta_1}{r_2 \operatorname{cis} \theta_2} = \frac{r_1}{r_2} \operatorname{cis} (\theta_1 - \theta_2).$$

EXAMPLE 10   Find the product of $3 \operatorname{cis} 40°$ and $7 \operatorname{cis} 20°$.

$$3 \operatorname{cis} 40° \cdot 7 \operatorname{cis} 20° = 3 \cdot 7 \operatorname{cis} (40° + 20°)$$
$$= 21 \operatorname{cis} 60°$$

EXAMPLE 11   Find the product of $2 \operatorname{cis} \pi$ and $3 \operatorname{cis} \left(-\dfrac{\pi}{2} \right)$.

$$2 \operatorname{cis} \pi \cdot 3 \operatorname{cis} \left(-\frac{\pi}{2} \right) = 2 \cdot 3 \operatorname{cis} \left(\pi - \frac{\pi}{2} \right)$$
$$= 6 \operatorname{cis} \frac{\pi}{2}.$$

EXAMPLE 12   Convert to polar notation and multiply: $(1 + i)(\sqrt{3} - i)$.

We first find polar notation (see Examples 8 and 9):

$$1 + i = \sqrt{2} \operatorname{cis} 45°, \qquad \sqrt{3} - i = 2 \operatorname{cis} 330°.$$

We now multiply, using Theorem 8:

$$(\sqrt{2} \operatorname{cis} 45°)(2 \operatorname{cis} 330°) = 2 \cdot \sqrt{2} \operatorname{cis} 375° \quad \text{or} \quad 2\sqrt{2} \operatorname{cis} 15°.$$

EXAMPLE 13   Divide $2 \operatorname{cis} \pi$ by $4 \operatorname{cis} \dfrac{\pi}{2}$.

$$\frac{2 \operatorname{cis} \pi}{4 \operatorname{cis} \dfrac{\pi}{2}} = \frac{2}{4} \operatorname{cis} \left(\pi - \frac{\pi}{2} \right)$$

$$= \frac{1}{2} \operatorname{cis} \frac{\pi}{2}$$

EXAMPLE 14   Convert to polar notation and divide: $(1 + i)/(1 - i)$.

We first convert to polar notation:

$$1 + i = \sqrt{2} \operatorname{cis} 45°, \qquad \text{See Example 8}$$
$$1 - i = \sqrt{2} \operatorname{cis} 315°.$$

We now divide, using Theorem 9:

$$\frac{1+i}{1-i} = \frac{\sqrt{2}\ \text{cis}\ 45°}{\sqrt{2}\ \text{cis}\ 315°} = 1 \cdot \text{cis}\ (45° - 315°)$$

$$= 1 \cdot \text{cis}\ (-270°) \quad \text{or} \quad 1 \cdot \text{cis}\ 90°.$$

DO EXERCISES 8–12.

THE PRINCIPLE OF ZERO PRODUCTS

In the system of complex numbers, when we multiply by 0 the result is 0. This is half of the principle of zero products. We have yet to show that the converse is true: that if a product is zero, at least one of the factors must be zero. With polar notation that is easy to do. The number 0 has polar notation $0\ \text{cis}\ \theta$, where θ can be any angle. Now consider a product $r_1\ \text{cis}\ \theta_1 \cdot r_2\ \text{cis}\ \theta_2$, where neither factor is 0. This means that $r_1 \neq 0$ and $r_2 \neq 0$. Since these are nonzero real numbers, their product is not zero, and the product of the complex numbers $r_1 r_2\ \text{cis}\ (\theta_1 + \theta_2)$ is not zero. Thus the principle of zero products holds in complex numbers.

Multiply.

8. $5\ \text{cis}\ 25° \cdot 4\ \text{cis}\ 30°$

9. $8\ \text{cis}\ \pi \cdot \frac{1}{2}\ \text{cis}\ \frac{\pi}{4}$

10. Convert to polar notation and multiply.

$$(1 + i)(2 + 2i)$$

11. Divide.

$$10\ \text{cis}\ \frac{\pi}{2} \div 5\ \text{cis}\ \frac{\pi}{4}$$

12. Convert to polar notation and divide.

$$\frac{\sqrt{3} - i}{1 + i}$$

EXERCISE SET 8.4

i In Exercises 1–8, graph each pair of complex numbers and their sum.

1. $3 + 2i, 2 - 5i$ **2.** $4 + 3i, 3 - 4i$ **3.** $-5 + 3i, -2 - 3i$ **4.** $-4 + 2i, -3 - 4i$

5. $2 - 3i, -5 + 4i$ **6.** $3 - 2i, -5 + 5i$ **7.** $-2 - 5i, 5 + 3i$ **8.** $-3 - 4i, 6 + 3i$

iii In Exercises 9–16, find rectangular notation.

9. $3(\cos 30° + i\sin 30°)$ **10.** $6(\cos 150° + i\sin 150°)$ **11.** $10\ \text{cis}\ 270°$ **12.** $5\ \text{cis}\ (-60°)$

13. $\sqrt{8}\left(\cos\frac{\pi}{4} + i\sin\frac{\pi}{4}\right)$ **14.** $5\left(\cos\frac{\pi}{3} + i\sin\frac{\pi}{3}\right)$ **15.** $\sqrt{8}\ \text{cis}\ \frac{5\pi}{4}$ **16.** $\sqrt{8}\ \text{cis}\left(-\frac{\pi}{4}\right)$

ii In Exercises 17–22, find polar notation.

17. $1 - i$ **18.** $\sqrt{3} + i$ **19.** $10\sqrt{3} - 10i$

20. $-10\sqrt{3} + 10i$ **21.** -5 **22.** $-5i$

iv In Exercises 23–30, convert to polar notation and then multiply or divide.

23. $(1 - i)(2 + 2i)$ **24.** $(1 + i\sqrt{3})(1 + i)$ **25.** $(2\sqrt{3} + 2i)(2i)$ **26.** $(3\sqrt{3} - 3i)(2i)$

27. $\frac{1 - i}{1 + i}$ **28.** $\frac{1 - i}{\sqrt{3} - i}$ **29.** $\frac{2\sqrt{3} - 2i}{1 + \sqrt{3}i}$ **30.** $\frac{3 - 3\sqrt{3}i}{\sqrt{3} - i}$

31. Show that for any complex number z, $|z| = |-z|$. (*Hint:* Let $z = a + bi$.)

32. Show that for any complex number z, $|z| = |\bar{z}|$. (*Hint:* Let $z = a + bi$.)

33. Show that for any complex number z, $|z\bar{z}| = |z^2|$.

34. Show that for any complex number z, $|z^2| = |z|^2$.

35. Show that for any complex numbers z and w, $|z \cdot w| = |z| \cdot |w|$. (*Hint:* Let $z = r_1\ \text{cis}\ \theta_1$ and $w = r_2\ \text{cis}\ \theta_2$.)

36. Show that for any complex number z and any nonzero complex number w, $\left|\frac{z}{w}\right| = \frac{|z|}{|w|}$. (Use the hint for Exercise 35.)

37. On a complex plane, graph $|z| = 1$.

38. On a complex plane, graph $z + \bar{z} = 3$.

39. Find polar notation for $(\cos\theta + i\sin\theta)^{-1}$.

40. Compute $\begin{bmatrix} i & 0 \\ 0 & -i \end{bmatrix}^3$.

OBJECTIVES

You should be able to:

i Use DeMoivre's theorem to raise complex numbers to powers.

ii Find the nth roots of a complex number.

1. Find $(1 - i)^{10}$.

2. Find $(\sqrt{3} + i)^4$.

8.5

DeMOIVRE'S THEOREM

i POWERS OF COMPLEX NUMBERS

An important theorem about powers and roots of complex numbers is named for the French mathematician DeMoivre (1667–1754). Let us consider a number $r \operatorname{cis} \theta$ and its square:

$$(r \operatorname{cis} \theta)^2 = (r \operatorname{cis} \theta)(r \operatorname{cis} \theta) = r \cdot r \operatorname{cis} (\theta + \theta) = r^2 \operatorname{cis} 2\theta.$$

Similarly, we see that

$$(r \operatorname{cis} \theta)^3 = r \cdot r \cdot r \operatorname{cis} (\theta + \theta + \theta) = r^3 \operatorname{cis} 3\theta.$$

The generalization of this is DeMoivre's theorem.

Theorem 10 DeMoivre's Theorem

> For any complex number $r \operatorname{cis} \theta$ and any natural number n, $(r \operatorname{cis} \theta)^n = r^n \operatorname{cis} n\theta$.

EXAMPLE 1 Find $(1 + i)^9$.

We first find polar notation: $1 + i = \sqrt{2} \operatorname{cis} 45°$. Then

$$
\begin{aligned}
(1 + i)^9 &= (\sqrt{2} \operatorname{cis} 45°)^9 \\
&= \sqrt{2}^9 \operatorname{cis} 9 \cdot 45° \\
&= 2^{9/2} \operatorname{cis} 405° \\
&= 16\sqrt{2} \operatorname{cis} 45°. \qquad \text{405° has the same terminal side as 45°.}
\end{aligned}
$$

DO EXERCISES 1 AND 2.

ii ROOTS OF COMPLEX NUMBERS

As we shall see, every nonzero complex number has two square roots. A number has three cube roots, four fourth roots, and so on. In general, a nonzero complex number has n different nth roots. They can be found by the formula that we now state and prove.

Theorem 11

> The nth roots of a complex number $r \operatorname{cis} \theta$ are given by
>
> $$r^{1/n} \operatorname{cis} \left(\frac{\theta}{n} + k \cdot \frac{360°}{n} \right),$$
>
> where $k = 0, 1, 2, \ldots, n - 1$.

We show that this formula gives us n different roots, using DeMoivre's theorem. We take the expression for the nth roots and raise it to the nth power, to show that we get $r \operatorname{cis} \theta$:

$$\left[r^{1/n} \operatorname{cis} \left(\frac{\theta}{n} + k \cdot \frac{360°}{n} \right) \right]^n = (r^{1/n})^n \operatorname{cis} \left(\frac{\theta}{n} \cdot n + k \cdot n \cdot \frac{360°}{n} \right)$$

$$= r \operatorname{cis} (\theta + k \cdot 360°) = r \operatorname{cis} \theta.$$

Thus we know that the formula gives us nth roots for any natural number k. Next we show that there are at least n different roots. To see this, consider substituting 0, 1, 2, and so on, for k. When $k = n$ the cycle begins to repeat, but from 0 to $n - 1$ the angles obtained and their sines and cosines are all different. There cannot be more than n different nth roots. That fact follows from the *fundamental theorem of algebra*, considered in the next chapter.

EXAMPLE 2 Find the square roots of $2 + 2\sqrt{3}i$.

We first find polar notation: $2 + 2\sqrt{3}i = 4$ cis $60°$. Then

$$(4 \text{ cis } 60°)^{1/2} = 4^{1/2} \text{ cis} \left(\frac{60°}{2} + k \cdot \frac{360°}{2} \right), \quad k = 0, 1,$$

$$= 2 \text{ cis} \left(30° + k \cdot \frac{360°}{2} \right), \quad k = 0, 1.$$

Thus the roots are 2 cis $30°$ and 2 cis $210°$, or

$$\sqrt{3} + i \quad \text{and} \quad -\sqrt{3} - i.$$

DO EXERCISES 3 AND 4.

In Example 2 it should be noted that the two square roots of a number were additive inverses of each other. The same is true of the square roots of any complex number. To see this, let us find the square roots of any complex number r cis θ:

$$(r \text{ cis } \theta)^{1/2} = r^{1/2} \text{ cis} \left(\frac{\theta}{2} + k \cdot \frac{360°}{2} \right), \quad k = 0, 1,$$

$$= r^{1/2} \text{ cis} \frac{\theta}{2} \quad \text{or} \quad r^{1/2} \text{ cis} \left(\frac{\theta}{2} + 180° \right).$$

Now let us look at the two numbers on a graph. They lie on a line, so if one number has binomial notation $a + bi$, the other has binomial notation $-a - bi$. Hence their sum is 0 and they are additive inverses of each other.

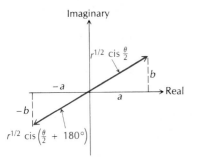

EXAMPLE 3 Find the cube roots of 1. Locate them on a graph.

$$1 = 1 \text{ cis } 0°$$

$$(1 \text{ cis } 0°)^{1/3} = 1^{1/3} \text{ cis} \left(\frac{0°}{3} + k \cdot \frac{360°}{3} \right), \quad k = 0, 1, 2.$$

The roots are 1 cis $0°$, 1 cis $120°$, and 1 cis $240°$, or

$$1, \quad -\frac{1}{2} + \frac{\sqrt{3}}{2}i, \quad \text{and} \quad -\frac{1}{2} - \frac{\sqrt{3}}{2}i.$$

Find the square roots.

3. $2i$

4. $10i$

5. Find and graph the cube roots of -1.

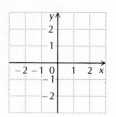

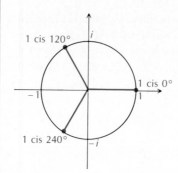

Note in the example that the graphs of the cube roots lie equally spaced about a circle. The same is true of the nth roots of any complex number.

DO EXERCISE 5.

EXERCISE SET 8.5

i In Exercises 1–6, raise the number to the power and write polar notation for the answer.

1. $\left(2 \operatorname{cis} \dfrac{\pi}{3}\right)^3$

2. $\left(3 \operatorname{cis} \dfrac{\pi}{2}\right)^4$

3. $\left(2 \operatorname{cis} \dfrac{\pi}{6}\right)^6$

4. $\left(2 \operatorname{cis} \dfrac{\pi}{5}\right)^5$

5. $(1+i)^6$

6. $(1-i)^6$

In Exercises 7–14, raise the number to the power and write rectangular notation for the answer.

7. $(2 \operatorname{cis} 240°)^4$

8. $(2 \operatorname{cis} 120°)^4$

9. $(1+\sqrt{3}i)^4$

10. $(-\sqrt{3}+i)^6$

11. $\left(\dfrac{1}{\sqrt{2}}+\dfrac{1}{\sqrt{2}}i\right)^{10}$

12. $\left(\dfrac{1}{\sqrt{2}}-\dfrac{1}{\sqrt{2}}i\right)^{12}$

13. $\left(\dfrac{\sqrt{3}}{2}+\dfrac{1}{2}i\right)^{12}$

14. $\left(\dfrac{\sqrt{3}}{2}-\dfrac{1}{2}i\right)^{14}$

ii

15. Solve $x^2=-1+\sqrt{3}i$. That is, find the square roots of $-1+\sqrt{3}i$.

16. Solve $x^2=-\sqrt{3}-i$. That is, find the square roots of $-\sqrt{3}-i$.

17. Solve $x^3=i$.

18. ▤ Solve $x^3=68.4321$.

19. Find and graph the fourth roots of 16.

20. Find and graph the fourth roots of i.

In Exercises 21 and 22, solve. Evaluate square roots.

21. $x^2+(1-i)x+i=0$

22. $3x^2+(1+2i)x+1-i=0$

SUMMARY AND REVIEW: CHAPTER 8

The following contains a summary of what you should be able to do after completing this chapter. The review exercises are for practice. Answers are at the back of the book. If you miss an exercise, restudy the section indicated alongside the answer.

You should be able to:

Express imaginary numbers in terms of i and simplify.

Simplify. Leave answers in terms of i.

1. $-\sqrt{-40}$

2. $\sqrt{-12}\cdot\sqrt{-20}$

Add, subtract, multiply, and divide complex numbers, expressing the answer as $a + bi$, and determine whether a complex number is a solution of an equation.

Simplify.

3. $(2 - 2i)(3 + 4i)$ **4.** $(3 - 5i) - (2 - i)$ **5.** $(6 + 2i) + (-4 - 3i)$ **6.** $\dfrac{2 - 3i}{1 - 3i}$

7. Determine whether $1 - 3i$ is a solution of $x^2 - 2x - 10 = 0$.

Find the reciprocal of a complex number and express it in the form $a + bi$, and use the fact that for equality of complex numbers, the real parts must be the same and the imaginary parts must be the same.

8. Find the reciprocal of $6 - 7i$ and express it in the form $a + bi$.

9. Solve for x and y: $4x + 2i = 8 - (2 + y)i$.

Given a polynomial in z, find a polynomial in $\bar{z}$ that is its conjugate.

10. Find a polynomial in $\bar{z}$ that is the conjugate of $3z^3 + z - 7$.

Solve linear and quadratic equations with complex coefficients and find an equation having specified complex numbers as solutions.

Solve.

11. $(6 - i)x + 4 - 8i = 3 - 3i + 2ix$ **12.** $5x^2 - 4x + 1 = 0$ **13.** $x^2 + 3ix - 1 = 0$

14. Find an equation having the solutions $1 - 2i, 1 + 2i$.

Find square roots of complex numbers.

15. Find the square roots of $4i$.

Graph complex numbers and sums of complex numbers.

16. Graph the pair of complex numbers $-3 - 2i, 4 + 7i$, and their sum.

Given rectangular, or binomial, notation for a complex number, find polar notation and vice versa, and use polar notation to multiply and divide complex numbers.

17. Find rectangular notation for $2(\cos 135° + i \sin 135°)$.

18. Find polar notation for $1 + i$.

19. Find the product of $7 \operatorname{cis} 18°$ and $10 \operatorname{cis} 32°$.

20. Convert to polar notation and then divide:

$$\frac{1 + i}{\sqrt{3} + i}.$$

Raise complex numbers to powers using DeMoivre's theorem, and find the nth roots of a complex number.

21. Find $(1 - i)^5$ and write polar notation for the answer.

22. Find $(3 \operatorname{cis} 120°)^4$ and write rectangular notation for the answer.

23. Find the cube roots of $1 + i$.

24. Solve $x^3 - 27 = 0$.

☐

25. Let $f(z) = 3z^2 - 2\bar{z} + 5i$. Find $f(2 - i)$.

26. Solve $3z + 2\bar{z} = 5 + 2i$.

27. Solve:

$$2x - 3y = 7 + 7i,$$
$$3x + 2y = 4 - 9i.$$

28. Graph:

$$|z - (1 + i)| = 1.$$

TEST: CHAPTER 8

Simplify.

1. i^{83} **2.** $-\sqrt{-3}\sqrt{-16}$ **3.** $(7 - i)(2 + 4i)$

4. $(1 + 6i) - (-3 - 8i)$ **5.** $\dfrac{5 - 2i}{2 - 3i}$ **6.** $(1 - i)(1 + i)$

7. Find the reciprocal of $4 + 2i$ and express it in the form $a + bi$.

8. Solve for x and y:

$$3x + 4 - i = 10 + (2 + y)i.$$

9. Find a polynomial in $\bar{z}$ that is the conjugate of

$$5z^3 + 6z^2 - 5.$$

10. Find an equation having the solutions $5i$, $-5i$.

Solve.

11. $6x^2 - 2x + 1 = 0$

12. $2x^2 - ix + 1 = 0$

13. $z^2 = 8i$

14. Graph the pair of complex numbers $2 - 7i$, $-4 + 2i$, and their sum.

15. Find rectangular notation for $3 \operatorname{cis}(-30°)$.

16. Find the polar notation for $-1 + \sqrt{3}\,i$.

17. Find the product of $2 \operatorname{cis} 50°$ and $5 \operatorname{cis} 40°$.

18. Divide $3 \operatorname{cis} 120°$ by $6 \operatorname{cis} 40°$.

19. Find $(1 + i)^6$. Write polar notation for the answer.

20. Find the cube roots of -8.

21. Subtract: $\dfrac{3 - i}{2 - i} - \dfrac{2 + i}{3 + i}.$

22. Find $f(2 - i)$, where $f(z) = \dfrac{(z - 1)^2}{\bar{z} + 2}.$

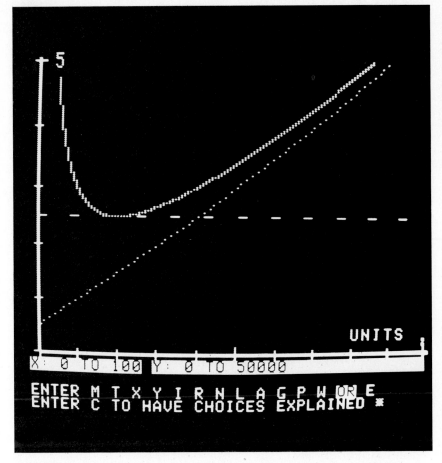

Polynomial and rational functions are the most widely used functions in elementary
mathematics, with many applications. Here a rational function is being graphed
by computer.

POLYNOMIALS AND RATIONAL FUNCTIONS

The most important kind of elementary function is the *polynomial
function*. A polynomial function is a function that can be defined by
a polynomial expression. A function that can be defined as a quotient
of two polynomials is known as a *rational function*. In this chapter
we will study both kinds of functions and their graphs.

The polynomial functions and rational functions, together with
the trigonometric, logarithmic, and exponential functions, comprise
the functions that we consider in elementary mathematics. Those
are the functions considered in calculus, for example, and are the
functions most used in the applications of mathematics.

9

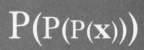

OBJECTIVES

You should be able to:

i Determine the degree of a polynomial.

ii Determine whether a given number is a root of a polynomial.

iii By division, determine whether one polynomial is a factor of another.

iv Given a polynomial $P(x)$ and a divisor $d(x)$, determine the quotient and remainder and express $P(x)$ as

$$P(x) = d(x) \cdot Q(x) + R(x).$$

Determine the degree of each polynomial.

1. $x^5 - x^3 + x^2 + 2i$

2. $3x - i$

3. $3x^2 - \sqrt{2}$

4. 0.5

5. 0

9.1

POLYNOMIALS AND POLYNOMIAL FUNCTIONS

i We have already considered polynomials. We now make a formal definition.

Definition

A *polynomial* in one variable is any expression of the type

$$a_n x^n + a_{n-1} x^{n-1} + \cdots + a_2 x^2 + a_1 x + a_0,$$

where n is a nonnegative integer and $a_0, \ldots, a_n$, are numbers.

The numbers mentioned in this definition are complex numbers, but in certain cases may be restricted to real numbers, rational numbers, or integers. Some or all of them may be 0.

Terminology

In a polynomial, the numbers $a_0, \ldots, a_n$ are called *coefficients*. The first nonzero coefficient, a_n, is called the *leading coefficient*. The number n is called the *degree* of the polynomial, except that the polynomial consisting only of the number 0 has no degree.

EXAMPLE 1 We list some polynomials and their degrees.

Polynomial	Degree
$5x^3 - 3x^2 + i$	3
$-4x + \sqrt{2}$	1
25	0
0	No degree

DO EXERCISES 1–5.

ii ZEROS OR ROOTS OF POLYNOMIALS

When a number is substituted for the variable in a polynomial, the result is some unique number. Thus every polynomial defines a function. We often refer to polynomials, therefore, using function notation, such as $P(x)$. If a number a makes a polynomial 0, that is, $P(a) = 0$, then a is said to be a *zero* of $P(x)$. Note that a is also a solution of the equation $P(x) = 0$. We often say that a is a *root* of the equation $P(x) = 0$ and also a *root* of the polynomial $P(x)$.*

EXAMPLE 2 $P(x) = x^3 + 2x^2 - 5x - 6$. Is 3 a *zero* of $P(x)$?

We substitute 3 into the polynomial:

$$P(3) = 3^3 + 2(3)^2 - 5 \cdot 3 - 6 = 24.$$

Since $P(3) \neq 0$, 3 is not a zero of the polynomial.

---

* Terminology varies from author to author. Traditionally, one referred to the *zeros* of a function and the *roots* of an equation. Subsequently it was found to be convenient in exposition to speak also of the "roots" of a function, especially a polynomial function. In some books the traditional distinction is still made today. We have found that the freer use of terminology facilitates simplification of the exposition.

EXAMPLE 3 Is -1 a root of $P(x)$ in Example 2?

$$P(-1) = (-1)^3 + 2(-1)^2 - 5(-1) - 6 = 0$$

Since $P(-1) = 0$, -1 is a root of the polynomial.

DO EXERCISES 6–8.

iii DIVISION OF POLYNOMIALS AND FACTORS

When we divide one polynomial by another, we obtain a quotient and a remainder. If the remainder is 0, then the divisor is a *factor* of the dividend.

EXAMPLE 4 Divide, to find whether $x^2 + 9$ is a factor of $x^4 - 81$.

$$
\begin{array}{r}
x^2 - 9 \\
x^2 + 9 \overline{)\, x^4 \qquad\qquad - 81} \\
\underline{x^4 \quad + 9x^2} \\
-9x^2 \quad - 81 \\
\underline{-9x^2 \quad - 81} \\
0
\end{array}
$$

Note how spaces have been left for missing terms in the dividend.

Since the remainder is 0, we know that $x^2 + 9$ is a factor.

EXAMPLE 5 Divide, to find whether $x^2 + 3x - 1$ is a factor of $x^4 - 81$.

$$
\begin{array}{r}
x^2 - 3x + 10 \\
x^2 + 3x - 1 \overline{)\, x^4 \qquad\qquad\qquad - 81} \\
\underline{x^4 + 3x^3 - \quad x^2} \\
-3x^3 + \quad x^2 \\
\underline{-3x^3 - \quad 9x^2 + \quad 3x} \\
10x^2 - \quad 3x - 81 \\
\underline{10x^2 + 30x - 10} \\
-33x - 71
\end{array}
$$

Since the remainder is not 0, we know that $x^2 + 3x - 1$ is not a factor of $x^4 - 81$.

DO EXERCISES 9 AND 10.

iv In general, when we divide a polynomial $P(x)$ by a divisor $d(x)$ we obtain some polynomial $Q(x)$ for a quotient and some polynomial $R(x)$ for a remainder. The remainder must either be 0 or have degree less than that of $d(x)$. To check a division we multiply the quotient by the divisor and add the remainder, to see if we get the dividend. Thus these polynomials are related as follows:

$$P(x) = d(x) \cdot Q(x) + R(x).$$

EXAMPLE 6 If $P(x) = x^4 - 81$ and $d(x) = x^2 + 9$, then $Q(x) = x^2 - 9$ and $R(x) = 0$, and

$$\underbrace{x^4 - 81}_{P(x)} = \underbrace{(x^2 + 9)}_{d(x)} \cdot \underbrace{(x^2 - 9)}_{Q(x)} + \underbrace{0}_{R(x)}$$

6. Determine whether the following numbers are roots of the polynomial $x^2 - 4x - 21$.

a) 7

b) 3

c) -7

7. Determine whether the following numbers are roots of the polynomial $x^4 - 16$.

a) 2

b) -2

c) -1

d) 0

8. Determine whether the following numbers are roots of the polynomial $x^2 + 1$.

a) 1

b) -1

c) i

d) $-i$

9. By division, determine whether the following polynomials are factors of the polynomial $x^4 - 16$.

a) $x - 2$

b) $x^2 + 3x - 1$

10. By division, determine whether the following polynomials are factors of the polynomial $x^3 + 2x^2 - 5x - 6$.

a) $x + 1$

b) $x - 3$

c) $x^2 + 3x - 1$

11. Divide $x^3 + 2x^2 - 5x - 6$ by $x - 3$. Then express the dividend as

$$P(x) = d(x) \cdot Q(x) + R(x).$$

EXAMPLE 7 If $P(x) = x^4 - 81$ and $d(x) = x^2 + 3x - 1$, then $Q(x) = x^2 - 3x + 10$ and $R(x) = -33x - 71$, and

$$\underbrace{x^4 - 81}_{P(x)} = \underbrace{(x^2 + 3x - 1)}_{d(x)} \cdot \underbrace{(x^2 - 3x + 10)}_{Q(x)} + \underbrace{(-33x - 71)}_{R(x)}$$

DO EXERCISE 11.

EXERCISE SET 9.1

i Determine the degree of each polynomial.

1. $x^4 - 3x^2 + 1$ **2.** $2x^5 - x^4 + \frac{1}{4}x - 7$ **3.** $-2x + 5$ **4.** $3x - \sqrt{\pi}$

5. $2x^2 - 3x + 4$ **6.** $\frac{1}{4}x^2 - 7$ **7.** 3 **8.** 0

ii

9. Determine whether 2, 3, and -1 are roots, or zeros, of

$$P(x) = x^3 + 6x^2 - x - 30.$$

10. Determine whether 2, 3, and -1 are roots, or zeros, of

$$P(x) = 2x^3 - 3x^2 + x - 1.$$

iii

11. For $P(x)$ in Exercise 9, which of the following are factors of $P(x)$?

  a) $x - 2$ b) $x - 3$ c) $x + 1$

12. For $P(x)$ in Exercise 10, which of the following are factors of $P(x)$?

  a) $x - 2$ b) $x - 3$ c) $x + 1$

iv In each of the following, a polynomial $P(x)$ and a divisor $d(x)$ are given. Find the quotient $Q(x)$ and the remainder $R(x)$ when $P(x)$ is divided by $d(x)$, and express $P(x)$ in the form $d(x) \cdot Q(x) + R(x)$.

13. $P(x) = x^3 + 6x^2 - x - 30$,
$d(x) = x - 2$

14. $P(x) = 2x^3 - 3x^2 + x - 1$,
$d(x) = x - 2$

15. $P(x)$ as in Exercise 13,
$d(x) = x - 3$

16. $P(x)$ as in Exercise 14,
$d(x) = x - 3$

17. $P(x) = x^3 - 8$,
$d(x) = x + 2$

18. $P(x) = x^3 + 27$,
$d(x) = x + 1$

19. $P(x) = x^4 + 9x^2 + 20$,
$d(x) = x^2 + 4$

20. $P(x) = x^4 + x^2 + 2$,
$d(x) = x^2 + x + 1$

21. $P(x) = 5x^7 - 3x^4 + 2x^2 - 3$,
$d(x) = 2x^2 - x + 1$

22. $P(x) = 6x^5 + 4x^4 - 3x^2 + x - 2$,
$d(x) = 3x^2 + 2x - 1$

23. For $P(x) = x^5 - 64$,

  a) find $P(2)$;

  b) find the remainder when $P(x)$ is divided by $x - 2$, and compare your answer to (a);

  c) find $P(-1)$;

  d) find the remainder when $P(x)$ is divided by $x + 1$, and compare your answer to (c).

24. For $P(x) = x^3 + x^2$,

  a) find $P(-1)$;

  b) find the remainder when $P(x)$ is divided by $x + 1$, and compare your answer to (a);

  c) find $P(2)$;

  d) find the remainder when $P(x)$ is divided by $x - 2$, and compare your answer to (c).

25. For $P(x) = 2x^2 - ix + 1$,

  a) find $P(-i)$;

  b) find the remainder when $P(x)$ is divided by $x + i$.

26. For $P(x) = 2x^2 + ix - 1$,

  a) find $P(i)$;

  b) find the remainder when $P(x)$ is divided by $x - i$.

27. Under what conditions is a polynomial function of real variables an even function, that is, $P(x) = P(-x)$ for all x? Does your answer still hold for complex variables? Explain.

28. Under what conditions is a polynomial function of real variables an odd function, that is, $-P(x) = P(-x)$ for all x? Does your answer still hold for complex variables?

29. Find a rule for determining the degree of the product of two polynomials with real coefficients.

30. Find a rule for determining the degree of the sum of two polynomials with real coefficients.

9.2

THE REMAINDER AND FACTOR THEOREMS

Some of the exercises in the preceding exercise set illustrate the following theorem.

Theorem 1 **The remainder theorem**

If a number r is substituted for x in the polynomial $P(x)$, then the result $P(r)$ is the remainder that would be obtained by dividing $P(x)$ by $x - r$.

> **Proof.** The equation $P(x) = d(x) \cdot Q(x) + R(x)$ is the basis of this proof. If we divide $P(x)$ by $x - r$, we obtain a quotient $Q(x)$ and a remainder $R(x)$ related as follows:
>
> $$P(x) = (x - r) \cdot Q(x) + R(x).$$
>
> The remainder $R(x)$ must either be 0 or have degree less than $x - r$. Thus $R(x)$ must be a constant. Let us call this constant R. In the above expression we get a true sentence whenever we replace x by any number. Let us replace x by r. We get
>
> $$P(r) = (r - r) \cdot Q(r) + R$$
> $$P(r) = \quad 0 \cdot Q(r) + R$$
> $$P(r) = R.$$
>
> This tells us that the function value $P(r)$ is the remainder obtained when we divide $P(x)$ by $x - r$.

Theorem 1 motivates us to find a rapid way of dividing by $x - r$, in order to find function values.

i SYNTHETIC DIVISION

To streamline division, we can arrange the work so that duplicate and unnecessary writing are avoided. Consider the following.

A.
$$
\begin{array}{r}
4x^2 + 5x + 11 \\
x - 2 \overline{\smash{)}\ 4x^3 - 3x^2 + x + 7} \\
\underline{4x^3 - 8x^2} \\
5x^2 + x \\
\underline{5x^2 - 10x} \\
11x + 7 \\
\underline{11x - 22} \\
29
\end{array}
$$

B.
$$
\begin{array}{r}
4 \quad 5 \quad 11 \\
1 - 2 \overline{\smash{)}\ 4 - 3 + 1 + 7} \\
\underline{4 - 8} \\
5 + 1 \\
\underline{5 - 10} \\
11 + 7 \\
\underline{11 - 22} \\
29
\end{array}
$$

The division in (B) is the same as that in (A), but we wrote only the coefficients. The color numerals are duplicated, so we look for an arrangement in which they are not duplicated. We can also simplify things by using the additive inverse of -2 and then adding instead of subtracting. When things are thus "collapsed," we have the algorithm known as *synthetic division*.

C. (Synthetic Division)

$$
\begin{array}{r|rrrr}
2 & 4 & -3 & 1 & 7 \\
  &   & 8 & 10 & 22 \\
\hline
  & 4 & 5 & 11 & 29
\end{array}
$$

OBJECTIVES

You should be able to:

i Use synthetic division to find the quotient and remainder when a polynomial is divided by $x - r$.

ii Use the remainder theorem to find a function value $P(r)$ when a polynomial $P(x)$ is divided by $x - r$.

iii Determine whether $x - r$ is a factor of $P(x)$ by determining whether $P(r) + 0$.

Use synthetic division to find the quotient and remainder.

1. $(x^3 + 6x^2 - x - 30) \div (x - 2)$

2. $(x^3 - 2x^2 + 5x - 4) \div (x + 2)$

3. $(y^3 + 1) \div (y + 1)$

4. Let $P(x) = x^5 - 2x^4 - 7x^3 + x^2 + 20$. Use synthetic division to find:

a) $P(10)$;

b) $P(-8)$.

5. Let $P(x) = x^3 + 6x^2 - x - 30$. Using synthetic division, determine whether the given numbers are roots of $P(x)$.

a) 2

b) 5

c) -3

We "bring down" the 4. Then we multiply it by the 2, to get 8, and add to get 5. We then multiply 5 by 2 to get 10, add, and so on. The last number, 29, is the remainder. The others are the coefficients of the quotient.

EXAMPLE 1 Use synthetic division to find the quotient and remainder:

$$(2x^3 + 7x^2 - 5) \div (x + 3).$$

First note that $x + 3 = x - (-3)$.

$$
\begin{array}{r|rrrr}
-3 & 2 & 7 & 0 & -5 \\
   &   & -6 & -3 & 9 \\
\hline
   & 2 & 1 & -3 & \;4 \\
\end{array}
$$
Note: We must write 0's for missing terms.

The quotient is $2x^2 + x - 3$. The remainder is 4.

DO EXERCISES 1–3.

ii FUNCTION VALUES FOR POLYNOMIALS

We can now apply synthetic division to the finding of function values for polynomials.

EXAMPLE 2 Given that $P(x) = 2x^5 - 3x^4 + x^3 - 2x^2 + x - 8$, find $P(10)$.

By Theorem 1, $P(10)$ is the remainder when $P(x)$ is divided by $x - 10$. We use synthetic division to find that remainder.*

$$
\begin{array}{r|rrrrrr}
10 & 2 & -3 & 1 & -2 & 1 & -8 \\
   &   & 20 & 170 & 1710 & 17{,}080 & 170{,}810 \\
\hline
   & 2 & 17 & 171 & 1708 & 17{,}081 & \;170{,}802 \\
\end{array}
$$

Thus $P(10) = 170{,}802$.

DO EXERCISE 4.

EXAMPLE 3 Determine whether -4 is a zero or root of $P(x)$, where $P(x) = x^3 + 8x^2 + 8x - 32$.

We use synthetic division and Theorem 1 to find $P(-4)$.

$$
\begin{array}{r|rrrr}
-4 & 1 & 8 & 8 & -32 \\
   &   & -4 & -16 & 32 \\
\hline
   & 1 & 4 & -8 & \;0 \\
\end{array}
$$

Since $P(-4) = 0$, the number -4 is a root of $P(x)$.

DO EXERCISE 5.

iii FINDING FACTORS OF POLYNOMIALS

We now consider the following useful corollary of the remainder theorem.

---

* Compare this with the work involved in a direct calculation! ▦ A calculator is most useful when finding polynomial function values by synthetic division. In Example 2, one would begin by entering 2. Then multiply by 10, add the result to -3, multiply that result by 10, and so on. The number 10 can be stored and recalled at each step where needed if the calculator has a memory. Since only the last result is needed, no intermediate values need be recorded.

Theorem 2 The factor theorem

For a polynomial $P(x)$, if $P(r) = 0$, then the polynomial $x - r$ is a factor of $P(x)$.

Proof. If we divide $P(x)$ by $x - r$, we obtain a quotient and remainder, related as follows:

$$P(x) = (x - r) \cdot Q(x) + P(r).$$

Then if $P(r) = 0$, we have

$$P(x) = (x - r) \cdot Q(x),$$

so $x - r$ is a factor of $P(x)$.

This theorem is very useful in factoring polynomials, and hence in the solving of equations.

EXAMPLE 4 Let $P(x) = x^3 + 2x^2 - 5x - 6$. Factor $P(x)$ and thus solve the equation $P(x) = 0$.

We look for linear factors of the form $x - r$. Let us try $x - 1$. We use synthetic division to see whether $P(1) = 0$.

$$\begin{array}{r|rrrr} 1\rfloor & 1 & 2 & -5 & -6 \\ & & 1 & 3 & -2 \\ \hline & 1 & 3 & -2\,\rfloor & -8 \end{array}$$

We know that $x - 1$ is not a factor of $P(x)$. We try $x + 1$.

$$\begin{array}{r|rrrr} -1\rfloor & 1+2 & -5 & -6 \\ & & -1 & -1 & 6 \\ \hline & 1 & 1 & -6 & 0 \end{array}$$

We know that $x + 1$ is one factor and the quotient, $x^2 + x - 6$, is another. Thus

$$P(x) = (x + 1)(x^2 + x - 6).$$

The trinomial is easily factored, so we have

$$P(x) = (x + 1)(x + 3)(x - 2).$$

To solve the equation $P(x) = 0$, we use the principle of zero products. The solutions are -1, -3, and 2.

DO EXERCISES 6–8.

6. Determine whether $x - \frac{1}{2}$ is a factor of $4x^4 + 2x^3 + 8x - 1$.

7. Determine whether $x + 5$ is a factor of $x^3 + 625$.

8. a) Let $P(x) = x^3 + 6x^2 - x - 30$. Determine whether $x - 2$ is a factor of $P(x)$.
   b) Find another factor of $P(x)$.
   c) Find a complete factorization of $P(x)$.

EXERCISE SET 9.2

i Use synthetic division to find the quotient and remainder.

1. $(2x^4 + 7x^3 + x - 12) \div (x + 3)$
2. $(x^3 - 7x^2 + 13x + 3) \div (x - 2)$
3. $(x^3 - 2x^2 - 8) \div (x + 2)$
4. $(x^3 - 3x + 10) \div (x - 2)$
5. $(x^4 - 1) \div (x - 1)$
6. $(x^5 + 32) \div (x + 2)$
7. $(2x^4 + 3x^2 - 1) \div (x - \frac{1}{2})$
8. $(3x^4 - 2x^2 + 2) \div (x - \frac{1}{4})$
9. $(x^4 - y^4) \div (x - y)$
10. $(x^3 + 3ix^2 - 4ix - 2) \div (x + i)$

ii Use synthetic division to find the function values.

11. $P(x) = x^3 - 6x^2 + 11x - 6$; find $P(1)$, $P(-2)$, $P(3)$.
12. $P(x) = x^3 + 7x^2 - 12x - 3$; find $P(-3)$, $P(-2)$, $P(1)$.
13. $P(x) = 2x^5 - 3x^4 + 2x^3 - x + 8$; find $P(20)$ and $P(-3)$.
14. $P(x) = x^5 - 10x^4 + 20x^3 - 5x - 100$; find $P(-10)$ and $P(5)$.

iii Using synthetic division, determine whether the numbers are roots of the polynomials.

15. $-3, 2; P(x) = 3x^3 + 5x^2 - 6x + 18$

16. $-4, 2; P(x) = 3x^3 + 11x^2 - 2x + 8$

17. $-3, \frac{1}{2}; P(x) = x^3 - \frac{7}{2}x^2 + x - \frac{3}{2}$

18. $i, -i, -2; P(x) = x^3 + 2x^2 + x + 2$

Factor the polynomial $P(x)$. Then solve the equation $P(x) = 0$.

19. $P(x) = x^3 + 4x^2 + x - 6$

20. $P(x) = x^3 + 5x^2 - 2x - 24$

21. $P(x) = x^3 - 6x^2 + 3x + 10$

22. $P(x) = x^3 + 2x^2 - 13x + 10$

23. $P(x) = x^3 - x^2 - 14x + 24$

24. $P(x) = x^3 - 3x^2 - 10x + 24$

25. $P(x) = x^4 - 22x^2 + 51x - 30$

26. $P(x) = x^4 + 11x^3 + 41x^2 + 61x + 30$

Solve.

27. $x^3 + 2x^2 - 13x + 10 > 0$

28. $x^4 - x^3 - 19x^2 + 49x - 30 < 0$

29. Find k so that $x + 2$ is a factor of $x^3 - kx^2 + 3x + 7k$.

30. ▦ Given that $f(x) = 2.13x^5 - 42.1x^3 + 17.5x^2 + 0.953x - 1.98$, find $f(3.21)$ (a) by synthetic division; (b) by substitution.

31. For what values of k will the remainder be the same when $x^2 + kx + 4$ is divided by $x - 1$ or $x + 1$?

32. When $x^2 - 3x + 2k$ is divided by $x + 2$ the remainder is 7. Find the value of k.

33. Devise a way to use the method of synthetic division when the divisor is a polynomial such as $bx - r$.

34. Prove that $x - a$ is a factor of $x^n - a^n$, for any natural number n.

OBJECTIVES

You should be able to:

i Factor polynomials and find their roots and their multiplicities.

ii Find a polynomial with specified roots.

iii In certain cases, given some of the roots of a polynomial, find such a polynomial and find the rest of its roots.

9.3

THEOREMS ABOUT ROOTS

THE FUNDAMENTAL THEOREM OF ALGEBRA

A linear, or first-degree, polynomial $ax + b$ (where $a \neq 0$, of course) has just one root, $-b/a$. From Chapter 8 we know that any quadratic polynomial with complex numbers for coefficients has at least one, and at most two, roots. The following theorem is a generalization. A proof is beyond this text.

Theorem 3 The fundamental theorem of algebra

Every polynomial of degree greater than 0, with complex coefficients, has at least one root in the system of complex numbers.*

This is a very powerful theorem. Note that although it guarantees that a root exists, it does not tell how to find it. We now develop some theory that can help in finding roots. First, we prove a corollary of the fundamental theorem of algebra.

Theorem 4

Every polynomial of degree n, where $n > 0$, having complex coefficients, can be factored into n linear factors.

---

* It is wise to keep in mind that when we speak of "roots," or "zeros," of a polynomial $P(x)$, we are also speaking about "roots" or "solutions" of the polynomial equation $P(x) = 0$.

Proof. Let us consider any polynomial of degree n, say $P(x)$. By the fundamental theorem, it has a root r_1. By the factor theorem, $x - r_1$ is a factor of $P(x)$. Thus we know that

$$P(x) = (x - r_1) \cdot Q_1(x),$$

where $Q_1(x)$ is the quotient that would be obtained by dividing $P(x)$ by $x - r_1$. Let the leading coefficient of $P(x)$ be a_n. By considering the actual division process, we see that the leading coefficient of $Q_1(x)$ is also a_n and that the degree of $Q_1(x)$ is $n - 1$. Now if the degree of $Q_1(x)$ is greater than 0, then it has a root r_2, and we have

$$Q_1(x) = (x - r_2) \cdot Q_2(x),$$

where the degree of $Q_2(x)$ is $n - 2$ and the leading coefficient is a_n. Thus we have

$$P(x) = (x - r_1)(x - r_2) \cdot Q_2(x).$$

This process can be continued until a quotient $Q_n(x)$ is obtained having degree 0. The leading coefficient will be a_n, so $Q_n(x)$ is actually the constant a_n. We now have the following:

$$P(x) = a_n(x - r_1)(x - r_2)(x - r_3) \cdots (x - r_n).$$

This completes the proof. We have actually shown a little more than what the theorem states. We see that $P(x)$ has been factored with one constant factor a_n and n linear factors having leading coefficient 1.

i FINDING ROOTS BY FACTORING

To find the roots of a polynomial, we can attempt to factor it.

EXAMPLES

1. $x^4 + x^3 - 13x^2 - x + 12$ can be factored as

$$(x - 3)(x + 4)(x + 1)(x - 1),$$

so the roots are 3, -4, -1, and 1.

2. $3x^4 - 15x^3 + 18x^2 + 12x - 24$ can be factored as

$$3(x - 2)(x - 2)(x - 2)(x + 1),$$

so the roots are 2 and -1.

In Example 2 the factor $x - 2$ occurs three times. In a case like this we sometimes say that the root we obtain, 2, has a *multiplicity* of three.

From Theorem 4 and the above examples, we have the following.

Theorem 5

Every polynomial of degree n, where $n > 0$, has at least one root and at most n roots.*

DO EXERCISES 1–5.

ii FINDING POLYNOMIALS WITH GIVEN ROOTS

Given several numbers, we can find a polynomial having them as its roots.

---

* Theorem 5 is often stated as follows: "Every polynomial of degree n, where $n > 0$, has *exactly* n roots." This statement is not incompatible with Theorem 5, as it first seems, because to make sense of the statement just quoted, one must take multiplicities into account. Theorem 5, as stated here, is simpler and more straightforward.

Find the roots of each polynomial and state the multiplicity of each.

1. $P(x) = (x - 5)(x - 5)(x + 6)$

2. $P(x) = 4(x + 7)^2(x - 3)$

3. $P(x) = (x + 2)^3(x^2 - 9)$

4. $P(x) = (x^2 - 7x + 12)^2$

5. $5x^2 - 5$

6. Find a polynomial of degree 3 that has -1, 2, and 5 as roots.

7. Find a polynomial of degree 3 that has -1, 2, and $-5i$ as roots.

8. Find a polynomial of degree 5 with -2 as a root of multiplicity 3 and 0 as a root of multiplicity 2.

9. Find a polynomial of degree 4 with 1 as a root of multiplicity 3 and -5 as a root of multiplicity 1.

EXAMPLE 3   Find a polynomial of degree 3, having the roots -2, 1, and $3i$.

By Theorem 2, such a polynomial has factors $x + 2$, $x - 1$, and $x - 3i$, so we have

$$P(x) = a_n(x + 2)(x - 1)(x - 3i).$$

The number a_n can be any nonzero number. The simplest polynomial will be obtained if we let it be 1. If we then multiply the factors we obtain

$$P(x) = x^3 + (1 - 3i)x^2 + (-2 - 3i)x + 6i.$$

EXAMPLE 4   Find a polynomial of degree 5 with -1 as a root of multiplicity 3, 4 as a root of multiplicity 1, and 0 as a root of multiplicity 1.

Proceeding as in Example 3, letting $a_n = 1$, we obtain

$$(x + 1)^3(x - 4)(x - 0), \quad \text{or} \quad x^5 - x^4 - 9x^3 - 11x^2 - 4x.$$

DO EXERCISES 6–9.

iii ROOTS OF POLYNOMIALS WITH REAL COEFFICIENTS

Consider the quadratic equation $x^2 - 2x + 2 = 0$, with real coefficients. Its roots are $1 + i$ and $1 - i$. Note that they are complex conjugates. This generalizes to any polynomial with real coefficients.

Theorem 6

If a complex number z is a root of a polynomial $P(x)$ of degree greater than or equal to 1 with real coefficients, then its conjugate $\bar{z}$ is also a root. (Nonreal roots occur in conjugate pairs.)

Proof.   Let

$$P(x) = a_n x^n + a_{n-1}x^{n-1} + \cdots + a_1 x + a_0,$$

where the coefficients are real numbers. Suppose z is a complex root of $P(x)$. Then $P(z) = 0$, or

$$a_n z^n + a_{n-1}z^{n-1} + \cdots + a_1 z + a_0 = 0.$$

Now let us find the conjugate of each side of the equation. First note that $\bar{0} = 0$, since 0 is a real number. Then we have the following.

$$
\begin{aligned}
0 = \bar{0} &= \overline{a_n z^n + a_{n-1}z^{n-1} + \cdots + a_1 z + a_0} \\
&= \overline{a_n z^n} + \overline{a_{n-1}z^{n-1}} + \cdots + \overline{a_1 z} + \overline{a_0} && \text{By Theorem 3, Chapter 8} \\
&= \overline{a_n} \cdot \overline{z^n} + \overline{a_{n-1}} \cdot \overline{z^{n-1}} + \cdots + \overline{a_1} \cdot \overline{z} + \overline{a_0} && \text{By Theorem 4, Chapter 8} \\
&= a_n \overline{z^n} + a_{n-1}\overline{z^{n-1}} + \cdots + a_1 \overline{z} + a_0 && \text{By Theorem 6, Chapter 8} \\
&= a_n \overline{z}^n + a_{n-1}\overline{z}^{n-1} + \cdots + a_1 \overline{z} + a_0. && \text{By Theorem 5, Chapter 8}
\end{aligned}
$$

Thus $P(\bar{z}) = 0$, so $\bar{z}$ is a root of the polynomial.

For Theorem 6, it is essential that the coefficients be real numbers. This can be seen by considering Example 3. In that polynomial the root $3i$ occurs, but its conjugate does not. This can happen because some of the coefficients of the polynomial are not real.

Rational Coefficients

When a polynomial has rational numbers for coefficients, certain irrational roots also occur in pairs, as described in the following theorem.

Theorem 7

Suppose $P(x)$ is a polynomial with rational coefficients and of degree greater than 0. Then if either of the following is a root, so is the other: $a + c\sqrt{b}$, $a - c\sqrt{b}$, a and c rational, b not a square.

This theorem can be proved in a manner analogous to Theorem 6, but we shall not do it here. The theorem can be used to help in finding roots.

EXAMPLE 5 Suppose a polynomial of degree 6 with rational coefficients has $-2 + 5i$, $-2i$, and $1 - \sqrt{3}$ as some of its roots. Find the other roots.

The other roots are $-2 - 5i$, $2i$, and $1 + \sqrt{3}$. There are no other roots since the degree is 6.

EXAMPLE 6 Find a polynomial of lowest degree with rational coefficients that has $1 - \sqrt{2}$ and $1 + 2i$ as some of its roots.

The polynomial must also have the roots $1 + \sqrt{2}$ and $1 - 2i$. Thus the polynomial is

$$[x - (1 - \sqrt{2})][x - (1 + \sqrt{2})][x - (1 + 2i)][x - (1 - 2i)],$$

or

$$(x^2 - 2x - 1)(x^2 - 2x + 5),$$

or

$$x^4 - 4x^3 + 8x^2 - 8x - 5.$$

DO EXERCISES 10–12.

EXAMPLE 7 Let $P(x) = x^4 - 5x^3 + 10x^2 - 20x + 24$. Find the other roots of $P(x)$, given that $2i$ is a root.

Since $2i$ is a root, we know that $-2i$ is also a root. Thus

$$P(x) = (x - 2i)(x + 2i) \cdot Q(x)$$

for some $Q(x)$. Since $(x - 2i)(x + 2i) = x^2 + 4$, we know that

$$P(x) = (x^2 + 4) \cdot Q(x).$$

We find, using division, that $Q(x) = x^2 - 5x + 6$, and since we can factor $x^2 - 5x + 6$, we get

$$P(x) = (x^2 + 4)(x - 2)(x - 3).$$

Thus the other roots are $-2i$, 2, and 3.

DO EXERCISE 13.

10. Suppose a polynomial of degree 5 with rational coefficients has -4, $7 - 2i$, and $3 + \sqrt{5}$ as roots. Find the other roots.

11. Find a polynomial of lowest degree with rational coefficients that has $2 + \sqrt{3}$ and $1 - i$ as some of its roots.

12. Find a polynomial of lowest degree with real coefficients that has $2i$ and 2 as some of its roots.

13. Find the other roots of

$$x^4 + x^3 - x^2 + x - 2,$$

given that i is a root.

EXERCISE SET 9.3

i In Exercises 1–4, find the roots of each polynomial, or polynomial equation, and state the multiplicity of each.

1. $(x + 3)^2(x - 1)$

2. $-8(x - 3)^2(x + 4)^3 x^4 = 0$

3. $x^3(x - 1)^2(x + 4) = 0$

4. $(x^2 - 5x + 6)^2$

ii In Exercises 5–9, find a polynomial of degree 3 with the given numbers as roots.

5. $-2, 3, 5$

6. $2, i, -i$

7. $-3, 2i, -2i$

8. $1 + 4i, 1 - 4i, -1$

9. $\sqrt{2}, -\sqrt{2}, \sqrt{3}$
Are the coefficients rational?

10. Find a polynomial equation of degree 4 with -2 as a root of multiplicity 1, 3 as a root of multiplicity 2, and -1 as a root of multiplicity 1.

iii In Exercises 11 and 12, suppose a polynomial or polynomial equation of degree 5 with rational coefficients has the given numbers as roots. Find the other roots.

11. $6, -3 + 4i, 4 - \sqrt{5}$

12. $-2, 3, 4, 1 - i$

In Exercises 13–18, find a polynomial of lowest degree with rational coefficients that has the given numbers as some of its roots.

13. $1 + i, 2$

14. $2 - i, -1$

15. $-4i, 5$

16. $2 - \sqrt{3}, 1 + i$

17. $\sqrt{5}, -3i$

18. $-\sqrt{2}, 4i$

In Exercises 19–22, given that the polynomial or polynomial equation has the given root, find the other roots.

19. $x^4 - 5x^3 + 7x^2 - 5x + 6; -i$

20. $x^4 - 16 = 0; 2i$

21. $x^3 - 6x^2 + 13x - 20 = 0; 4$

22. $x^3 - 8; 2$

In Exercises 23–25, solve the equations. Use synthetic division, the quadratic formula, the theorems of this section, or whatever else you think might help.

23. $x^3 - 4x^2 + x - 4 = 0$

24. $x^3 - x^2 - 7x + 15 = 0$

25. $x^4 - 2x^3 - 2x - 1 = 0$

26. The equation $x^2 + 2ax + b = 0$ has a double root. Find it.

27. What does the fundamental theorem of algebra tell you about the following equation?

$$2(\log x)^5 - 2(\log x)^3 + \log x = \tfrac{1}{8}$$

28. Prove that a polynomial with positive coefficients cannot have a positive root.

29. Prove that every polynomial of odd degree, with real coefficients, has at least one real root.

30. Prove that every polynomial with real coefficients has a factorization into linear and quadratic factors (with real coefficients).

OBJECTIVES

You should be able to:

i Given a polynomial with integer coefficients, find the rational roots and find the other roots, if possible.

ii Do the same for polynomials with rational coefficients.

9.4

RATIONAL ROOTS

i INTEGER COEFFICIENTS

It is not always easy to find the roots of a polynomial. However, if a polynomial has integer coefficients, there is a procedure that will yield all the rational roots.

Theorem 8 The rational roots theorem

Let

$$P(x) = a_n x^n + a_{n-1} x^{n-1} + \cdots + a_1 x + a_0,$$

where all the coefficients are integers. Consider a rational number denoted by c/d, where c and d are relatively prime (having no common factor besides 1 and -1). If c/d is a root of $P(x)$, then c is a factor of a_0 and d is a factor of a_n.

Proof. Since c/d is a root of $P(x)$, we know that

$$a_n\left(\frac{c}{d}\right)^n + a_{n-1}\left(\frac{c}{d}\right)^{n-1} + \cdots + a_1\left(\frac{c}{d}\right) + a_0 = 0. \qquad (1)$$

We multiply by d^n and get the equation

$$a_n c^n + a_{n-1} c^{n-1} d + \cdots + a_1 c d^{n-1} + a_0 d^n = 0. \qquad (2)$$

Then we have

$$a_n c^n = (-a_{n-1} c^{n-1} - \cdots - a_1 c d^{n-2} - a_0 d^{n-1}) d.$$

Note that d is a factor of $a_n c^n$. Now d is not a factor of c because c and d are relatively prime. Thus d is not a factor of c^n. So d is a factor of a_n.

In a similar way we can show from Eq. (2) that

$$a_0 d^n = (-a_n c^{n-1} - a_{n-1} c^{n-2} d - \cdots - a_1 d^{n-1}) c.$$

Thus c is a factor of $a_0 d^n$. Again, c is not a factor of d^n, so it must be a factor of a_0.

EXAMPLE 1   Let $P(x) = 3x^4 - 11x^3 + 10x - 4$. Find the rational roots of $P(x)$. If possible, find the other roots.

By the rational roots theorem, if c/d is a root of $P(x)$, then c must be a factor of -4 and d must be a factor of 3. Thus the possibilities for c and d are

$$c: \quad 1, -1, 4, -4, 2, -2; \qquad d: \quad 1, -1, 3, -3.$$

Then the resulting possibilities for c/d are

$$\frac{c}{d}: \quad 1, -1, 4, -4, \frac{1}{3}, -\frac{1}{3}, \frac{4}{3}, -\frac{4}{3}, \frac{2}{3}, -\frac{2}{3}, 2, -2.$$

Of these 12 possibilities, we know that at most 4 of them could be roots because $P(x)$ is of degree 4. To find which are roots we could use substitution, but synthetic division is usually more efficient.

We try 1:

$$
\begin{array}{r|rrrrr}
1 & 3 & -11 & 0 & 10 & -4 \\
  &   & 3 & -8 & -8 & 2 \\
\hline
  & 3 & -8 & -8 & 2 & -2. \\
\end{array}
$$

We try -1:

$$
\begin{array}{r|rrrrr}
-1 & 3 & -11 & 0 & 10 & -4 \\
   &   & -3 & 14 & -14 & 4 \\
\hline
   & 3 & -14 & 14 & -4 & 0. \\
\end{array}
$$

Since $P(1) = -2$, 1 is not a root; but $P(-1) = 0$, so -1 is a root. We use synthetic division again, to see whether -1 is a double root.

$$
\begin{array}{r|rrrr}
-1 & 3 & -14 & 14 & -4 \\
   &   & -3 & 17 & -31 \\
\hline
   & 3 & -17 & 31 & -35 \\
\end{array}
$$

It is not. Using the results of the second synthetic division above, we can express $P(x)$ as follows:

$$P(x) = (x + 1)(3x^3 - 14x^2 + 14x - 4).$$

1. Let

$$P(x) = 2x^4 - 7x^3 - 35x^2 + 13x + 3.$$

If c/d is a rational root of $P(x)$, then:

a) What are the possibilities for c?

b) What are the possibilities for d?

c) What are the possibilities for c/d?

d) Find the rational roots.

e) If possible, find the other roots.

2. Let $P(x) = x^3 + 7x^2 + 4x + 28$. If c/d is a rational root of $P(x)$, then:

a) What are the possibilities for c?

b) What are the possibilities for d?

c) What are the possibilities for c/d?

d) How can you tell without substitution or synthetic division that there are no positive roots?

e) Find the rational roots of $P(x)$.

f) Find the other roots if they exist.

We now use $3x^3 - 14x^2 + 14x - 4$ and check the other possible roots. We try $\frac{2}{3}$:

$$\begin{array}{r|rrrr} \tfrac{2}{3} & 3 & -14 & 14 & -4 \\ & & 2 & -8 & 4 \\ \hline & 3 & -12 & 6 & |\ 0. \end{array}$$

Since $P(\frac{2}{3}) = 0$, $\frac{2}{3}$ is a root. Again using the results of the synthetic division, we can express $P(x)$ as

$$P(x) = (x + 1)(x - \tfrac{2}{3})(3x^2 - 12x + 6).$$

Since the factor $3x^2 - 12x + 6$ is quadratic, we can use the quadratic formula to find that the other roots are $2 + \sqrt{2}$ and $2 - \sqrt{2}$. These are irrational numbers. Thus the rational roots are -1 and $\frac{2}{3}$.

DO EXERCISE 1.

EXAMPLE 2   Let $P(x) = x^3 + 6x^2 + x + 6$. Find the rational roots of $P(x)$. If possible, find the other roots.

By the rational roots theorem, if c/d is a root of $P(x)$, then c must be a factor of 6 and d must be a factor of 1. Thus the possibilities for c and d are

$$c: \quad 1, -1, 2, -2, 3, -3, 6, -6; \qquad d: \quad 1, -1.$$

Then the resulting possibilities for c/d are

$$\frac{c}{d}: \quad 1, -1, 2, -2, 3, -3, 6, -6.$$

Note that these are the same as the possibilities for c. If the leading coefficient is 1, we need only check the factors of the last coefficient as possibilities for rational roots.

There is another aid in eliminating possibilities for rational roots. Note that all coefficients of $P(x)$ are positive. Thus when any positive number is substituted in $P(x)$, we get a positive value, never 0. Therefore, no positive number can be a root. Thus the only possibilities for roots are

$$-1, -2, -3, -6.$$

We try -6:

$$\begin{array}{r|rrrr} -6 & 1 & 6 & 1 & 6 \\ & & -6 & 0 & -6 \\ \hline & 1 & 0 & 1 & |\ 0. \end{array}$$

Thus $P(-6) = 0$, so -6 is a root. We could divide again, to determine whether -6 is a double root, but since we can now factor $P(x)$ as a product of a linear and a quadratic polynomial, it is preferable to proceed that way. We have

$$P(x) = (x + 6)(x^2 + 1).$$

Now $x^2 + 1$ has the complex roots i and $-i$. Thus the only rational root of $P(x)$ is -6.

DO EXERCISE 2.

EXAMPLE 3   Find the rational roots of $x^4 + 2x^3 + 2x^2 - 4x - 8$.

Since the leading coefficient is 1, the only possibilities for rational roots are the factors of the last coefficient -8:

$$1, -1, 2, -2, 4, -4, 8, -8.$$

But, using substitution or synthetic division, we find that none of the possibilities is a root. We leave it to the student to verify this. Thus there are no rational roots.

The polynomial given in Example 3 has no rational roots. We can approximate the irrational roots of such a polynomial by graphing it and determining the x-intercepts (see Section 9.6).

DO EXERCISES 3 AND 4.

ii   RATIONAL COEFFICIENTS

Suppose some (or all) of the coefficients of a polynomial are rational, but not integers. After multiplying on both sides by the LCM of the denominators, we can then find the rational roots.

EXAMPLE 4   Let $P(x) = \frac{1}{12}x^3 - \frac{1}{12}x^2 - \frac{2}{3}x + 1$. Find the rational roots of $P(x)$.

The LCM of the denominators is 12. When we multiply on both sides by 12, we get

$$12P(x) = x^3 - x^2 - 8x + 12.$$

This equation is equivalent to the first, and all coefficients on the right are integers. Thus any root of $12P(x)$ is a root of $P(x)$. We leave it to the student to verify that 2 and -3 are the rational roots, in fact the only roots, of $P(x)$.

DO EXERCISE 5.

3. Let $P(x) = x^4 + x^2 + 2x + 6$.

   a) How do you know at the outset that this polynomial has no positive roots?

   b) Find the rational roots of $P(x)$.

4. a) Find the rational roots of $x^2 + 3x + 3$.

   b) Can you find the other roots of this polynomial? What are they? Why can you find them?

5. Let

$$P(x) = x^4 - \frac{1}{6}x^3 - \frac{4}{3}x^2 + \frac{1}{6}x + \frac{1}{3}.$$

   a) Which, if any, of the coefficients are *not* integers?

   b) What is the LCM of the denominators?

   c) Multiply by the LCM.

   d) Find the rational roots of the resulting polynomial.

   e) Are they rational roots of $P(x)$? Why?

EXERCISE SET 9.4

i   In Exercises 1–4, use Theorem 8 to list all *possible* rational roots.

1. $x^5 - 3x^2 + 1$

2. $x^7 + 37x^5 - 6x^2 + 12$

3. $15x^6 + 47x^2 + 2$

4. $10x^{25} + 3x^{17} - 35x + 6$

In Exercises 5–16, find the rational roots, if they exist, of each polynomial or equation. If possible, find the other roots. Then write the equation or polynomial in factored form.

5. $x^3 + 3x^2 - 2x - 6$

6. $x^3 - x^2 - 3x + 3 = 0$

7. $5x^4 - 4x^3 + 19x^2 - 16x - 4 = 0$

8. $3x^4 - 4x^3 + x^2 + 6x - 2$

9. $x^4 - 3x^3 - 20x^2 - 24x - 8$

10. $x^4 + 5x^3 - 27x^2 + 31x - 10$

11. $x^3 + 3x^2 - x - 3 = 0$

12. $x^3 + 5x^2 - x - 5$

13. $x^3 + 8$

14. $x^3 - 8 = 0$

ii

15. $\frac{1}{3}x^3 - \frac{1}{2}x^2 - \frac{1}{6}x + \frac{1}{6}$

16. $\frac{2}{3}x^3 - \frac{1}{2}x^2 + \frac{2}{3}x - \frac{1}{2}$

i   In Exercises 17–24, find only the rational roots.

17. $x^4 + 32$

18. $x^6 + 8 = 0$

19. $x^3 - x^2 - 4x + 3 = 0$

20. $2x^3 + 3x^2 + 2x + 3$

21. $x^4 + 2x^3 + 2x^2 - 4x - 8 = 0$

22. $x^4 + 6x^3 + 17x^2 + 36x + 66 = 0$

23. $x^5 - 5x^4 + 5x^3 + 15x^2 - 36x + 20$

24. $x^5 - 3x^4 - 3x^3 + 9x^2 - 4x + 12$

25. The volume of a cube is 64 cm^3. Find the length of a side. (*Hint:* Solve $x^3 - 64 = 0$.)

26. The volume of a cube is 125 cm^3. Find the length of a side.

27. An open box of volume 48 cm^3 can be made from a piece of tin 10 cm on a side by cutting a square from each corner and folding up the edges. What is the length of a side of the squares?

28. An open box of volume 500 cm^3 can be made from a piece of tin 20 cm on a side by cutting a square from each corner and folding up the edges. What is the length of a side of the squares?

29. Show that $\sqrt{5}$ is irrational, by considering the equation $x^2 - 5 = 0$.

30. Generalize the result of Exercise 29 to find which integers have rational square roots.

OBJECTIVES

You should be able to:

i Use Descartes' rule of signs to find information about the number of real roots of a polynomial with real coefficients.

ii Find upper and lower bounds of the roots of a polynomial with real coefficients, using synthetic division.

Determine the number of variations of sign in each polynomial.

1. $3x^5 - 2x^3 - x^2 + x - 2$

2. $4p^7 + 6p^5 + 2p^3 - 5p^2 + 3$

9.5

FURTHER HELPS IN FINDING ROOTS

i DESCARTES' RULE OF SIGNS

A rule that helps determine the number of positive real roots of a polynomial is due to Descartes. To use the rule we must have the polynomial arranged in descending or ascending order, with no zero terms written in. Then we determine the number of *variations of sign*, that is, the number of times, in going through the polynomial, that successive coefficients are of different sign.

EXAMPLE 1 Determine the number of variations of sign in the polynomial $2x^6 - 3x^2 + x + 4$.

$$\underline{2x^6 - 3x^2} + \underline{x + 4}$$

From positive to negative. A variation

Both positive. No variation

From negative to positive. A variation

The number of variations of sign is two.

DO EXERCISES 1 AND 2.

We now state Descartes' rule, without proof.

Theorem 9 Descartes' rule of signs

The number of positive real roots of a polynomial with real coefficients is either

1. The same as the number of its variations of sign, or

2. Less than the number of its variations of sign by a positive even integer.

A root of multiplicity *m* must be counted *m* times.

EXAMPLES In each case, what does Descartes' rule of signs tell you about the number of positive real roots?

2. $2x^5 - 5x^2 + 3x + 6$

The number of variations of sign is two. Therefore, the number of positive real roots is either 2 or less than 2 by 2, 4, 6, etc. Thus the number of positive roots is either 2 or 0, since a negative number of roots has no meaning.

3. $5x^4 - 3x^3 + 7x^2 - 12x + 4$

There are four variations of sign. Thus the number of positive real roots is either

$$4 \quad \text{or } 4 - 2 \quad \text{or } 4 - 4.$$

That is, the number of roots is 4, 2, or 0.

4. $6x^5 - 2x - 5 = 0$

The number of variations of sign is 1. Therefore, there is exactly one positive real root.

DO EXERCISES 3–5.

Negative Roots

Descartes' rule can also be used to help determine the number of negative roots of a polynomial. Consider the following graphs of a polynomial equation $y = P(x)$ and its reflection across the y-axis, that is, $y = P(-x)$. The points at which the graphs cross the x-axis are the roots of the polynomials.

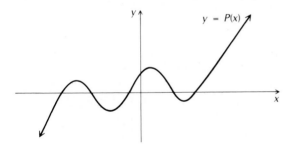

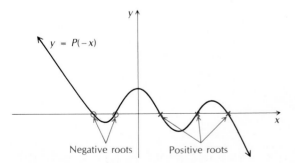

Negative roots | Positive roots

From the graphs we see that the number of positive roots of $P(-x)$ is the same as the number of negative roots of $P(x)$.

In each case, what does Descartes' rule of signs tell you about the number of positive real roots?

3. $5x^3 - 4x - 5$

4. $6p^6 - 5p^4 + 3p^3 - 7p^2 + p - 2$

5. $3x^2 - 2x + 4$

In each case, what does Descartes' rule of signs tell you about the number of negative real roots?

6. $5x^3 - 4x - 5$

7. $6p^6 - 5p^4 + 3p^3 - 7p^2 + p - 2$

8. $3x^2 - 2x + 4$

Theorem 10 Corollary to Descartes' rule of signs

> The number of negative real roots of a polynomial $P(x)$ with real coefficients is either
>
> 1. The number of variations of sign of $P(-x)$, or
> 2. Less than the number of variations of sign of $P(-x)$ by a positive even integer.

To apply Theorem 10, we construct $P(-x)$ by replacing x by $-x$ wherever it occurs and then count the variations of sign.

EXAMPLES   In each case, what does Descartes' rule of signs tell you about the number of negative real roots?

5. $5x^4 + 3x^3 + 7x^2 + 12x + 4$

$$P(-x) = 5(-x)^4 + 3(-x)^3 + 7(-x)^2 + 12(-x) + 4 \quad \text{Replacing } x \text{ by } -x$$

$$= 5x^4 \quad - 3x^3 \quad + 7x^2 \quad - 12x \quad + 4 \quad \text{Simplifying*}$$

There are four variations of sign, so the number of negative roots is either 4 or 2 or 0.

6. $2x^5 - 5x^2 + 3x + 6$

$$P(-x) = 2(-x)^5 - 5(-x)^2 + 3(-x) + 6 \quad \text{Replacing } x \text{ by } -x$$

$$= -2x^5 \quad - 5x^2 \quad - 3x \quad + 6 \quad \text{Simplifying}$$

There is one variation of sign, so there is exactly one negative root.

DO EXERCISES 6–8.

ii UPPER BOUNDS ON ROOTS

Recall that when a polynomial $P(x)$ is divided by $x - a$, we obtain a quotient $Q(x)$ and a remainder R, related as follows:

$$P(x) = (x - a) \cdot Q(x) + R.$$

Let us consider the case in which a is positive, R is positive or zero, and all the coefficients of $Q(x)$ are nonnegative. Then $P(b)$ will be positive for any positive number b that is greater than a. We show this as follows.

$$P(b) = \underbrace{(b - a)}_{\text{positive}} \cdot \underbrace{Q(b)}_{\text{positive}} + \underbrace{R}_{\text{nonnegative}}$$

$b - a > 0$ because we assumed that $b > a$;

$Q(b) > 0$ because all coefficients are nonnegative and the number b we are substituting is positive. (We know that $Q(x)$ is not the zero polynomial, so it has some nonzero coefficients.)

R is assumed to be nonnegative.

---

* It is simple to construct $P(-x)$ mechanically. If $P(x)$ is in descending order with no missing terms, we change the sign of every second term beginning with the second term from the right.

Change signs.

This shows that there can be no root of $P(x)$ greater than the positive number a. If there were such a root r it would make $P(r) = 0$ and this cannot occur, because all values of $P(x)$ are positive when x is greater than a. In terms of synthetic division, this tells us the following.

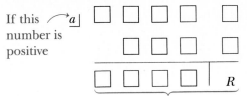

If this $a|$
number is
positive

and all of the numbers in the bottom row are nonnegative,

then there is no root greater than a. In other words, the number a is an *upper bound* to all the roots of $P(x)$. Of course, if the number R is 0, we know that a is actually a root, as well as an upper bound. We state this result as a theorem.

Theorem 11

> If when a polynomial is divided by $x - a$, where a is positive, the remainder and all coefficients of the quotient are nonnegative, the number a is an upper bound to the roots of the polynomial (all of the roots are less than or equal to a).

EXAMPLE 7   Determine an upper bound to the roots of

$$3x^4 - 11x^3 + 10x - 4.$$

We choose some *positive* number a and divide by $x - a$, using synthetic division. Let's try 1 for a.

$$
\begin{array}{r|rrrrr}
1 & 3 & -11 & 0 & 10 & -4 \\
  &   & 3 & -8 & -8 & 2 \\
\hline
  & 3 & -8 & -8 & 2 & -2
\end{array}
$$

Since some of the coefficients are negative, the theorem does not guarantee that 1 is an upper bound. We choose a larger number, this time 4.

$$
\begin{array}{r|rrr|rr}
4 & 3 & -11 & 0 & 10 & -4 \\
  &   & 12 & 4 & 16 & 104 \\
\hline
  & 3 & 1 & 4 & 26 & 100
\end{array}
$$

One reason for choosing 4 is that we know *this* number must be at least 11, in order for the addition to give a nonnegative result.

Since there are no negative numbers in the bottom row, 4 is an upper bound.

> CAUTION!   Remember, in applying Theorem 11 the number a must be a *positive* number.

DO EXERCISES 9 AND 10.

LOWER BOUNDS

A corollary of the above theorem can be used to find lower bounds of roots of a polynomial. Consider the following graphs.

In each case, determine an upper bound to the roots.

9. $5x^4 - 18x^2 + 3x - 2$

10. $x^3 - 75x^2 + 3$

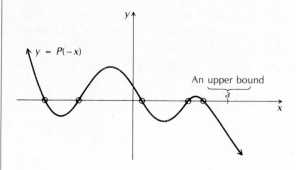

Here we have an upper bound, a, to the roots of the polynomial $P(-x)$. Now let us consider the reflection across the y-axis, $y = P(x)$.

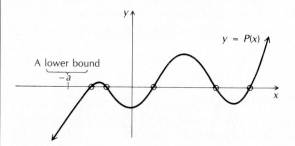

It is easy to see that the negative number $-a$ is a lower bound to the roots of $P(x)$.

Theorem 12

> The number $-a$, where a is positive, is a lower bound to the roots of the polynomial $P(x)$, if when $P(-x)$ is divided by $x - a$ the remainder and all coefficients of the quotient are nonnegative.

EXAMPLE 8 Determine a lower bound to the roots of $P(x)$, where

$$P(x) = 3x^4 + 11x^3 + 10x - 4.$$

We first construct $P(-x)$ by replacing x by $-x$:

$$3(-x)^4 + 11(-x)^3 + 10(-x) - 4$$
$$3x^4 - 11x^3 - 10x - 4. \qquad \text{Simplifying}$$

Next, we use synthetic division, dividing by $x - 4$:*

$$
\begin{array}{r|rrrrr}
4 & 3 & -11 & 0 & -10 & -4 \\
  &   & 12 & 4 & 16 & 24 \\
\hline
  & 3 & 1 & 4 & 6 & \:|\: 20
\end{array}
$$

---

* To use a shortcut, we can set up the synthetic division using $P(x)$.

$$
\begin{array}{r|rrrrr}
4 & 3 & 11 & 0 & 10 & -4
\end{array}
$$

Then we change the sign of every second number, beginning at the second from the right, and then divide.

$$
\begin{array}{r|rrrrr}
4 & 3 & \boxed{-11} & 0 & \boxed{-10} & -4
\end{array}
$$

Change signs.

Since none of the numbers in the bottom row is negative, 4 is an upper bound to the roots of $P(-x)$. Thus -4 is a lower bound to the roots of $3x^4 + 11x^3 + 10x - 4$, which is $P(x)$.

EXAMPLE 9   Determine a lower bound to the roots of $P(x)$, where

$$P(x) = 5x^3 - 12x^2 + 2x + 3.$$

We construct $P(-x)$ by replacing x by $-x$ (or by simply changing the sign of every other term, beginning at the second from the right).

$$\begin{aligned} P(-x) &= 5(-x)^3 - 12(-x)^2 + 2(-x) + 3 \\ &= -5x^3 \quad - 12x^2 \quad - 2x \quad + 3. \end{aligned}$$

We try synthetic division.

$$\begin{array}{r|rrrr} a| & -5 & -12 & -2 & 3 \\ & & & b & \\ \hline & -5 & c & & \end{array}$$

Recall that the number a must be positive in this procedure. It is clear that no positive number a will give us a positive number in position b. Therefore, it will be impossible to get a nonnegative number in position c.

In a case like this we will use $-P(-x)$ instead of $P(-x)$, since the two polynomials have the same roots. To change to $-P(-x)$ all we need do is change the sign of every coefficient in $P(-x)$. Then we proceed as before. We try 1 for a.

$$\begin{array}{r|rrrr} 1| & 5 & 12 & 2 & -3 \\ & & 5 & 17 & 19 \\ \hline & 5 & 17 & 19 & | \; 16 \end{array}$$

The number -1 is a lower bound.

In determining bounds for roots, if the leading coefficient of $P(x)$ is negative, we simply convert to $-P(x)$ by changing the sign of every term. Then we proceed as before.

DO EXERCISES 11 AND 12.

A SHORTCUT

On p. 345 you were cautioned to use only positive numbers when seeking upper bounds. Likewise, in looking for lower bounds we have used only positive numbers, in accordance with Theorem 11. In looking for lower bounds there is a way to use negative numbers in the synthetic division that may save a small amount of time. We illustrate, using the polynomial of Example 8. The following is what we have already done.

$$\begin{array}{r|rrrrr} 4| & 3 & -11 & 0 & -10 & -4 \\ & & 12 & 4 & 16 & 24 \\ \hline & 3 & 1 & 4 & 6 & | \; 20 \end{array} \quad \longleftarrow \text{This is } P(-x).$$

Now let us use -4 with $P(x)$ and compare.

$$\begin{array}{r|rrrrr} -4| & 3 & 11 & 0 & 10 & -4 \\ & & -12 & 4 & -16 & 24 \\ \hline & 3 & -1 & 4 & -6 & | \; 20 \end{array} \quad \longleftarrow \text{This is } P(x).$$

Note that the only difference is a change of sign in the indicated columns. This illustrates the following corollary to Theorem 12.

Determine a lower bound to the roots.

11. $5x^4 + 18x^3 + 3x - 2$

12. $4x^3 + 7x^2 + 3x + 5$

13. Determine a lower bound to the roots of $5x^4 + 18x^3 - 3x - 3$.

Theorem 13

When a polynomial is divided by $x - a$, where a is negative, a will be a lower bound to the roots if in the result of the synthetic division the first number is nonnegative, the second nonpositive, the third nonnegative, the fourth nonpositive, and so on.

EXAMPLE 10 Determine a lower bound to the roots of $P(x)$, where

$$P(x) = 5x^3 - 12x^2 + 2x + 3.$$

We try -2:

$$
\begin{array}{r|rrrr}
-2 & 5 & -12 & 2 & 3 \\
   &   & -10 & 44 & -92 \\
\hline
   & 5 & -22 & 46 & -89.
\end{array}
$$

Since the odd-numbered coefficients (from left to right) are nonnegative and the even-numbered ones are nonpositive, -2 is a lower bound to the roots.

DO EXERCISE 13.

EXAMPLE 11 What does Descartes' rule of signs tell you about the roots of $P(x)$? Find upper and lower bounds to the real roots.

$$P(x) = 4x^4 + 3x^3 + x - 1$$

a) There is one variation of sign, so there is just one real positive root.

b) $P(-x) = 4(-x)^4 + 3(-x)^3 + (-x) - 1$
$$= 4x^4 \quad - 3x^3 \quad - x \quad - 1$$

There is just one variation of sign, so there is just one negative root.

c) Since the equation is of degree 4, it has four roots. Therefore, there are two nonreal roots.

d) We look for an upper bound.

$$
\begin{array}{r|rrrrr}
1 & 4 & 3 & 0 & 1 & -1 \\
  &   & 4 & 7 & 7 & 8 \\
\hline
  & 4 & 7 & 7 & 8 & 7
\end{array}
$$

The number 1 is an upper bound. Let's try $\frac{1}{4}$.

$$
\begin{array}{r|rrrrr}
\frac{1}{4} & 4 & 3 & 0 & 1 & -1 \\
  &   & 1 & 1 & \frac{1}{4} & \frac{5}{16} \\
\hline
  & 4 & 4 & 1 & \frac{5}{4} & -\frac{11}{16}
\end{array}
$$

We do not know whether or not $\frac{1}{4}$ is an upper bound,* but we look no further.

e) We look for a lower bound, using Theorem 12.

$$
\begin{array}{r|rrrrr}
-1 & 4 & 3 & 0 & 1 & -1 \\
   &   & -4 & 1 & -1 & 0 \\
\hline
   & 4 & -1 & 1 & 0 & -1
\end{array}
$$

---

* The converse of Theorem 10 is not true. That is, if the positive number a is an upper bound to the roots, it is not necessarily true that when we divide by $x - a$ all of the coefficients of the quotient and the remainder will be nonnegative. Thus when we divide and find negative numbers in the bottom row, this does not mean that the number a is *not* an upper bound. We simply do not know whether it is or not.

We do not know whether or not -1 is a lower bound.

$$\begin{array}{r|rrrrr} -2 & 4 & 3 & 0 & 1 & -1 \\ & & -8 & 10 & -20 & 38 \\ \hline & 4 & -5 & 10 & -19 & 37 \end{array}$$

The number -2 is a lower bound and not a root (because $R \neq 0$).

To summarize, we have learned that $P(x)$ has two real roots, one between 0 and 1 and one between -2 and 0. It also has two nonreal roots.

From the information obtained in Example 11, we can actually say a little more. In (d) we found that $P(1) = 7$ (a positive number) and $P(\frac{1}{4}) = -\frac{11}{16}$ (a negative number). Thus $P(x)$ must be 0 for some number between $\frac{1}{4}$ and 1. One of the roots is therefore between $\frac{1}{4}$ and 1.

DO EXERCISES 14 AND 15.

In each case, what does Descartes' rule of signs tell you about the roots of the polynomial? Find upper and lower bounds to the real roots.

14. $x^4 - 6x^3 + 7x^2 + 6x - 2$

15. $x^4 + 2x^2 + x - 2$

EXERCISE SET 9.5

i What does Descartes' rule of signs tell you about the number of positive real roots?

1. $3x^5 - 2x^2 + x - 1$

2. $5x^6 - 3x^3 + x^2 - x$

3. $6x^7 + 2x^2 + 5x + 4 = 0$

4. $-3x^5 - 7x^3 - 4x - 5 = 0$

5. $3p^{18} + 2p^4 - 5p^2 + p + 3$

6. $5t^{12} - 7t^4 + 3t^2 + t + 1$

What does Descartes' rule of signs tell you about the number of negative real roots?

7. $3x^5 - 2x^2 + x - 1$

8. $5x^6 - 3x^3 + x^2 - x$

9. $6x^7 + 2x^2 + 5x + 4 = 0$

10. $-3x^5 - 7x^3 - 4x - 5 = 0$

11. $3p^{18} + 2p^3 - 5p^2 + p + 3$

12. $5t^{11} - 7t^4 + 3t^2 + t + 1$

ii Determine an upper bound to the roots.

13. $3x^4 - 15x^2 + 2x - 3$

14. $4x^4 - 14x^2 + 4x - 2$

15. $6x^3 - 17x^2 - 3x - 1$

16. $5x^3 - 15x^2 + 5x - 4$

Determine a lower bound to the roots.

17. $3x^4 - 15x^3 + 2x - 3$

18. $4x^4 - 17x^3 + 3x - 2$

19. $6x^3 + 15x^2 + 3x - 1$

20. $6x^3 + 12x^2 + 5x - 3$

i, **ii** What does Descartes' rule of signs tell you about the roots of the polynomial or equation? Find upper and lower bounds to the real roots.

21. $x^4 - 2x^2 + 12x - 8$

22. $x^4 - 6x^2 + 20x - 24$

23. $x^4 - 2x^2 - 8 = 0$

24. $3x^4 - 5x^2 - 4 = 0$

25. $x^4 - 9x^2 - 6x + 4$

26. $x^4 - 21x^2 + 4x + 6$

27. $x^4 + 3x^2 + 2 = 0$

28. $x^4 + 5x^2 + 6 = 0$

29. Prove that for n a positive even integer, $x^n - 1$ has only two real roots.

30. Prove that for n an odd positive integer, $x^n - 1$ has only one real root.

31. Show that $x^4 + ax^2 + bx - c$, where a, b, and c are positive, has just two nonreal roots.

32. In the three-body problem that arises in astronomy the following equation occurs:

$$r^5 + (3 - k)r^4 + (3 - 2k)r^3 - kr^2 - 2kr - k = 0,$$

where $0 < k < 1$. Show that this equation has just one positive real solution.

9.6

GRAPHS OF POLYNOMIAL FUNCTIONS

i Graphs of first-degree polynomial functions are lines; graphs of second-degree, or quadratic, functions are parabolas. We now consider polynomials of higher degree. We begin with some general principles to be kept in mind while graphing. We consider only polynomials with real coefficients.

1. Every polynomial function is a continuous function, whose domain is the set of all real numbers. The graph of any function must pass the vertical-line test, so no polynomial function can have a graph like the following.

2. Unless a polynomial function is linear, no part of its graph is straight. A common error is to make a graph look like the black one below, rather than the color one below.

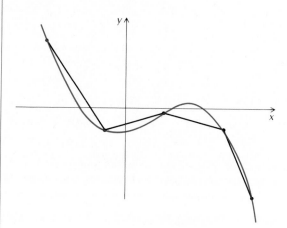

3. A polynomial of degree n cannot have more than n real roots. This means that the graph cannot cross the x-axis more than n times, so we know something about how the curve "wiggles." Third-degree, or *cubic*, functions have graphs like the following.

(a)

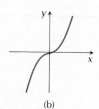

(b)

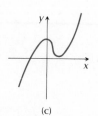

(c)

In (a) the graph crosses the axis three times, so there are three real roots. In (b) and (c) there is only one x-intercept, so there is only one real root in each case. The graph of a cubic cannot look like the following, because there would be a possibility that it might cross the x-axis more than three times.

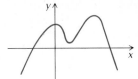

Graphs of fourth-degree, or *quartic*, polynomials look like these.

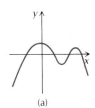 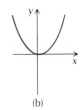 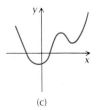

(a)   (b)   (c)

In (a) there are four real roots, in (b) there is one, and in (c) there are two.

Multiple roots occur at points like the following.

 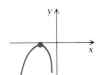

When a graph looks like the one below, missing the x-axis at one of its opportunities, a pair of nonreal roots occurs. There is no easy way to find them from the graph.

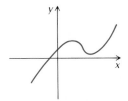

4. The leading term of a polynomial tells a lot about how the graph looks for values of x with large absolute value. This is so because for such values of x the contributions of the other terms are relatively minor. First, let us suppose that the coefficient is positive. Then for large positive values of x, the function value will be positive and increasing. Thus we know that as we move far to the right the graph looks like (a) below.

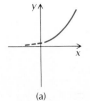 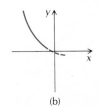 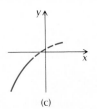

(a)   (b)   (c)

If the exponent of the leading term is even, the same thing happens as we move far to the left, as in (b). If the exponent is odd, then the function values are negative and they decrease as we move far to the left, as in (c) above. If the leading coefficient is negative, the graphs will of course be reflected across the x-axis.

To graph polynomials, keep in mind previously established results and proceed somewhat as follows.

To graph polynomials

a) Look at the degree of the polynomial and its leading coefficient. This gives a lot of information about the general shape of the graph.

b) Look for symmetries, as covered in Chapter 3.

c) Make a table of values using synthetic division.

d) Find the y-intercept and as many x-intercepts as possible (the latter are roots of the polynomial). In doing this, recall the theorems about roots, including Descartes' rule of signs.

e) Plot the points and connect them appropriately.

EXAMPLE 1   Graph $P(x) = 2x^3 - x + 2$.

a) This polynomial is of degree 3 with leading coefficient positive. Thus as we move far to the right, function values will increase beyond bound. As we move far left, they will be negative and decrease beyond bound.

   The curve will have the general shape of a cubic, one of those shown on p. 350.

b) The function is not odd or even. However, $2x^3 - x$ is an odd function, with the origin as a point of symmetry. $P(x)$ is a translation of this upward 2 units; hence the point $(0, 2)$ is a point of symmetry.

c) We make a table of values. The first entry is merely the y-intercept. The other rows contain the bottom line in synthetic division (which is really all we need to write ordinarily).

	2	0	-1	2	(x, y)	
0				2	$(0, 2)$	The y-intercept
1	2	2	1	3	$(1, 3)$	Since all numbers are positive, 1 is an upper bound to the roots.
2	2	4	7	16	$(2, 16)$	
-1	2	-2	1	1	$(-1, 1)$	
-2	2	-4	7	-12	$(-2, -12)$	Since signs alternate, -2 is a lower bound to the roots.

d) Because we know that the curve is symmetric with respect to the point $(0, 2)$, we do not really need to calculate $P(x)$ for negative values of x. We can also see from this table that the graph will cross the x-axis between -1 and -2. This is because $P(-1)$ is positive (and the graph there is above the axis), whereas $P(-2)$ is negative (and the graph there is below the axis). Somewhere between -1 and -2, then, $P(x)$ must be 0. Thus there is a real root of $P(x)$ between -1 and -2.

By Descartes' rule of signs, there are either 2 or 0 positive real roots. From our table of values, we see that $P(x)$ is positive and increasing rapidly. These facts would lead us to *suspect* that there are *no* positive roots. If there are any, they must be between 0 and 1 because 1 is an upper bound. Now $P(-x) = -2x^3 + x + 2$, so Descartes' rule tells us there is just one negative root. Thus the one we have already eyeballed is the only one. We also know that if there are no positive roots, there will be two nonreal ones, and they will be conjugates of each other.

e) We now plot points and connect them appropriately.*

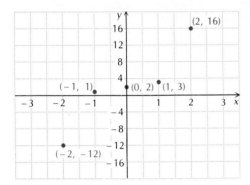

In this example, it is not easy to see how to draw the curve. It might look like any of the following so far as we can tell from the points plotted.

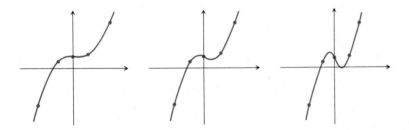

We therefore need to plot some more points. Because of symmetry, it will be sufficient to consider positive values of x.

	2	0	-1		(x, y)
0.1	2	0.2	0.98	1.9	$(0.1, 1.9)$
0.3	2	0.6	-0.82	1.75	$(0.3, 1.75)$
0.5	2	1.0	-0.5	1.75	$(0.5, 1.75)$
0.7	2	1.4	-0.02	1.99	$(0.7, 1.99)$
0.9	2	1.8	0.62	2.56	$(0.9, 2.56)$

---

* In this graph we are using different scales on the x-axis and the y-axis. We are reluctant to do so, but it is sometimes necessary.

1. Graph.

$$P(x) = x^3 - 4x^2 - 3x + 12$$

We now have enough points to determine the shape of the graph.

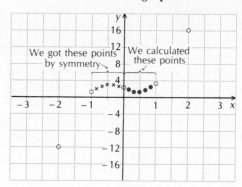

We got these points by symmetry

We calculated these points

We can now draw the graph confidently.

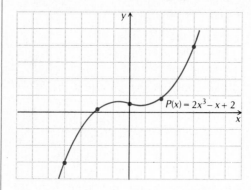

$$P(x) = 2x^3 - x + 2$$

DO EXERCISE 1.

ii   SOLVING EQUATIONS

Whenever we find the roots, or zeros, of a function $P(x)$ we have solved the equation $P(x) = 0$. One means of finding approximate solutions of polynomial equations is by graphing. We graph the function $y = P(x)$ and note where the graph crosses the x-axis. In Example 1, there is a real solution of the equation $2x^3 - x + 2 = 0$ at about -1.1.

We can approximate roots more precisely by doing further calculation.

▥ **EXAMPLE 2**   Find a better approximation of the solution of $2x^3 - x + 2 = 0$, which is near -1.1.

We know that $P(-1) = 1$ and $P(-2) = -12$, so there must be a root between -1 and -2. To find a better approximation we use synthetic division.

	2	0	-1	2
-1.1	2	-2.2	1.42	0.44
-1.2	2	-2.4	1.88	-0.26
.	.	.	.	.
.	.	.	.	.
.	.	.	.	.
-1.15	2	-2.30	1.65	0.11
-1.16	2	-2.32	1.69	0.04
-1.17	2	-2.34	1.74	-0.03

Since $P(-1.1) > 0$, there is no root between -1 and -1.1.

Since $P(-1.2) < 0$ and $P(-1.1) > 0$, there is a root between -1.1 and -1.2. We'll try -1.15, halfway between.

We have a root between -1.16 and -1.17.

We now have an approximation to hundredths, -1.16. In this manner the approximation can be made as close as we please.

DO EXERCISE 2.

2. a) Graph $P(x) = x^3 - 3x^2 + 1$.

   b) Use the graph to approximate the roots to tenths.

EXERCISE SET 9.6

i In Exercises 1–6, graph.

1. $P(x) = x^3 - 3x^2 - 2x - 6$

2. $P(x) = x^3 + 4x^2 - 3x - 12$

3. $P(x) = 2x^4 + x^3 - 7x^2 - x + 6$

4. $P(x) = 3x^4 + 5x^3 + 5x^2 - 5x - 6$

5. $P(x) = x^5 - 2x^4 - x^3 + 2x^2$

6. $P(x) = x^5 + 4x^4 - 5x^3 - 14x^2 - 8x$

ii In Exercises 7–16, graph the corresponding polynomial functions, in order to find approximate solutions of the equations.

7. $x^3 - 3x - 2 = 0$

8. $x^3 - 3x^2 + 3 = 0$

9. $x^3 - 3x - 4 = 0$

10. $x^3 - 3x^2 + 5 = 0$

11. $x^4 + x^2 + 1 = 0$

12. $x^4 + 2x^2 + 2 = 0$

13. $x^4 - 6x^2 + 8 = 0$

14. $x^4 - 4x^2 + 2 = 0$

15. $x^5 + x^4 - x^3 - x^2 - 2x - 2 = 0$

16. $x^5 - 2x^4 - 2x^3 + 4x^2 - 3x + 6 = 0$

17. ▦ The following equation has a solution between 0 and 1. Approximate it, to hundredths.

$$2x^5 + 2x^3 - x^2 - 1 = 0$$

18. ▦ The following equation has a solution between 1 and 2. Approximate it, to hundredths.

$$x^4 - 2x^3 - 3x^2 + 4x + 2$$

In each of the following, graph and then approximate the irrational roots, to hundredths.

19. ▦ $P(x) = x^3 - 2x^2 - x + 4$

20. ▦ $P(x) = x^3 - 4x^2 + x + 3$

---

21. ▦ Graph $P(x) = 5.8x^4 - 2.3x^2 - 6.1$.

22. (*Nested evaluation*). A procedure for evaluating a polynomial is as follows. Given a polynomial, such as $3x^4 - 5x^3 + 4x^2 - 5$, successively factor out x, as shown:

$$x(x(x(3x - 5) + 4)) - 5.$$

Given a value for x, substitute it in the innermost parentheses and work your way out, at each step multiplying, then adding or subtracting. Show that this process is identical to synthetic division.

9.7

RATIONAL FUNCTIONS

i A *rational function* is a function definable as the quotient of two polynomials. Here are some examples:

$$y = \frac{x^2 + 3x - 5}{x + 4}, \qquad y = \frac{5}{x^2 + 3}, \qquad y = \frac{3x^5 - 5x + 2}{4}.$$

Definition

A *rational function* is a function that can be defined as $y = P(x)/Q(x)$, where $P(x)$ and $Q(x)$ are polynomials having no common factors other than 1 and -1, and with $Q(x)$ not the zero polynomial.

GRAPHS

Polynomial functions are themselves special kinds of rational functions, since $Q(x)$ can be the polynomial 1. Here we are interested in rational functions in which the denominator is not a constant. We begin with the simplest such function.

EXAMPLE 1   Graph the function $y = 1/x$.

a)   Note that the domain of this function consists of all real numbers except 0.

b)   For nonzero x or y, the above equation is equivalent to $xy = 1$. Now it is easy to see that the graph is symmetric with respect to the line $y = x$ (because interchanging x and y produces an equivalent equation).

c)   The graph is also symmetric with respect to the origin (because replacing x with $-x$ and y with $-y$ produces an equivalent equation).

d)   Now we find some values, keeping in mind the two symmetries.

x	1	2	3	4	5
y	1	$\frac{1}{2}$	$\frac{1}{3}$	$\frac{1}{4}$	$\frac{1}{5}$

e)   We plot these points. We use one symmetry to get other points in the first quadrant. We use the other symmetry to get the points in the third quadrant.

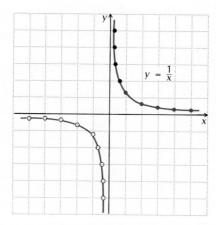

$$y = \frac{1}{x}$$

The points indicated by ● are obtained from the table. Those marked ● are obtained by reflection across the line $y = x$. Following this, the points marked ○ are obtained by reflection across the origin.

ASYMPTOTES

Note that this curve does not touch either axis, but comes very close. As $|x|$ becomes very large, the curve comes very near to the x-axis. In fact, we can find points as close to the x-axis as we please by choosing x large enough. We say that the curve approaches the x-axis *asymptotically*, and we say that the x-axis is an *asymptote* to the curve. The y-axis is also an asymptote to this curve.

Using the ideas of transformations we can easily graph certain variations of the above function.

EXAMPLE 2 $y = -1/x$ is a reflection across the x-axis (or the y-axis; the result is the same).

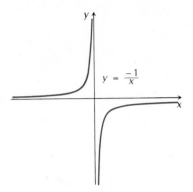

EXAMPLE 3 $y = 1/(x-2)$ is a translation of $y = 1/x$ two units to the right.

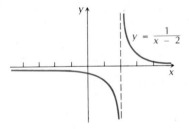

DO EXERCISES 1–4.

EXAMPLE 4 Graph the function $y = 1/x^2$.

a) Note that this function is defined for all x except 0. Therefore, the line $x = 0$ is an asymptote.

b) As $|x|$ gets very large, y approaches 0. Therefore, the x-axis is also an asymptote, for both positive and negative values of x.

c) This function is even. Therefore, it is symmetric with respect to the y-axis.

d) All function values are positive. Therefore, the entire graph is above the x-axis.

e) With this much information, we can already sketch a rough graph of the function. However, a table of values will help.

x	1	2	3	4	$\frac{1}{2}$	$\frac{1}{3}$
y	1	$\frac{1}{4}$	$\frac{1}{9}$	$\frac{1}{16}$	4	9

The graph is as follows. Points marked ● are obtained from the table. Points marked ● are obtained by reflection across the y-axis.

Graph these equations. (Use graph paper.)

1. $y = \dfrac{1}{x + 5}$

2. $y = \dfrac{3x + 1}{x}$

$\left(Hint: \dfrac{3x + 1}{x} = 3 + \dfrac{1}{x}.\right)$

3. $y = \dfrac{2}{x}$

4. $y = \dfrac{1}{3x + 15}$

[*Hint:* $3x + 15 = 3(x + 5)$.] Compare with Exercise 1.

Graph.

5. $y = -\dfrac{1}{x^2}$

6. $y = \dfrac{-3x^2 + 1}{x^2}$

(*Hint:* Divide numerator by denominator.)

7. $y = \dfrac{1}{(x-3)^2}$

8. Find the vertical asymptotes.

$$y = \dfrac{x+3}{x^3 - x^2 - 6x}$$

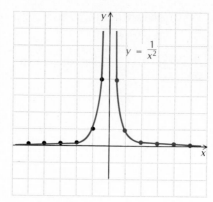

$y = \dfrac{1}{x^2}$

DO EXERCISES 5–8.

Occurrence of Asymptotes

It is important in graphing rational functions to determine where the asymptotes, if any, occur. Vertical asymptotes are easy to locate when a denominator is factored. The *x*-values that make a denominator 0 but do not make the numerator 0 are those of the vertical asymptotes.

EXAMPLES Determine the vertical asymptotes.

5. $y = \dfrac{3x - 2}{x(x-5)(x+3)}$

The vertical asymptotes are the lines $x = 0$, $x = 5$, and $x = -3$.

6. $y = \dfrac{x-2}{x^3 - x}$

We factor the denominator:

$$x^3 - x = x(x+1)(x-1).$$

The vertical asymptotes are $x = 0$, $x = -1$, and $x = 1$.

Horizontal asymptotes occur when the degree of the numerator is the same as or less than that of the denominator. Let us first consider a function for which the degree of the numerator is less than that of the denominator:

$$y = \dfrac{2x + 3}{x^3 - 2x^2 + 4}.$$

We multiply by $\dfrac{1/x^3}{1/x^3}$, to obtain

$$y = \dfrac{\dfrac{2}{x^2} + \dfrac{3}{x^3}}{1 - \dfrac{2}{x} + \dfrac{4}{x^3}}.$$

Let us now consider what happens to the function values as $|x|$ becomes very large. Each expression with *x* in its denominator takes on smaller and smaller values, approaching 0. Thus the numerator approaches 0 and the denominator approaches 1; hence the entire expression takes on values closer and closer to 0. Therefore, the *x*-axis is an asymptote. Whenever the degree of a numerator is less than that of the denominator, the x-axis will be an asymptote.

DO EXERCISE 9.

Next, we consider a function for which the numerator and denominator have the same degree:

$$y = \frac{3x^2 + 2x - 4}{2x^2 - x + 1} = \frac{3x^2 + 2x - 4}{2x^2 - x + 1} \cdot \frac{\dfrac{1}{x^2}}{\dfrac{1}{x^2}}$$

$$= \frac{3 + \dfrac{2}{x} - \dfrac{4}{x^2}}{2 - \dfrac{1}{x} + \dfrac{1}{x^2}}.$$

As $|x|$ gets very large, the numerator approaches 3 and the denominator approaches 2. Therefore, the function values get very close to $\frac{3}{2}$, and thus the line $y = \frac{3}{2}$ is an asymptote. From this example, we can see that the asymptote in such cases can be determined by dividing the leading coefficients of the two polynomials.

EXAMPLES Determine the horizontal asymptotes.

7. $y = \dfrac{5x^3 - x^2 + 7}{3x^3 + x - 10}$

The line $y = \frac{5}{3}$ is an asymptote.

8. $y = \dfrac{-7x^4 - 10x^2 + 1}{11x^4 + x - 2}$

The line $y = -\frac{7}{11}$ is an asymptote.

DO EXERCISES 10 AND 11.

There are asymptotes that are neither vertical nor horizontal. They are called *oblique*, and they occur when the degree of the numerator is greater than that of the denominator by 1. To find the asymptote we divide the numerator by the denominator. Consider

$$y = \frac{2x^2 - 3x - 1}{x - 2}.$$

When we divide the numerator by the denominator, we obtain a quotient of $2x + 1$ and a remainder of 1. Thus,

$$y = 2x + 1 + \frac{1}{x - 2}.$$

Now we can see that when $|x|$ becomes very large, $1/(x - 2)$ approaches 0, and the y-values thus approach $2x + 1$. This means that the graph comes closer and closer to the straight line $y = 2x + 1$.

EXAMPLE 9 Find and draw the asymptotes of the function

$$y = \frac{2x^2 - 11x - 10}{x - 4}.$$

9. For which of the following is the x-axis an asymptote?

a) $y = \dfrac{3x^4 - x^2 + 4}{18x^4 - x^3 + 44}$

b) $y = \dfrac{17x^2 + 14x - 5}{x^3 - 2x - 1}$

c) $y = \dfrac{135x^5 - x^2}{x^7}$

Find the horizontal asymptotes.

10. $y = \dfrac{3x^3 + 4x - 9}{6x^3 - 7x^2 + 3}$

11. $y = \dfrac{9x^4 - 7x^2 - 9}{3x^4 + 7x^2 + 9}$

Find the oblique asymptotes.

12. $y = \dfrac{3x^2 - 7x + 2}{x - 2}$

a) We first note that $x = 4$ is a vertical asymptote.

b) We divide, to obtain

$$y = 2x - 3 - \frac{22}{x - 4}.$$

Thus the line $y = 2x - 3$ is an asymptote.

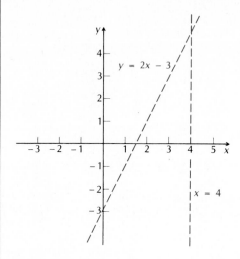

DO EXERCISES 12 AND 13.

The following summarizes the conditions under which asymptotes occur.

Asymptotes of a rational function occur as follows.

1. *Vertical asymptotes.* Have the same x-values that make the denominator zero, but the numerator nonzero.

2. *The x-axis an asymptote.* When the degree of the denominator is greater than that of the numerator.

3. *Horizontal asymptotes other than the x-axis.* Occur when numerator and denominator have the same degree.

4. *Oblique asymptotes.* Occur when the degree of the numerator is greater than that of the denominator by 1.

---

13. $y = \dfrac{5x^3 + 2x + 1}{x^2 - 4}$

ZEROS OR ROOTS

Zeros, or roots, of a rational function occur when the numerator is 0 but the denominator is not 0. The zeros of a function occur at points where the graph crosses the x-axis. Therefore, knowing the zeros helps in making a graph. If the numerator can be factored, the zeros are easy to determine.

EXAMPLE 10 Find the zeros of the function

$$y = \frac{x^3 - x^2 - 6x}{x^2 - 3x + 2}.$$

We factor numerator and denominator:

$$y = \frac{x(x + 2)(x - 3)}{(x - 1)(x - 2)}.$$

The values making the numerator zero are 0, -2, and 3. Since none of these makes the denominator zero, they are the zeros of the function.

DO EXERCISES 14 AND 15.

The following is an outline of a procedure to be followed in graphing rational functions.

> To graph a rational function
>
> a) Determine any symmetries.
> b) Determine any horizontal or oblique asymptotes and sketch them.
> c) Factor the denominator and the numerator.
>   i) Determine any vertical asymptotes and sketch them.
>   ii) Determine the zeros, if possible, and plot them.
> d) Make a table, showing where the function values are positive and where negative.
> e) Make a table of values and plot them.
> f) Sketch the curve.

EXAMPLE 11 Graph

$$y = \frac{x^3 - x^2 - 6x}{x^2 - 3x + 2}.$$

We follow the outline above.

a) The function is neither even nor odd and no symmetries are apparent.

b) The degree of the numerator is one greater than that of the denominator. Dividing numerator by denominator will show that the line $y = x + 2$ is an oblique asymptote.

c) The numerator and denominator are easily factorable. We have

$$y = \frac{x(x + 2)(x - 3)}{(x - 1)(x - 2)}.$$

The zeros are 0, -2, and 3 and there are vertical asymptotes $x = 1$ and $x = 2$.

d) The zeros and asymptotes divide the x-axis into intervals in a natural way, pictured as follows. We tabulate signs in these intervals.

Interval	$x + 2$	x	$x - 1$	$x - 2$	$x - 3$	y
$x < -2$	$-$	$-$	$-$	$-$	$-$	$-$
$-2 < x < 0$	$+$	$-$	$-$	$-$	$-$	$+$
$0 < x < 1$	$+$	$+$	$-$	$-$	$-$	$-$
$1 < x < 2$	$+$	$+$	$+$	$-$	$-$	$+$
$2 < x < 3$	$+$	$+$	$+$	$+$	$-$	$-$
$3 < x$	$+$	$+$	$+$	$+$	$+$	$+$

Find the zeros of these functions.

14. $y = \dfrac{x(x - 3)(x + 5)}{(x + 2)(x - 7)}$

15. $y = \dfrac{x^3 + 2x^2 - 3x}{x^2 + 5}$

e) We next make a table of function values and plot them.

x	$\frac{1}{2}$	$\frac{3}{2}$	$\frac{5}{2}$	4	5	-1	-3	-5
y	$-\frac{25}{6}$	$\frac{63}{2}$	$-\frac{15}{2}$	4	$\frac{35}{6}$	$\frac{2}{3}$	$-\frac{9}{10}$	$-\frac{20}{7}$

f) Using all available information, we draw the graph.

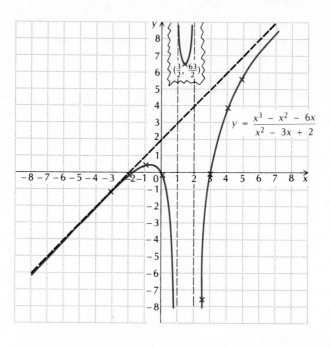

$$y = \frac{x^3 - x^2 - 6x}{x^2 - 3x + 2}$$

The lower left-hand part of the graph is as shown below. The curve crosses the oblique asymptote at $(-2, 0)$, and then, moving to the left, comes back close to the asymptote from above.*

$(-2, 0)$

---

* Mathematical folklore often has it that no curve ever crosses its asymptote. This example shows the folklore to be false.

EXAMPLE 12   Graph

$$y = \frac{1}{x^2 + 1}.$$

We follow the outline above.

a) The function is even, so the graph is symmetric with respect to the y-axis.

b) The degree of the denominator is greater than that of the numerator. Thus the x-axis is an asymptote.

c) The denominator is not factorable. Neither is the numerator. Therefore, there are no vertical asymptotes and there are no zeros.

d) The numerator is a positive constant and the denominator can never be negative, so this function has only positive values. Thus no table of signs is necessary.

e) We tabulate some values.

x	0	1	2	3
y	1	$\frac{1}{2}$	$\frac{1}{5}$	$\frac{1}{10}$

f) We sketch the graph.

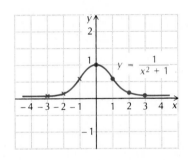

DO EXERCISE 16.

16. Graph this function.

$$y = \frac{x^2 - 4}{x^2 - x - 12}$$

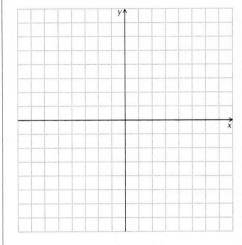

EXERCISE SET 9.7

i Graph these functions.

1. $y = \dfrac{1}{x - 3}$

2. $y = \dfrac{1}{x - 5}$

3. $y = \dfrac{-2}{x - 5}$

4. $y = \dfrac{-3}{x - 3}$

5. $y = \dfrac{2x + 1}{x}$

6. $y = \dfrac{3x - 1}{x}$

7. $y = \dfrac{1}{(x - 2)^2}$

8. $y = \dfrac{-2}{(x - 3)^2}$

9. $y = \dfrac{2}{x^2}$

10. $y = \dfrac{1}{3x^2}$

11. $y = \dfrac{1}{x^2 + 3}$

12. $y = \dfrac{-1}{x^2 + 2}$

13. $y = \dfrac{x - 1}{x + 2}$

14. $y = \dfrac{x - 2}{x + 1}$

15. $y = \dfrac{3x}{x^2 + 5x + 4}$

16. $y = \dfrac{x + 3}{2x^2 - 5x - 3}$

17. $y = \dfrac{x^2 - 4}{x - 1}$

18. $y = \dfrac{x^2 - 9}{x + 1}$

19. $y = \dfrac{x^2 + x - 2}{2x^2 + 1}$

20. $y = \dfrac{x^2 - 2x - 3}{3x^2 + 2}$

21. $y = \dfrac{x - 1}{x^2 - 2x - 3}$

22. $y = \dfrac{x + 2}{x^2 + 2x - 15}$

23. $y = \dfrac{x + 2}{(x - 1)^3}$

24. $y = \dfrac{x - 3}{(x + 1)^3}$

25. $y = \dfrac{x^3 + 1}{x}$

26. $y = \dfrac{x^3 - 1}{x}$

27. $y = \dfrac{x^3 + 2x^2 - 15x}{x^2 - 5x - 14}$

28. $y = \dfrac{x^3 + 2x^2 - 3x}{x^2 - 25}$

29. $y = \dfrac{1}{x^2 + 3} - 5$

30. $y = \dfrac{-1}{x^2 + 2} + 4$

31. $y = \dfrac{1}{|x + 2|}$

32. $y = \left| \dfrac{1}{x} - 2 \right|$

33. $y = \left| \dfrac{1}{x - 2} - 3 \right|$

34. $y = \dfrac{3x}{|x^2 + 5x + 4|}$

OBJECTIVE

You should be able to:

i Decompose fractional expressions into partial fractions.

9.8

PARTIAL FRACTIONS

i There are situations in which it is helpful to do the reverse of adding fractional expressions, that is, to *decompose* a fractional expression into a sum of several fractional expressions. The following illustrates:

$$\frac{3}{x + 2} - \frac{2}{2x - 3} = \frac{4x - 13}{2x^2 + x - 6}.$$

Adding is straightforward. The present problem is to take the sum and find the fractions that were added to get it. That procedure turns out to be rather straightforward also, and it is based on the following theorem, whose proof we omit.

Theorem 14

> Any rational expression $P(x)/Q(x)$, where the degree of the numerator is less than that of the denominator, can be decomposed into partial fractions. To each linear factor $ax + b$ of $Q(x)$ corresponds a fraction $A/(ax + b)$. To each linear factor $ax + b$ occurring twice in $Q(x)$, there correspond two fractions: $B_1/(ax + b)$ and $B_2/(ax + b)^2$. If $ax + b$ occurs three times, there corresponds an additional fraction, $B_3/(ax + b)^3$, and so on. To each quadratic factor $ax^2 + bx + c$ of $Q(x)$ corresponds a fraction $(Ax + B)/(ax^2 + bx + c)$. If the quadratic factor occurs twice, there is an additional fraction, $(Cx + D)/(ax^2 + bx + c)^2$, and so on.* The expressions in the numerators $A, B, \ldots$ are constants, hence they do not depend on x.

For our first example, we use the fractional expression in the opening paragraph.

EXAMPLE 1 Decompose into partial fractions:

$$\frac{4x - 13}{2x^2 + x - 6}.$$

We begin by factoring the denominator: $(x + 2)(2x - 3)$. By Theorem 14 we know that there are constants A and B such that

$$\frac{4x - 13}{(x + 2)(2x - 3)} = \frac{A}{x + 2} + \frac{B}{2x - 3}.$$

---

* This theorem covers all situations, because any polynomial $Q(x)$ with real coefficients has a factorization into linear and quadratic factors.

To determine A and B, we add on the right:

$$\frac{4x - 13}{(x + 2)(2x - 3)} = \frac{A(2x - 3) + B(x + 2)}{(x + 2)(2x - 3)}.$$

Next, we equate the numerators:

$$4x - 13 = A(2x - 3) + B(x + 2) = 2Ax - 3A + Bx + 2B \qquad (1)$$

$$4x - 13 = (2A + B)x + (2B - 3A). \qquad (2)$$

Then we equate corresponding coefficients of Eq. (2):*

$$4 = 2A + B \qquad \text{The coefficients of the } x\text{-terms}$$

$$-13 = 2B - 3A. \qquad \text{The constant terms}$$

We have two equations in A and B. We solve, to obtain

$$A = 3 \quad \text{and} \cdot \quad B = -2.$$

The decomposition is as follows:

$$\frac{3}{x + 2} - \frac{2}{2x - 3}.$$

To check, we would add, to see if we get the original expression.

DO EXERCISE 1.

EXAMPLE 2   Decompose into partial fractions:

$$\frac{7x^2 - 29x + 24}{(2x - 1)(x - 2)^2}.$$

By Theorem 14, the decomposition looks like the following:

$$\frac{A}{2x - 1} + \frac{B}{x - 2} + \frac{C}{(x - 2)^2}.$$

As in Example 1, we add and equate numerators. This gives us

$$7x^2 - 29x + 24 = A(x - 2)^2 + B(2x - 1)(x - 2) + C(2x - 1).$$

This time we use a timesaving approach to determine A, B, and C. Since the equation containing A, B, and C is true for all x, we can substitute any value of x whatever and still have a true equation. We let $2x - 1 = 0$, or $x = \frac{1}{2}$. Then, we get

$$7\left(\frac{1}{2}\right)^2 - 29 \cdot \frac{1}{2} + 24 = A\left(\frac{1}{2} - 2\right)^2 + 0.$$

Solving, we obtain $A = 5$. Next, we let $x - 2 = 0$, or $x = 2$. Substituting, we get

$$7(2)^2 - 29(2) + 24 = 0 + C(2 \cdot 2 - 1).$$

Solving, we obtain $C = -2$. To find B we now equate the coefficients of x^2:

$$7 = A + 2B.$$

---

* To see that this is valid, note that Eq. (2) is equivalent to $(2A + B - 4)x + (2B - 3A + 13) = 0$. By Theorem 14, A and B do not depend on x. Thus we have a polynomial function $P(x)$ that has the value 0 for all x. Since a polynomial $P(x)$ has no more than n real roots, where n is the degree of $P(x)$, it follows that $P(x)$ is the zero polynomial. This means that every coefficient must be 0.

1. Decompose into partial fractions.

$$\frac{13x + 5}{3x^2 - 7x - 6}$$

2. Decompose into partial fractions.

$$\frac{3x^2 - 3x - 2}{(x+1)(x-1)^2}$$

Substituting for A and solving gives us $B = 1$. The decomposition is as follows:

$$\frac{5}{(2x-1)} + \frac{1}{(x-2)} - \frac{2}{(x-2)^2}.$$

DO EXERCISE 2.

EXAMPLE 3 Decompose into partial fractions:

$$\frac{x^2 - 17x + 35}{(x^2 + 1)(x-4)}.$$

By Theorem 14, the decomposition looks like the following:

$$\frac{Ax + B}{x^2 + 1} + \frac{C}{x - 4}.$$

Adding and equating numerators, we get

$$x^2 - 17x + 35 = (Ax + B)(x - 4) + C(x^2 + 1).$$

Letting $x = 4$, we get

$$4^2 - 17 \cdot 4 + 35 = 0 + C(4^2 + 1).$$

Solving, we obtain $C = -1$. Equating the coefficients of x^2, we get $1 = A + C$. Thus we know that $A = 2$. Equating the constant terms, we get $35 = -4B + C$. This gives us $B = -9$. The decomposition is as follows:

$$\frac{2x - 9}{x^2 + 1} - \frac{1}{x - 4}.$$

Theorem 14 refers only to rational expressions $P(x)/Q(x)$ in which the degree of the numerator is less than the degree of the denominator. As we show in the following example, this is not much of a restriction. For an expression not satisfying the condition, we can divide numerator by denominator. The result will be some polynomial plus a fraction $R(x)/Q(x)$, where $R(x)$ is the remainder. The latter fraction meets the conditions of Theorem 14, hence it can be decomposed.

DO EXERCISE 3.

3. Decompose into partial fractions.

$$\frac{2x^2 + 4x + 5}{(x^2 + 1)(x+2)}$$

EXAMPLE 4 Decompose into partial fractions:

$$\frac{6x^3 + 5x^2 - 7}{3x^2 - 2x - 1}.$$

Since the degree of the numerator is not less than that of the denominator, we divide:

$$
\begin{array}{r}
2x + 3 \\
3x^2 - 2x - 1 \overline{)\, 6x^3 + 5x^2 - 7} \\
\underline{6x^3 - 4x^2 - 2x } \\
9x^2 + 2x \\
\underline{9x^2 - 6x - 3} \\
8x - 4
\end{array}
$$

The original expression is thus equivalent to the following:

$$2x + 3 + \frac{8x - 4}{3x^2 - 2x - 1}.$$

We decompose the fraction. The decomposition is as follows:

$$\frac{8x - 4}{(3x + 1)(x - 1)} = \frac{5}{3x + 1} + \frac{1}{x - 1}.$$

The final result is the following:

$$2x + 3 + \frac{5}{3x + 1} + \frac{1}{x - 1}.$$

DO EXERCISE 4.

4. Decompose into partial fractions.

$$\frac{6x^3 + 29x^2 - 6x - 5}{2x^2 + 9x - 5}$$

EXERCISE SET 9.8

i Decompose into partial fractions.

1. $\dfrac{x + 7}{(x - 3)(x + 2)}$

2. $\dfrac{2x}{(x + 1)(x - 1)}$

3. $\dfrac{7x - 1}{6x^2 - 5x + 1}$

4. $\dfrac{13x + 46}{12x^2 - 11x - 15}$

5. $\dfrac{3x^2 - 11x - 26}{(x^2 - 4)(x + 1)}$

6. $\dfrac{5x^2 + 9x - 56}{(x - 4)(x - 2)(x + 1)}$

7. $\dfrac{9}{(x + 2)^2(x - 1)}$

8. $\dfrac{x^2 - x - 4}{(x - 2)^3}$

9. $\dfrac{2x^2 + 3x + 1}{(x^2 - 1)(2x - 1)}$

10. $\dfrac{x^2 - 10x + 13}{(x^2 - 5x + 6)(x - 1)}$

11. $\dfrac{x^4 - 3x^3 - 3x^2 + 10}{(x + 1)^2(x - 3)}$

12. $\dfrac{10x^3 - 15x^2 - 35x}{x^2 - x - 6}$

13. $\dfrac{2x^2 - 11x + 5}{(x - 3)(x^2 + 2x - 5)}$

14. $\dfrac{26x^2 + 208x}{(x^2 + 1)(x + 5)}$

15. $\dfrac{6 + 26x - x^2}{(2x - 1)(x + 2)^2}$

16. $\dfrac{5x^3 + 6x^2 + 5x}{(x^2 - 1)(x + 1)^3}$

17. Decompose into partial fractions:

$$\frac{x}{x^4 - a^4}.$$

18. Decompose into partial fractions:

$$\frac{9x^3 - 24x^2 + 48x}{(x - 2)^4(x + 1)}.$$

[*Hint:* Let the expression equal

$$\frac{A}{x + 1} + \frac{P(x)}{(x - 2)^4}$$

and find $P(x)$.]

Decompose into partial fractions and then graph by addition of ordinates.

19. $y = \dfrac{3x}{x^2 + 5x + 4}$

20. $y = \dfrac{x - 1}{x^2 - 2x - 3}$

Decompose into partial fractions.

21. $\dfrac{1 + \ln x^2}{(\ln x + 2)(\ln x - 3)^2}$

22. $\dfrac{1}{e^{-x} + 3 + 2e^x}$

SUMMARY AND REVIEW: CHAPTER 9

The following contains a summary of what you should be able to do after completing this chapter. The review exercises are for practice. Answers are at the back of the book. If you miss an exercise, restudy the section indicated alongside the answer.

You should be able to:

Use synthetic division to find the quotient and remainder when a polynomial is divided by $x - r$.

1. Find the remainder when $x^4 + 3x^3 + 3x^2 + 3x + 2$ is divided by $x + 2$.

2. Use synthetic division to find the quotient and remainder:

$$(2x^4 - 6x^3 + 7x^2 - 5x + 1) \div (x + 2).$$

Use the remainder theorem to find a function value $P(r)$ when a polynomial $P(x)$ is divided by $x - r$ and determine whether $x - r$ is a factor of $P(x)$ by determining whether $P(r) = 0$.

3. Use synthetic division to find $P(3)$:

$$P(x) = 2x^4 - 3x^3 + x^2 - 3x + 7.$$

4. Factor the polynomial $P(x)$. Then solve the equation $P(x) = 0$.

$$P(x) = x^3 + 7x^2 + 7x - 15$$

5. Determine whether $x + 1$ is a factor of $x^3 + 6x^2 + x + 30$.

Factor polynomials finding their roots and their multiplicities and find a polynomial with specified roots.

6. Find the roots of $P(x) = x^2(x^2 + x - 12)^2(x^2 - 16)$ and state the multiplicity of each.

7. Find a polynomial of degree 3 with roots 0, 1, and 2.

8. Find a polynomial of lowest degree having roots 1 and -1, and having 2 as a root of multiplicity 2 and -3 as a root of multiplicity 3.

Given some of the roots of a polynomial, find such a polynomial and find the rest of its roots.

9. The equation $x^4 - 81 = 0$ has $3i$ for a root. Find the other roots.

10. A polynomial of degree 4 with rational coefficients has roots $-8 - 7i$ and $10 + \sqrt{5}$. Find the other roots.

Given a polynomial with rational coefficients, find the rational roots and find the other roots, if possible.

11. List all possible rational roots of $2x^6 - 12x^4 + 17x^2 + 12$.

12. Let $P(x) = x^4 - x^3 - 3x^2 - 9x - 108$. Find the rational roots of $P(x)$. If possible, find the other roots.

Use Descartes' rule of signs to find information about the number of real roots of a polynomial with real coefficients, and find upper and lower bounds of those roots using synthetic division.

13. What does Descartes' rule of signs tell you about the number of positive real roots of

$$3x^{12} + 3x^4 - 7x^2 + x + 5?$$

14. What does Descartes' rule of signs tell you about the number of negative real roots of

$$6x^8 - 12x^4 + 5x^2 + x + 2?$$

15. Find an upper bound to the roots of

$$2x^4 - 7x^2 + 2x - 1.$$

16. Find a lower bound to the roots of

$$12x^3 + 24x^2 + 10x - 6.$$

Graph polynomial functions and approximate roots of polynomial equations by graphing.

17. The equation $x^5 + x^4 - x^3 - x^2 - 2x - 2$ has a root between 1 and 2. Approximate this root to hundredths.

18. Graph $P(x) = x^3 - 3x^2 + 3$. Use the graph to approximate the irrational roots.

Graph rational functions.

19. Graph $y = \dfrac{x^2 + x - 6}{x^2 - x - 20}$.

Decompose fractional expressions into partial fractions.

20. Decompose into partial fractions: $\dfrac{5}{(x + 2)^2(x + 1)}$.

21. Find a complete factorization of $x^3 - 1$.

22. Find k such that $x + 3$ is a factor of $x^3 + kx^2 + kx - 15$.

23. The equation $x^2 - 8x + c = 0$ has a double root. Find it.

24. When $x^2 - 4x + 3k$ is divided by $x + 5$, the remainder is 33. Find the value of k.

25. Graph: $y = 1 - \dfrac{1}{x^2 + 4}$.

TEST: CHAPTER 9

1. Find the remainder when $3x^4 - 6x^3 + x - 2$ is divided by $x - 3$.

2. Determine whether $x + i$ is a factor of $x^3 + x$.

3. Use synthetic division to find the quotient and remainder. Show all your work.

$$(5x^3 - x^2 + 4x - 3) \div (x + 1)$$

4. Use synthetic division to find $P(-5)$.

$$P(x) = 2x^4 - 7x^3 + 8x^2 - 10$$

5. Factor the polynomial $P(x)$. Then solve the equation $P(x) = 0$.

$$P(x) = x^3 - 4x^2 + x + 6$$

6. Find a polynomial of degree 4 with roots $2, -2, 2 + i, 2 - i$.

7. Find a polynomial of lowest degree having roots 0 and -1, and having 2 as a root of multiplicity 2, and 1 as a root of multiplicity 3.

8. Find a polynomial of lowest degree with rational coefficients that has $3 - i$ and 2 as two of its roots.

9. List all possible rational roots of

$$3x^5 - 4x^3 - 2x + 6.$$

10. What does Descartes' rule of signs tell you about the number of negative real roots of

$$5x^7 - 2x^6 + x^4 + 2x - 6?$$

11. Find an upper bound to the roots of

$$3x^4 - 2x^3 - 15x - 10.$$

12. Graph $P(x) = x^4 - x - 2$. Use the graph to approximate the irrational roots.

13. Graph $y = \dfrac{x - 2}{x^2 - 2x - 15}$.

14. Decompose into partial fractions: $\dfrac{-8x + 23}{2x^2 + 5x - 12}$.

15. Graph $y = \left| 2 - \dfrac{1}{3x} \right|$.

16. Solve $x^4 - 2x^3 + 3x^2 - 2x + 2 = 0$.

Some reflections are made in shapes given by equations of second degree.

Consider the most general kind of second-degree equation in two variables, x and y. There can be an x^2-term, a y^2-term, an xy-term, and then possibly first-degree terms in x and in y, besides a possible constant term:

$$ax^2 + bxy + cy^2 + dx + ey + f = 0.$$

When we consider graphs of such equations, we get a pleasant surprise. Almost all of them have graphs that are cross-sections of cones! In this section we study such graphs and their equations.

Curiously, there are many applications of second-degree equations and their graphs that have nothing to do with cones. In fact, most of them do not. We will have time to consider some of the more interesting applications of second-degree equations, but there are many, many more.

10

10.1

CONIC SECTIONS

CONES

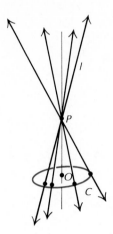 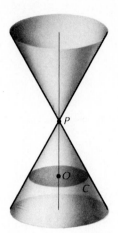

Suppose C is a circle with center O and P is a point not in the same plane as C such that the line $\overleftrightarrow{OP}$ is perpendicular to the plane of the circle C. The set of points on all lines through P and a point of the circle form a *right circular cone* (or *conical surface*). Any line contained in the surface is called a *surface element*. Note that there are two parts, or *nappes*, of a cone. Point P is called the *vertex* and line $\overleftrightarrow{OP}$ is called the *axis*.

Conic Sections

The nonempty intersection of any plane with a cone is a *conic section*. Some conic sections are shown below and at the top of the following page.

In this chapter we study equations of second degree and their graphs. Graphs of most such equations are conic sections.

(a) Ellipse   (b) Parabola   (c) Hyperbola

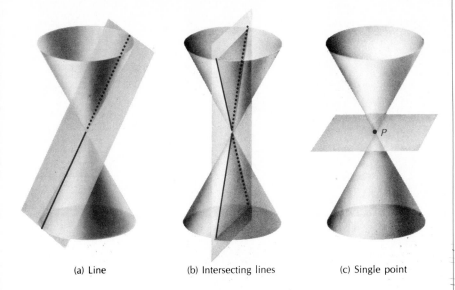

(a) Line   (b) Intersecting lines   (c) Single point

i LINES

Some second-degree equations have graphs consisting of a line. Some have graphs consisting of two lines. The graphs are conic sections except when the two lines are parallel.

EXAMPLE 1   Graph $3x^2 + 2xy - y^2 = 0$.

We factor and use the principle of zero products:

$$3x^2 + 2xy - y^2 = 0$$
$$(3x - y)(x + y) = 0$$
$$3x - y = 0 \quad \text{or} \quad x + y = 0$$
$$y = 3x \quad \text{or} \quad y = -x.$$

The graph consists of two intersecting lines.

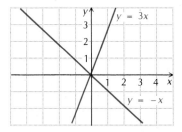

EXAMPLE 2   Graph $(y - x)(y - x - 1) = 0$.

$$(y - x)(y - x - 1) = 0$$
$$y - x = 0 \quad \text{or} \quad y - x - 1 = 0 \qquad \text{Using the principle of zero products}$$
$$y = x \quad \text{or} \quad y = x + 1$$

Graph. (Use graph paper.)

1. $x^2 - 4y^2 = 0$

The graph consists of two parallel lines.

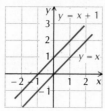

2. $y^2 = 4$

EXAMPLE 3 Graph $y^2 + x^2 = 2xy$.

$$y^2 + x^2 = 2xy$$
$$y^2 - 2xy + x^2 = 0$$
$$(y - x)(y - x) = 0$$
$$y - x = 0 \quad \text{or} \quad y - x = 0$$
$$y = x \quad \text{or} \quad \quad y = x$$

The graph consists of a single line.

3. $y^2 + 9x^2 = 6xy$

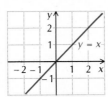

Generally a second-degree equation with 0 on one side and a factorable expression on the other has a graph consisting of one or two lines. A special case is $xy = 0$, in which the graph consists of the coordinate axes.

DO EXERCISES 1–3.

Graph. (Use graph paper.)

4. $x^2 + 2y^2 = 0$

ii SINGLE POINTS OR NO POINTS

When a plane intersects only the vertex of a cone, the result is a single point. The following is an equation for such a conic section.

EXAMPLE 4 Graph $x^2 + 4y^2 = 0$.

The expression $x^2 + 4y^2$ is not factorable in the real-number system. The only real-number solution of the equation is $(0, 0)$.

5. $x^2 + y^2 = -3$

EXAMPLE 5 Graph $3x^2 + 7y^2 = -2$.

Since squares of numbers are never negative, the left side of the equation can never be negative. The equation has no real-number solutions, hence has no graph.

DO EXERCISES 4 AND 5.

CIRCLES

Some equations of second degree have graphs that are circles. Circles are defined as follows.

Definition

A *circle* is the locus or set of all points in a plane that are at a fixed distance from a fixed point in that plane.

When a plane intersects a cone, perpendicular to the axis of the cone, as shown, a circle is formed. We shall prove this later.

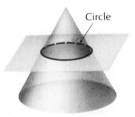

Circle

iii Equations of Circles

We first obtain an equation for a circle centered at the origin.

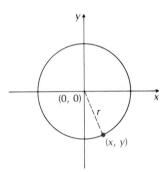

We let r represent the radius. If a point (x, y) is on the circle, then by the definition of a circle and the distance formula, we have

$$r^2 = x^2 + y^2.$$

We have shown that if a point (x, y) is on the circle, then $x^2 + y^2 = r^2$. We now prove the converse, that is, if a point (x, y) satisfies the equation $x^2 + y^2 = r^2$, then it is on the circle. We assume that (x, y) satisfies $x^2 + y^2 = r^2$. Then we have

$$(x - 0)^2 + (y - 0)^2 = r^2.$$

Taking the principal square root, we obtain

$$\sqrt{(x - 0)^2 + (y - 0)^2} = r.$$

Thus the distance from (x, y) to $(0, 0)$ is r, and the point (x, y) is on the circle. The two parts of this proof together show that the equation $x^2 + y^2 = r^2$ gives *all* the points of the circle *and no others*.

When a circle is translated so that its center is the point (h, k), an equation for it can be found by replacing x by $x - h$ and y by $y - k$.

6. Find an equation of the circle having center $(-3, 7)$ and radius 5.

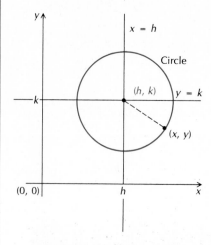

The equation, in standard form, of a circle with center (h, k) and radius r is

$$(x - h)^2 + (y - k)^2 = r^2.$$

EXAMPLE 6 Find an equation of the circle having center $(4, -5)$ and radius 6.

Using the standard form, we obtain

$$(x - 4)^2 + [y - (-5)]^2 = 6^2,$$

or

$$(x - 4)^2 + (y + 5)^2 = 36.$$

DO EXERCISE 6.

We now prove that a conic section obtained when the plane is perpendicular to the axis of a cone is actually a circle.

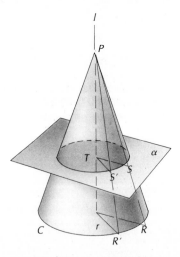

Consider a plane α that cuts a right circular cone perpendicular to its axis, and of course parallel to its base. Then look at the cross sections on any two planes containing the axis of the cone l. Angles PTS and PTS' are right angles because α is perpendicular to l. Thus $\triangle PTS \simeq \triangle PTS'$. Then $\overline{TS} \simeq \overline{TS'}$, for any two points S and S' on the intersection of the cone and the plane α. Thus the intersection consists of all points that are a distance TS from T and is thus a circle, centered at T.

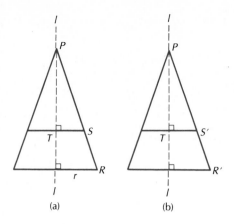

(a)   (b)

iv Finding Center and Radius

EXAMPLE 7 Find the center and radius of $(x - 2)^2 + (y + 3)^2 = 16$. Then graph the circle.

We can first write standard form: $(x - 2)^2 + [y - (-3)]^2 = 4^2$. Then the *center* is $(2, -3)$ and the *radius* is 4. The graph is then easy to draw, as shown, using a compass.

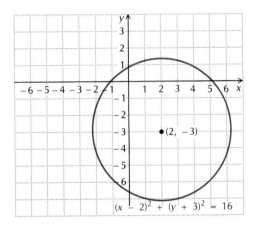

DO EXERCISE 7.

v Standard Form by Completing the Square

Completing the square allows us to find the standard form for the equation of a circle.

EXAMPLE 8 Find the center and radius of the circle

$$x^2 + y^2 + 8x - 2y + 15 = 0.$$

We complete the square twice to get standard form:

$$(x^2 + 8x + \quad) + (y^2 - 2y + \quad) = -15.$$

We take half the coefficient of the x-term and square it, obtaining 16. We add $16 - 16$ in the first parentheses. Similarly, we add $1 - 1$ in the second parentheses:

$$(x^2 + 8x + 16 - 16) + (y^2 - 2y + 1 - 1) = -15.$$

7. Find the center and radius of

$$(x + 1)^2 + (y - 3)^2 = 4.$$

Then graph the circle.

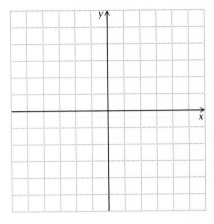

8. Find the center and radius of the circle $x^2 + y^2 - 14x + 4y - 11 = 0$.

Next we do some rearranging and factoring:
$$(x^2 + 8x + 16) + (y^2 - 2y + 1) - 16 - 1 = -15$$
$$(x + 4)^2 + (y - 1)^2 = 2. \quad \text{This is standard form.}$$
The center is $(-4, 1)$ and the radius is $\sqrt{2}$.

DO EXERCISE 8.

vi Circle Determined by Center and One Point

EXAMPLE 9 Find an equation of a circle with center $(-2, -3)$ that passes through the point $(1, 1)$.

Since $(-2, -3)$ is the center, we have
$$(x + 2)^2 + (y + 3)^2 = r^2.$$

The circle passes through $(1, 1)$. We find r by substituting in the above equation:
$$(1 + 2)^2 + (1 + 3)^2 = r^2$$
$$9 + 16 = r^2$$
$$25 = r^2$$
$$5 = r.$$
Then $(x + 2)^2 + (y + 3)^2 = 25$ is an equation of the circle.

DO EXERCISE 9.

9. Find an equation of the circle with center $(-1, 4)$ that passes through $(3, -1)$.

EXERCISE SET 10.1

i , ii Graph.

1. $x^2 - y^2 = 0$

2. $x^2 - 9y^2 = 0$

3. $3x^2 + xy - 2y^2 = 0$

4. $x^2 - xy - 2y^2 = 0$

5. $2x^2 + y^2 = 0$

6. $5x^2 + y^2 = -3$

iii Find an equation of a circle with center and radius as given.

7. Center: $(0, 0)$
Radius: 7

8. Center: $(-2, 7)$
Radius: $\sqrt{5}$

iv Find the center and radius of each circle. Then graph the circle.

9. $(x + 1)^2 + (y + 3)^2 = 4$

10. $(x - 2)^2 + (y + 3)^2 = 1$

Find the center and radius of each circle.

11. $(x - 8)^2 + (y + 3)^2 = 40$

12. $(x + 5)^2 + (y - 1)^2 = 75$

v

13. $x^2 + y^2 + 8x - 6y - 15 = 0$

14. $x^2 + y^2 + 25x + 10y + 12 = 0$

15. $x^2 + y^2 - 4x = 0$

16. $x^2 + y^2 + 10y - 75 = 0$

17. ▦ $x^2 + y^2 + 8.246x - 6.348y - 74.35 = 0$

18. ▦ $x^2 + y^2 + 25.074x + 10.004y + 12.054 = 0$

19. $9x^2 + 9y^2 = 1$

20. $16x^2 + 16y^2 = 1$

vi Find an equation of a circle satisfying the given conditions.

21. Center $(0, 0)$, passing through $(-3, 4)$

22. Center $(3, -2)$, passing through $(11, -2)$

Find an equation of a circle satisfying the given conditions.

23. Center $(2, 4)$ and tangent (touching at one point) to the x-axis

24. Center $(-3, -2)$ and tangent to the y-axis

25. 🔲 A circular swimming pool is being constructed in the corner of a lot as shown. In laying out the pool the contractor wishes to know the distances a_1 and a_2. Find them.

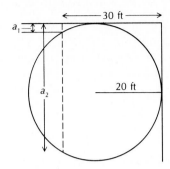

26. Find an equation of a circle such that the endpoints of a diameter are $(5, -3)$ and $(-3, 7)$.

27. a) Graph $x^2 + y^2 = 4$. Is this relation a function?
b) Solve $x^2 + y^2 = 4$ for y.
c) Graph $y = \sqrt{4 - x^2}$ and determine whether it is a function. Find the domain and range.
d) Graph $y = -\sqrt{4 - x^2}$ and determine whether it is a function. Find the domain and range.

28. Show that the following equation is an equation of a circle with center (h, k) and radius r.

$$\begin{vmatrix} x - h & -(y - k) \\ y - k & x - h \end{vmatrix} = r^2$$

Determine whether each of the following points lies on the *unit circle* $x^2 + y^2 = 1$.

29. $(0, -1)$

30. $\left(\dfrac{\sqrt{3}}{2}, -\dfrac{1}{2}\right)$

31. $(\sqrt{2} + \sqrt{3}, 0)$

32. $\left(\dfrac{\pi}{4}, \dfrac{4}{\pi}\right)$

33. $\left(\dfrac{1}{2}, \dfrac{\sqrt{3}}{2}\right)$

34. $\left(\dfrac{\sqrt{2}}{2}, \dfrac{1}{2}\right)$

35. $(\sqrt{3}, 2)$

36. (e^x, e^{-x})

---

37. Prove that ABC is a right angle. Assume that point B is on the circle whose radius is a and whose center is at the origin. (*Hint:* Use slopes and an equation of the circle.)

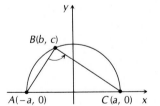

10.2

ELLIPSES

Some equations of second degree have graphs that are ellipses. Ellipses are defined as follows.

Definition

An *ellipse* is the locus or set of all points *P* in a plane such that the sum of the distances from *P* to two fixed points F_1 and F_2 in the plane is constant. F_1 and F_2 are called *foci* (singular, *focus*) of the ellipse.

OBJECTIVES

You should be able to:

i Given an equation of an ellipse centered at the origin, find the vertices and foci, and graph the ellipse.

ii Given an equation of an ellipse centered at other than the origin, complete the square if necessary and then find the vertices and foci, and graph the ellipse.

When a plane intersects a cone as shown, not perpendicular to the axis of the cone, an ellipse is formed. See the figure on p. 372. We shall prove this later.

Here is a way to draw an ellipse. Stick two tacks in a piece of cardboard. These will be the foci F_1 and F_2. Attach a piece of string to the tacks. The length of the string will be the constant sum of the distances from the foci to points on the ellipse. Take a pencil and pull the string tight. Now swing the pencil around, keeping the string tight.

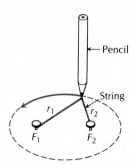

i EQUATIONS OF ELLIPSES

We first consider an equation of an ellipse whose center is at the origin and whose foci lie on one of the coordinate axes.

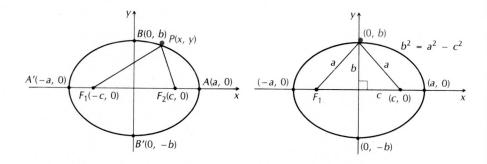

Suppose we have an ellipse with foci $F_1(-c, 0)$ and $F_2(c, 0)$. If $P(x, y)$ is a point on the ellipse, then $F_1P + F_2P$ is the given constant distance. We will call it $2a$:

$$F_1P + F_2P = 2a.$$

By the distance formula,

$$\sqrt{(x + c)^2 + y^2} + \sqrt{(x - c)^2 + y^2} = 2a,$$

or

$$\sqrt{(x + c)^2 + y^2} = 2a - \sqrt{(x - c)^2 + y^2}.$$

Squaring, we get

$$x^2 + 2cx + c^2 + y^2 = 4a^2 - 4a\sqrt{(x - c)^2 + y^2} + x^2 - 2cx + c^2 + y^2,$$

or

$$-4a^2 + 4cx = -4a\sqrt{(x - c)^2 + y^2}$$
$$-a^2 + cx = -a\sqrt{(x - c)^2 + y^2}.$$

Squaring again, we get

$$a^4 - 2a^2cx + c^2x^2 = a^2x^2 - 2cxa^2 + a^2c^2 + a^2y^2,$$

or

$$x^2(a^2 - c^2) + a^2 y^2 = a^2(a^2 - c^2).$$

It follows from when P is at $(0, b)$ that $b^2 = a^2 - c^2$. Substituting b^2 for $a^2 - c^2$ in the last equation, we have the equation of the ellipse $b^2 x^2 + a^2 y^2 = a^2 b^2$, or the following.

$$\frac{x^2}{a^2} + \frac{y^2}{b^2} = 1$$ **Standard form of an equation of an ellipse, center at the origin**

We have proved that if a point is on the ellipse, then its coordinates satisfy this equation. We also need to know the converse, that is, if the coordinates of a point satisfy this equation, then the point is on the ellipse. The proof of the latter will be omitted here. In the above, the longer axis of symmetry $\overline{A'A}$ is called the *major axis*. The shorter axis of symmetry $\overline{B'B}$ is called the *minor axis*. The intersection of these axes is called the *center*. The points A, A', B, and B' are called *vertices*. If the center of an ellipse is at the origin, the vertices are also the intercepts.

Ellipses as Stretched Circles

Let us consider a unit circle centered at the origin:

$$x^2 + y^2 = 1.$$

If we replace x by x/a and y by y/b, we get an equation of an ellipse:

$$\left(\frac{x}{a}\right)^2 + \left(\frac{y}{b}\right)^2 = 1 \quad \text{or} \quad \frac{x^2}{a^2} + \frac{y^2}{b^2} = 1.$$

It follows that an ellipse is a circle transformed by a stretch or shrink in the x-direction and also in the y-direction. If $a = 2$, for example, the unit circle is stretched in the x-direction by a factor of 2. In any case, the x-intercepts become a and $-a$ and the y-intercepts become b and $-b$.

EXAMPLE 1   For the ellipse $x^2 + 16y^2 = 16$, find the vertices and the foci. Then graph the ellipse.

a)   We first multiply by $\frac{1}{16}$ to find standard form:

$$\frac{x^2}{16} + \frac{y^2}{1} = 1 \quad \text{or} \quad \frac{x^2}{4^2} + \frac{y^2}{1^2} = 1.$$

Thus $a = 4$ and $b = 1$. Then two of the vertices are $(-4, 0)$ and $(4, 0)$. These are also x-intercepts. The other vertices are $(0, 1)$ and $(0, -1)$. These are also y-intercepts. Since we know that $c^2 = a^2 - b^2$, we have $c^2 = 16 - 1$, so $c = \sqrt{15}$ and the foci are $(-\sqrt{15}, 0)$ and $(\sqrt{15}, 0)$.

b)   The graph is as follows.

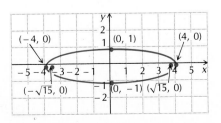

DO EXERCISES 1–4.

---

1. Suppose a circle has been distorted to get the ellipse below.

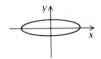

a)   Has the circle been stretched or shrunk in the x-direction?

b)   Has the circle been stretched or shrunk in the y-direction?

For each ellipse find the vertices and the foci, and draw a graph.

2. $x^2 + 9y^2 = 9$

3. $9x^2 + 25y^2 = 225$

4. $2x^2 + 4y^2 = 8$

For each ellipse find the center, the vertices, and the foci, and draw a graph.

5. $9x^2 + y^2 = 9$

6. $25x^2 + 9y^2 = 225$

7. $4x^2 + 2y^2 = 8$

EXAMPLE 2 Graph this ellipse and its foci: $9x^2 + 2y^2 = 18$.

a) We first multiply by $\frac{1}{18}$:

$$\frac{x^2}{2} + \frac{y^2}{9} = 1 \quad \text{or} \quad \frac{x^2}{(\sqrt{2})^2} + \frac{y^2}{3^2} = 1.$$

Thus $a = \sqrt{2}$ and $b = 3$.

b) Since $b > a$, the foci are on the y-axis and the major axis lies along the y-axis. To find c in this case we proceed as follows:

$$c^2 = b^2 - a^2 = 9 - 2 = 7$$
$$c = \sqrt{7}.$$

c) The graph is as follows.

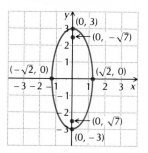

DO EXERCISES 5–7.

ⅱ STANDARD FORM BY COMPLETING THE SQUARE

If the center of an ellipse is not at the origin but at some point (h, k), then the standard form of the equation is as follows.

> $$\frac{(x - h)^2}{a^2} + \frac{(y - k)^2}{b^2} = 1$$
> **Standard form of an equation of an ellipse, center at (h, k)**

EXAMPLE 3 For the ellipse

$$16x^2 + 4y^2 + 96x - 8y + 84 = 0,$$

find the center, vertices, and foci. Then graph the ellipse.

a) We first complete the square to get standard form:

$$16(x^2 + 6x + \quad) + 4(y^2 - 2y + \quad) = -84$$
$$16(x^2 + 6x + 9 - 9) + 4(y^2 - 2y + 1 - 1) = -84$$
$$16(x^2 + 6x + 9) + 4(y^2 - 2y + 1) = -84 + 144 + 4$$
$$16(x + 3)^2 + 4(y - 1)^2 = 64$$

$$\frac{(x + 3)^2}{2^2} + \frac{(y - 1)^2}{4^2} = 1.$$

The center is $(-3, 1)$, $a = 2$, and $b = 4$.

b) The vertices of the ellipse $x^2/2^2 + y^2/4^2 = 1$ are $(2, 0)$, $(-2, 0)$, $(0, 4)$, and $(0, -4)$; and since $c^2 = 16 - 4 = 12$, $c = 2\sqrt{3}$, and its foci are $(0, 2\sqrt{3})$ and $(0, -2\sqrt{3})$.

c) Then the vertices and foci of the translated ellipse are found by translation in the same way in which the center has been translated. Thus the vertices are

$$(-3 + 2, 1), \quad (-3 - 2, 1), \quad (-3, 1 + 4), \quad (-3, 1 - 4),$$

or

$$(-1, 1), \quad (-5, 1), \quad (-3, 5), \quad (-3, -3).$$

The foci are $(-3, 1 + 2\sqrt{3})$ and $(-3, 1 - 2\sqrt{3})$.

d) The graph is as follows.

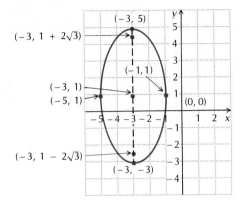

DO EXERCISES 8 AND 9.

APPLICATIONS

Ellipses have many applications. Earth satellites travel in elliptical orbits. The planets travel around the sun in elliptical orbits with the sun at one focus.

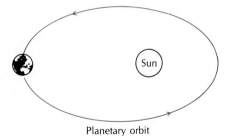

Planetary orbit

An interesting attraction found in museums is the *whispering gallery*. It is elliptical. Persons with their heads at the foci can whisper and hear each other clearly, while persons at other positions cannot hear them. This happens because sound waves emanating from one focus are reflected to the other focus, being concentrated there.

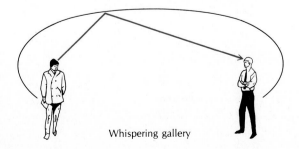

Whispering gallery

For each ellipse find the center, the vertices, and the foci, and graph the ellipse.

8. $25x^2 + 9y^2 + 150x - 36y + 260 = 0$

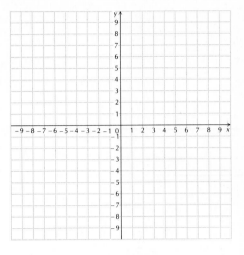

9. $9x^2 + 25y^2 - 36x + 150y + 260 = 0$

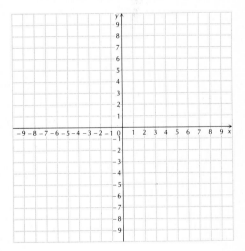

A PROOF

We now prove that a conic section obtained when the plane is not perpendicular to the axis of a cone is actually an ellipse.

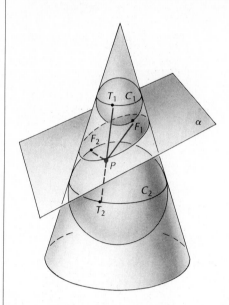

Consider a cone cut by a plane α not perpendicular to the axis. Now consider two spheres tangent to the cones at circles C_1 and C_2 and to plane α at points F_1 and F_2, respectively. Then consider any point P on the intersection of α with the cone. $\overline{PF_2}$ is tangent to the large sphere at F_2, and $\overline{PT_2}$, a segment on an element of the cone, is tangent to the large sphere at T_2. Hence $PF_2 = PT_2$.

Similarly, $\overline{PF_1}$ and $\overline{PT_1}$ are tangent to the small sphere, hence $PF_1 = PT_1$. Thus $PF_1 + PF_2 = PT_1 + PT_2$. But $PT_1 + PT_2$ is constant, being the distance between the circles C_1 and C_2. Therefore, for all points P in the intersection of α with the cone, $PF_1 + PF_2$ is constant. The curve is thus an ellipse.

EXERCISE SET 10.2

i For each ellipse find the vertices and the foci, and draw a graph.

1. $\dfrac{x^2}{4} + \dfrac{y^2}{1} = 1$ **2.** $\dfrac{x^2}{1} + \dfrac{y^2}{4} = 1$ **3.** $16x^2 + 9y^2 = 144$ **4.** $9x^2 + 16y^2 = 144$

5. $2x^2 + 3y^2 = 6$ **6.** $5x^2 + 7y^2 = 35$ **7.** $4x^2 + 9y^2 = 1$ **8.** $25x^2 + 16y^2 = 1$

ii Find the center, vertices, and foci, and draw a graph.

9. $\dfrac{(x-1)^2}{4} + \dfrac{(y-2)^2}{1} = 1$ **10.** $\dfrac{(x-1)^2}{1} + \dfrac{(y-2)^2}{4} = 1$

11. $\dfrac{(x+3)^2}{25} + \dfrac{(y-2)^2}{16} = 1$ **12.** $\dfrac{(x-2)^2}{25} + \dfrac{(y+3)^2}{16} = 1$

13. $3(x+2)^2 + 4(y-1)^2 = 192$ **14.** $4(x-5)^2 + 3(y-5)^2 = 192$

15. $4x^2 + 9y^2 - 16x + 18y - 11 = 0$ **16.** $x^2 + 2y^2 - 10x + 8y + 29 = 0$

17. $4x^2 + y^2 - 8x - 2y + 1 = 0$ **18.** $9x^2 + 4y^2 + 54x - 8y + 49 = 0$

For each ellipse find the center and vertices.

19. ▦ $4x^2 + 9y^2 - 16.025x + 18.0927y - 11.346 = 0$

20. ▦ $9x^2 + 4y^2 + 54.063x - 8.016y + 49.872 = 0$

Find equations of the ellipses with the following vertices. (*Hint:* Graph the vertices.)

21. $(2, 0)$, $(-2, 0)$, $(0, 3)$, $(0, -3)$

22. $(1, 0)$, $(-1, 0)$, $(0, 4)$, $(0, -4)$

23. $(1, 1)$, $(5, 1)$, $(3, 6)$, $(3, -4)$

24. $(-1, -1)$, $(-1, 5)$, $(-3, 2)$, $(1, 2)$

Find equations of the ellipses satisfying the given conditions.

25. Center at $(-2, 3)$ with major axis of length 4 and parallel to the y-axis, minor axis of length 1.

26. Vertices $(3, 0)$ and $(-3, 0)$ and containing the point $(2, \frac{22}{3})$.

27. a) Graph $9x^2 + y^2 = 9$. Is this relation a function?
  b) Solve $9x^2 + y^2 = 9$ for y.
  c) Graph $y = 3\sqrt{1 - x^2}$ and determine whether it is a function. Find the domain and range.
  d) Graph $y = -3\sqrt{1 - x^2}$ and determine whether it is a function. Find the domain and range.

28. Describe the graph of

$$\frac{x^2}{a^2} + \frac{y^2}{b^2} = 1$$

when $a^2 = b^2$.

29. ▦ Draw a large-scale precise graph of

$$\frac{x^2}{25} + \frac{y^2}{16} = 1$$

by calculating and plotting a large number of points.

30. The maximum distance of the earth from the sun is 9.3×10^7 miles. The minimum distance is 9.1×10^7 miles. The sun is at one focus of the elliptical orbit. Find the distance from the sun to the other focus.

31. A point moves so that its x-coordinate is given by $x = a \cos t$ and its y-coordinate is given by $y = b \sin t$, where a and b are constants and t is time. Show that the path of the point is an ellipse.

32. The unit square on the left is transformed to the rectangle on the right by a stretch or shrink in the x-direction and a stretch or shrink in the y-direction.

33. The toy pictured here is called a "vacuum grinder." It consists of a rod hinged to two blocks A and B that slide in perpendicular grooves. One grasps the knob at C and grinds. Determine (and prove) whether or not the path of the handle C is an ellipse.

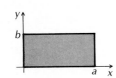

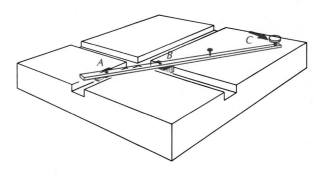

a) Use the above result to develop a formula for the area of the ellipse

$$\frac{x^2}{a^2} + \frac{y^2}{b^2} = 1.$$

(*Hint:* The area of the circle $x^2 + y^2 = r^2$ is $\pi \cdot r \cdot r$.)

b) Use the result of (a) to find the area of the ellipse

$$\frac{x^2}{16} + \frac{y^2}{25} = 1.$$

c) Use the result of (a) to find the area of the ellipse

$$\frac{x^2}{4} + \frac{y^2}{3} = 1.$$

OBJECTIVES

You should be able to:

i Given an equation of a hyperbola with center at the origin, find the vertices, foci, and asymptotes, and graph.

ii Given an equation of a hyperbola with center not at the origin, complete the square if necessary, and then find the center, vertices, foci, and asymptotes, and graph.

iii Graph hyperbolas having an equation $xy = c$.

10.3

HYPERBOLAS

Some equations of second degree have graphs that are hyperbolas. Hyperbolas are defined as follows.

Definition

> A *hyperbola* is a locus or set of all points P in a plane such that the absolute value of the difference of the distances from P to two fixed points F_1 and F_2 in the plane is constant. The points F_1 and F_2 are called *foci* (singular, *focus*) and the midpoint of the segment joining them is called the *center*.

When a plane intersects a cone as shown, parallel to the axis of the cone, a hyperbola is formed. See p. 372. Note that a hyperbola has two parts. These are called *branches*.

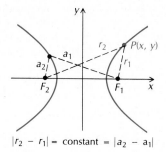

$|r_2 - r_1| = \text{constant} = |a_2 - a_1|$

i EQUATIONS OF HYPERBOLAS

Now let us find equations for hyperbolas. We first consider an equation of a hyperbola whose center is at the origin and whose foci lie on one of the coordinate axes. Suppose we have a hyperbola as shown with foci $F_1(c, 0)$ and $F_2(-c, 0)$ on the x-axis. We consider a point $P(x, y)$ in the first quadrant. The proof for the other quadrants is similar to the proof that follows.

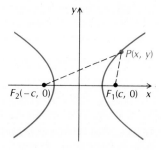

We know that $PF_2 > PF_1$, so $PF_2 - PF_1 > 0$ and $|PF_2 - PF_1| = PF_2 - PF_1$. Let the constant difference be $2a$. Then

$$|PF_2 - PF_1| = PF_2 - PF_1 = 2a.$$

In the triangle F_2PF_1, $PF_2 - PF_1 < F_1F_2$, or $2a < 2c$; therefore $a < c$. Using the distance formula, we have

$$\sqrt{(x + c)^2 + y^2} - \sqrt{(x - c)^2 + y^2} = 2a,$$

or

$$\sqrt{(x+c)^2 + y^2} = 2a + \sqrt{(x-c)^2 + y^2}.$$

Squaring, we get

$$x^2 + 2xc + c^2 + y^2 = 4a^2 + 4a\sqrt{(x-c)^2 + y^2} + x^2 - 2xc + c^2 + y^2,$$

which simplifies to

$$4cx - 4a^2 = 4a\sqrt{(x-c)^2 + y^2},$$

or

$$cx - a^2 = a\sqrt{(x-c)^2 + y^2}.$$

Squaring again, we get

$$c^2 x^2 - 2a^2 cx + a^4 = a^2 x^2 - 2a^2 cx + a^2 c^2 + a^2 y^2,$$

or

$$x^2(c^2 - a^2) - a^2 y^2 = a^2(c^2 - a^2).$$

Since $c > a$, $c^2 > a^2$, so $c^2 - a^2$ is positive. We represent $c^2 - a^2$ by b^2. The previous equation then becomes

$$x^2 b^2 - a^2 y^2 = a^2 b^2,$$

or the following.

$$\frac{x^2}{a^2} - \frac{y^2}{b^2} = 1 \qquad \text{**Standard equation of a hyperbola, center at the origin, foci on the x-axis**}$$

We have shown that if a point is on the hyperbola, it satisfies this equation. We also need to know the converse: If a point satisfies the equation, then it is on the hyperbola. We omit the proof.

The following figure is a hyperbola.

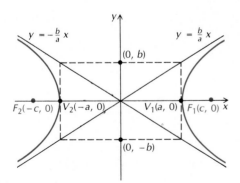

Points $V_1(a, 0)$ and $V_2(-a, 0)$ are called the *vertices*, and the line segment $\overline{V_1 V_2}$ is called the *transverse axis*. The line segment from $(0, b)$ to $(0, -b)$ is called the *conjugate axis*. Note that when we replace either or both of x by $-x$ and/or y by $-y$, we get an equivalent equation. Thus the hyperbola is symmetric with respect to the origin, and the x- and y-axes are lines of symmetry.

The lines $y = (b/a)x$ and $y = -(b/a)x$ are *asymptotes*. They have slopes b/a and $-b/a$.

For each hyperbola find the vertices, the foci, and the asymptotes. Then draw a graph.

1. $4x^2 - 9y^2 = 36$

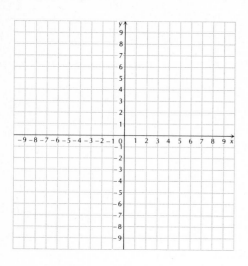

2. $x^2 - y^2 = 16$

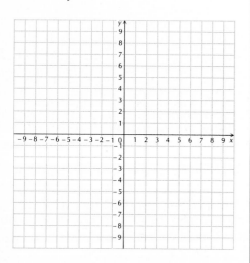

EXAMPLE 1 For the hyperbola $9x^2 - 16y^2 = 144$, find the vertices, the foci, and the asymptotes. Then graph the hyperbola.

a) We first multiply by $\frac{1}{144}$ to find the standard form:

$$\frac{x^2}{16} - \frac{y^2}{9} = 1.$$

Thus $a = 4$ and $b = 3$. The vertices are $(4, 0)$ and $(-4, 0)$. Since $b^2 = c^2 - a^2$, $c = \sqrt{a^2 + b^2} = \sqrt{4^2 + 3^2} = 5$. Thus the foci are $(5, 0)$ and $(-5, 0)$. The asymptotes are $y = \frac{3}{4}x$ and $y = -\frac{3}{4}x$.

b) To graph the hyperbola it is helpful to first graph the asymptotes. An easy way to do this is to draw the rectangle shown below. Then draw the branches of the hyperbola outward from the vertices toward the asymptotes.

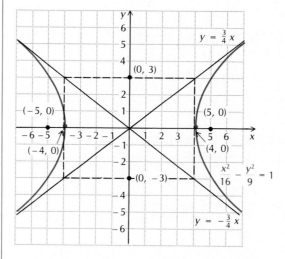

Why are $y = (b/a)x$ and $y = -(b/a)x$ asymptotes? To answer this, we solve $x^2/16 - y^2/9 = 1$ for y^2:

$$-16y^2 = 144 - 9x^2$$
$$16y^2 = 9x^2 - 144$$
$$y^2 = \frac{1}{16}(9x^2 - 144) = \frac{9x^2 - 144}{16}.$$

From this last equation, we see that as $|x|$ gets larger, the term -144 is very small compared to $9x^2$, so y^2 gets close to $9x^2/16$. That is, when $|x|$ is large,

$$y^2 \approx \frac{9x^2}{16},$$

so $\qquad\qquad y \approx |\tfrac{3}{4}x| \quad \text{or} \quad y \approx \pm\tfrac{3}{4}x.$

Thus the lines $y = \frac{3}{4}x$ and $y = -\frac{3}{4}x$ are asymptotes.

DO EXERCISES 1 AND 2.

iii ASYMPTOTES ON THE COORDINATE AXES

If a hyperbola has its center at the origin and its axes at $45°$ to the coordinate axes, it has a simple equation. The coordinate axes are its asymptotes.

$xy = c$, c a nonzero constant	Hyperbola, asymptotes the coordinate axes

If c is positive, the branches of the hyperbola lie in the first and third quadrants. If c is negative, the branches lie in the second and fourth quadrants. In either case, the asymptotes are the x-axis and the y-axis.

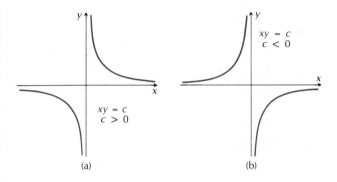

(a)   (b)

DO EXERCISES 7 AND 8.

APPLICATIONS

Hyperbolas also have many applications. A jet breaking the sound barrier creates a sonic boom whose wave front has the shape of a cone. The cone intersects the ground in one branch of a hyperbola.

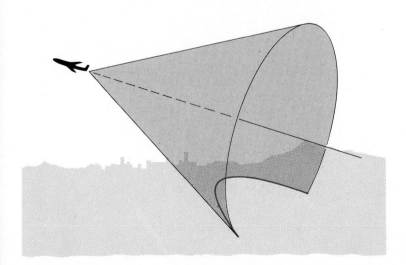

Some comets travel in hyperbolic orbits. A cross section of an amphitheater may be half of one branch of a hyperbola.

Graph.

7. $xy = 3$

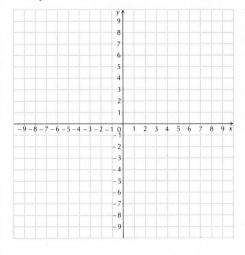

8. $xy = -12$

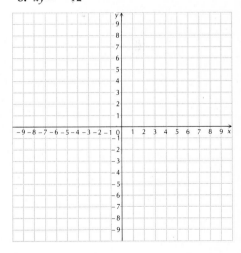

EXERCISE SET 10.3

i For each hyperbola find the center, the vertices, the foci, and the asymptotes, and graph the hyperbola.

1. $\dfrac{x^2}{9} - \dfrac{y^2}{1} = 1$ **2.** $\dfrac{x^2}{1} - \dfrac{y^2}{9} = 1$ **3.** $\dfrac{(x-2)^2}{9} - \dfrac{(y+5)^2}{1} = 1$ **4.** $\dfrac{(x-2)^2}{1} - \dfrac{(y+5)^2}{9} = 1$

5. $\dfrac{(y+3)^2}{4} - \dfrac{(x+1)^2}{16} = 1$ **6.** $\dfrac{(y+3)^2}{25} - \dfrac{(x+1)^2}{16} = 1$ **7.** $x^2 - 4y^2 = 4$ **8.** $4x^2 - y^2 = 4$

9. $4y^2 - x^2 = 4$ **10.** $y^2 - 4x^2 = 4$ **11.** $x^2 - y^2 = 2$ **12.** $x^2 - y^2 = 3$

13. $x^2 - y^2 = \frac{1}{4}$ **14.** $x^2 - y^2 = \frac{1}{9}$

ii

15. $x^2 - y^2 - 2x - 4y - 4 = 0$ **16.** $4x^2 - y^2 + 8x - 4y - 4 = 0$

17. $36x^2 - y^2 - 24x + 6y - 41 = 0$ **18.** $9x^2 - 4y^2 + 54x + 8y + 45 = 0$

iii Graph.

19. $xy = 1$ **20.** $xy = -4$ **21.** $xy = -8$ **22.** $xy = 3$

23. Find the center, the vertices, and the asymptotes.

$$x^2 - y^2 - 2.046x - 4.088y - 4.228 = 0$$

Find an equation of a hyperbola having:

24. Vertices at $(1, 0)$ and $(-1, 0)$; and foci at $(2, 0)$ and $(-2, 0)$.

25. Asymptotes $y = \frac{3}{2}x$ and $y = -\frac{3}{2}x$ and one vertex $(2, 0)$.

26. a) Graph $x^2 - 4y^2 = 4$. Is this relation a function?
  b) Solve $x^2 - 4y^2 = 4$ for y.
  c) Graph $y = \frac{1}{2}\sqrt{x^2 - 4}$ and determine whether it is a function. Find the domain and range.
  d) Graph $y = \frac{1}{2}\sqrt{x^2 - 4}$ and determine whether it is a function. Find the domain and range.

27. Show that the equation

$$\begin{vmatrix} \dfrac{x-h}{a} & \dfrac{y-k}{b} \\[2mm] \dfrac{y-k}{b} & \dfrac{x-h}{a} \end{vmatrix} = 1$$

is an equation of a hyperbola with center (h, k).

28. A rifle at A fires a bullet, which hits a target at B. A person at C hears the sound of the rifle shot and the sound of the bullet hitting the target simultaneously. Describe the set of all such points C.

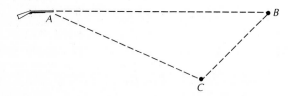

29. In a navigation system called Loran, a radio transmitter at M (the *master* station) sends out pulses. Each pulse triggers another transmitter at S (the *slave* station), which then also transmits a pulse. A ship or airplane at A receives pulses from both M and S and a device measures the difference in the time at which they arrive at A. Knowing this difference in time, the navigator can locate the vessel as being somewhere along a curve predrawn on a chart. What is the shape of that curve?

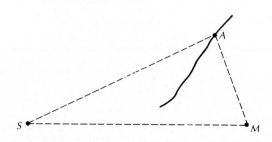

10.4

PARABOLAS

Some equations of second degree have graphs that are parabolas. Parabolas are defined as follows.

Definition

A *parabola* is a locus or set of all points *P* in a plane equidistant from a fixed line and a fixed point in the plane. The fixed line is called the *directrix* and the fixed point is called the *focus*.

When a plane intersects a cone parallel to an element of the cone, as shown in p. 372, a parabola is formed.

i EQUATIONS OF PARABOLAS

Now let us find equations for parabolas. Given the focus *F* and the directrix *l*, we place coordinate axes as shown. The *y*-axis contains *F* and is perpendicular to *l*. The *x*-axis is halfway between *F* and *l*. We shall call the distance from *F* to the *x*-axis *p*. Then *F* has coordinates $(0, p)$ and *l* has the equation $y = -p$.

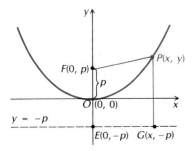

Let $P(x, y)$ be any point of the parabola and consider $\overline{PG}$ perpendicular to the line $y = -p$. The coordinates of *G* are $(x, -p)$. By definition of a parabola,

$$PF = PG.$$

Then using the distance formula, we have

$$\sqrt{(x - 0)^2 + (y - p)^2} = \sqrt{(x - x)^2 + (y + p)^2}.$$

Squaring, we get

$$x^2 + y^2 - 2py + p^2 = y^2 + 2py + p^2$$

$$x^2 = 4py.$$

Thus we have the following.

$$x^2 = 4py$$

Standard equation of a parabola with focus at $(0, p)$ and directrix $y = -p$. The vertex is $(0, 0)$ and the *y*-axis is the only line of symmetry.

OBJECTIVES

You should be able to:

i Given an equation of a parabola, find the vertex, focus, and directrix, and graph the parabola.

ii Given the focus and directrix of a parabola, find an equation of the parabola.

iii Find standard form of equations of parabolas by completing the square if necessary, and graph the parabolas.

For each parabola find the vertex, the focus, and the directrix, and draw a graph.

1. $8y = x^2$

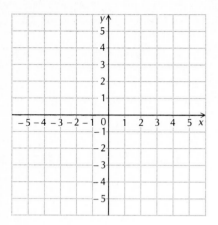

2. $y = 2x^2$

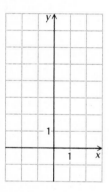

3. $y^2 = -6x$

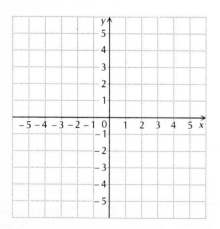

We have shown that if $P(x, y)$ is on the parabola, then its coordinates satisfy this equation. The converse is also true, but we omit the proof.

Note that if $p > 0$, as above, the graph opens upward. If $p < 0$, the graph opens downward and the focus and directrix exchange sides of the x-axis.

The inverse of the above parabola is described as follows.

$$y^2 = 4px$$

Standard equation of a parabola with focus at $(p, 0)$ and directrix $x = -p$. The vertex is $(0, 0)$ and the x-axis is the only line of symmetry.

EXAMPLE 1 For the parabola $y = x^2$, find the vertex, the focus, and the directrix, and draw a graph.

We first write

$$x^2 = 4py$$
$$x^2 = 4(\tfrac{1}{4})y.$$

Vertex: $(0, 0)$

Focus: $(0, \tfrac{1}{4})$

Directrix: $y = -\tfrac{1}{4}$

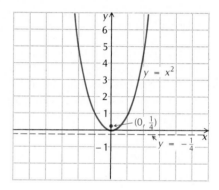

EXAMPLE 2 For the parabola $y^2 = -12x$, find the vertex, the focus, and the directrix, and draw a graph.

We first write

$$y^2 = 4px$$
$$y^2 = 4(-3)x.$$

Vertex: $(0, 0)$

Focus: $(-3, 0)$

Directrix: $x = -(-3)$
$\qquad\quad = 3$

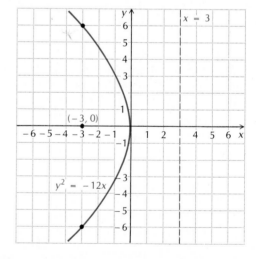

DO EXERCISES 1–3.

ii Focus and Directrix Known

EXAMPLE 3 Find an equation of a parabola with focus $(5, 0)$ and directrix $x = -5$.

The focus is on the x-axis and $x = -5$ is the directrix, so the line of symmetry is the x-axis. Thus the equation is of the type

$$y^2 = 4px.$$

Since $p = 5$, the equation is $y^2 = 20x$.

EXAMPLE 4 Find an equation of a parabola with focus $(0, -7)$ and directrix $y = 7$.

The focus is on the y-axis and $y = 7$ is the directrix, so the line of symmetry is the y-axis. Thus the equation is of the type

$$x^2 = 4py.$$

Since $p = -7$, we obtain $x^2 = -28y$.

DO EXERCISES 4–7.

iii STANDARD FORM BY COMPLETING THE SQUARE

If a parabola is translated so that its vertex is (h, k) and its axis of symmetry is parallel to the y-axis, it has an equation as follows:

$$(x - h)^2 = 4p(y - k),$$

where the vertex is (h, k), the focus is $(h, k + p)$, and the directrix is $y = k - p$.

If a parabola is translated so that its vertex is (h, k) and its axis of symmetry is parallel to the x-axis, it has an equation as follows:

$$(y - k)^2 = 4p(x - h),$$

where the vertex is (h, k), the focus is $(h + p, k)$, and the directrix is $x = h - p$.

EXAMPLE 5 For the parabola

$$x^2 + 6x + 4y + 5 = 0,$$

find the vertex, the focus, and the directrix, and graph the parabola.

We complete the square:

$$\begin{aligned} x^2 + 6x \qquad\quad &= -4y - 5 \\ x^2 + 6x + 9 - 9 &= -4y - 5 \\ x^2 + 6x + 9 &= -4y + 4 \\ (x + 3)^2 &= -4y + 4 = 4(-1)(y - 1). \end{aligned}$$

Vertex: $(-3, 1)$

Focus: $(-3, 1 + (-1))$ or $(-3, 0)$

Directrix: $y = 2$

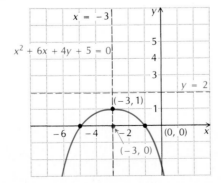

Find an equation of a parabola satisfying the given conditions.

4. Focus $(3, 0)$, directrix $x = -3$

5. Focus $(0, \frac{1}{2})$, directrix $y = -\frac{1}{2}$

6. Focus $(-6, 0)$, directrix $x = 6$

7. Focus $(0, -1)$, directrix $y = 1$

For each parabola find the vertex, the focus, and the directrix, and graph the parabola.

8. $x^2 + 2x - 8y - 3 = 0$

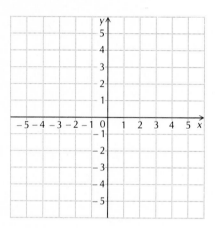

9. $y^2 + 2y + 4x - 7 = 0$

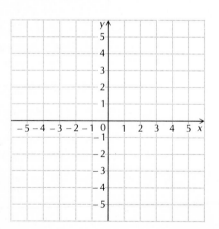

EXAMPLE 6   For the parabola

$$y^2 + 6y - 8x - 31 = 0,$$

find the vertex, the focus, and the directrix, and draw a graph.

We complete the square:

$$
\begin{aligned}
y^2 + 6y &= 8x + 31 \\
y^2 + 6y + 9 - 9 &= 8x + 31 \\
y^2 + 6y + 9 &= 8x + 40 \\
(y + 3)^2 &= 8x + 40 = 8(x + 5) = 4(2)(x + 5).
\end{aligned}
$$

Vertex: $(-5, -3)$

Focus: $(-5 + 2, -3)$ or
$\quad\quad (-3, -3)$

Directrix: $x = -5 - 2 = -7$

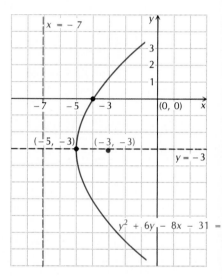

DO EXERCISES 8 AND 9.

APPLICATIONS

Parabolas have many applications. Cross sections of headlights are parabolas. The bulb is located at the focus. All light from that point is reflected outward, parallel to the axis of symmetry.

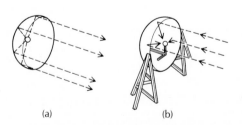

(a)   (b)

Radar and ratio antennas may have cross sections that are parabolas. Incoming radio waves are reflected and concentrated at the focus. Cables hung between structures to form suspension bridges form parabolas. When a cable supports only its own weight, it does not form a parabola, but rather a curve called a *catenary*.

EXERCISE SET 10.4

i For each parabola find the vertex, the focus, and the directrix, and graph the parabola.

1. $x^2 = 8y$

2. $x^2 = 16y$

3. $y^2 = -6x$

4. $y^2 = -2x$

5. $x^2 - 4y = 0$

6. $y^2 + 4x = 0$

7. $y = 2x^2$

8. $y = \frac{1}{2}x^2$

ii Find an equation of a parabola satisfying the given conditions.

9. Focus $(4, 0)$, directrix $x = -4$

10. Focus $(0, \frac{1}{4})$, directrix $y = -\frac{1}{4}$

11. Focus $(-\sqrt{2}, 0)$, directrix $x = \sqrt{2}$

12. Focus $(0, -\pi)$, directrix $y = \pi$

13. Focus $(3, 2)$, directrix $x = -4$

14. Focus $(-2, 3)$, directrix $y = -3$

iii Find the vertex, focus, and directrix, and graph.

15. $(x + 2)^2 = -6(y - 1)$

16. $(y - 3)^2 = -20(x + 2)$

17. $x^2 + 2x + 2y + 7 = 0$

18. $y^2 + 6y - x + 16 = 0$

19. $x^2 - y - 2 = 0$

20. $x^2 - 4x - 2y = 0$

21. $y = x^2 + 4x + 3$

22. $y = x^2 + 6x + 10$

23. $4y^2 - 4y - 4x + 24 = 0$

24. $4y^2 + 4y - 4x - 16 = 0$

For each parabola find the vertex, the focus, and the directrix.

25. ▦ $x^2 - 8056.25y$

26. ▦ $y^2 = -7645.88x$

27. Graph each of the following using the same set of axes.
$$x^2 - y^2 = 0, \qquad x^2 - y^2 = 1,$$
$$x^2 + y^2 = 1, \qquad y = x^2.$$

28. Graph each of the following using the same set of axes.
$$x^2 - 4y^2 = 0, \qquad x^2 - 4y^2 = 1,$$
$$x^2 + 4y^2 = 1, \qquad x = 4y^2.$$

29. Find equations of the following parabola: Line of symmetry parallel to the y-axis, vertex $(-1, 2)$, and passing through $(-3, 1)$.

30. a) Graph $(y - 3)^2 = -20(x + 1)$. Is this relation a function?
b) In general, is $(y - k)^2 = 4p(x - h)$ a function?

31. Show that the equation
$$\begin{vmatrix} y - k & x - h \\ 4p & y - k \end{vmatrix} = 0$$
is an equation of a parabola with vertex (h, k).

32. The cables of a suspension bridge are 50 ft above the roadbed at the ends of the bridge and 10 ft above it in the center of the bridge. The roadbed is 200 ft long. Vertical cables are to be spaced every 20 ft along the bridge. Calculate the lengths of these vertical cables.

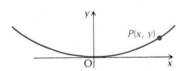

33. Prove that when a cable supports a load distributed uniformly horizontally, it hangs in the shape of a parabola. (*Hint:* Proceed as follows.)

a) Place a coordinate system as shown to the right, with the origin at the lowest point of the cable.
b) For a point $P(x, y)$ on the cable, write an equation of rotational equilibrium. The forces involved are the tensions in the cable, F_1 and F_2, and the weight supported, W (which is a function of x). The weight of the cable is essentially neglected. Use point P as the center of rotation.
c) Solve for y.

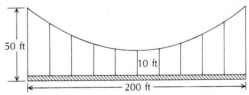

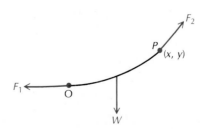

OBJECTIVES

You should be able to:

i Solve a system of one first-degree and one second-degree equation graphically or using the substitution method.

ii Solve applied problems involving the solution of one first-degree and one second-degree equation.

Solve these systems graphically.

1. $x^2 + y^2 = 25, \quad y - x = -1$

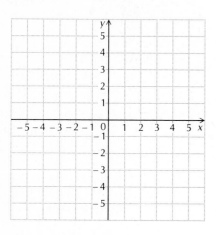

2. $y = x^2 - 2x - 1, \quad y = x + 3$

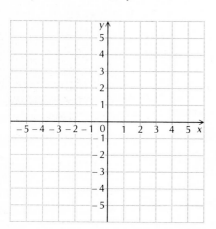

10.5

SYSTEMS OF FIRST-DEGREE AND SECOND-DEGREE EQUATIONS

When we studied systems of linear equations, we solved them both graphically and algebraically. Here we study systems in which one equation is of first degree and one is of second degree. We will use graphical and then algebraic methods of solving.

i **SOLVING GRAPHICALLY**

We consider a system of equations, an equation of a circle and an equation of a line. Let us think about the possible ways in which a circle and a line can intersect. The three possibilities are shown in the figure. For L_1 there is no point of intersection, hence the system of equations has no real solution. For L_2 there is one point of intersection, hence one real solution. For L_3 there are two points of intersection, hence two real solutions.

EXAMPLE 1 Solve this system graphically:

$$x^2 + y^2 = 25,$$
$$3x - 4y = 0.$$

We graph the two equations, using the same axes. The points of intersection have coordinates that must satisfy both equations. The solutions seem to be $(4, 3)$ and $(-4, -3)$. We check.

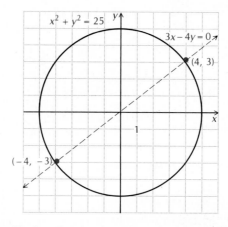

Check: For $(-4, -3)$. You should do the check for $(4, 3)$.

$x^2 + y^2 = 25$		$3x - 4y = 0$	
$(-4)^2 + (-3)^2$	25	$3(-4) - 4(-3)$	0
$16 + 9$		$-12 + 12$	
25		0	

DO EXERCISES 1–3. (NOTE THAT EXERCISE 3 IS ON THE FOLLOWING PAGE.)

SOLVING BY SUBSTITUTION

Remember that we used the *addition* and *substitution* methods to solve systems of linear equations. In solving systems where one equation is of first degree and one is of second degree, it is preferable to use the *substitution* method.

EXAMPLE 2   Solve this system algebraically:

$$x^2 + y^2 = 25, \tag{1}$$

$$3x - 4y = 0. \tag{2}$$

First solve the linear equation (2) for x:

$$x = \tfrac{4}{3}y.$$

Then substitute $\tfrac{4}{3}y$ for x in Eq. (1) and solve for y:

$$(\tfrac{4}{3}y)^2 + y^2 = 25$$
$$\tfrac{16}{9}y^2 + y^2 = 25$$
$$\tfrac{16}{9}y^2 + \tfrac{9}{9}y^2 = 25$$
$$\tfrac{25}{9}y^2 = 25$$
$$\tfrac{9}{25} \cdot \tfrac{25}{9}y^2 = \tfrac{9}{25} \cdot 25$$
$$y^2 = 9$$
$$y = \pm 3.$$

Now substitute these numbers into the linear equation and solve for x:

$$x = \tfrac{4}{3}(3), \quad \text{or } 4;$$
$$x = \tfrac{4}{3}(-3), \quad \text{or } -4.$$

The pairs $(4, 3)$ and $(-4, -3)$ check, hence are solutions.

DO EXERCISES 4–6.

Sometimes an equation may not take on a familiar form like those studied previously in this chapter. We can still rely on the substitution method for solution.

EXAMPLE 3   Solve the system

$$y + 3 = 2x, \tag{1}$$
$$x^2 + 2xy = -1. \tag{2}$$

First solve the linear equation (1) for y:

$$y = 2x - 3.$$

Then substitute $2x - 3$ for y in Eq. (2) and solve for x:

$$x^2 + 2x(2x - 3) = -1$$
$$x^2 + 4x^2 - 6x = -1$$
$$5x^2 - 6x + 1 = 0$$
$$(5x - 1)(x - 1) = 0$$
$$5x - 1 = 0 \quad \text{or} \quad x - 1 = 0$$
$$x = \tfrac{1}{5} \quad \text{or} \qquad x = 1.$$

3. $y = \dfrac{x^2}{4}, \quad x + 2y = 4$

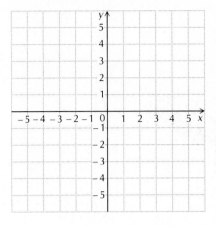

4. Solve the system in Exercise 1 algebraically.

5. Solve the system in Exercise 2 algebraically.

6. Solve the system in Exercise 3 algebraically.

7. Solve the system.

$$y + 3x = 1,$$
$$x^2 - 2xy = 5$$

Now substitute these numbers into the linear equation and solve for y:

$$y = 2(\tfrac{1}{5}) - 3, \quad \text{or} \ -\tfrac{13}{5};$$
$$y = 2(1) - 3, \quad \text{or} \ -1.$$

The pairs $(\tfrac{1}{5}, -\tfrac{13}{5})$ and $(1, -1)$ check and are solutions.

DO EXERCISE 7.

ii APPLIED PROBLEMS

EXAMPLE 4   The perimeter of a rectangular field is 204 m and the area is 2565 m^2. Find the dimensions of the field.

We first translate the conditions of the problem to equations, using w for the width and l for the length:

8. The difference of two numbers is 4 and the difference of their squares is 72. What are the numbers?

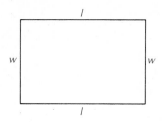

9. The perimeter of a rectangular field is 34 ft and the length of a diagonal is 13 ft. Find the dimensions of the field.

Perimeter: $2w + 2l = 204,$
Area: $lw = 2565.$

Now we solve the system

$$2w + 2l = 204,$$
$$lw = 2565$$

and get the solution $(45, 57)$. Now check in the original problem: The perimeter is $2 \cdot 45 + 2 \cdot 57$, or 204. The area is $45 \cdot 57$, or 2565. The numbers check, so the answer is $l = 57$ m, $w = 45$ m.

DO EXERCISES 8 AND 9.

EXERCISE SET 10.5

i Solve each system graphically. Then solve algebraically.

1. $x^2 + y^2 = 25,$
$y - x = 1$

2. $x^2 + y^2 = 100,$
$y - x = 2$

3. $y^2 - x^2 = 9,$
$2x - 3 = y$

4. $x + y = -6,$
$xy = -7$

5. $4x^2 + 9y^2 = 36,$
$3y + 2x = 6$

6. $9x^2 + 4y^2 = 36,$
$3x + 2y = 6$

7. $y^2 = x + 3,$
$2y = x + 4$

8. $y = x^2,$
$3x = y + 2$

Solve.

9. $x^2 + 4y^2 = 25,$
$x + 2y = 7$

10. $y^2 - x^2 = 16,$
$2x - y = 1$

11. $x^2 - xy + 3y^2 = 27,$
$x - y = 2$

12. $2y^2 + xy + x^2 = 7,$
$x - 2y = 5$

13. $3x + y = 7,$
$4x^2 + 5y = 56$

14. $2y^2 + xy = 5,$
$4y + x = 7$

ii Applied problems

15. The sum of two numbers is 14 and the sum of their squares is 106. What are the numbers?

16. The sum of two numbers is 15 and the difference of their squares is also 15. What are the numbers?

17. A rectangle has perimeter 28 cm and the length of a diagonal is 10 cm. What are its dimensions?

18. A rectangle has perimeter 6 m and the length of a diagonal $\sqrt{5}$ m. What are its dimensions?

19. A rectangle has area 20 in^2 and perimeter 18 in. Find its dimensions.

20. A rectangle has area 2 yd^2 and perimeter 6 yd. Find its dimensions.

i Solve.

21. ▦ $x^2 + y^2 = 19{,}380{,}510.36,$
$27{,}942.25x - 6.125y = 0$

22. ▦ $2x + 2y = 1660,$
$xy = 35{,}325$

23. Given the area A and the perimeter P of a rectangle, show that the length L and the width W are given by the formulas

$$L = \frac{1}{4}\left(P + \sqrt{P^2 - 16A}\right),$$

$$W = \frac{1}{4}\left(P - \sqrt{P^2 - 16A}\right).$$

24. Show that a hyperbola does not intersect its asymptotes. That is, solve the system

$$\frac{x^2}{a^2} - \frac{y^2}{b^2} = 1,$$

$$y = \frac{b}{a}x \quad \left(\text{or, } y = -\frac{b}{a}x\right).$$

25. Find an equation of a circle that passes through the points $(2, 4)$ and $(3, 3)$ and whose center is on the line $3x - y = 3$.

26. Find an equation of a circle that passes through the points $(7, 3)$ and $(5, 5)$ and whose center is on the line $y - 4x = 1$.

10.6

SYSTEMS OF SECOND-DEGREE EQUATIONS

We now consider systems of two second-degree equations. The following figure shows ways in which a circle and a hyperbola can intersect.

4 real solutions   3 real solutions   2 real solutions   1 real solution   0 real solutions

i SOLVING GRAPHICALLY

EXAMPLE 1 Solve this system graphically:

$$x^2 + y^2 = 25,$$

$$\frac{x^2}{25} - \frac{y^2}{25} = 1.$$

We graph the two equations using the same axes. The points of intersection have coordinates that must satisfy both equations; the solutions seem to be $(5, 0)$ and $(-5, 0)$.

Solve these systems graphically.

1. $x^2 + y^2 = 4$,

$\dfrac{x^2}{4} - \dfrac{y^2}{4} = 1$

2. $x^2 + y^2 = 16$,

$\dfrac{x^2}{16} + \dfrac{y^2}{9} = 1$

3. $x^2 + y^2 = 4$,

$\dfrac{x^2}{25} + \dfrac{y^2}{4} = 1$

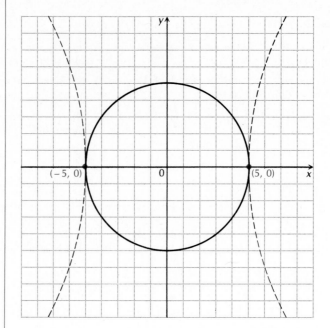

Check: Since $(5)^2 = 25$ and $(-5)^2 = 25$, we can do both checks at once.

$$
\begin{array}{c|c}
x^2 + y^2 = 25 & \\
\hline
(\pm 5)^2 + 0^2 & 25 \\
25 + 0 & \\
25 &
\end{array}
\qquad
\begin{array}{c|c}
\dfrac{x^2}{25} - \dfrac{y^2}{25} = 1 & \\
\hline
\dfrac{(\pm 5)^2}{25} - \dfrac{0^2}{25} & 1 \\[2mm]
\dfrac{25}{25} - 0 & \\[2mm]
1 &
\end{array}
$$

DO EXERCISES 1–3.

ii   THE ADDITION METHOD

To solve systems of two second-degree equations, we can use either the substitution or the addition method.

EXAMPLE 2   Solve this system:

$$2x^2 + 5y^2 = 22, \tag{1}$$
$$3x^2 - \ y^2 = -1. \tag{2}$$

Here we use the addition method:

$$2x^2 + 5y^2 = 22.$$

From the second equation, we get

$$
\begin{array}{ll}
\underline{15x^2 - 5y^2 = -5} & \text{Multiplying by 5} \\
17x^2 = 17 & \text{Adding} \\
x^2 = 1 & \\
x = \pm 1. &
\end{array}
$$

If $x = 1$, $x^2 = 1$, and if $x = -1$, $x^2 = 1$, so substituting 1 or -1 for x in Eq. (2) we have

$$3 \cdot 1^2 - y^2 = -1$$
$$y^2 = 4$$
$$y = \pm 2.$$

Thus, if $x = 1$, $y = 2$ or $y = -2$, and if $x = -1$, $y = 2$ or $y = -2$. The possible solutions are $(1, 2)$, $(1, -2)$, $(-1, 2)$, and $(-1, -2)$.

Check: Since $(2)^2 = 4$, $(-2)^2 = 4$, $(1)^2 = 1$, and $(-1)^2 = 1$, we can check all four pairs at one time.

$$\frac{2x^2 + 5y^2 = 22}{\begin{array}{c|c} 2(\pm 1)^2 + 5(\pm 2)^2 & 22 \\ 2 + 20 & \\ 22 & \end{array}} \qquad \frac{3x^2 - y^2 = -1}{\begin{array}{c|c} 3(\pm 1)^2 - (\pm 2)^2 & -1 \\ 3 - 4 & \\ -1 & \end{array}}$$

DO EXERCISE 4.

iii THE SUBSTITUTION METHOD

EXAMPLE 3   Solve the system

$$x^2 + 4y^2 = 20, \qquad (1)$$
$$xy = 4. \qquad (2)$$

Here we use the substitution method. First solve Eq. (2) for y:

$$y = \frac{4}{x}.$$

Then substitute $4/x$ for y in Eq. (1) and solve for x:

$$x^2 + 4\left(\frac{4}{x}\right)^2 = 20$$

$$x^2 + \frac{64}{x^2} = 20$$

$$x^4 + 64 = 20x^2 \qquad \text{Multiplying by } x^2$$
$$x^4 - 20x^2 + 64 = 0$$
$$u^2 - 20u + 64 = 0 \qquad \text{Letting } u = x^2$$
$$(u - 16)(u - 4) = 0.$$

Then $x^2 = 4$ or $x^2 = -4$ or $x = 2$ or $x = -2$. Since $y = 4/x$, if $x = 4$, $y = 1$; if $x = -4$, $y = -1$; if $x = 2$, $y = 2$; if $x = -2$, $y = -2$. The solutions are $(4, 1)$, $(-4, -1)$, $(2, 2)$, and $(-2, -2)$.

DO EXERCISE 5.

iv AN APPLIED PROBLEM

EXAMPLE 4   The area of a rectangle is 300 yd^2 and the length of a diagonal is 25 yd. Find the dimensions.

4. Solve this system.

$$2y^2 - 3x^2 = 6,$$
$$5y^2 + 2x^2 = 53$$

5. Solve this system.

$$x^2 + xy + y^2 = 19,$$
$$xy = 6$$

6. The area of a rectangle is 2 ft^2 and the length of a diagonal is $\sqrt{5}$ ft. Find the dimensions of the rectangle.

First make a drawing.

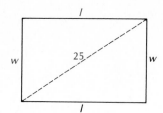

We use l for the length and w for the width and translate to equations:

From the Pythagorean theorem: $\quad l^2 + w^2 = 25^2,$

$$\text{Area:} \qquad lw = 300.$$

Now we *solve* the system

$$l^2 + w^2 = 625,$$
$$lw = 300$$

and get the solutions $(15, 20)$ and $(-15, -20)$. Then we check in the original problem: $15^2 + 20^2 = 25^2$ and $15 \cdot 20 = 300$, so $(15, 20)$ is a solution of the problem. Lengths of sides cannot be negative so $(-15, -20)$ is not a solution. The answer is $l = 20$ yd and $w = 15$ yd.

DO EXERCISE 6.

EXERCISE SET 10.6

i , **ii** , **iii** Solve each system graphically. Then solve algebraically.

1. $x^2 + y^2 = 25,$
$\quad y^2 = x + 5$

2. $y = x^2,$
$\quad x = y^2$

3. $x^2 + y^2 = 9,$
$\quad x^2 - y^2 = 9$

4. $y^2 - 4x^2 = 4,$
$\quad 4x^2 + y^2 = 4$

5. $x^2 + y^2 = 25,$
$\quad xy = 12$

6. $x^2 - y^2 = 16,$
$\quad x + y^2 = 4$

7. $x^2 + y^2 = 4,$
$\quad 16x^2 + 9y^2 = 144$

8. $x^2 + y^2 = 25,$
$\quad 25x^2 + 16y^2 = 400$

ii , **iii** Solve.

9. $x^2 + y^2 = 16,$
$\quad y^2 - 2x^2 = 10$

10. $x^2 + y^2 = 14,$
$\quad x^2 - y^2 = 4$

11. $x^2 + y^2 = 5,$
$\quad xy = 2$

12. $x^2 + y^2 = 20,$
$\quad xy = 8$

13. $x^2 + y^2 = 13,$
$\quad xy = 6$

14. $x^2 + 4y^2 = 20,$
$\quad xy = 4$

15. $x^2 + y^2 + 6y + 5 = 0,$
$\quad x^2 + y^2 - 2x - 8 = 0$

16. $2xy + 3y^2 = 7,$
$\quad 3xy - 2y^2 = 4$

17. ▦ $18.465x^2 + 788.723y^2 = 6408,$
$\quad 106.535x^2 - 788.723y^2 = 2692$

18. ▦ $0.319x^2 + 2688.7y^2 = 56,548,$
$\quad 0.306x^2 - 2688.7y^2 = 43,452$

iv

19. Find two numbers whose product is 156 if the sum of their squares is 313.

20. Find two numbers whose product is 60 if the sum of their squares is 136.

21. The area of a rectangle is $\sqrt{3}$ m^2 and the length of a diagonal is 2 m. Find the dimensions.

22. The area of a rectangle is $\sqrt{2}$ m^2 and the length of a diagonal is $\sqrt{3}$ m. Find the dimensions.

23. A garden contains two square peanut beds. Find the length of each bed if the sum of their areas is 823 ft^2, and the difference of their areas is 320 ft^2.

24. A certain amount of money saved for 1 yr at a certain interest rate yielded $7.50. If the principal had been $25 more and the interest rate 1% less, the interest would have been the same. Find the principal and the rate.

25. Find an equation of the circle that passes through the points $(4, 6)$, $(-6, 2)$, and $(1, -3)$.

26. Find an equation of the circle that passes through the points $(2, 3)$, $(4, 5)$, and $(0, -3)$.

27. Find k such that the following equations have two roots in common.

$$(x - 2)^4 - (x - 2) = 0,$$
$$x^2 - kx + k = 0$$

SUMMARY AND REVIEW: CHAPTER 10

The following contains a summary of what you should be able to do after completing this chapter. The review exercises are for practice. Answers are at the back of the book. If you miss an exercise, restudy the section indicated alongside the answer.

You should be able to:

Graph equations whose graph consists of a union of lines.

1. Graph $2x^2 - 3xy - 2y^2 = 0$.

Given the center and radius of a circle, find an equation for the circle, and given an equation of a circle, find its center and radius.

2. Find an equation of the circle with center $(-2, 6)$ and radius $\sqrt{13}$.

3. Find the center and radius of the circle

$$(x + 1)^2 + (y - 3)^2 = \tfrac{9}{4}.$$

Then graph the circle.

4. Find the center and radius of the circle $2x^2 + 2y^2 - 3x - 5y + 3 = 0$.

Given the center and a point that a circle passes through, find an equation of the circle.

5. Find an equation of the circle having its center at $(3, 4)$ and passing through the origin.

6. Find an equation of the circle having a diameter with endpoints $(-3, 5)$ and $(7, 3)$.

Given an equation of an ellipse, find the vertices and foci and graph the ellipse.

7. Find the center, vertices, and foci of the ellipse

$$16x^2 + 25y^2 - 64x + 50y - 331 = 0.$$

Then graph the ellipse.

8. Find an equation of the ellipse having vertices $(3, 0)$ and $(0, 4)$ and centered at the origin.

Given an equation of a hyperbola, find the center, vertices, foci, and asymptotes, and graph.

9. Find the center, vertices, foci, and asymptotes of the hyperbola $x^2 - 2y^2 + 4x + y - \tfrac{1}{8} = 0$.

Graph.

10. $xy = -2$

11. $4y^2 - x^2 = 16$

Given an equation of a parabola, find the vertex, focus, and directrix, and graph the parabola, and given the focus and directrix of a parabola, find an equation of the parabola.

12. Find an equation of the parabola with directrix $y = \tfrac{3}{2}$ and focus $(0, -\tfrac{3}{2})$.

13. Find the focus, vertex, and directrix of the parabola $y^2 = -12x$.

14. Find the vertex, focus, and directrix of the parabola $x^2 + 10x + 2y + 9 = 0$.

Solve systems of one first-degree and one second-degree equation and systems of second-degree equations.

Solve.

15. $x^2 - 16y = 0,$
$\quad x^2 - y^2 = 64$

16. $4x^2 + 4y^2 = 65,$
$\quad 6x^2 - 4y^2 = 25$

17. The sides of a triangle are 8, 10, and 14. Find the altitude to the longest side.

18. The sum of two numbers is 11 and the sum of their squares is 65. Find the numbers.

19. Find an equation of the ellipse that contains the point $\left(-\dfrac{1}{2}, \dfrac{3\sqrt{3}}{2}\right)$ and two of whose vertices are $(-1, 0)$ and $(1, 0)$.

20. Classify the graph of each of the following as a circle, an ellipse, a parabola, or a hyperbola.

   a) $x^2 + y^2 - 8x = 0$
   b) $4x^2 = y - 3$
   c) $\dfrac{x^2}{3} + \dfrac{y^2}{3} = 1$
   d) $1 - 5y = 4y^2 - x$
   e) $9x^2 - 4y^2 - 90x - 16y - 173 = 0$
   f) $x^2 + 4y^2 - 6x + 8y - 23 = 0$

21. Find two numbers whose product is 4 and the sum of whose reciprocals is $\frac{65}{56}$.

22. Find an equation of the circle that passes through the points $(10, 7)$, $(-6, 7)$, and $(-8, 1)$.

TEST: CHAPTER 10

1. Graph $3x^2 - 5xy - 2y^2 = 0$.

2. Find an equation of the circle with center at $(-2, 4)$ and radius 4.

3. Find the center and the radius of the circle

$$x^2 + y^2 - 2x + 6y + 5 = 0.$$

4. Find the center, vertices, and foci of the ellipse

$$9x^2 + y^2 - 36x - 8y + 43 = 0.$$

Then graph the ellipse.

5. Find an equation of the ellipse with vertices $(-7, 0)$, $(7, 0)$, $(0, -2)$, and $(0, 2)$.

6. Find the center, vertices, foci, and asymptotes of the hyperbola

$$9y^2 - 25x^2 = 225.$$

7. Graph $xy = 6$.

8. Find an equation of the parabola with directrix $y = 5$ and focus $(0, -5)$.

9. Find the vertex, focus, and directrix of the parabola $y^2 + 6y + 8x - 7 = 0$.

Solve.

10. $x^2 + y^2 = 100,$
$2y + x = 20$

11. $x^2 + y^2 = 5,$
$xy = 2$

12. The sum of two numbers is 4 and the difference of their squares is 32. What are the numbers?

13. The area of a rectangle is 10 yd^2 and the length of a diagonal is $\sqrt{101}$ yd. Find the dimensions.

14. Find an equation of the ellipse with vertices $(-3, -4)$ and $(-3, 2)$ and containing the point $(-1, 1)$.

15. Find an equation of the circle through points $(-3, 8)$, $(4, 1)$, and $(-3, -4)$.

Under certain conditions, the value of an office machine depreciates according to an arithmetic sequence. How can we find the value after *n* years?

SEQUENCES, SERIES, AND MATHEMATICAL INDUCTION

11

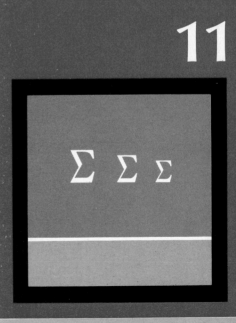

In the first section of this chapter we shall define precisely, in a mathematical way, what we mean by a *sequence*. The idea is a familiar one. For example, you might draw straws to form a batting order in a softball game. In so doing, you associate each batter with a natural number; that is, you define a function. The precise definition of a sequence uses the notion of a function.

When the members of a sequence are numbers, we can think of adding them. Such a sum is known as a *series*. Most of the chapter is devoted to the study of sequences and series. The ideas are, for most people, interesting in themselves, although there are important applications, too.

In this chapter, we also study a method of proof called *mathematical induction*. This approach is a clever one, and it enables us to prove many important formulas, as well as other mathematical results.

OBJECTIVES

You should be able to:

i Given a formula for the nth term (general term) of a sequence, find any term in the sequence, given a value for n.

ii Given a sequence, look for a pattern and try to guess a rule or formula for the general term.

iii Given a recursively defined sequence, construct its terms.

iv Find sums of sequences.

v Convert between sigma (Σ) notation and other notation for the sum of a sequence.

1. A sequence is given by
$$a(n) = 2n - 1.$$

a) Find the first 3 terms.

b) Find the 34th term.

2. A sequence is given by
$$a_n = \frac{(-1)^n}{n - 1}, \quad n \geqslant 2.$$

a) Find the first 4 terms.

b) Find the 48th term.

11.1

SEQUENCES AND SUMS

i SEQUENCES

A sequence is a set in which the order of the elements is specified. It can be defined precisely as a certain kind of function.

Definition

A *sequence* is a function having for its domain some set of natural numbers.

As an example, consider the sequence given by
$$a(n) = 2^n, \quad \text{or} \quad a_n = 2^n,$$

where the domain is the entire set of natural numbers $\{1, 2, 3, 4, 5, \ldots\}$. The notation a_n means the same as $a(n)$, and is more commonly used with sequences. Let us find some function values (also known as *terms* of the sequence):

$$a_1 = 2^1 = 2,$$
$$a_2 = 2^2 = 4,$$
$$a_3 = 2^3 = 8,$$
$$a_6 = 2^6 = 64.$$

The first term of the sequence is a_1, the fifth term is a_5, and the nth term, or *general* term, is a_n. This sequence can also be denoted in the following ways.

a) $2, 4, 8, \ldots$

b) $2, 4, 8, \ldots, 2^n, \ldots$

The preceding was an example of an *infinite sequence*. The sequence 2, 4, 8, 16 is a finite sequence. Its nth term is 2^n and its domain is the set $\{1, 2, 3, 4\}$.

EXAMPLE 1 Find the first 4 terms and the 57th term of the sequence whose general term is given by $a_n = (-1)^n/(n + 1)$.

$$a_1 = \frac{(-1)^1}{1 + 1} = -\frac{1}{2}, \qquad a_2 = \frac{(-1)^2}{2 + 1} = \frac{1}{3},$$

$$a_3 = \frac{(-1)^3}{3 + 1} = -\frac{1}{4}, \qquad a_4 = \frac{(-1)^4}{4 + 1} = \frac{1}{5},$$

$$a_{57} = \frac{(-1)^{57}}{57 + 1} = \frac{-1}{58}.$$

DO EXERCISES 1 AND 2.

ii FINDING A GENERAL TERM

When a sequence is described merely by naming the first few terms, we do not know for sure what the general term is, but the reader is expected to make a guess by looking for a pattern.

EXAMPLES   For each sequence make a guess at the general term.

2. 1, 4, 9, 16, 25, . . .

These are squares of numbers, so the general term may be n^2.

3. $\sqrt{1}, \sqrt{2}, \sqrt{3}, \sqrt{4}, \ldots$

These are square roots of numbers, so the general term may be $\sqrt{n}$.

4. $-1, 2, -4, 8, -16, \ldots$

These are powers of 2 with alternating signs, so the general term may be $(-1)^n 2^{n-1}$.

5. 2, 4, 8, . . .

If we see the pattern of powers of 2, we will see 16 as the next term and guess 2^n for the general term. If we see that we can get the second term by adding 2, the third term by adding 4, and the next term by adding 6, and so on, we will see 14 as the next term. The general term is then $n^2 - n + 2$.

DO EXERCISES 3–7.

iii RECURSIVE DEFINITIONS

A sequence may be defined by a *recursive definition*. Such a definition lists the first term and tells how to get the $(k+1)$st term from the kth term.

EXAMPLE 6   Find the first 5 terms of the sequence defined by

$$a_1 = 5,$$
$$a_{k+1} = 2a_k - 3, \quad \text{for } k \geqslant 1.$$

We have

$$a_1 = 5,$$
$$a_2 = 2a_1 - 3 = 2 \cdot 5 - 3 = 7,$$
$$a_3 = 2a_2 - 3 = 2 \cdot 7 - 3 = 11,$$
$$a_4 = 2a_3 - 3 = 2 \cdot 11 - 3 = 19,$$
$$a_5 = 2a_4 - 3 = 2 \cdot 19 - 3 = 35.$$

DO EXERCISE 8.

iv SUMS

With any sequence there is associated another sequence, obtained by taking sums of terms of the given sequence. The general term of this sequence of "partial sums" is denoted S_n. Consider the sequence

$$3, 5, 7, 9, \ldots, 2n + 1.$$

We construct some terms of the sequence of partial sums.

$S_1 = 3$ This is the first term of the given sequence.

$S_2 = 3 + 5 = 8$ This is the sum of the first two terms.

$S_3 = 3 + 5 + 7 = 15$ The sum of the first three terms

$S_4 = 3 + 5 + 7 + 9 = 24$ The sum of the first four terms

A sum like any of the above is also known as a *series*.

For each sequence try to find a rule for finding the general term, or the nth term. Answers may vary.

3. 2, 4, 6, 8, 10, . . .

4. 1, 2, 3, 4, 5, 6, . . .

5. 1, 8, 27, 64, 125, . . .

6. $x, \dfrac{x^2}{2}, \dfrac{x^3}{3}, \dfrac{x^4}{4}, \dfrac{x^5}{5}, \ldots$

7. 1, 2, 4, 8, 16, 32, . . .

8. Find the first 5 terms of each recursively defined sequence.

 a) $a_1 = 2$
   $a_{k+1} = 3a_k - 4, \quad k \geqslant 1$

 b) $a_1 = -3$
   $a_{k+1} = (-1) \cdot a_k^2, \quad k \geqslant 1$

9. Find the sum of the sequence

$$\frac{1}{2}, \frac{1}{4}, \frac{1}{8}, \frac{1}{16}, \frac{1}{32}.$$

Name each series without using Σ.

10. $\displaystyle\sum_{k=1}^{3} 2 + \frac{1}{k}$

11. $\displaystyle\sum_{k=0}^{4} 5^k$ (Sometimes a sequence can start with a 0th term.)

12. $\displaystyle\sum_{k=8}^{11} k^3$

Name each series with sigma notation. Answers may vary.

13. $2 + 4 + 6 + 8 + 10$

14. $1 + 8 + 27 + 64$

15. $x + \dfrac{x^2}{2} + \dfrac{x^3}{3} + \dfrac{x^4}{4} + \dfrac{x^5}{5} + \dfrac{x^6}{6}$

16. $2 + 3 + 4 + 5$

EXAMPLE 7   Find the sum of the sequence $-2, 4, -6, 8, -10$.

$$S = -2 + 4 + (-6) + 8 + (-10)$$
$$= -6$$

DO EXERCISE 9.

v SIGMA (Σ) NOTATION

Sums, or series, like those above can be handily denoted when the general term is known, using the Greek letter Σ (sigma).* The sum of the first four terms (S_4) of the sequence $3, 5, 7, 9, \ldots$ can be named as follows:

$$\sum_{k=1}^{4} (2k + 1).$$

This is read "the sum as k goes from 1 to 4 of $(2k + 1)$."

EXAMPLES   Rename the following sums without using sigma notation.

8. $\displaystyle\sum_{k=1}^{5} k^2 = 1^2 + 2^2 + 3^2 + 4^2 + 5^2$

9. $\displaystyle\sum_{k=1}^{4} (-1)^k(2k) = (-1)^1(2 \cdot 1) + (-1)^2(2 \cdot 2)$
$$+ (-1)^3(2 \cdot 3) + (-1)^4(2 \cdot 4)$$
$$= -2 + 4 - 6 + 8$$

10. $\displaystyle\sum_{k=10}^{12} \log k = \log 10 + \log 11 + \log 12$

To find sigma notation for a sum we must find a formula for the general term.

EXAMPLES   Write sigma notation for these sums.

11. $x + x^2 + x^3 + x^4 + x^5 + x^6 = \displaystyle\sum_{k=1}^{6} x^k$

12. $\log \dfrac{\pi}{2} - \log \dfrac{3\pi}{2} + \log \dfrac{5\pi}{2} - \log \dfrac{7\pi}{2} = \displaystyle\sum_{k=1}^{4} (-1)^{k+1} \log\left(\dfrac{2k-1}{2}\right)\pi$

DO EXERCISES 10–16.

EXERCISE SET 11.1

i   In each of the following, the general, or nth, term of a sequence is given. In each case find the first 4 terms, the 10th term, and the 15th term.

1. $a_n = 3n + 1$

2. $a_n = (n-1)(n-2)(n-3)$

3. $a_n = \dfrac{1}{n}$

4. $a_n = \dfrac{(-1)^n}{n}$

5. $a_n = \dfrac{n}{n+1}$

6. $a_n = \left(-\dfrac{1}{2}\right)^{n-1}$

---

* The letter Σ corresponds to S, for "sum."

ii For each of the following sequences, find the general, or nth, term. Answers may vary.

7. $1, 3, 5, 7, 9, \ldots$

8. $3, 9, 27, 81, 243, \ldots$

9. $\frac{2}{3}, \frac{3}{4}, \frac{4}{5}, \frac{5}{6}, \frac{6}{7}, \ldots$

10. $\sqrt{2}, \sqrt{4}, \sqrt{6}, \sqrt{8}, \sqrt{10}, \ldots$

11. $\sqrt{3}, 3, 3\sqrt{3}, 9, 9\sqrt{3}, \ldots$

12. $1 \cdot 2, 2 \cdot 3, 3 \cdot 4, 4 \cdot 5, \ldots$

13. $\log 1, \log 10, \log 100, \log 1000, \ldots$

14. $2, 4, 6, 8, 10, 12, 14, 16$

iii Find the first four terms of each recursively defined sequence.

15. $a_1 = 4$, $a_{k+1} = 1 + \dfrac{1}{a_k}$

16. $a_1 = 16$, $a_{k+1} = \sqrt{a_k}$

17. $a_1 = 64$, $a_{k+1} = \sqrt{a_k}$

18. $a_1 = e^Q$, $a_{k+1} = \ln a_k$

iv For each sequence find the associated sum.

19. $1, 2, 3, 4, 5, 6, 7, \ldots$. Find S_7.

20. $1, -3, 5, -7, 9, -11, \ldots$. Find S_8.

21. $2, 4, 6, 8, \ldots$. Find S_5.

22. $1, \frac{1}{4}, \frac{1}{9}, \frac{1}{16}, \frac{1}{25}, \ldots$. Find S_5.

v Rename without using Σ notation.

23. $\displaystyle\sum_{k=1}^{5} \frac{1}{2k}$

24. $\displaystyle\sum_{k=1}^{6} \frac{1}{2k+1}$

25. $\displaystyle\sum_{k=0}^{5} 2^k$

26. $\displaystyle\sum_{k=4}^{7} \sqrt{2k-1}$

27. $\displaystyle\sum_{k=7}^{10} \log k$

28. $\displaystyle\sum_{k=0}^{4} \pi k$

iv, **v** Write Σ notation for each sum. Then compute the sum.

29. $\dfrac{1}{1^2} + \dfrac{1}{2^2} + \dfrac{1}{3^2} + \dfrac{1}{4^2}$

30. $\dfrac{1}{2} + \dfrac{2}{4} + \dfrac{3}{8} + \dfrac{4}{16}$

31. $-2 + 4 - 8 + 16 - 32 + 64$

32. $\dfrac{1}{2^1} - \dfrac{2}{2^2} + \dfrac{3}{2^3} - \dfrac{4}{2^4} + \dfrac{5}{2^5}$

Find the first five terms of each sequence.

33. $a_n = \dfrac{1}{2n} \log 1000^n$

34. $a_n = i^n$, $i = \sqrt{-1}$

35. $a_n = \ln e^n$

36. $a_n = \ln (1 \cdot 2 \cdot 3 \cdots n)$

37. $a_n = e^{\ln n}$

38. $a_n = \log |-10|^n$

Find decimal notation, rounded to six decimal places, for the sum of the first six terms of each sequence.

39. $a_n = \left(1 + \dfrac{1}{n}\right)^n$

40. $a_n = \sqrt{n+1} - \sqrt{n}$

41. $a_1 = 2$, $a_{k+1} = \sqrt{1 + \sqrt{a_k}}$

42. $a_1 = 2$, $a_{k+1} = \dfrac{1}{2}\left(a_k + \dfrac{2}{a_k}\right)$

For each sequence, find a formula for S_n.

43. $a_n = \ln n$

44. $a_n = \dfrac{1}{n} - \dfrac{1}{n+1}$

OBJECTIVES

You should be able to:

i Use the formula of Theorem 1 to find, for any arithmetic sequence,

   a) the *n*th term when *n* is given;
   b) *n*, when an *n*th term is given.
   c) Given two terms of a sequence, find the common difference and construct the sequence.

ii Use Theorems 2 and 3 for arithmetic sequences to find the sum of any arithmetic sequence.

iii Insert arithmetic means between two numbers.

1. Find the 13th term of the sequence 2, 6, 10, 14, ...

2. In the sequence given in Exercise 1, what term would be 298? That is, what is *n* if $a_n = 298$?

11.2

ARITHMETIC SEQUENCES AND SERIES

i ARITHMETIC SEQUENCES

Consider this sequence:

$$2, 5, 8, 11, 14, 17, \ldots .$$

Note that adding 3 to any term produces the following term. In other words, the difference between any term and the preceding one is 3. Sequences in which the difference between successive terms is constant are called *arithmetic sequences.**

We now find a formula for the general, or *n*th, term of any arithmetic sequence. Let us denote the difference between successive terms (called the *common difference*) by *d*, and write out the first few terms:

$$a_1,$$
$$a_2 = a_1 + d,$$
$$a_3 = a_2 + d = (a_1 + d) + d = a_1 + 2d,$$
$$a_4 = a_3 + d = (a_1 + 2d) + d = a_1 + 3d.$$

Generalizing, we obtain the following.

Theorem 1

> **The *n*th term of an arithmetic sequence is given by**
> $$a_n = a_1 + (n - 1)d,$$
> **where *d* is the common difference between successive terms.**

EXAMPLE 1 Find the 14th term of the arithmetic sequence 4, 7, 10, 13, ...

First note that $a_1 = 4$, $d = 3$, and $n = 14$. Using the formula of Theorem 1, we obtain

$$a_{14} = 4 + (14 - 1) \cdot 3$$
$$= 4 + 39$$
$$= 43.$$

DO EXERCISE 1.

EXAMPLE 2 In the sequence given in Example 1, which term is 301? That is, what is *n* if $a_n = 301$?

We substitute into the formula of Theorem 1 and solve for *n*:

$$a_n = a_1 + (n - 1)d$$
$$301 = 4 + (n - 1) \cdot 3$$
$$301 = 4 + 3n - 3$$
$$300 = 3n$$
$$100 = n.$$

The 100th term is 301.

DO EXERCISE 2.

---

* Sometimes called *arithmetic progressions*.

Given two terms and their places in an arithmetic sequence, we can construct the sequence.

EXAMPLE 3 The 3rd term of an arithmetic sequence is 8 and the 16th term is 47. Find a_1 and d and construct the sequence.

We know that $a_3 = 8$ and $a_{16} = 47$. Thus we would have to add d 13 times to get from 8 to 47. That is,

$$13d = 47 - 8 = 39.$$

Solving, we obtain

$$d = 3.$$

Since $a_3 = 8$, we subtract d twice to get to a_1. Thus

$$a_1 = 2.$$

The sequence is 2, 5, 8, 11, ...

DO EXERCISE 3.

ii SUMS

We now look for a formula for the sum of the first n terms of an arithmetic sequence. Let us denote the terms as follows:

$$a_1, (a_1 + d), (a_1 + 2d), \ldots, (a_n - 2d), (a_n - d), a_n.$$

Then the sum of the first n terms is given by

$$S_n = a_1 + (a_1 + d) + (a_1 + 2d) + \cdots + (a_n - 2d) + (a_n - d) + a_n.$$

Reversing the right side, we obtain

$$S_n = a_n + (a_n - d) + (a_n - 2d) + \cdots + (a_1 + 2d) + (a_1 + d) + a_1.$$

Adding these two equations, we obtain

$$2S_n = (a_1 + a_n) + (a_1 + a_n) + \cdots + (a_1 + a_n),$$

a total of n terms on the right. Thus $2S_n = n(a_1 + a_n)$. Dividing by 2 gives us the formula we seek.

Theorem 2

The sum of the first n terms of an arithmetic sequence is given by

$$S_n = \frac{n}{2}(a_1 + a_n).$$

We now derive a variant of this formula, which we can use in case a_n is not known. We substitute the expression for a_n given in Theorem 1, $a_n = a_1 + (n - 1)d$, into the formula of Theorem 2:

$$S_n = \frac{n}{2}(a_1 + [a_1 + (n - 1)d]).$$

We now have the following.

Theorem 3

The sum of the first n terms of an arithmetic sequence is given by

$$S_n = \frac{n}{2}[2a_1 + (n - 1)d].$$

3. The 7th term of an arithmetic sequence is 79 and the 13th term is 151. Find a_1 and d. Construct the sequence.

4. Find the sum of the first 200 natural numbers.

EXAMPLE 4 Find the sum of the first 100 natural numbers.

The series is $1 + 2 + 3 + \cdots + 100$, so $a_1 = 1$, $a_n = 100$, and $n = 100$. We substitute into the formula of Theorem 2:

$$S_n = \frac{n}{2}(a_1 + a_n)$$

$$S_{100} = \frac{100}{2}(1 + 100) = 5050.$$

EXAMPLE 5 Find the sum of the first 14 terms of the arithmetic sequence $2, 5, 8, 11, 14, \ldots$.

We note that $a_1 = 2$, $d = 3$, and $n = 14$. We use the formula of Theorem 3:

$$S_n = \frac{n}{2}[2a_1 + (n - 1)d]$$

$$S_{14} = \frac{14}{2} \cdot [2 \cdot 2 + (14 - 1)3]$$

$$= 7 \cdot [4 + 13 \cdot 3]$$

$$= 7 \cdot 43$$

$$S_{14} = 301.$$

5. Find the sum of the first 15 terms of the arithmetic sequence

$$1, 3, 5, 7, 9, \ldots.$$

EXAMPLE 6 Find the sum $\displaystyle\sum_{k=1}^{13}(4k + 5)$.

It is helpful to write out a few terms:

$$9 + 13 + 17 + \cdots.$$

We see that the sequence is arithmetic, with $a_1 = 9$, $d = 4$, and $n = 13$. We use the formula of Theorem 3:

$$S_n = \frac{n}{2}[2a_1 + (n - 1)d]$$

$$S_{13} = \frac{13}{2}[2 \cdot 9 + (13 - 1)4]$$

$$= \frac{13}{2}[18 + 12 \cdot 4]$$

$$= \frac{13}{2} \cdot 66$$

$$S_{13} = 429.$$

6. Find the sum.

$$\sum_{k=1}^{10}(9k - 4)$$

DO EXERCISES 4–6.

EXAMPLE 7 A family saves money in an arithmetic sequence. They save $500 one year, $600 the next, $700 the next, and so on, for 10 years. How much do they save in all, exclusive of interest?

The amount saved is the sum of the numbers and is given by the series

$$\$500 + \$600 + \$700 + \cdots.$$

We have $a_1 = 500$, $n = 10$, and $d = 100$. We use the formula of Theorem 3:

$$S_n = \frac{n}{2}[2a_1 + (n-1)d]$$

$$S_{10} = \frac{10}{2}[2 \cdot 500 + (10-1)100]$$

$$S_{10} = 9500.$$

They save $9500.

DO EXERCISES 7 AND 8.

iii ARITHMETIC MEANS

If p, m, and q form an arithmetic sequence, it can be shown (see Exercise 35) that $m = (p+q)/2$. We call m the *arithmetic mean* of p and q. Given two numbers p and q, if we find k other numbers $m_1, m_2, m_3, \ldots, m_k$ such that the following is an arithmetic sequence,

$$p, m_1, m_2, \ldots, m_k, q,$$

we say that we have "inserted k arithmetic means between p and q."

EXAMPLE 8 Insert three arithmetic means between 4 and 13.

We look for m_1, m_2, and m_3 such that 4, m_1, m_2, m_3, 13 is an arithmetic sequence. In this case, $a_1 = 4$, $n = 5$, and $a_5 = 13$. We use the formula of Theorem 1:

$$a_n = a_1 + (n-1)d$$
$$13 = 4 + (5-1)d.$$

Then $d = 2\frac{1}{4}$, so we have

$$m_1 = a_1 + d$$
$$= 4 + 2\frac{1}{4} = 6\frac{1}{4},$$

$$m_2 = m_1 + d$$
$$= 6\frac{1}{4} + 2\frac{1}{4} = 8\frac{1}{2},$$

$$m_3 = m_2 + d$$
$$= 8\frac{1}{2} + 2\frac{1}{4} = 10\frac{3}{4}.$$

DO EXERCISE 9.

7. Find the sum of all two-digit natural numbers divisible by 3.

8. A cheerleader pyramid has 15 girls on the bottom row, 14 on the next row, and so on, until there is just one girl on top. How many cheerleaders are in the pyramid?

9. Insert three arithmetic means between 4 and 16.

EXERCISE SET 11.2

i Exercises 1–10 refer to arithmetic sequences.

1. Find the 12th term of 2, 6, 10, ...

2. Find the 11th term of 0.07, 0.12, 0.17, ...

3. In the sequence of Exercise 1, which term is 106?

4. In the sequence of Exercise 2, which term is 1.67?

5. If $a_1 = 5$ and $d = 6$, what is a_{17}?

6. If $a_1 = 14$ and $d = -3$, what is a_{20}?

7. If $d = 4$ and $a_8 = 33$, what is a_1?

8. If $a_1 = 8$ and $a_{11} = 26$, what is d?

9. If $a_1 = 5$, $d = -3$, and $a_n = -76$, what is n?

10. If $a_1 = 25$, $d = -14$, and $a_n = -507$, what is n?

11. In an arithmetic sequence $a_{17} = -40$ and $a_{28} = -73$. Find a_1 and d. Find the first 5 terms of the sequence.

12. A bomb drops from an airplane and falls 16 ft the first sec, 48 ft the second sec, and so on, forming an arithmetic sequence. How many feet will the bomb fall during the 20th sec?

ii

13. Find the sum of the first 20 terms of the sequence

$$5, 8, 11, 14, \ldots .$$

14. Find the sum of the first 15 terms of the sequence

$$5, \tfrac{55}{7}, \tfrac{75}{7}, \tfrac{95}{7}, \ldots .$$

15. Find the sum of the odd numbers from 1 to 99 inclusive.

16. Find the sum of the multiples of 7, from 7 to 98 inclusive.

17. An arithmetic sequence has $a_1 = 2$, $d = 5$, and $n = 20$. What is S_{20}?

18. Find the sum of all multiples of 4 that are between 14 and 523.

19. Find the sum $\displaystyle\sum_{k=1}^{16} (7k - 76)$.

20. Find the sum $\displaystyle\sum_{k=1}^{12} (6k - 3)$.

21. How many poles will be in a pile of telephone poles if there are 30 in the first layer, 29 in the second, and so on, until there is one in the last layer?

22. If a student saved 10¢ on October 1, 20¢ on October 2, 30¢ on October 3, etc., how much would be saved during October? (October has 31 days.)

iii

23. Insert four arithmetic means between 4 and 13.

24. Insert three arithmetic means between -3 and 5.

25. Find a formula for the sum of the first n odd natural numbers.

26. Find three numbers in an arithmetic sequence such that the sum of the first and third is 10 and the product of the first and second is 15.

27. The zeros of this polynomial form an arithmetic sequence. Find them.

$$x^4 + 4x^3 - 84x^2 - 176x + 640$$

28. Insert enough arithmetic means between 1 and 50 so the sum of the resulting arithmetic series will be 459.

29. Suppose that the lengths of the sides of a right triangle form an arithmetic sequence. Prove that the triangle is similar to a right triangle whose sides have lengths 3, 4, 5.

(*Straight-line depreciation*). A company buys an office machine for $5200 on January 1 of a given year. It is expected to last for 8 years, at which time its *trade-in*, or *salvage*, *value* will be $1100. If the company figures the decline in value to be the same each year, then the *book value*, or *salvage value*, after n years, $0 \leqslant n \leqslant 8$, is given by an arithmetic sequence:

$$V_n = C - n\left(\frac{C - S}{L}\right),$$

where $C =$ the original cost of the item ($5200), $L =$ years of expected life (8), and $S =$ the salvage value ($1100).

30. Find the general term for the straight-line depreciation of the office machine.

31. Find the salvage value after 0 years, 1 year, 2 years, 3 years, 4 years, 5 years, 6 years, 7 years, 8 years.

32. ▦ Find the first 10 terms of the arithmetic sequence for which $a_1 = \$8760$ and $d = -\$798.23$.

33. ▦ Find the sum of the first 10 terms of the sequence given in Exercise 32.

34. Prove that an expression for the nth term of an arithmetic sequence defines a linear function.

35. Prove that if p, m, and q form an arithmetic sequence, then

$$m = \frac{p + q}{2}.$$

OBJECTIVES

You should be able to:

i Use Theorem 4 for geometric sequences to find a given term.

ii Find sums of geometric sequences.

11.3

GEOMETRIC SEQUENCES AND SERIES

i **GEOMETRIC SEQUENCES**

Consider this sequence:

$$3, 6, 12, 24, 48, 96, \ldots .$$

Note that multiplying any term by 2 produces the following term. In other words, the ratio of any term and the preceding one is 2. Sequences in which the ratio of successive terms is constant but different from 1 are called *geometric sequences.**

We now find a formula for the general, or *n*th, term of any geometric sequence. Let us denote the ratio of successive terms (called the *common ratio*) by r, and write out the first few terms:

$$a_1,$$
$$a_2 = a_1 r,$$
$$a_3 = a_2 r = (a_1 r) r = a_1 r^2,$$
$$a_4 = a_3 r = (a_1 r^2) r = a_1 r^3.$$

Generalizing, we obtain the following.

Theorem 4

> **The *n*th term of a geometric sequence is given by**
>
> $$a_n = a_1 r^{n-1},$$
>
> **where r is the common ratio of successive terms.**

EXAMPLE 1   Find the 6th term of the geometric sequence $4, 20, 100, \ldots$.

First note that $a_1 = 4$, $n = 6$, and $r = 5$. We use the formula given in Theorem 4:

$$a_n = a_1 r^{n-1}$$
$$a_6 = 4 \cdot 5^{6-1} = 4 \cdot 5^5$$
$$= 12{,}500.$$

DO EXERCISE 1.

EXAMPLE 2   Find the 11th term of the geometric sequence $64, -32, 16, -8, \ldots$.

Note that $a_1 = 64$, $n = 11$, and $r = -\frac{1}{2}$. We use the formula given in Theorem 4:

$$a_n = a_1 r^{n-1}$$
$$a_{11} = 64 \cdot \left(-\frac{1}{2}\right)^{11-1}$$
$$= 64 \cdot \left(-\frac{1}{2}\right)^{10}$$
$$= 2^6 \cdot \frac{1}{2^{10}}$$
$$= \frac{1}{2^4}, \quad \text{or} \quad \frac{1}{16}.$$

DO EXERCISE 2.

---

* Sometimes called *geometric progressions*.

1. Find the 8th term of the geometric sequence

$$2, 4, 8, 16, \ldots.$$

2. Find the 6th term of the geometric sequence

$$3, -15, 75, \ldots.$$

3. A college student borrows \$400 at 6% interest compounded annually. She pays off the loan at the end of 3 yr. How much does she pay?

EXAMPLE 3   A college student borrows \$600 at 12% interest compounded annually. The loan is paid off at the end of 3 years. How much is paid back?

For principal P, at 12% interest, $P + 0.12P$ will be owed at the end of a year. This is also $1.12P$, and is the principal for the second year, so at the end of that year $1.12(1.12P)$ is owed. Thus the principal at the beginnings of successive years is

$$P, \ 1.12P, \ 1.12^2 P, \ 1.12^3 P, \quad \text{and so on.}$$

We have a geometric sequence with $a_1 = P = 600$, $n = 4$, and $r = 1.12$. We use the formula given in Theorem 4:

$$\begin{aligned} a_n &= a_1 r^{n-1} \\ a_4 &= 600 \cdot (1.12)^{4-1} \\ &= 600(1.12)^3 \\ &= 600(1.404928) \\ &= 842.96. \end{aligned}$$

The amount paid back is \$842.96.

Note that the equation $a_4 = \$600(1.12)^3$, used in Example 3, conforms to the compound interest formula $A = P(1 + i)^t$, developed in Chapter 2. If we consult that development we see that the numbers A_1, A_2, A_3, and so on, form a geometric sequence, where $A_1 = P$, $A_2 = P(1 + i)$, $r = 1 + i$, and $n - 1 = t$. (Note that "after t years" and "at the beginning of the nth year" have the same meaning here.)

DO EXERCISE 3.

ii SUMS

We now look for a formula for the sum of the first n terms of a geometric sequence.* Let us denote the terms as follows:

$$a_1, \ a_1 r, \ a_1 r^2, \ a_1 r^3, \dots, a_1 r^{n-1}.$$

Then the sum of the first n terms is given by

$$S_n = a_1 + a_1 r + a_1 r^2 + \cdots + a_1 r^{n-2} + a_1 r^{n-1}.$$

We next multiply on both sides of this equation by $-r$, obtaining

$$-rS_n = -a_1 r - a_1 r^2 - \cdots - a_1 r^{n-1} - a_1 r^n.$$

We now add the two equations, obtaining

$$S_n - rS_n = a_1 - a_1 r^n.$$

Solving for S_n gives us the formula we seek.

Theorem 5

> The sum of the first n terms of a geometric sequence is given by
>
> $$S_n = \frac{a_1 - a_1 r^n}{1 - r}, \quad r \neq 1.$$

---

* It is not uncommon to see or hear the redundant terminology "sum of a geometric series."

We now derive a variant of this formula, which we can use when a_n is known. Using Theorem 4 we substitute a_n for $a_1 r^{n-1}$ into the formula of Theorem 5. We then have the following.

Theorem 6

> The sum of the first n terms of a geometric sequence is given by
>
> $$S_n = \frac{a_1 - ra_n}{1 - r}, \quad r \neq 1.$$

EXAMPLE 4 Find the sum of the first 6 terms of the geometric sequence $3, 6, 12, 24, \ldots$.

Note that $a_1 = 3$, $n = 6$, and $r = 2$. We use the formula given in Theorem 5:

$$S_n = \frac{a_1 - a_1 r^n}{1 - r}$$

$$S_6 = \frac{3 - 3 \cdot 2^6}{1 - 2}$$

$$= \frac{3 - 192}{-1}, \quad \text{or } 189.$$

EXAMPLE 5 Find the sum $\sum_{k=1}^{11} (0.3)^k$.

This is a geometric series. The first term is 0.3, $r = 0.3$, and $n = 11$. We use the formula given in Theorem 5:

$$S_n = \frac{a_1 - a_1 r^n}{1 - r}$$

$$S_{11} = \frac{0.3 - 0.3 \cdot (0.3)^{11}}{1 - 0.3}$$

$$= \frac{0.3 - (0.3)^{12}}{0.7}$$

$$= 0.42857 \ldots.$$

DO EXERCISES 4–6.

EXAMPLE 6 Someone offers you a job for 30 days, the pay being 1¢ the first day, 2¢ the second day, and doubled every day thereafter. Would you take the job? Calculate your pay for the 30 days.

1. *Familiarize:* Your pay is the sum of the geometric sequence (in dollars)

$$0.01, 2(0.01), 2^2(0.01), 2^3(0.01), \ldots, 2^{29}(0.01).$$

2. *Translate:* We note that $a_1 = 0.01$, $n = 30$, and $r = 2$. We use the formula given in Theorem 5:

$$S_n = \frac{a_1 - a_1 r^n}{1 - r}$$

$$S_{30} = \frac{0.01 - 0.01(2)^{30}}{1 - 2}.$$

4. Find the sum of the first 6 terms of the geometric sequence

$$3, 15, 75, 375, \ldots.$$

5. Find the sum of the first 10 terms of the geometric sequence

$$2, -1, \tfrac{1}{2}, -\tfrac{1}{4}, \ldots.$$

6. Find the sum.

$$\sum_{k=1}^{5} 3^k$$

7. In Example 6, how much would you earn in 20 days?

3. *Carry out:* We carry out the computation on the right:

$$S_{30} = \frac{0.01(1 - 2^{30})}{-1} = \frac{0.01(1 - 1073741824)}{-1}$$

$$= 10,737,418.23.$$

Your pay: $10,737,418.23.

DO EXERCISE 7.

EXERCISE SET 11.3

1. Find the 10th term of the geometric sequence

$$\frac{8}{243}, \frac{4}{81}, \frac{2}{27}, \ldots.$$

3. Find the sum of the first 8 terms of the geometric sequence

$$5, 10, 20, \ldots.$$

5. Find the sum of the first 7 terms of the geometric sequence

$$\frac{1}{18}, -\frac{1}{6}, \frac{1}{2}, \ldots.$$

7. Find the sum.

$$\sum_{k=1}^{6} \left(\frac{1}{2}\right)^{k-1}$$

9. A college student borrows $800 at 8% interest compounded annually. She pays off the loan in full at the end of 2 yr. How much does she pay?

11. A ping-pong ball is dropped from a height of 16 ft and always rebounds $\frac{1}{4}$ of the distance of the previous fall. What distance does it rebound the 6th time?

13. ▦ Find the sum of the first 5 terms of the geometric sequence

$$\$1000, \$1000(1.08), \$1000(1.08)^2, \ldots.$$

Round to the nearest cent.

2. Find the 5th term of the geometric sequence

$$2, -10, 50, \ldots.$$

4. Find the sum of the first 6 terms of the geometric sequence

$$16, -8, 4, \ldots.$$

6. Find the sum of the geometric sequence

$$-8, 4, -2, \ldots -\frac{1}{32}.$$

8. Find the sum.

$$\sum_{k=1}^{10} 2^k$$

10. A college student borrows $1000 at 8% interest compounded annually. He pays off the loan in full at the end of 4 yr. How much does he pay?

12. Gaintown has a population of 100,000 now and the population is increasing 10% every year. What will be the population in 5 yr?

14. ▦ Find the sum of the first 6 terms of the geometric sequence

$$\$200, \$200(1.13), \$200(1.13)^2, \ldots.$$

Round to the nearest cent.

For Exercises 15–17, assume that $a_1, a_2, a_3, \ldots$, is a geometric sequence.

15. Show that

$$a_1^2, a_2^2, a_3^2, \ldots,$$

is a geometric sequence.

17. Show that

$$\ln a_1, \ln a_2, \ln a_3, \ldots,$$

is an arithmetic sequence.

16. Show that

$$a_1^{-3}, a_2^{-3}, a_3^{-3}, \ldots,$$

is a geometric sequence.

18. Show that

$$5^{a_1}, 5^{a_2}, 5^{a_3}, \ldots,$$

is a geometric sequence, if $a_1, a_2, a_3, \ldots$, is an arithmetic sequence.

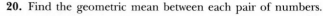

19. Show that if the positive numbers p, G, q form a geometric sequence, then

$$G = \sqrt{pq}.$$

G is called the *geometric mean* of the numbers p and q.

20. Find the geometric mean between each pair of numbers.
  a) 4 and 9
  b) 2 and 6
  c) $\frac{1}{2}$ and $\frac{1}{3}$
  d) $\sqrt{5} + \sqrt{2}$ and $\sqrt{5} - \sqrt{2}$

21. 🖩 A piece of paper is 0.01 in. thick. It is folded in such a way that its thickness is doubled each time for 20 times. How thick is the result?

22. 🖩 A superball dropped from the top of the Washington Monument (556 ft high) always rebounds $\frac{3}{4}$ of the distance of the previous fall. How far up and down has it traveled when it hits the ground for the 6th time?

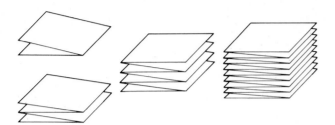

23. (*An annuity*). A person decides to save money in a savings account for retirement. At the beginning of each year $1000 is invested at 8% compounded annually. How much is in the retirement fund at the end of 40 years?

24. A square has sides 1 m in length. Another square is formed whose vertices are the midpoints of the original square. Inside this square another is formed whose vertices are the midpoints of the second square, and so on.

  a) Find the first 5 terms of the sequence of areas of the squares.
  b) Find the nth term.
  c) Find a formula for the sum of the areas of the first n squares.

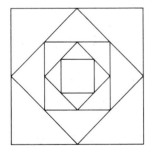

11.4

INFINITE GEOMETRIC SEQUENCES AND SERIES

Recall the definition of a *sequence* as a function having for its domain some set of natural numbers. If that domain is an infinite (unending) set of natural numbers, then the sequence is called an *infinite sequence*. In the examples that follow, the domain will be the set of *all* natural numbers, even though this is not a requirement of the definition.

Let us look at two examples of infinite geometric sequences and their sequences of partial sums.

OBJECTIVES

You should be able to:

| i | Determine whether the sum of an infinite geometric sequence exists, and if so, use the formula of Theorem 7 to find it. |

| ii | Convert from repeating decimal notation to fractional notation. |

A. $2, 4, 8, 16, \ldots, 2^n, \ldots$

$$S_1 = 2$$
$$S_2 = 6$$
$$S_3 = 14$$
$$S_4 = 30$$
$$S_5 = 62$$

The terms of the sequence S_n get larger and larger, without bound. Now let us look at the next sequence.

B. $\dfrac{1}{2}, \dfrac{1}{4}, \dfrac{1}{8}, \dfrac{1}{16}, \ldots, \dfrac{1}{2^n}, \ldots$

$$S_1 = \tfrac{1}{2}$$
$$S_2 = \tfrac{1}{2} + \tfrac{1}{4} = \tfrac{3}{4}$$
$$S_3 = S_2 + \tfrac{1}{8} = \tfrac{3}{4} + \tfrac{1}{8} = \tfrac{7}{8}$$
$$S_4 = S_3 + \tfrac{1}{16} = \tfrac{15}{16}$$
$$S_5 = S_4 + \tfrac{1}{32} = \tfrac{31}{32}$$

It appears that we can write a formula for S_n as follows:

$$S_n = \frac{2^n - 1}{2^n}.$$

Later (Example 5, p. 428) we will prove that this is correct. In this formula the numerator is less than the denominator for all values of n. Yet as n increases the numerator gets very close to the denominator. Thus the values of S_n approach the number 1 more and more closely. We say that S_n "approaches 1 as a limit."

i SUMS

The sequence (B) in the above example is infinite. For any finite number of terms, the sum is given by

$$S_n = \sum_{k=1}^{n} \frac{1}{2^k}$$

$$= \frac{2^n - 1}{2^n}.$$

Since S_n approaches a limit, we *define* that limit to be the "sum" of the infinite sequence of terms, and use the following notation:*

$$S_\infty = \sum_{k=1}^{\infty} \frac{1}{2^k} = 1.$$

Not all infinite sequences have sums (for example, the sequence (A) at the beginning of the section). In those cases the notation $\sum_{k=1}^{\infty} a_k$ is meaningless. An infinite sum is also called an *infinite series*.

We now find a formula for the "sum" of an infinite geometric sequence, by looking at values of S_n as n becomes very large. Recall that

$$S_n = \frac{a_1 - a_1 r^n}{1 - r}$$

---

* $\sum_{k=1}^{\infty} \dfrac{1}{2^k}$ is read "the sum, as k goes from 1 to infinity, of $1/2^k$."

or, equivalently,

$$S_n = \frac{a_1}{1-r} - \frac{a_1 r^n}{1-r}.$$

If $r = 1$, the formula does not hold, since 1 is a nonsensible replacement for r. However, in that case the sequence is not geometric. For any other value of r, the first fractional expression can be evaluated. It remains to look at the second fractional expression as n becomes large.

If $r > 1$, r^n will become very large and the second fractional expression will become large, hence no sum exists. If r is negative and $|r| > 1$, then r^n will alternate between positive and negative numbers having larger and larger absolute value. In either case, S_n will not approach a limit.

If $|r| < 1$, then the values of r^n approach 0 as n becomes very large, hence the second fractional expression also approaches 0, and we have the following.

Theorem 7

An infinite geometric sequence has a sum if and only if $|r| < 1$. That sum is given by

$$S_\infty = \frac{a_1}{1-r}.$$

EXAMPLE 1 Find the sum of the infinite geometric sequence $5, \frac{5}{2}, \frac{5}{4}, \frac{5}{8}, \ldots$, if it exists. That is, find

$$\sum_{k=1}^{\infty} \frac{5}{2^{k-1}}.$$

Note that $a_1 = 5$. The sum exists because $r = \frac{1}{2}$, hence $|r| < 1$. We use the formula of Theorem 7:

$$S_\infty = \frac{a_1}{1-r}$$

$$= \frac{5}{1 - \frac{1}{2}}$$

$$= 10.$$

DO EXERCISES 1–6.

ii **REPEATING DECIMALS**

Any repeating decimal represents a geometric series. For example,

$$0.33333\ldots = 0.3 + 0.03 + 0.003 + 0.0003 + \cdots.$$

In this series the common ratio is 0.1. It is a common series and the sum is known to be $\frac{1}{3}$. Consider the following:

$$0.35\overline{35}^* = 0.35 + 0.0035 + 0.000035 + \cdots.$$

In this case the common ratio is 0.01. In any repeating decimal the common ratio will be 0.1, 0.01, 0.001, or 10^{-n} for some natural number n. Thus in any repeating decimal the common ratio is less than 1, so the infinite sum exists.

---

* The bar indicates the repeating cycle.

Determine which geometric sequences have sums.

1. 4, 16, 64, ...

2. 5, -30, 180, ...

3. 1, $\frac{1}{3}$, $\frac{1}{9}$, $\frac{1}{27}$, ...

Find these infinite sums. The series are geometric.

4. $1 + \frac{1}{3} + \frac{1}{9} + \frac{1}{27} + \cdots$

5. $\sum_{n=0}^{\infty} \frac{1}{2^n}$

6. $4 - 1 + \frac{1}{4} - \frac{1}{16}, \ldots$

Find the fractional notation for each
number.

7. $0.45\overline{45}$

8. $5.36\overline{36}$

9. Some states gave veterans a bonus
after World War II. In one such
state, the bonus was \$500. Assuming
that each veteran spent 85% of the
bonus and that each person receiving
the money thus put into circulation
also spent 85% of it, how much money
was effectively put into the economy
by each veteran?

EXAMPLE 2   Find fractional notation for $0.27\overline{27}$.

We can think of this as $0.27 + 0.0027 + 0.000027 + 0.00000027 + \cdots$.
Note that $a_1 = 0.27$ and that $r = 0.01$. We use the formula given in Theorem 7:

$$S_\infty = \frac{a_1}{1 - r}$$

$$S_\infty = \frac{0.27}{1 - 0.01} = \frac{0.27}{0.99}$$

$$= \frac{3}{11}.$$

Fractional notation for $0.27\overline{27}$ is $\frac{3}{11}$. This can be checked by dividing 3
by 11.

EXAMPLE 3   Find fractional notation for $0.4166\overline{6}$.

We first split this up into a repeating and a nonrepeating part as follows: $0.41 + 0.00666$. The repeating part can be considered as $0.006 +
0.0006 + 0.00006 + 0.000006 + \cdots$. Thus the repeating part is a geometric
series with $a_1 = 0.006$ and $r = 0.1$. We find fractional notation for this sum
first:

$$S_\infty = \frac{a_1}{1 - r}$$

$$= \frac{0.006}{1 - 0.1}$$

$$= \frac{0.006}{0.9}$$

$$= \frac{6}{900}.$$

The number in question is then $0.41 + \frac{6}{900}$, or $\frac{41}{100} + \frac{6}{900}$. Adding and simplifying, we obtain $\frac{5}{12}$. The reader should check this by dividing.

DO EXERCISES 7 AND 8.

EXAMPLE 4   (*The economic multiplier*). A recent change in the tax laws
reduced the tax paid by a family by \$1000. Each family then had an extra
\$1000 to spend. Let us suppose that they spend 90% of this amount, that
the people who get it spend 90% of it, the people who get that spend 90%
of it, and so on.

According to certain economic theory, the money that this effectively
puts into the economy can be calculated as the sum of an infinite geometric
sequence, as follows:

$$\$1000 + \$1000(0.90) + \$1000(0.90)^2 + \$1000(0.90)^3 + \cdots.$$

Using the formula of Theorem 7 we find this amount to be

$$\frac{\$1000}{1 - 0.90}, \quad \text{or } \$10,000.$$

DO EXERCISE 9.

EXERCISE SET 11.4

i Determine which of the following infinite geometric sequences have sums.

1. $5, 10, 20, 40, \ldots$

2. $16, 8, 4, 2, \ldots$

3. $6, 2, \frac{2}{3}, \frac{2}{9}, \ldots$

4. $2, -4, 8, -16, 32, \ldots$

5. $1, 0.1, 0.01, 0.001, 0.0001, \ldots$

6. $-\frac{5}{3}, -\frac{10}{9}, -\frac{20}{27}, \ldots$

7. $1, -\frac{1}{5}, \frac{1}{25}, -\frac{1}{125}, \ldots$

8. $6, \frac{42}{5}, \frac{294}{25}, \ldots$

Find these infinite sums. (The series are geometric.)

9. $4 + 2 + 1 + \cdots$

10. $7 + 3 + \frac{9}{7} + \cdots$

11. $25 + 20 + 16 + \cdots$

12. $12 + 9 + \frac{27}{4} + \cdots$

13. $\sum_{k=1}^{\infty} \frac{1}{2^{k-1}}$

14. $\sum_{k=1}^{\infty} \frac{8}{3}\left(\frac{1}{2}\right)^{k-1}$

15. $\sum_{k=1}^{\infty} 16(0.1)^{k-1}$

16. $\sum_{k=1}^{\infty} 4(0.6)^{k-1}$

17. 🔳 $\$1000(1.08)^{-1} + \$1000(1.08)^{-2} + \$1000(1.08)^{-3} + \cdots$

18. 🔳 $\$500(1.12)^{-1} + \$500(1.12)^{-2} + \$500(1.12)^{-3} + \cdots$

ii Find fractional notation for these numbers.

19. $0.7\overline{77}$

20. $0.5\overline{33}$

21. $0.2121\overline{21}$

22. $0.6444\overline{4}$

23. $5.1515\overline{15}$

24. $0.4125\overline{125}$

25. The government makes an $8,000,000,000 expenditure for a new type of aircraft. If 85% of this gets spent again, and 85% of that gets spent again, and so on, what is the total effect on the economy?

26. Repeat Exercise 25 for $9,400,000,000 and 99%.

---

27. (*Advertising effect*). A company is marketing a new product in a city of 5,000,000 people. They plan an advertising campaign that they think will induce 40% of the people to buy the product. They estimate that if those people like the product, they will induce 40% (of the 40% of 5,000,000) more to buy the product, and those will induce 40%, and so on. In all, how many people will buy the product as a result of the advertising campaign? What percentage of the population is this?

28. How far (up and down) will a ball travel before stopping if it is dropped from a height of 12 ft, and each rebound is $\frac{1}{3}$ of the previous distance? (*Hint:* Use an infinite geometric series.)

---

29. 🔳 The function e^x is given by the infinite series

$$e^x = 1 + x + \frac{x^2}{2!} + \frac{x^2}{3!} + \cdots$$

(n! is read "n factorial," and is defined to be the product $1 \cdot 2 \cdot 3 \cdot \cdots \cdot n$). Approximate e to three decimal places using the first 6 terms of the series.

30. 🔳 An infinite sequence is defined recursively by

$$a_1 = 2, \quad a_{k+1} = \frac{1}{2}\left(a_k + \frac{2}{a_k}\right).$$

Find decimal notation, rounded to six decimal places, for a_1, $a_2, a_3, a_4, a_5,$ and a_6. Make a conjecture about the limit of the sequence.

31. 🔳 The infinite sequence

$$2, \quad \frac{1}{2}, \quad \frac{1}{2 \cdot 3}, \quad \frac{1}{2 \cdot 3 \cdot 4}, \quad \frac{1}{2 \cdot 3 \cdot 4 \cdot 5}, \quad \frac{1}{2 \cdot 3 \cdot 4 \cdot 5 \cdot 6}, \ldots$$

is not geometric, but does have a sum. Find $S_1, S_2, S_3, S_4, S_5,$ and S_6. Make a conjecture about the value of S_∞.

OBJECTIVES

You should be able to:

i List the statements of an infinite sequence that is defined by a formula.

ii Do proofs by mathematical induction.

1. List the first 5 statements in the sequence obtainable from
$$n^2 + 1 > n + 1.$$

2. List the first 5 statements in the sequence obtainable from
$$1 + 2 + \cdots + n = \frac{n(n + 1)}{2}.$$

11.5

MATHEMATICAL INDUCTION

i SEQUENCES OF STATEMENTS

Infinite sequences of statements occur often in mathematics. In an infinite sequence of statements there is, of course, a statement for each natural number. For example, consider the sentence

$$\text{For } x \text{ between 0 and 1,} \quad 0 < x^n < 1.$$

Let us think of this as $S(n)$ or S_n. Substituting natural numbers for n gives a sequence of statements. We list a few of them.

Statement 1 (S_1):*   For x between 0 and 1, $0 < x^1 < 1$.

Statement 2 (S_2):   For x between 0 and 1, $0 < x^2 < 1$.

S_3:   For x between 0 and 1, $0 < x^3 < 1$.

S_4:   For x between 0 and 1, $0 < x^4 < 1$.

EXAMPLE 1   List the first few statements in the sequence obtainable from $\log n < n$.

This time S_n is "$\log n < n$."

$$S_1: \quad \log 1 < 1$$
$$S_2: \quad \log 2 < 2$$
$$S_3: \quad \log 3 < 3$$

Many sequences of statements concern sums.

EXAMPLE 2   List the first few statements in the sequence obtainable from
$$1 + 3 + 5 + \cdots + (2n - 1) = n^2.$$

This time the entire equation is the statement S_n.

$$S_1: \quad 1 = 1^2$$
$$S_2: \quad 1 + 3 = 2^2$$
$$S_3: \quad 1 + 3 + 5 = 3^2$$
$$S_4: \quad 1 + 3 + 5 + 7 = 4^2$$

DO EXERCISES 1 AND 2.

ii PROVING INFINITE SEQUENCES OF STATEMENTS

The method of proof of this section, called *mathematical induction*, allows us to prove infinite sequences of statements. They are usually verbalized somewhat as follows:

$$\text{For all natural numbers } n, \ S_n,$$

where S_n is some sentence such as those in the preceding examples. Of course, we cannot prove each statement of an infinite sequence individ-

---

* Note that S_1, S_2, and so on do *not* represent sums in this context.

ually. Instead, we try to show that whenever S_k holds, then S_{k+1} will have to hold. We abbreviate this as $S_k \to S_{k+1}$. (This is read "S_k *implies* S_{k+1}.") If we can establish that this holds for all natural numbers k, we then have the following:

$S_1 \to S_2$ meaning whenever S_1 holds, S_2 must hold;

$S_2 \to S_3$ meaning whenever S_2 holds, S_3 must hold;

$S_3 \to S_4$ meaning whenever S_3 holds, S_4 must hold;

and so on, indefinitely.

At this stage we do not yet know whether there is *any* k for which S_k holds. All we know is that *if* S_k holds, then S_{k+1} must hold. So we now show that S_k holds for some k, usually $k = 1$. We are then in good condition because we have the following.

S_1 is true.   We have verified, or proved, this.

$S_1 \to S_2$ This means that whenever S_1 holds,
Therefore S_2 is true. S_2 must hold.

$S_2 \to S_3$ This means that whenever S_2 holds,
Therefore S_3 is true, S_3 must hold.

and so on.

We conclude that S_n is true for all natural numbers n.*
    We now state the principle of mathematical induction.

Principle of Mathematical Induction

We can prove an infinite sequence of statements S_n by showing the following.

a) S_1 is true. (This is called the *basis step*.)

b) For all natural numbers k, $S_k \to S_{k+1}$. (This is called the *induction step*.)

In learning to do proofs by mathematical induction, it is helpful to first write out S_n, S_1, S_k, and S_{k+1}. This helps to identify what is to be assumed and what is to be deduced.

EXAMPLE 3   Prove: For every natural number n, $n < 2^n$.

We first list S_n, S_1, S_k, and S_{k+1}.

$$S_n: \qquad n < 2^n$$
$$S_1: \qquad 1 < 2^1$$
$$S_k: \qquad k < 2^k$$
$$S_{k+1}: \qquad k + 1 < 2^{k+1}$$

a) *Basis step.* S_1 as listed is obviously true.

b) *Induction step.* We assume S_k as hypothesis and try to show that it implies S_{k+1}. For any k,

$$k < 2^k \qquad \text{By hypothesis (this is } S_k)$$
$$2k < 2 \cdot 2^k \qquad \text{Multiplying on both sides by 2}$$
$$2k < 2^{k+1}. \qquad \text{Adding exponents on the right}$$

---

*This is reminiscent of knocking over dominoes. The first one hits the second one, the second hits the third, and so on.

3. Prove that if x is a number greater than 1, then for any natural number n,

$$x \leqslant x^n.$$

Now, since k is any natural number,

$$1 \leqslant k$$
$$k + 1 \leqslant k + k \qquad \text{Adding } k \text{ on both sides}$$
$$k + 1 \leqslant 2k.$$

Thus we have

$$k + 1 \leqslant 2k < 2^{k+1}$$

and

$$k + 1 < 2^{k+1}. \qquad \text{This is } S_{k+1}.$$

We have now shown that $S_k \to S_{k+1}$ for any natural number k. Thus the induction step and the proof are complete. We have proved that for every natural number n, $n < 2^n$.

DO EXERCISE 3.

EXAMPLE 4   Prove: For every natural number n,

$$1 + 3 + 5 + 7 + \cdots + (2n - 1) = n^2.$$

We first list S_n, S_1, S_k, and S_{k+1}.

S_n: $1 + 3 + 5 + \cdots + (2n - 1) = n^2$

S_1: $1 = 1^2$

S_k: $1 + 3 + 5 + \cdots + (2k - 1) = k^2$

S_{k+1}: $1 + 3 + 5 + \cdots + (2k - 1) + [2(k + 1) - 1] = (k + 1)^2$

a)   *Basis step.* S_1 as listed is obviously true.

4. Consider

$$2 + 4 + 6 + \cdots + 2n = n(n + 1).$$

a) List S_1 and S_2.

b) List S_k.

c) List S_{k+1}.

d) Complete the basis step, that is, verify that S_1 is true.

e) Complete the proof that the formula holds for all n, by proving that S_k implies S_{k+1} for all natural numbers k.

b)   *Induction step.* We assume S_k as hypothesis and try to show that it implies S_{k+1}.

$$1 + 3 + 5 + \cdots + (2k - 1) = k^2 \quad \text{for any natural number } k.$$

This is hypothesis (it is S_k). Let us add $[2(k + 1) - 1]$ on both sides. This gives us

$$1 + 3 + \cdots + (2k - 1) + [2(k + 1) - 1] = k^2 + [2(k + 1) - 1]$$
$$= k^2 + 2k + 1$$
$$= (k + 1)^2.$$

We have arrived at S_{k+1}. Thus we have shown that for all natural numbers k, $S_k \to S_{k+1}$. This completes the induction step. The proof is complete. We have proved that the equation is true for all natural numbers n.

DO EXERCISE 4.

EXAMPLE 5   Prove

$$\sum_{p=1}^{n} \frac{1}{2^p} = \frac{2^n - 1}{2^n},$$

for all natural numbers n. In other words, prove that for all natural numbers n,

$$\frac{1}{2} + \frac{1}{4} + \frac{1}{8} + \cdots + \frac{1}{2^n} = \frac{2^n - 1}{2^n}.$$

We first list S_n, S_1, S_k, and S_{k+1}.

S_n: $\displaystyle\sum_{p=1}^{n} \frac{1}{2^p} = \frac{1}{2} + \frac{1}{4} + \frac{1}{8} + \cdots + \frac{1}{2^n} = \frac{2^n - 1}{2^n}$

S_1: $\displaystyle\frac{1}{2} = \frac{2^1 - 1}{2^1}$

S_k: $\displaystyle\sum_{p=1}^{k} \frac{1}{2^p} = \frac{1}{2} + \frac{1}{4} + \cdots + \frac{1}{2^k} = \frac{2^k - 1}{2^k}$

S_{k+1}: $\displaystyle\sum_{p=1}^{k+1} \frac{1}{2^p} = \frac{1}{2} + \frac{1}{4} + \frac{1}{8} + \cdots + \frac{1}{2^k} + \frac{1}{2^{k+1}} = \frac{2^{k+1} - 1}{2^{k+1}}$

a) *Basis step.* Since

$$\frac{2^1 - 1}{2^1} = \frac{2 - 1}{2} = \frac{1}{2},$$

S_1 is true.

b) *Induction step.* Using S_k as hypothesis, we have, for all natural numbers k,

$$\frac{1}{2} + \frac{1}{4} + \frac{1}{8} + \cdots + \frac{1}{2^k} = \frac{2^k - 1}{2^k}.$$

Let us add $1/2^{k+1}$ on both sides. Then we have

$$\frac{1}{2} + \frac{1}{4} + \cdots + \frac{1}{2^k} + \frac{1}{2^{k+1}} = \frac{2^k - 1}{2^k} + \frac{1}{2^{k+1}} = \frac{2^k - 1}{2^k} \cdot \frac{2}{2} + \frac{1}{2^{k+1}}$$

$$= \frac{(2^k - 1) \cdot 2 + 1}{2^{k+1}}$$

$$= \frac{2^{k+1} - 1}{2^{k+1}}.$$

We have arrived at S_{k+1}. Thus we have shown that for all natural numbers k, $S_k \to S_{k+1}$. This completes the induction and the proof is complete. We have proved that the equation is true for all natural numbers n.

DO EXERCISE 5.

5. Prove that

$$\sum_{p=1}^{n} (3p - 1) = \frac{n(3n + 1)}{2}$$

for all natural numbers n.

EXERCISE SET 11.5

i List the first 5 statements in the sequence obtainable from each of the following.

1. $n^2 < n^3$

2. $n^2 + n + 40$ is prime.

3. A polygon of n sides has $[n(n - 3)]/2$ diagonals.

4. The sum of the angles of a polygon of n sides is $(n - 2) \cdot 180°$.

ii In Exercises 5–22, use mathematical induction to prove each, for every natural number n.

5. $1 + 2 + 3 + \cdots + n = \dfrac{n(n + 1)}{2}$

6. $4 + 8 + 12 + \cdots + 4n = 2n(n + 1)$

7. $1 + 5 + 9 + \cdots + (4n - 3) = n(2n - 1)$

8. $3 + 6 + 9 + \cdots + 3n = \dfrac{3n(n + 1)}{2}$

9. $\dfrac{1}{1 \cdot 2} + \dfrac{1}{2 \cdot 3} + \cdots + \dfrac{1}{n(n + 1)} = \dfrac{n}{n + 1}$

10. $2 + 4 + 8 + \cdots + 2^n = 2(2^n - 1)$

11. $n < n + 1$

12. $2 \leqslant 2^n$

13. $3^n < 3^{n+1}$

14. $2n \leqslant 2^n$

15. $1^3 + 2^3 + 3^3 + \cdots + n^3 = \dfrac{n^2(n + 1)^2}{4}$

16. $\dfrac{1}{1 \cdot 2 \cdot 3} + \dfrac{1}{2 \cdot 3 \cdot 4} + \dfrac{1}{3 \cdot 4 \cdot 5} + \cdots + \dfrac{1}{n(n + 1)(n + 2)} = \dfrac{n(n + 3)}{4(n + 1)(n + 2)}$

---

17. $\left(1 + \dfrac{1}{1}\right)\left(1 + \dfrac{1}{2}\right)\left(1 + \dfrac{1}{3}\right) \cdots \left(1 + \dfrac{1}{n}\right) = n + 1$

18. $a_1 + (a_1 + d) + (a_2 + d) + \cdots + [a_1 + (n - 1)d] = \dfrac{n}{2}[2a_1 + (n - 1)d]$

(This is a formula for the sum of the first n terms of an *arithmetic* sequence (Theorem 3).)

19. $a_1 + a_1 r + a_1 r^2 + \cdots + a_1 r^{n-1} = \dfrac{a_1 - a_1 r^n}{1 - r}$

(This is a formula for the sum of the first n terms of a *geometric* sequence (Theorem 5).)

In Exercises 20 and 21, prove using mathematical induction.

20. For every natural number $n \geqslant 2$,

$$\log_a (b_1 b_2 \ldots b_n) = \log_a b_1 + \log_a b_2 + \cdots + \log_a b_n.$$

21. For every natural number $n \geqslant 2$,

$$\left(1 - \dfrac{1}{2^2}\right)\left(1 - \dfrac{1}{3^2}\right) \cdots \left(1 - \dfrac{1}{n^2}\right) = \dfrac{n + 1}{2n}.$$

Prove the following for any complex numbers $z_1, \ldots, z_n$, where $i^2 = -1$ and $\bar{z}$ is the conjugate of z (see Section 8.2).

22. $\overline{z^n} = \bar{z}^n$

23. $\overline{z_1 + z_2 + \cdots + z_n} = \bar{z}_1 + \bar{z}_2 + \cdots + \bar{z}_n$

24. $\overline{z_1 \cdot z_2 \cdots z_n} = \bar{z}_1 \cdot \bar{z}_2 \cdots \bar{z}_n$

25. i^n is either 1, -1, i, or $-i$.

For any integers a and b, b is a factor of a if there exists an integer c such that $a = bc$. Prove the following for any natural number n.

26. 3 is a factor of $n^3 + 2n$.

27. 2 is a factor of $n^2 + n$.

---

28. For every natural number $n \geqslant 2$,

$$\frac{1}{\sqrt{1}} + \frac{1}{\sqrt{2}} + \frac{1}{\sqrt{3}} + \cdots + \frac{1}{\sqrt{n}} > \sqrt{n}.$$

29. (*The Tower of Hanoi problem*). There are three pegs on a board. On one peg are n disks, each smaller than the one on which it rests. The problem is to move this pile of disks to another peg. The final order must be the same, but you can move only one disk at a time and you can never place a larger disk on a smaller one.

a) What is the *least* number of moves it takes to move 3 disks?
b) What is the *least* number of moves it takes to move 4 disks?
c) What is the *least* number of moves it takes to move 2 disks?
d) What is the *least* number of moves it takes to move 1 disk?
e) Conjecture a formula for the *least* number of moves it takes to move n disks. Prove it by mathematical induction.

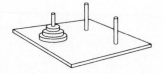

SUMMARY AND REVIEW: CHAPTER 11

The following contains a summary of what you should be able to do after completing this chapter. The review exercises are for practice. Answers are at the back of the book. If you miss an exercise, restudy the section indicated alongside the answer.

You should be able to:

Use the formula $a_n = a_1 + (n-1)d$, for arithmetic sequences, to find (a) the nth term, when n is given; (b) n, when an nth term is given. Given two terms of a sequence, find the common difference and construct the sequence. Use either the formula $S_n = \dfrac{n}{2}(a_1 + a_n)$ or $S_n = \dfrac{n}{2}[2a_1 + (n-1)d]$ for arithmetic sequences to find the sum of any arithmetic sequence.

1. Find the 10th term in the arithmetic sequence $\frac{3}{4}, \frac{13}{12}, \frac{17}{12}, \ldots$.

2. Find the 6th term in the arithmetic sequence $a - b$, $a, a + b, \ldots$.

3. Find the sum of the first 18 terms of the arithmetic sequence $4, 7, 10, \ldots$.

4. Find the sum of the first 30 positive integers.

5. The first term of an arithmetic sequence is 5. The 17th term is 53. Find the 3rd term.

6. The common difference in an arithmetic sequence is 3. The 10th term is 23. Find the first term.

Use the formula $a_n = a_1 r^{n-1}$ for geometric sequences to find a given term, or n when an nth term is given, and find sums of geometric sequences using the formula $S_n = \dfrac{a_1 - a_1 r^n}{1 - r}$ or $S_n = \dfrac{a_1 - ra_n}{1 - r}$.

7. For a geometric sequence, $a_1 = -2$, $r = 2$, and $a_n = -64$. Find n and S_n.

8. For a geometric sequence, $r = \frac{1}{2}$, $n = 5$, and $S_n = \frac{31}{2}$. Find a_1 and a_n.

Determine whether the sum of an infinite geometric sequence exists, and use the formula $S_\infty = a_1/(1-r)$ to find it. Convert from repeating decimal notation to fractional notation.

9. Determine whether this geometric sequence has a sum.

$$25, 27.5, 30.25, 33.275, \ldots$$

10. Determine whether this geometric sequence has a sum.

$$0.27, 0.0027, 0.000027, \ldots$$

11. Find this infinite sum. The series is geometric.

$$\tfrac{1}{2} - \tfrac{1}{6} + \tfrac{1}{18} - \cdots$$

12. Find fractional notation for $2.\overline{13}$.

Insert arithmetic means between two numbers.

13. Insert four arithmetic means between 5 and 9.

Solve applied problems involving sequences and series.

14. A golf ball is dropped from a height of 30 ft to the pavement, and the rebound is $\frac{1}{4}$ the distance it drops. If, after each descent, it continues to rebound $\frac{1}{4}$ the distance dropped, what is the total distance the ball has traveled when it reaches the pavement on its 10th descent?

15. You receive 10¢ on the first day of the year, 12¢ on the 2nd day, 14¢ on the 3rd day, and so on. How much will you receive on the 365th day? What is the sum of all these 365 gifts?

16. The present population of a city is 30,000. Its population is supposed to double every 10 yr. What will its population be at the end of 80 yr?

17. The sides of a square are each 16 in. long. A second square is inscribed by joining the midpoints of the sides, successively. In the second square we repeat the process, inscribing a third square. If this process is continued indefinitely, what is the sum of the perimeters of all of the squares? (*Hint:* Use an infinite geometric series.)

18. A pendulum is moving back and forth in such a way that it traverses an arc 10 cm in length and thereafter arcs that are $\frac{4}{7}$ the length of the previous arc. What is the sum of the arc lengths that the pendulum traverses?

Do proofs by mathematical induction.

Use mathematical induction.

19. Prove: For all natural numbers n,

$$1 + 4 + 7 + \cdots + (3n - 2) = \frac{n(3n-1)}{2}.$$

20. Prove: For all natural numbers n,

$$1 + 3 + 3^2 + \cdots + 3^{n-1} = \frac{3^n - 1}{2}.$$

21. Prove: For every natural number $n \geqslant 2$,

$$\left(1 - \frac{1}{2}\right) \cdot \left(1 - \frac{1}{3}\right) \cdots \left(1 - \frac{1}{n}\right) = \frac{1}{n}.$$

Given a recursively defined sequence, construct its terms. Convert between sigma (Σ) notation and other notation for the sum of a sequence.

22. Find the first 4 terms of this recursively defined sequence.

$$a_1 = 5, \quad a_{k+1} = 2a_k^2 + 1$$

23. Write Σ notation for this sequence.

$$0 + 3 + 8 + 15 + 24 + 35 + 48$$

24. Explain why the following cannot be proved by mathematical induction: For every natural number n,

a) $3 + 5 + \cdots + (2n + 1) = (n + 1)^2$;

b) $1 + 3 + \cdots + (2n - 1) = n^2 + 3$.

25. Suppose a and b are geometric sequences. Prove that the sequence c is a geometric sequence where $c_n = a_n b_n$.

26. Suppose a is an arithmetic sequence. Prove that c is a geometric sequence, where $c_n = b^{a_n}$, for some positive number b.

27. Suppose a is an arithmetic sequence. Under what conditions is b an arithmetic sequence where:

a) $b_n = |a_n|$? b) $b_n = a_n + 8$? c) $b_n = 7a_n$?

d) $b_n = \dfrac{1}{a_n}$? e) $b_n = \log a_n$? f) $b_n = a_n^3$?

28. The zeros of this polynomial form an arithmetic sequence. Find them.

$$x^4 - 4x^3 - 4x^2 + 16x$$

29. Write the first 3 terms of the infinite geometric sequence with $S_\infty = \frac{3}{11}$ and $r = 0.01$.

30. Write the first 3 terms of the infinite geometric sequence with $r = -\frac{1}{3}$ and $S_\infty = \frac{3}{8}$.

TEST: CHAPTER 11

1. Find the 18th term of the arithmetic sequence $\frac{1}{4}, 1, \frac{7}{4}, \frac{5}{2}, \ldots$.

2. The 2nd term of an arithmetic sequence is 9, and the 9th term is 37. Find the common difference.

3. Which term of the arithmetic sequence $1, \frac{3}{2}, 2, \frac{5}{2}, \ldots$ is $\frac{31}{2}$?

4. Insert three arithmetic means between 3 and 14.

5. Find the 8th term of the geometric sequence

$$0.2, 0.6, 1.8, \ldots.$$

6. Find the sum

$$\sum_{k=1}^{5} \left(\frac{1}{2}\right)^{k+1}.$$

7. Which of the following infinite geometric sequences have sums?

a) $2, 0.2, 0.02, 0.002, \ldots$

b) $3, -6, 12, -24, 48, \ldots$

c) $\frac{1}{20}, \frac{1}{10}, \frac{1}{5}, \frac{2}{5}, \ldots$

8. Find the sum of the infinite geometric sequence

$$25, -5, 1, -\frac{1}{5}, \ldots.$$

9. A student made deposits in a savings account as follows: $20.50 the first month, $26 the second month, $31.50 the third month, and so on for 2 years. What was the sum of the deposits?

10. A publishing company prints only $\frac{3}{5}$ as many books with each new printing of a book. If 100,000 copies of a book are printed originally, how many will be printed in the 5th printing?

11. Find fractional notation for $0.1288\overline{8}$.

12. Use mathematical induction. Prove for every natural number n,

$$5 + 10 + 15 + \cdots + 5n = \frac{5n(n + 1)}{2}.$$

13. Find the first 4 terms of this recursively defined sequence.

$$a_1 = 5, \quad a_{k+1} = 4a_k + 3$$

14. Find the geometric mean between 0.3 and 0.6.

15. Find 4 numbers in an arithmetic sequence such that twice the second minus the fourth is 1 and the sum of the first and third is 14.

Has a standard typewriter keyboard been designed to locate the most frequently occurring letters in the most convenient place? How can probability help to answer this question?

"What are your chances?" Whenever you hear this question your mind may be ready to delve into the realm of a study that we call *probability*. If you flip pennies all day long, about half of them will fall heads. Thus we assign the number $\frac{1}{2}$ as the *probability* of a coin falling heads. That number doesn't tell you what will happen if you flip just one coin, but we still say to ourselves that the chances are 1 out of 2, or that the probability is $\frac{1}{2}$.

In this chapter, we give you a brief introduction to probability theory. We assign, as a probability, a number from 0 to 1. In effect, we construct a probability *function* and then look at its properties and how we can use it to solve problems.

A great many of the applications of probability are to games of chance, such as dice and cards, but there are also many other applications, such as predicting elections or designing the timing of traffic lights or the makeup of telephone systems.

In order to study probability we need to learn methods of counting, which we consider in the first two sections. These methods are then applied to a way of expanding expressions of the type $(a + b)^n$. Finally, we use counting methods to study probability.

12

OBJECTIVE

You should be able to:

[i] **Evaluate factorial and permutation notation and solve related counting problems.**

12.1

PERMUTATIONS

[i] COMBINATORICS

We shall develop means of determining the number of ways in which a set of objects can be arranged or combined, certain objects can be chosen, or a succession of events can occur. This study is called *combinatorics*.

EXAMPLE 1   How many 3-letter code symbols can be formed with the letters A, B, and C *without* repetition?

Consider placing letters in these frames.

We can select any of the 3 letters for the first letter in the symbol. Once this letter has been selected, the second is selected from the remaining 2 letters. Then the third is already determined since there is only 1 letter left. The possibilities can be arrived at with a *tree diagram*.

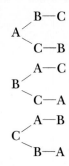

There are $3 \cdot 2 \cdot 1$ possibilities. The set of all of them is as follows:

$$\{ABC, ACB, BAC, BCA, CAB, CBA\}.$$

Suppose we perform an experiment such as selecting letters (as in the preceding example), flipping a coin, or drawing a card. The results are called *outcomes*. An *event* is a set of outcomes. When several events occur together, we say that the event is *compound*. The following principle is concerned with compound events.

Fundamental counting principle

> In a compound event in which the first event can occur independently in n_1 ways, the second can occur in n_2 ways, and so on, the kth event in n_k ways, the total number of ways in which the compound event can occur is
>
> $$n_1 \cdot n_2 \cdot n_3 \cdot \cdots \cdot n_k.$$

EXAMPLE 2   How many 3-letter code symbols can be formed with the letters A, B, and C *with* repetition?

There are 3 choices for the first letter, and since we allow repetition, 3 choices for the second, and 3 for the third. Thus by the fundamental counting principle there are $3 \cdot 3 \cdot 3$, or 27, choices.

DO EXERCISES 1–3.

Permutations

Definition

A *permutation* of a set of *n* objects is an ordered arrangement of all *n* objects.

Consider, for example, a set of 4 objects

$$\{A, B, C, D\}.$$

To find the number of ordered arrangements of the set, we select a first letter; there are 4 choices. Then we select a second letter; there are 3 choices. Then we select a third letter; there are 2 choices. Finally there is 1 choice for the last selection. Thus by the fundamental counting principle, there are $4 \cdot 3 \cdot 2 \cdot 1$, or 24, permutations of a set of 4 objects.

DO EXERCISE 4.

We can find a formula for the total number of permutations of all objects in a set of *n* objects. We have *n* choices for the first selection, $n - 1$ for the second, $n - 2$ for the third, and so on. For the *n*th selection there is only 1 choice.

Theorem 1

The total number of permutations of a set of *n* objects, denoted $_nP_n$, is given by

$$_nP_n = n(n - 1)(n - 2) \cdots (3)(2)(1).$$

EXAMPLE 3   Find (a) $_4P_4$ and (b) $_7P_7$.

a) $_4P_4 = 4 \cdot 3 \cdot 2 \cdot 1 = 24$

b) $_7P_7 = 7 \cdot 6 \cdot 5 \cdot 4 \cdot 3 \cdot 2 \cdot 1 = 5040$

EXAMPLE 4   In how many different ways can 9 letters be placed in 9 mailboxes, one letter to a box?

$$_9P_9 = 9 \cdot 8 \cdot 7 \cdot 6 \cdot 5 \cdot 4 \cdot 3 \cdot 2 \cdot 1 = 362{,}880$$

DO EXERCISES 5–10.

Factorial Notation

Products of successive natural numbers, such as $7 \cdot 6 \cdot 5 \cdot 4 \cdot 3 \cdot 2 \cdot 1$ and $9 \cdot 8 \cdot 7 \cdot 6 \cdot 5 \cdot 4 \cdot 3 \cdot 2 \cdot 1$, are used so often that it is convenient to adopt a notation for them.

1. How many 3-digit numbers can be named using all the digits 5, 6, 7 without repetition? with repetition?

2. A man is planning a date. He will first put on one of 4 suits, then call one of 2 girlfriends, and then select one of 5 restaurants. How many possible arrangements can he make for his date (assuming either girl would accept)?

3. In how many ways can 5 different cars be parked in a row in a parking lot?

4. How many permutations are there of a set of 5 objects? Consider a set $\{A, B, C, D, E\}$.

Compute.

5. $_3P_3$

6. $_5P_5$

7. $_6P_6$

8. In how many different ways can 4 horses be lined up for a race?

9. In how many different ways can 6 people line up at a ticket window?

10. In how many different ways can the 9-person batting order of a baseball team be made up, if you assume the pitcher bats last?

11. Find 9!.

12. Find 8!.

13. Using factorial notation only, represent the number of permutations of 18 objects.

14. Represent each in the form $n(n-1)!$

a) 10!

b) 20!

For the product $7 \cdot 6 \cdot 5 \cdot 4 \cdot 3 \cdot 2 \cdot 1$ we write 7!, read "7-factorial."

Definition

$$n! = n(n-1)(n-2) \cdots (3)(2)(1).$$

Here are some examples.

$$7! = 7 \cdot 6 \cdot 5 \cdot 4 \cdot 3 \cdot 2 \cdot 1 = 5040$$
$$6! = 6 \cdot 5 \cdot 4 \cdot 3 \cdot 2 \cdot 1 = 720$$
$$5! = 5 \cdot 4 \cdot 3 \cdot 2 \cdot 1 = 120$$
$$4! = 4 \cdot 3 \cdot 2 \cdot 1 = 24$$
$$3! = 3 \cdot 2 \cdot 1 = 6$$
$$2! = 2 \cdot 1 = 2$$
$$1! = 1 = 1$$

DO EXERCISES 11 AND 12.

We also define 0! to be 1. We do this so that certain formulas and theorems can be stated concisely.

We can now simplify the formula of Theorem 1 as follows:

$$_nP_n = n!.$$

DO EXERCISE 13.

Note that $8! = 8 \cdot 7!$ We can see this as follows: By definition of factorial notation,

$$8! = 8 \cdot 7 \cdot 6 \cdot 5 \cdot 4 \cdot 3 \cdot 2 \cdot 1$$
$$= 8 \cdot (7 \cdot 6 \cdot 5 \cdot 4 \cdot 3 \cdot 2 \cdot 1)$$
$$= 8 \cdot 7!.$$

Generalizing, we get the following.

For any natural number n, $n! = n(n-1)!$

By using this result repeatedly, we can further manipulate factorial notation.

EXAMPLE 5 Rewrite 7! with a factor of 5!.

$$7! = 7 \cdot 6 \cdot 5!$$

DO EXERCISE 14.

Permutations of n Objects Taken r at a Time

Consider a set of 6 objects. How many ordered arrangements are there having 3 members? We can select the first object in 6 ways. There are then 5 choices for the second and then 4 choices for the third. By the fundamental counting principle, there are then $6 \cdot 5 \cdot 4$ ways to construct the subset. In other words, there are $6 \cdot 5 \cdot 4$ permutations of a set of 6 objects taken 3 at a time. Note that

$$6 \cdot 5 \cdot 4 = \frac{6 \cdot 5 \cdot 4 \cdot 3 \cdot 2 \cdot 1}{3 \cdot 2 \cdot 1}, \quad \text{or} \quad \frac{6!}{3!}.$$

Definition

A *permutation* of a set of *n* objects taken *r* at a time is an ordered arrangement of *r* objects taken from the set.

Consider a set of n objects and the selecting of an ordered arrangement of r objects. The first object can be selected in n ways. The second can be selected in $n - 1$ ways, and so on. The rth can be selected in $n - (r - 1)$ ways. By the fundamental counting principle, the total number of permutations is

$$n(n - 1)(n - 2) \cdots [n - (r - 1)].$$

We now multiply by 1:

$$n(n - 1)(n - 2) \cdots [n - (r - 1)] \frac{(n - r)!}{(n - r)!}$$

$$= \frac{n(n - 1)(n - 2)(n - 3) \cdots [n - (r - 1)](n - r)!}{(n - r)!}.$$

The numerator is now the product of all natural numbers from n to 1, hence is $n!$. Thus the total number of permutations is

$$\frac{n!}{(n - r)!}.$$

This gives us the following theorem.

Theorem 2

The number of permutations of a set of *n* objects taken *r* at a time, denoted $_nP_r$, is given by

$$_nP_r = n(n - 1)(n - 2) \cdots [n - (r - 1)] \qquad (1)$$

$$= \frac{n!}{(n - r)!}. \qquad (2)$$

Formula (1) is most useful in application, but formula (2) will be important in a later development.

EXAMPLE 6   Compute $_6P_4$ using both formulas of Theorem 2.

Using formula (1), we have

$$_6P_4 = \overbrace{6 \cdot 5 \cdot 4 \cdot 3}$$ Note that the 6 in $_6P_4$ shows where to start and the 4 shows how many factors there are.

$$= 360.$$

Using formula (2) of Theorem 2, we have

$$_6P_4 = \frac{6!}{(6 - 4)!} = \frac{6!}{2!} = \frac{6 \cdot 5 \cdot 4 \cdot 3 \cdot 2 \cdot 1}{2 \cdot 1} = 6 \cdot 5 \cdot 4 \cdot 3 = 360.$$

DO EXERCISES 15 AND 16.

15. Compute $_7P_3$.

16. Compute.

a) $_{10}P_4$

b) $_8P_2$

c) $_{11}P_5$

d) $_nP_1$

e) $_nP_2$

17. In how many ways can a 5-woman starting unit be selected from a 12-woman basketball squad and arranged in a straight line?

18. A professor wants to write a 6-item test from a pool of 10 questions. In how many ways can he do this?

19. How many 7-digit numbers can be named, without repetition, using the digits 2, 3, 4, 5, 6, 7, and 8, if the even digits come first?

EXAMPLE 7 In how many ways can letters of the set {A, B, C, D, E, F, G} be arranged to form code words of (a) 7 letters? (b) 5 letters? (c) 4 letters? (d) 2 letters?

a) $_7P_7 = 7 \cdot 6 \cdot 5 \cdot 4 \cdot 3 \cdot 2 \cdot 1 = 5040$

b) $_7P_5 = 7 \cdot 6 \cdot 5 \cdot 4 \cdot 3 \qquad = 2520$

c) $_7P_4 = 7 \cdot 6 \cdot 5 \cdot 4 \qquad\quad = \ 840$

d) $_7P_2 = 7 \cdot 6 \qquad\qquad\qquad = \quad 42$

EXAMPLE 8 A baseball manager arranges the batting order as follows: The 4 infielders will bat first, then the outfielders, catcher, and pitcher will follow, not necessarily in that order. How many different batting orders are possible?

The infielders can bat in 4! different ways; the rest in 5! different ways. Then by the fundamental counting principle we have $_4P_4 \cdot {}_5P_5 = 4! \cdot 5!$, or 2880, possible batting orders.

DO EXERCISES 17–19.

Permutations of Sets with Nondistinguishable Objects

Consider a set of 7 marbles, 4 of which are colored and 3 of which are black. Although the marbles are all different, when they are lined up, one black marble will look just like any other black marble. In this sense we say that the marbles are nondistinguishable.

We know that there are 7! permutations of this set. Many of them will look alike, however. We develop a formula for finding the number of distinguishable permutations.

Consider a set of n objects in which n_1 are of one kind, n_2 are of a second kind, . . . , n_k are of the kth kind. By Theorem 1 the total number of permutations of the set is $n!$. Let P be the number of distinguishable permutations. For each of these P permutations there are $n_1!$ actual permutations, obtained by permuting the objects of the first kind. For each of these $P_{n_1}!$ permutations there are $n_2!$ actual permutations, obtained by permuting the objects of the second kind. And so on. By the fundamental counting principle, the total number of actual permutations is

$$P \cdot n_1! \cdot n_2! \cdots n_k!.$$

Then we have $P \cdot n_1! \cdot n_2! \cdots n_k! = n!$. Solving for P, we obtain

$$P = \frac{n!}{n_1! n_2! \cdots n_k!}.$$

This proves the following theorem.

Theorem 3

For a set of n objects in which n_1 are of one kind, n_2 are of another kind, . . . , n_k are of a kth kind, the number of distinguishable permutations is

$$\frac{n!}{n_1! \cdot n_2! \cdots n_k!}.$$

EXAMPLE 9 In how many distinguishable ways can the letters of the word CINCINNATI be arranged?

Note: There are 2 C's, 3 I's, 3 N's, 1 A, and 1 T, for a total of 10. Thus

$$P = \frac{10!}{2! \cdot 3! \cdot 3! \cdot 1! \cdot 1!}, \quad \text{or} \quad 50,400.$$

DO EXERCISES 20–22.

Circular Arrangements

Suppose we arrange the 4 letters A, B, C, and D in a circular arrangement.

Note that the arrangements ABCD, BCDA, CDAB, and DABC are not distinguishable. For each circular arrangement there are 4 distinguishable arrangements on a line. Thus if there are P circular arrangements, these yield $4 \cdot P$ arrangements on a line, which we know is 4!. Thus $4 \cdot P = 4!$, so $P = 4!/4$, or 3!. To generalize:

Theorem 4

> The number of distinct (different) circular arrangements of n objects is $(n-1)!$.

EXAMPLE 10 In how many ways can 9 different foods be arranged around a lazy Susan?

$$(9-1)! = 8! = 8 \cdot 7 \cdot 6 \cdot 5 \cdot 4 \cdot 3 \cdot 2 \cdot 1 = 40,320$$

DO EXERCISES 23 AND 24.

Repeated Use of the Same Object

EXAMPLE 11 How many 5-letter code symbols can be formed with the letters A, B, C, and D if we allow a letter to occur more than once?

We have five spaces:

We can select the first letter in 4 ways, the second in 4 ways, and so on. Thus there are 4^5, or 1024, arrangements.

Generalizing, we have the following.

Theorem 5

> The number of distinct arrangements of n objects taken r at a time, allowing repetition, is n^r.

DO EXERCISE 25.

---

20. In how many distinguishable ways can the letters of the word MISSISSIPPI be arranged?

21. How many 6-digit numbers can be named with all the digits 3, 3, 3, 4, 4, and 5?

22. How many vertical signal flag arrangements can be formed with 3 solid red, 3 solid green, and 2 solid yellow flags?

23. In how many ways can 7 men be arranged around a round table?

24. In how many ways can the numbers on a clock face be arranged?

25. How many 5-letter code symbols can be formed by repeated use of the letters of the alphabet?

EXERCISE SET 12.1

| i | Evaluate.

1. $_4P_3$ **2.** $_7P_5$ **3.** $_{10}P_7$ **4.** $_{10}P_3$

Table 5 at the back of the book can help with many of these problems.

5. How many 5-digit numbers can be named using the digits 5, 6, 7, 8, and 9 without repetition? with repetition?

6. How many 4-digit numbers can be named using the digits 2, 3, 4, and 5 without repetition? with repetition?

7. In how many ways can 5 students be arranged in a straight line? in a circle?

8. In how many ways can 7 athletes be arranged in a straight line? in a circle?

9. In how many distinguishable ways can the letters of the word DIGIT be arranged?

10. In how many distinguishable ways can the letters of the word RABBIT be arranged?

11. How many 7-digit phone numbers can be formed with the digits 0, 1, 2, 3, 4, 5, 6, 7, 8, and 9, assuming that no digit is used more than once and the first digit is not 0?

12. A program is planned and is to have 5 rock numbers and 4 speeches. In how many ways can this be done if a rock number and a speech are to alternate and the rock numbers come first?

13. Suppose the expression $a^2b^3c^4$ is rewritten without exponents. In how many ways can this be done?

14. Suppose the expression a^3bc^2 is rewritten without exponents. In how many ways can this be done?

15. In how many ways could King Arthur and his 12 knights sit at his Round Table?

16. In how many ways can 4 people be seated at a bridge table?

17. A penny, nickel, dime, quarter, and half dollar (if you have one) are arranged in a straight line. a) Considering just the coins, in how many ways can they be lined up? b) Considering the coins and heads and tails, in how many ways can they be lined up?

18. A penny, nickel, dime, and quarter are arranged in a straight line. a) Considering just the coins, in how many ways can they be lined up? b) Considering the coins and heads and tails, in how many ways can they be lined up?

19. | Compute $_{52}P_4$.

20. | Compute $_{50}P_5$.

21. A professor is going to grade her 24 students on a curve. She will give 3 A's, 5 B's, 9 C's, 4 D's, and 3 F's. In how many ways can she do this?

22. A professor is planning to grade his 20 students on a curve. He will give 2 A's, 5 B's, 8 C's, 3 D's, and 2 F's. In how many ways can he do this?

23. | How many distinguishable code symbols can be formed from the letters of the word MATH? BUSINESS? PHILOSOPHICAL?

24. | How many distinguishable code symbols can be formed from the letters of the word ORANGE? BIOLOGY? MATHEMATICS?

25. | A state forms its license plates by first listing a number that corresponds to the county in which the car owner lives (the names of the counties are alphabetized and the number is its location in the order). Then the plate lists a letter of the alphabet and this is followed by a number from 1 to 9999. How many such plates are possible if there are 80 counties?

26. How many code symbols can be formed using 4 out of 5 letters of A, B, C, D, E if the letters

   a) are not repeated?
   b) can be repeated?
   c) are not repeated but must begin with D?
   d) are not repeated but must end with DE?

```
29 B 7480
INDIANA
```

Solve for n.

27. $_nP_5 = 7 \cdot {_nP_4}$ **28.** $_nP_4 = 8 \cdot {_{n-1}P_3}$ **29.** $_nP_5 = 9 \cdot {_{n-1}P_4}$ **30.** $_nP_4 = 8 \cdot {_nP_3}$

31. In a single-elimination sports tournament consisting of n teams, a team is eliminated when it loses one game. How many games are required to complete the tournament?

32. In a double-elimination softball tournament consisting of n teams, a team is eliminated when it loses two games. At most how many games are required to complete the tournament?

33. Find a formula for the total number of arrangements of n keys on a key ring.

12.2

COMBINATIONS

i Permutations of a set are ordered arrangements. *Combinations* of a set are simply subsets and are not considered with regard to order.

EXAMPLE 1 How many combinations are there of the set {A, B, C, D} taken 3 at a time?

The combinations are the subsets {A, B, C}, {A, C, D}, {B, C, D}, and {A, B, D}. There are four of them.

Definition

> The number of combinations taken *r* at a time from a set of *n* objects, denoted $_nC_r$, is the number of subsets containing exactly *r* objects.

DO EXERCISE 1.

We develop a formula for finding $_nC_r$, but we first define some convenient symbolism.

Definition

> $\binom{n}{r}$ **is defined to mean** $\dfrac{n!}{r!(n-r)!}$.

The notation $\binom{n}{r}$ is read "*n* over *r*," or (for reasons we see later) "binomial coefficient *n* over *r*."

CAUTION!

$\binom{n}{r}$ does *not* mean $n \div r$, or $\dfrac{n}{r}$.

EXAMPLE 2 Evaluate $\binom{5}{2}$.

$$\binom{5}{2} = \frac{5!}{2!(5-2)!} = \frac{5!}{2!3!} = \frac{5 \cdot 4 \cdot 3!}{2!3!} = \frac{5 \cdot 4}{2} = 10$$

We now find a formula for $_nC_r$. Consider a set of *n* objects. The number of subsets with *r* members is $_nC_r$. Each subset with *r* members can be arranged in *r*! ways. By the fundamental counting principle we have $_nC_r \cdot r!$ ordered subsets with *r* members. This gives us

$$_nP_r = {}_nC_r \cdot r!.$$

Then

$$_nC_r = \frac{_nP_r}{r!}.$$

OBJECTIVE

You should be able to:

i Evaluate combination notation and solve related problems.

1. Consider the set

$$\{A, B, C, D\}.$$

How many combinations are there taken

a) 2 at a time?

b) 4 at a time?

c) 1 at a time?

d) 0 at a time?

2. Consider the set of 5 objects

$$\{A, B, C, D, E\}.$$

a) Calculate the number of permutations of this set taken 3 at a time.

b) List these permutations.

c) Calculate the number of combinations of this set taken 3 at a time.

d) List these combinations.

3. Compute.

a) $\binom{10}{8}$

b) $\binom{10}{2}$

c) $_7C_5$

d) $_7C_7$

4. Compute $\binom{9}{7}$ and $\binom{9}{2}$.

5. In how many ways can a 5-woman starting unit be selected from a 12-woman basketball squad?

6. A committee is to be chosen from 12 men and 8 women and is to consist of 3 men and 2 women. How many committees can be formed?

Since

$$_nP_r = \frac{n!}{(n-r)!}, \qquad \text{Theorem 2}$$

we get

$$_nC_r = \frac{n!}{r!(n-r)!},$$

or

$$_nC_r = \binom{n}{r}.$$

This proves the following.

Theorem 6

The number of combinations of a set of n objects taken r at a time is given by

$$_nC_r = \binom{n}{r} = \frac{n!}{r!(n-r)!}.$$

EXAMPLE 3 Compute $_7C_4$.

By Theorem 6 we have $_7C_4 = \binom{7}{4}$.

$$\binom{7}{4} = \frac{7!}{4!3!} = \frac{7 \cdot 6 \cdot 5 \cdot 4!}{3! \cdot 4!} = \frac{7 \cdot 6 \cdot 5}{3 \cdot 2} = 35$$

DO EXERCISES 2–4.

EXAMPLE 4 How many committees can be formed from a set of 5 governors and 7 senators if each committee contains 3 governors and 4 senators?

The 3 governors can be selected in $_5C_3$ ways and the 4 senators can be selected in $_7C_4$ ways. Using the fundamental counting principle, it follows that the number of possible committees is

$$_5C_3 \cdot {_7C_4} = 10 \cdot 35 = 350.$$

DO EXERCISES 5 AND 6.

EXERCISE SET 12.2

i Compute. Table 5 at the back of the book or a calculator can be of help with these problems.

1. $_9C_5$

2. $_{14}C_2$

3. $\binom{50}{2}$

4. $\binom{40}{3}$

5. $_nC_3$

6. $_nC_2$

7. There are 23 students in a fraternity. How many sets of 4 officers can be selected?

8. How many basketball games can be played in a 9-team league if each team plays all other teams once? twice?

9. On a test a student is to select 6 out of 10 questions. In how many ways can he do this?

10. On a test a student is to select 7 out of 11 questions. In how many ways can she do this?

11. How many lines are determined by 8 points, no 3 of which are collinear? How many triangles are determined by the same points?

12. How many lines are determined by 7 points, no 3 of which are collinear? How many triangles are determined by the same points?

13. Of the first 10 questions on a test, a student must answer 7. On the second 5 questions, she must answer 3. In how many ways can this be done?

14. Of the first 8 questions on a test, a student must answer 6. On the second 4 questions, he must answer 3. In how many ways can this be done?

15. Suppose the Senate of the United States consisted of 58 Democrats and 42 Republicans. How many committees consisting of 6 Democrats and 4 Republicans could be formed? You need not simplify the expression.

16. Suppose the Senate of the United States consisted of 63 Republicans and 37 Democrats. How many committees consisting of 8 Republicans and 12 Democrats could be formed? You need not simplify the expression.

17. How many 5-card poker hands consisting of 3 aces and 2 cards that are not aces are possible with a 52-card deck? (See Section 15.4 for a description of a 52-card deck.)

18. How many 5-card poker hands consisting of 2 kings and 3 cards that are not kings are possible with a 52-card deck?

---

19. ▦ How many 5-card poker hands are possible with a 52-card deck?

20. ▦ How many 13-card bridge hands are possible with a 52-card deck?

21. There are 8 points on a circle. How many triangles can be inscribed with these points as vertices?

22. There are n points on a circle. How many quadrilaterals can be inscribed with these points as vertices?

23. A set of 5 parallel lines crosses another set of 8 parallel lines at angles that are not right angles. How many parallelograms are formed?

24. Prove: For any natural numbers n and $r \leqslant n$,

$$\binom{n}{r} = \binom{n}{n-r}.$$

Solve for n.

25. $\binom{n+1}{3} = 2 \cdot \binom{n}{2}$

26. $\binom{n}{n-2} = 6$

27. $\binom{n+2}{4} = 6 \cdot \binom{n}{2}$

28. $\binom{n}{3} = 2 \cdot \binom{n-1}{2}$

---

29. How many line segments are determined by the 5 vertices of a pentagon? Of these, how many are diagonals?

30. How many line segments are determined by the 6 vertices of a hexagon? Of these, how many are diagonals?

31. How many line segments are determined by the n vertices of an n-gon? Of these, how many are diagonals? Use mathematical induction to prove the result for the diagonals.

32. Prove: For any natural numbers n and $r \leqslant n$,

$$\binom{n}{r-1} + \binom{n}{r} = \binom{n+1}{r}.$$

12.3

BINOMIAL SERIES

i   EXPANDING POWERS OF BINOMIALS

Our goal in this section is to find an efficient way to expand expressions of the type $(a + b)^n$.

Consider the following powers of $a + b$, where $a + b$ is any binomial. Look for patterns.

$$
\begin{aligned}
(a+b)^0 &= 1 \\
(a+b)^1 &= a + b \\
(a+b)^2 &= a^2 + 2ab + b^2 \\
(a+b)^3 &= a^3 + 3a^2b + 3ab^2 + b^3 \\
(a+b)^4 &= a^4 + 4a^3b + 6a^2b^2 + 4ab^3 + b^4 \\
(a+b)^5 &= a^5 + 5a^4b + 10a^3b^2 + 10a^2b^3 + 5ab^4 + b^5
\end{aligned}
$$

OBJECTIVES

You should be able to:

i Expand a binomial power and find the rth term of such an expression.

ii Find the total number of subsets of a set of n objects.

Each expansion is a polynomial that can be regarded as a series. There are some patterns to be noted in the expansions.

1. In each term, the sum of the exponents is n.

2. The exponents of a start with n and decrease, to 0. The last term has no factor of a. The first term has no factor of b. The exponents of b start in the second term with 1 and increase to n.

3. There is one more term than the degree of the polynomial. The expansion of $(a + b)^n$ has $n + 1$ terms.

We now find a way to determine the coefficients. Let us consider the nth power of a binomial $(a + b)$:

$$(a + b)^n = \underbrace{(a + b)(a + b)(a + b) \cdots (a + b)}_{n \text{ factors}}.$$

When we multiply, we will find all possible products of a's and b's. For example, when we multiply all the first terms we will get n factors of a, or a^n. Thus the first term in the expansion is a^n, and the first coefficient is 1. Similarly, the coefficient of the last term is 1.

To get a term such as the $a^{n-r}b^r$ term, we will take a's from $n - r$ factors and b's from r factors. Thus we take n objects, $n - r$ of them a's and r of them b's. The number of ways in which we can do this is

$$\frac{n!}{(n - r)!r!}.$$

This is $\binom{n}{r}$. Thus the $(r + 1)$st term in the expansion is $\binom{n}{r}a^{n-r}b^r$. We now have a theorem.

Theorem 7 The binomial theorem

For any binomial $(a + b)$ and any natural number n,

$$(a + b)^n = \binom{n}{0}a^n + \binom{n}{1}a^{n-1}b + \binom{n}{2}a^{n-2}b^2 + \cdots + \binom{n}{n}b^n.$$

The binomial theorem can be proved by mathematical induction but we will not do that here.

Sigma notation for a binomial series is as follows:

$$(a + b)^n = \sum_{r=0}^{n} \binom{n}{r}a^{n-r}b^r.$$

Because of Theorem 7, $\binom{n}{r}$ is called a *binomial coefficient*. It should now be apparent why 0! is defined to be 1. In the binomial expansion we want $\binom{n}{0}$ to equal 1 and we also want the definition

$$\binom{n}{r} = \frac{n!}{(n - r)!r!}$$

to hold for all whole numbers n and r. Thus we must have

$$\binom{n}{0} = \frac{n!}{(n - 0)!0!} = \frac{n!}{n!0!} = 1.$$

This will be satisfied if 0! is defined to be 1.

EXAMPLE 1 Expand $(x^2 - 2y)^5$.

Note that $a = x^2$, $b = -2y$, and $n = 5$. Then using the binomial theorem we have

$$(x^2 - 2y)^5 = \binom{5}{0}(x^2)^5 + \binom{5}{1}(x^2)^4(-2y) + \binom{5}{2}(x^2)^3(-2y)^2$$

$$+ \binom{5}{3}(x^2)^2(-2y)^3 + \binom{5}{4}x^2(-2y)^4 + \binom{5}{5}(-2y)^5$$

$$= \frac{5!}{0!5!}x^{10} + \frac{5!}{1!4!}x^8(-2y) + \frac{5!}{2!3!}x^6(-2y)^2$$

$$+ \frac{5!}{3!2!}x^4(-2y)^3 + \frac{5!}{4!1!}x^2(-2y)^4 + \frac{5!}{5!0!}(-2y)^5$$

$$= x^{10} - 10x^8 y + 40x^6 y^2 - 80x^4 y^3 + 80x^2 y^4 - 32y^5.$$

DO EXERCISE 1.

EXAMPLE 2 Expand $(2/x + 3\sqrt{x})^4$.

Note that $a = 2/x$, $b = 3\sqrt{x}$, and $n = 4$. Then using the binomial theorem, we have

$$\left(\frac{2}{x} + 3\sqrt{x}\right)^4 = \binom{4}{0}\cdot\left(\frac{2}{x}\right)^4 + \binom{4}{1}\cdot\left(\frac{2}{x}\right)^3(3\sqrt{x}) + \binom{4}{2}\cdot\left(\frac{2}{x}\right)^2(3\sqrt{x})^2$$

$$+ \binom{4}{3}\left(\frac{2}{x}\right)(3\sqrt{x})^3 + \binom{4}{4}(3\sqrt{x})^4$$

$$= \frac{4!}{0!4!}\cdot\frac{16}{x^4} + \frac{4!}{1!3!}\cdot\frac{8}{x^3}3\sqrt{x} + \frac{4!}{2!2!}\cdot\frac{4}{x^2}\cdot 9x$$

$$+ \frac{4!}{3!1!}\cdot\frac{2}{x}\cdot 27x^{3/2} + \frac{4!}{4!0!}\cdot 81x^2$$

$$= \frac{16}{x^4} + \frac{96}{x^{5/2}} + \frac{216}{x} + 216\sqrt{x} + 81x^2.$$

DO EXERCISES 2 AND 3.

EXAMPLE 3 Find the 7th term of $(4x - y^3)^9$.

We let $r = 6$, $n = 9$, $a = 4x$, and $b = -y^3$ in the formula $\binom{n}{r}a^{n-r}b^r$. Then

$$\binom{9}{6}(4x)^3(-y^3)^6 = \frac{9!}{6!3!}(4x)^3(-y^3)^6$$

$$= \frac{9\cdot 8\cdot 7\cdot 6!}{3!\cdot 6!}64x^3 y^{18}$$

$$= 5376x^3 y^{18}.$$

DO EXERCISES 4–7.

1. Expand $(x^2 - 1)^5$.

2. Expand $\left(2x + \dfrac{1}{y}\right)^4$.

3. Expand $(x - \sqrt{2})^6$.

4. Find the 4th term of $(x - 3)^8$.

5. Find the 5th term of $(x - 3)^8$.

6. Find the 6th term of $(y + 2)^{10}$.

7. Find the 1st term of $(3x + 5)^5$.

8. How many subsets are there in the set $\{a, b, c, d, e, f\}$?

9. How many subsets are there of the set of all states of the United States?

ii SUBSETS

Suppose a set has n objects. The number of subsets containing r members is $\binom{n}{r}$, by Theorem 6. The total number of subsets of a set is the number with 0 elements, plus the number with 1 element, plus the number with 2 elements, and so on. The total number of subsets of a set with n members is

$$\binom{n}{0} + \binom{n}{1} + \binom{n}{2} + \cdots + \binom{n}{n}.$$

Now let us expand $(1 + 1)^n$:

$$(1 + 1)^n = \binom{n}{0} + \binom{n}{1} + \binom{n}{2} + \cdots + \binom{n}{n}.$$

Thus the total number of subsets is $(1 + 1)^n$, or 2^n. We have proved the following theorem.

Theorem 8

> The total number of subsets of a set with n members is 2^n.

EXAMPLE 4 The set $\{A, B, C, D, E\}$ has how many subsets?

The set has 5 members, so the number of subsets is 2^5, or 32.

EXAMPLE 5 A fast-food restaurant can make hamburgers 256 ways using combinations of 8 seasonings. Show why.

The total number of combinations is

$$\binom{8}{0} + \binom{8}{1} + \cdots + \binom{8}{8} = 2^8$$
$$= 256.$$

DO EXERCISES 8 AND 9.

EXERCISE SET 12.3

i Find the indicated term of the binomial expression.

1. 3rd, $(a + b)^6$

2. 6th, $(x + y)^7$

3. 12th, $(a - 2)^{14}$

4. 11th, $(x - 3)^{12}$

5. 5th, $(2x^3 - \sqrt{y})^8$

6. 4th, $\left(\dfrac{1}{b^2} + \dfrac{b}{3}\right)^7$

7. Middle, $(2u - 3v^2)^{10}$

8. Middle two, $(\sqrt{x} + \sqrt{3})^5$

Expand.

9. $(m + n)^5$

10. $(a - b)^4$

11. $(x^2 - 3y)^5$

12. $(3c - d)^6$

13. $(x^{-2} + x^2)^4$

14. $\left(\dfrac{1}{\sqrt{x}} - \sqrt{x}\right)^6$

15. $(1 - 1)^n$

16. $(1 + 3)^n$

17. $(\sqrt{2} + 1)^6 - (\sqrt{2} - 1)^6$

18. $(1 - \sqrt{2})^4 + (1 + \sqrt{2})^4$

19. $(\sqrt{3} - t)^4$

20. $(\sqrt{5} + t)^6$

ii Determine the number of subsets:

21. Of a set of 7 members.

22. Of a set of 6 members.

23. ▦ Of the set of letters of the English alphabet, which contains 26 letters.

24. ▦ Of the set of letters of the Greek alphabet, which contains 24 letters.

☐

Expand.

25. $(\sqrt{2} - i)^4$, where $i^2 = -1$

26. $(1 + i)^6$, where $i^2 = -1$

27. $(e^x - e^{-x})^7$

28. $(2^x + 2^{-x})^{11}$

29. Find a formula for

$$(a - b)^n.$$

Use sigma notation.

30. Expand and simplify

$$\frac{(x + h)^n - x^n}{h}.$$

Use sigma notation.

Solve for x.

31. $\sum_{r=0}^{8} \binom{8}{r} x^{8-r} 3^r = 0$

32. $\sum_{r=0}^{4} \binom{4}{r} 5^{4-r} x^r = 64$

33. $\sum_{r=0}^{5} \binom{5}{r} (-1)^r x^{5-r} 3^r = 32$

34. $\sum_{r=0}^{9} \binom{9}{r} (-1)^{9-r} e^{rx} = 0$

35. A money clip contains one each of the following bills: \$1, \$2, \$5, \$10, \$20, \$50, and \$100. How many different sums of money can be formed using the bills?

■

Simplify.

36. $\sum_{r=0}^{23} \binom{23}{r} (\log_a x)^{23-r} (\log_a t)^r$

37. Use mathematical induction and the property $\binom{k}{r-1} + \binom{k}{r} = \binom{k+1}{r}$ to prove the binomial theorem.

12.4

PROBABILITY

We say that when a coin is tossed the chances that it will fall heads are 1 out of 2, or that the *probability* that it will fall heads is $\frac{1}{2}$. Of course this does not mean that if a coin is tossed ten times it will necessarily fall heads exactly five times. If the coin is tossed a great number of times, however, it will fall heads very nearly half of them.

EXPERIMENTAL AND THEORETICAL PROBABILITY

If we toss a coin a large number of times, say 1000, and count the number of heads, we can determine the probability. If there are 503 heads, we would calculate the probability to be

$$\frac{503}{1000}, \quad \text{or} \quad 0.503.$$

This is an *experimental* determination of probability. Such determination of probability is quite common. Here, for example, are some probabilities

OBJECTIVE

You should be able to:

i Compute the probability of a simple event.

that have been determined *experimentally*:

1. If you kiss someone who has a cold, the probability of your catching a cold is 0.07.

2. A person just released from prison has an 80% probability of returning.

If we consider a coin and reason that it is just as likely to fall heads as tails, we would calculate the probability to be $\frac{1}{2}$. This is a *theoretical* determination of probability. Here, for example, are some probabilities that have been determined *theoretically*:

1. If you have 30 people in a room, the probability that two of them have the same birthday (excluding year of birth) is 0.706.

2. You are on a trip. You meet someone, and after a period of conversation, discover that you have a common acquaintance. The typical reaction, "It's a small world!" is actually not appropriate because the probability of such an occurrence is quite high, just over 22%.

It is results like these that lend credence to the value of a study of probability. You might ask, "What is the *true* probability?" In fact there is none. Experimentally we can determine probabilities within certain limits. These may or may not agree with what we obtain theoretically.

i COMPUTING PROBABILITIES

Experimental Probabilities

We first consider experimental determination of probability. The basic principle we use in computing such probabilities is as follows.

Principle *P* (Experimental)

An experiment is performed in which *n* observations are made. If a situation E, or event, occurs *m* times out of the *n* observations, then we say that the probability of that event is given by

$$P(E) = \frac{m}{n}.$$

EXAMPLE 1 (*Sociological survey*). An actual experiment was conducted to determine the number of people who are left-handed, right-handed, or both. The results are shown in the following graph.

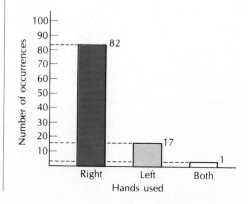

a) Determine the probability that a person is left-handed.

b) Determine the probability that a person is ambidexterous (uses both hands with equal ability).

1. In reference to Example 1, what is the probability that a person is right-handed?

a) The number of people who are right-handed was 82, the number who are left-handed was 17, and there was 1 person who is ambidexterous. The total number of observations was $82 + 17 + 1$, or 100. Thus the probability that a person is left-handed is P, where

$$P = \frac{17}{100}.$$

b) The probability that a person is ambidexterous is P, where

$$P = \frac{1}{100}.$$

DO EXERCISE 1.

EXAMPLE 2 (*Quality control*). It is very important to a manufacturer to maintain the quality of its products. The goal is to produce as few defective products as possible. But can a company afford to check every product to see if it is defective if it is producing thousands of them every day? Most often, it is not profitable. To find out about the quality of its production, that is, what percentage of its products are defective, the company checks a smaller sample.

The U.S. Department of Agriculture requires that 80% of the seeds that a company produces must sprout. To find out about the quality of the seeds a company has produced, it takes 500 seeds from those it has produced and plants them. It finds that 417 of the seeds sprout.

2. With another large batch of seeds the company of Example 2 plants 500 seeds and 367 of them sprout. What is the probability that a seed will sprout? Did this batch of seeds pass government inspection?

a) What is the probability that a seed will sprout?

b) Did the seeds pass government standards?

a) We know that 500 seeds were planted and 417 sprouted. The probability of a seed sprouting is P, where

$$P = \frac{417}{500} = 0.834, \quad \text{or} \quad 83.4\%.$$

b) Since the percentage of seeds exceeded the 80% requirement, the company deduces that it is producing quality seeds.

DO EXERCISE 2.

EXAMPLE 3 (*Estimating wildlife populations*). To determine the number of fish in a lake, a conservationist catches 225 fish, tags them, and puts them back into the lake. Later, 108 fish are caught, and 15 of them are found to be tagged. Estimate how many fish are in the lake.

Let $F =$ the number of fish in the lake. If there are F fish in the lake and 225 of them were tagged, then the probability that a fish is tagged is the ratio

$$\frac{225}{F}.$$

Later, 108 fish were caught and it was found that 15 of them are tagged. The ratio of fish tagged to fish caught is

$$\frac{15}{108}.$$

3. To determine the number of deer in a forest, a conservationist catches 612 deer, tags them, and lets them loose. Later, 244 deer are caught, and 72 of them are tagged. Estimate how many deer are in the forest.

This can also be thought of as an estimate, at least, of the probability that a fish is tagged. Thus we assume that the two ratios are the same. This gives us a proportion:

$$\frac{225}{F} = \frac{15}{108}.$$

We solve the proportion for F. We multiply by the LCM, which is $108F$:

$$108F \cdot \frac{225}{F} = 108F \cdot \frac{15}{108}$$

$$108 \cdot 225 = F \cdot 15$$

$$\frac{108 \cdot 225}{15} = F$$

$$1620 = F.$$

Thus we estimate that there are about 1620 fish in the lake.

DO EXERCISE 3.

EXAMPLE 4 (*TV ratings*). The major television networks and others such as cable TV are always concerned about the percentages of homes that have TVs and are watching their programs. It is too costly and unmanageable to contact every home in the country so a sample, or portion, of the homes are contacted. This is done by an electronic device attached to the TVs of about 1400 homes across the country. Viewing information is fed into a computer. The following are the results of a recent survey.

4. In reference to Example 4, what is the probability that a home was tuned to NBC? What is the probability that a home was tuned to a network other than CBS, ABC, or NBC, or was not tuned in at all?

Network	CBS	ABC	NBC	Other or not watching
Number of homes watching	258	231	206	705

What is the probability that a home was tuned to CBS during the time period considered? to ABC?

The probability that a home was tuned to CBS is P, where

$$P = \frac{258}{1400} \approx 0.184 = 18.4\%.$$

The probability that a home was tuned to ABC is P, where

$$P = \frac{231}{1400} \approx 0.165 = 16.5\%.$$

DO EXERCISE 4.

The numbers that we found either in Example 4 or in Margin Exercise 4 (18.4 for CBS, 16.5 for ABC, and 14.7 for NBC) are called the *ratings*.

Theoretical Probabilities

We need some terminology before we can continue. Suppose we perform an experiment such as flipping a coin, throwing a dart, drawing a card from a deck, or checking an item off an assembly line for quality. The results of an experiment are called *outcomes*. The set of all possible outcomes is called the *sample space*. An *event* is a set of outcomes, that is, a subset of the sample space. For example, for the experiment "throwing a dart," suppose the dartboard is as follows.

Then one event is

$$\{\text{hitting black}\}, \qquad \text{(an event)}$$

which is a subset of the sample space

$$\{\text{black, white, gray}\}, \qquad \text{(sample space)}$$

assuming the dart must hit the target somewhere.

We denote the probability that an event E occurs $P(E)$. For example, "getting a head" may be denoted by H. Then $P(H)$ represents the probability of getting a head. When the outcomes of an experiment all have the same probability of occurring, we say that they are *equally likely*. To see the distinction between events that are equally likely and those that are not, consider the dartboards.

For dartboard A, the events *hitting black*, *hitting white*, and *hitting gray* are equally likely, but for board B they are not. A sample space that can be expressed as a union of equally likely events can allow us to calculate probabilities of other events.

Principle *P* (Theoretical)

> If an event E can occur *m* ways out of *n* possible equally likely outcomes of a sample space *S*, the probability of that event is given by
>
> $$P(E) = \frac{m}{n}.$$

A die (pl., dice) is a cube, with six faces, each containing a number of dots from 1 to 6.

EXAMPLE 5   What is the probability of rolling a 3 on a die?

On a fair die there are 6 equally likely outcomes and there is 1 way to get a 3. By Principle *P*, $P(3) = \frac{1}{6}$.

5. What is the probability of rolling a prime number on a die?

6. Suppose we draw a card from a well-shuffled deck of 52 cards.

 a) What is the probability of drawing a king?

 b) What is the probability of drawing a spade?

 c) What is the probability of drawing a black card?

 d) What is the probability of drawing a jack or a queen?

7. Suppose we select, without looking, one marble from a bag containing 5 red marbles and 6 green marbles. What is the probability of selecting a green marble?

EXAMPLE 6 What is the probability of rolling an even number on a die?

The event is getting an *even* number. It can occur in 3 ways (getting 2, 4, or 6). The number of equally likely outcomes is 6. By Principle P, $P(\text{even}) = \frac{3}{6}$, or $\frac{1}{2}$.

DO EXERCISE 5.

We use a number of examples related to a standard bridge deck of 52 cards. Such a deck is made up as shown in the following figure.

A DECK OF 52 CARDS:

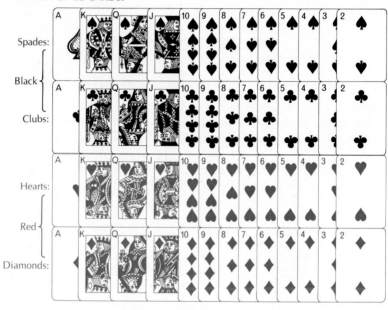

EXAMPLE 7 What is the probability of drawing an ace from a well-shuffled deck of 52 cards?

Since there are 52 outcomes (cards in the deck) and they are equally likely (from a well-shuffled deck) and there are 4 ways to obtain an ace, by Principle P we have

$$P(\text{drawing an ace}) = \tfrac{4}{52}, \quad \text{or} \quad \tfrac{1}{13}.$$

EXAMPLE 8 Suppose we select, without looking, one marble from a bag containing 3 red marbles and 4 green marbles. What is the probability of selecting a red marble?

There are 7 equally likely ways of selecting any marble, and since the number of ways of getting a red marble is 3,

$$P(\text{selecting a red marble}) = \tfrac{3}{7}.$$

DO EXERCISES 6 AND 7.

The following are some results that follow from Principle P.

Theorem 9

If an event E cannot occur, then $P(\text{E}) = 0.$

For example in coin tossing, the event that a coin would land on its edge has probability 0.

Theorem 10

> If an event E is certain to occur (every trial is a success), then $P(E) = 1$.

For example in coin tossing, the event that a coin falls either heads or tails has probability 1.

In general,

Theorem 11

> The probability that an event E will occur is a number from 0 to 1:
>
> $$0 \leqslant P(E) \leqslant 1.$$

DO EXERCISES 8 AND 9.

EXAMPLE 9 Suppose 2 cards are drawn from a well-shuffled deck of 52 cards. What is the probability that both of them are spades?

The number n of ways of drawing 2 cards from a deck of 52 is $_{52}C_2$. Now 13 of the 52 cards are spades, so the number m of ways of drawing 2 spades is $_{13}C_2$. Thus,

$$P(\text{getting 2 spades}) = \frac{m}{n} = \frac{_{13}C_2}{_{52}C_2} = \frac{78}{1326} = \frac{1}{17}.$$

EXAMPLE 10 Suppose 2 people are selected at random from a group that consists of 6 men and 4 women. What is the probability that both of them are women?

The number of ways of selecting 2 people from a group of 10 is $_{10}C_2$. The number of ways of selecting 2 women from a group of 4 is $_4C_2$. Thus the probability of selecting 2 women from the group of 10 is P, where

$$P = \frac{_4C_2}{_{10}C_2} = \frac{6}{45} = \frac{2}{15}.$$

EXAMPLE 11 Suppose 3 people are selected at random from a group that consists of 6 men and 4 women. What is the probability that 1 man and 2 women are selected?

The number of ways of selecting 3 people from a group of 10 is $_{10}C_3$. One man can be selected in $_6C_1$ ways, and 2 women can be selected in $_4C_2$ ways. By the fundamental counting principle, the number of ways of selecting 1 man and 2 women is $_6C_1 \cdot _4C_2$. Thus the probability is

$$\frac{_6C_1 \cdot _4C_2}{_{10}C_3}, \quad \text{or} \quad \frac{3}{10}.$$

DO EXERCISES 10–12.

EXAMPLE 12 What is the probability of getting a total of 8 on a roll of a pair of dice? (Assume the dice are different, say one red and one black.)

On each die there are 6 possible outcomes. The outcomes are paired so there are $6 \cdot 6$, or 36 possible ways in which the two can fall.

8. On a single roll of a die, what is the probability of getting a 7?

9. On a single roll of a die, what is the probability of getting a 1, 2, 3, 4, 5, or 6?

10. Suppose 3 cards are drawn from a well-shuffled deck of 52 cards. What is the probability that all 3 of them are spades?

11. Suppose 2 people are selected at random from a group that consists of 8 men and 6 women. What is the probability that both of them are women?

12. Suppose 3 people are selected at random from a group that consists of 8 men and 6 women. What is the probability that 2 men and 1 woman are selected?

13. What is the probability of getting a total of 7 on a roll of a pair of dice?

Red die

6	(1, 6)	(2, 6)	(3, 6)	(4, 6)	(5, 6)	(6, 6)
5	(1, 5)	(2, 5)	(3, 5)	(4, 5)	(5, 5)	(6, 5)
4	(1, 4)	(2, 4)	(3, 4)	(4, 4)	(5, 4)	(6, 4)
3	(1, 3)	(2, 3)	(3, 3)	(4, 3)	(5, 3)	(6, 3)
2	(1, 2)	(2, 2)	(3, 2)	(4, 2)	(5, 2)	(6, 2)
1	(1, 1)	(2, 1)	(3, 1)	(4, 1)	(5, 1)	(6, 1)
	1	2	3	4	5	6 Black die

The pairs that total 8 are as shown. Thus there are 5 possible ways of getting a total of 8, so the probability is $\frac{5}{36}$.

DO EXERCISE 13.

ORIGIN AND USE OF PROBABILITY

A desire to calculate odds in games of chance gave rise to the theory of probability. Today the theory of probability and its closely related field, mathematical statistics, have many applications, most of them not related to games of chance. Opinion polls, with such uses as predicting elections, are a familiar example. Quality control, in which a prediction about the percentage of faulty items manufactured is made without testing them all, is an important application, among many, in business. Still other applications are in the areas of genetics, medicine, and the kinetic theory of gases.

EXERCISE SET 12.4

1. In an actual survey 100 people were polled to determine the probability of a person wearing either glasses or contact lenses. Of those polled, 57 wore either glasses or contacts. What is the probability that a person wears either glasses or contacts? What is the probability that a person wears neither?

2. In another survey, 100 people were polled and asked to "select a number from 1 to 5." The results are shown in the following table.

Number choices	1	2	3	4	5
Number of people who selected that number	18	24	23	23	12

What is the probability that the number is selected is 1? 2? 3? 4? 5? What general conclusion might a psychologist make from this experiment?

(*Linguistics*). An experiment was conducted to determine the relative occurrence of various letters of the English alphabet. A paragraph from a newspaper, one from a textbook, and one from a magazine were considered. In all there were a total of 1044 letters. The number of occurrences of each letter of the alphabet is listed in the following table.

Letter	A	B	C	D	E	F	G	H	I	J	K	L	M
Number of occurrences	78	22	33	33	140	24	22	63	60	2	9	35	30

Letter	N	O	P	Q	R	S	T	U	V	W	X	Y	Z
Number of occurrences	74	74	27	4	67	67	95	31	10	22	8	13	1

Round answers to each of the following to three decimal places.

3. What is the probability of the occurrence of the letter A? E? I? O? U?

4. What is the probability of a vowel occurring?

5. What is the probability of a consonant occurring?

6. What letter has the least probability of occurring? What is the probability of this letter not occurring?

7. To determine the number of deer in a game preserve, a conservationist catches 318 deer, tags them, and lets them loose. Later, 168 deer are caught; 56 of them are tagged. How many deer are in the preserve?

8. To determine the number of trout in a lake, a conservationist catches 112 trout, tags them, and throws them back into the lake. Later, 82 trout are caught; 32 of them are tagged. How many trout are in the lake?

Suppose we draw a card from a well-shuffled deck of 52 cards.

9. How many equally likely outcomes are there?

10. What is the probability of drawing a queen?

11. What is the probability of drawing a heart?

12. What is the probability of drawing a club?

13. What is the probability of drawing a 4?

14. What is the probability of drawing a red card?

15. What is the probability of drawing a black card?

16. What is the probability of drawing an ace or a deuce?

17. What is the probability of drawing a 9 or a king?

Suppose we select, without looking, one marble from a bag containing 4 red marbles and 10 green marbles.

18. What is the probability of selecting a red marble?

19. What is the probability of selecting a green marble?

20. What is the probability of selecting a purple marble?

21. What is the probability of selecting a white marble?

Suppose 4 cards are drawn from a well-shuffled deck of 52 cards.

22. What is the probability that all 4 are spades?

23. What is the probability that all 4 are hearts?

24. If 4 marbles are drawn at random all at once from a bag containing 8 white marbles and 6 black marbles, what is the probability that 2 will be white and 2 will be black?

25. From a group of 8 men and 7 women, a committee of 4 is chosen. What is the probability that 2 men and 2 women will be chosen?

26. What is the probability of getting a total of 6 on a roll of a pair of dice?

27. What is the probability of getting a total of 3 on a roll of a pair of dice?

28. What is the probability of getting snake eyes (a total of 2) on a roll of a pair of dice?

29. What is the probability of getting box-cars (a total of 12) on a roll of a pair of dice?

30. From a bag containing 5 nickels, 8 dimes, and 7 quarters, 5 coins are drawn at random all at once. What is the probability of getting 2 nickels, 2 dimes, and 1 quarter?

31. From a bag containing 6 nickels, 10 dimes, and 4 quarters, 6 coins are drawn at random all at once. What is the probability of getting 3 nickels, 2 dimes, and 1 quarter?

(*Roulette*). A roulette wheel contains slots numbered 00, 0, 1, 2, 3, ..., 35, 36. Eighteen of the slots numbered 1 through 36 are colored red and eighteen are colored black. The 00 and 0 slots are uncolored. The wheel is spun, and a ball is rolled around the rim until it falls into a slot. What is the probability that the ball falls in

32. a black slot?

33. a red slot?

34. a red or black slot?

35. the 00 slot?

36. the 0 slot?

37. either the 00 or 0 slot? (Here the house always wins.)

38. an odd-numbered slot?

(*Five-card poker hands and probabilities*). In part (a) of each problem, give a reasoned expression as well as the answer. Read all the problems before beginning.

39. How many 5-card poker hands can be dealt from a standard 52-card deck?

40. A *royal flush* consists of a 5-card hand with A-K-Q-J-10 of the same suit.

  a) How many royal flushes are there?
  b) What is the probability of getting a royal flush?

41. A *straight flush* consists of 5 cards in sequence in the same suit, but excludes royal flushes. An ace can be used low, before a two.

  a) How many straight flushes are there?
  b) What is the probability of getting a straight flush?

42. *Four of a kind* is a 5-card hand in which 4 of the cards are of the same denomination, such as J-J-J-J-6, 7-7-7-7-A, or 2-2-2-2-5.

  a) How many are there?
  b) What is the probability of getting four of a kind?

43. ▦ A *full house* consists of a pair and three of a kind, such as Q-Q-Q-4-4.

   a) How many are there?
   b) What is the probability of getting a full house?

45. ▦ *Three of a kind* consists of a 5-card hand in which 3 of the cards are the same denomination, such as Q-Q-Q-10-7.

   a) How many are there?
   b) What is the probability of getting three of a kind?

47. ▦ *Two pairs* is a hand such as Q-Q-3-3-A.

   a) How many are there?
   b) What is the probability of getting two pairs?

44. ▦ A *pair* is a 5-card hand in which just 2 of the cards are the same denomination, such as Q-Q-8-A-3.

   a) How many are there?
   b) What is the probability of getting a pair?

46. ▦ A *flush* is a 5-card hand in which all the cards are the same suit, but not all in sequence (not a straight flush or royal flush).

   a) How many are there?
   b) What is the probability of getting a flush?

48. ▦ A *straight* is any 5 cards in sequence, but not in the same suit—for example, 4 of spades, 5 of spades, 6 of diamonds, 7 of hearts, and 8 of clubs.

   a) How many are there?
   b) What is the probability of getting a straight?

49. Suppose a dartboard is a mosaic made up of circles and squares like those in the figure. Also assume that each time you throw the dart you hit the board.

   a) Find the probability, *p*, of hitting inside a circle.
   b) Find the probability of hitting inside a square but outside a circle.
   c) Solve the answer to (a) for π. Suppose you throw a dart at the board 100 times and you hit the circle 78 times. Explain how you can use the result to find an estimate of π.

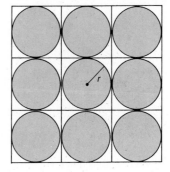

SUMMARY AND REVIEW: CHAPTER 12

The following contains a summary of what you should be able to do after completing this chapter. The review exercises are for practice. Answers are at the back of the book. If you miss an exercise, restudy the section indicated alongside the answer.

You should be able to:

Evaluate factorial, permutation, and combination notation and solve related problems.

1. In how many different ways can 6 books be arranged on a shelf?

3. The winner of a contest can choose any 8 of 15 prizes. How many different selections can be made?

5. In how many distinguishable ways can the letters of the word TENNESSEE be arranged?

7. How many code symbols can be formed using 5 out of 6 of the letters of G, H, I, J, K, L if the letters

   a) are not repeated?
   b) can be repeated?
   c) are not repeated but must begin with K?
   d) are not repeated but must end with IGH?

2. If 9 different signal flags are available, how many different displays are possible using 4 flags in a row?

4. The Greek alphabet contains 24 letters. How many fraternity or sorority names can be formed using 3 different letters?

6. A manufacturer of houses has one floor plan but achieves variety by having 3 different colored roofs, 4 different ways of attaching the garage, and 3 different types of entrance. Find the number of different houses that can be produced.

8. In how many different ways can 7 people be seated at a round table?

Expand a binomial power and find the rth term of such an expansion.

9. Find the 4th term of $(a + x)^{12}$.

10. Find the 12th term of $(a + x)^{18}$. Do not multiply out the factorials.

Expand.

11. $(m + n)^7$

12. $(x^2 + 3y)^4$

13. $(5i + 1)^6$, where $i^2 = -1$

14. $(\cos t + \sec t)^8$

Compute the probability of a simple event.

15. Before an election a poll was conducted to see which candidate was favored. Three people were running for a particular office. During the polling, 86 favored A, 97 favored B, and 23 favored C. Assuming that the poll is a valid indicator of the election, what is the probability that the election will be won by A? B? C?

16. What is the probability of rolling a 10 on a roll of a pair of dice? on a roll of one die?

17. From a deck of 52 cards 1 card is drawn. What is the probability that it is a club?

18. From a deck of 52 cards 3 are drawn at random without replacement. What is the probability that 2 are aces and 1 is a king?

19. Simplify: $\displaystyle\sum_{r=0}^{10} (-1)^r \binom{10}{r} (\log x)^{10-r} (\log y)^r$.

Solve for n.

20. $\dbinom{n}{n-1} = 36$

21. $26 \cdot \dbinom{n}{1} = \dbinom{n}{3}$

TEST: CHAPTER 12

1. How many code symbols can be formed using 4 out of 6 of the letters of D, E, F, G, H, I if the letters

   a) can be repeated?
   b) are not repeated?
   c) are not repeated but must begin with FH?

2. In how many different ways can 6 people be seated at a round table?

3. On a test a student must answer 4 out of 7 questions. In how many ways can this be done?

4. From a group of 20 seniors and 14 juniors, how many committees consisting of 3 seniors and 2 juniors are possible?

5. In how many distinguishable ways can the letters of the word ARKANSAS be arranged?

6. Solve for n:

$$\binom{n}{6} = 3 \cdot \binom{n-1}{5}.$$

7. Determine the number of subsets of a set of 8 members.

8. Find the 3rd term of $(2a + b)^7$.

9. Expand $(x - \sqrt{2})^5$.

10. What is the probability of getting a total of 6 on a roll of a pair of dice?

11. From a deck of 52 cards 1 card is drawn. What is the probability of drawing a 3 or a queen?

12. If 3 marbles are drawn at random all at once from a bag containing 5 green marbles, 7 red marbles, and 4 white marbles, what is the probability that 2 will be green and 1 white?

13. Solve for a:

$$\sum_{r=0}^{5} \binom{5}{r} 9^{5-r} a^r = 160.$$

14. How many diagonals does a dodecagon (12-sided polygon) have?

Tables are often prepared giving function values for a continuous function. Suppose the table gives four-digit precision. By using a procedure called *interpolation*, we can estimate values between those listed in the table, obtaining four-digit precision. Interpolation can be done in various ways, the simplest and most common being *linear interpolation*. We describe it now in relation to Table 2 for common logarithms. Remember that what we say applies to a table for *any* continuous function.

LINEAR INTERPOLATION

Let us consider how a table of values for any function is made. We select members of the domain x_1, x_2, x_3, and so on. Then we compute or somehow determine the corresponding function values $f(x_1)$, $f(x_2)$, $f(x_3)$, and so on. Then we tabulate the results. We might also graph the results.

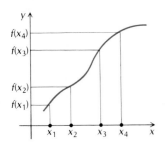

x	x_1	x_2	x_3	x_4	$\cdots$
$f(x)$	$f(x_1)$	$f(x_2)$	$f(x_3)$	$f(x_4)$	$\cdots$

Suppose we want to find the function value $f(x)$ for an x not in the table. If x is halfway between x_1 and x_2, then we can take the number halfway between $f(x_1)$ and $f(x_2)$ as an approximation to $f(x)$. If x is one fifth of the way between x_2 and x_3, we take the number that is one fifth of the way between $f(x_2)$ and $f(x_3)$ as an approximation to $f(x)$. What we do is divide the length from x_2 to x_3 in a certain ratio, and then divide the length from $f(x_2)$ to $f(x_3)$ in the same ratio. This is *linear interpolation*.

We can show this geometrically. The length from x_1 to x_2 is divided in a certain ratio by x. The length from $f(x_1)$ to $f(x_2)$ is divided in the same ratio by y. The number y approximates $f(x)$ with the noted error. Note the slanted line in the figure. The approximation y comes from this line. This explains the use of the term *linear interpolation*.

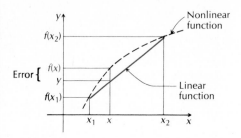

INTERPOLATION IN LOG TABLES

We now illustrate interpolation using Table 2.

EXAMPLE 1 Find log 34870.

a) We write scientific notation for the number:

$$3.487 \times 10^4.$$

Then $\log (3.487 \times 10^4) = \log 3.487 + 4$.

b) We find log 3.487. From Table 2, we have the following.

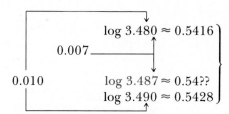

The tabular difference (the difference between consecutive values in the table) is 0.0012. Now 3.487 is $\frac{7}{10}$ of the way from 3.480 to 3.490. So we take 0.7 of 0.0012, which is 0.00084, and round it to 0.0008. We add this to 0.5416. Thus log $3.487 \approx 0.5424$.

c) We add the integer 4 from step (a):

$$\log (3.487 \times 10^4) \approx 4.5424.$$

EXAMPLE 2 Find log 0.009543.

a) We write scientific notation for the number:

$$9.543 \times 10^{-3}.$$

Then $\log (9.543 \times 10^{-3}) = \log 9.543 - 3$.

b) We find log 9.543. From Table 2 we have the following.

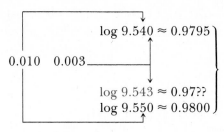

1. Find log 4793.

The tabular difference is 0.0005.

Now 9.543 is $\frac{3}{10}$ of the way from 9.540 to 9.550, so we take 0.3 of 0.0005, which is 0.00015, and round it to 0.0002. We add this to 0.9795. The result is 0.9797. Thus log 9.543 ≈ 0.9797.

c) We add the integer −3 from step (a):

$$\log (9.543 \times 10^{-3}) \approx -2.0203.$$

DO EXERCISES 1 AND 2.

ANTILOGARITHMS

We interpolate when finding antilogarithms, using the table in reverse.

EXAMPLE 3 Find antilog 4.9164.

a) We are setting out to find $10^{4.9164}$, which is

$10^{0.9164} \times 10^4$.   Separating the exponent into an integer and a number between 0 and 1

b) We use Table 2 to find $10^{0.9164}$.

2. Find log 0.0001732.

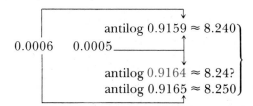

The difference is 0.010.

The difference between 0.9159 and 0.9165 is 0.0006. Thus 0.9164 is $\frac{0.0005}{0.0006}$, or $\frac{5}{6}$, of the way between 0.9159 and 0.9165. Then antilog 0.9164 is $\frac{5}{6}$ of the way between 8.240 and 8.250, so we take $\frac{5}{6}(0.010)$, which is 0.00833 . . . , and round it to 0.008. Thus $10^{0.9164} \approx 8.248$.

c) We now have

$$\text{antilog } 4.9164 \approx 10^{0.9164} \times 10^4$$
$$\approx 8.248 \times 10^4, \quad \text{or} \quad 82{,}480.$$

EXAMPLE 4 Find antilog (−2.5878).

a) We are setting out to find $10^{-2.5878}$. We need to separate −2.5878 into two parts, one of which is an integer and one of which is between 0 and 1. We can do that by adding and subtracting 3 (or any integer greater than 2):

$$3 - 2.5878 - 3 = 0.4122 - 3 \quad \text{Doing the first subtraction}$$

Now we have

$$\text{antilog} -2.5878 = \text{antilog } (0.4122 - 3)$$
$$= 10^{(0.4122-3)}$$
$$= 10^{0.4122} \times 10^{-3}.$$

3. Find antilog 5.5471.

4. Find antilog (-4.3690).

b)   Next, we use Table 2.

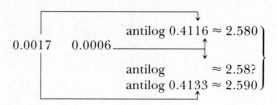

The difference between 0.4116 and 0.4133 is 0.0017. Thus 0.4122 i
$\frac{0.0006}{0.0017}$, or $\frac{6}{17}$, of the way between 0.4116 and 0.4133. Then antilog 0.412
is $\frac{6}{17}$ of the way between 2.580 and 2.590, so we take $\frac{6}{17}(0.010)$, whic
is 0.0035 to four places. We round it to 0.004. Therefore, antilo
$10^{0.4122} \approx 2.584$.

c)   We now have

$$\text{antilog } (-2.5878) \approx 10^{0.4122} \times 10^{-3}$$
$$\approx 2.584 \times 10^{-3}, \quad \text{or} \quad 0.002584.$$

DO EXERCISES 3 AND 4.

EXERCISES

Find the following logarithms using interpolation and Table 2.

1. log 41.63

2. log 472.1

3. log 2.944

4. log 21.76

5. log 650.2

6. log 37.37

7. log 0.1425

8. log 0.09045

9. log 0.004257

10. log 4518

11. log 0.1776

12. log 0.08356

13. log 600.6

14. log 800.1

Find the following antilogarithms using interpolation.

15. antilog 1.6350

16. antilog 2.3512

17. antilog 0.6478

18. antilog 1.1624

19. antilog 0.0342

20. antilog 4.8453

21. antilog $9.8564 - 10$

22. antilog $8.9659 - 10$

23. antilog $7.4128 - 10$

24. antilog $9.7278 - 10$

25. antilog $8.2010 - 10$

26. antilog $7.8630 - 10$

TABLES

TABLE 1 Powers, Roots, and Reciprocals

n	n^2	n^3	$\sqrt{n}$	$\sqrt[3]{n}$	$\sqrt{10n}$	$\frac{1}{n}$	n	n^2	n^3	$\sqrt{n}$	$\sqrt[3]{n}$	$\sqrt{10n}$	$\frac{1}{n}$
1	1	1	1.000	1.000	3.162	1.0000	51	2,601	132,651	7.141	3.708	22.583	.0196
2	4	8	1.414	1.260	4.472	.5000	52	2,704	140,608	7.211	3.733	22.804	.0192
3	9	27	1.732	1.442	5.477	.3333	53	2,809	148,877	7.280	3.756	23.022	.0189
4	16	64	2.000	1.587	6.325	.2500	54	2,916	157,464	7.348	3.780	23.238	.0185
5	25	125	2.236	1.710	7.071	.2000	55	3,025	166,375	7.416	3.803	23.452	.0182
6	36	216	2.449	1.817	7.746	.1667	56	3,136	175,616	7.483	3.826	23.664	.0179
7	49	343	2.646	1.913	8.367	.1429	57	3,249	185,193	7.550	3.849	23.875	.0175
8	64	512	2.828	2.000	8.944	.1250	58	3,364	195,112	7.616	3.871	24.083	.0172
9	81	729	3.000	2.080	9.487	.1111	59	3,481	205,379	7.681	3.893	24.290	.0169
10	100	1,000	3.162	2.154	10.000	.1000	60	3,600	216,000	7.746	3.915	24.495	.0167
11	121	1,331	3.317	2.224	10.488	.0909	61	3,721	226,981	7.810	3.936	24.698	.0164
12	144	1,728	3.464	2.289	10.954	.0833	62	3,844	238,328	7.874	3.958	24.900	.0161
13	169	2,197	3.606	2.351	11.402	.0769	63	3,969	250,047	7.937	3.979	25.100	.0159
14	196	2,744	3.742	2.410	11.832	.0714	64	4,096	262,144	8.000	4.000	25.298	.0156
15	225	3,375	3.873	2.466	12.247	.0667	65	4,225	274,625	8.062	4.021	25.495	.0154
16	256	4,096	4.000	2.520	12.648	.0625	66	4,356	287,496	8.124	4.041	25.690	.0152
17	289	4,913	4.123	2.571	13.038	.0588	67	4,489	300,763	8.185	4.062	25.884	.0149
18	324	5,832	4.243	2.621	13.416	.0556	68	4,624	314,432	8.246	4.082	26.077	.0147
19	361	6,859	4.359	2.668	13.784	.0526	69	4,761	328,509	8.307	4.102	26.268	.0145
20	400	8,000	4.472	2.714	14.142	.0500	70	4,900	343,000	8.367	4.121	26.458	.0143
21	441	9,261	4.583	2.759	14.491	.0476	71	5,041	357,911	8.426	4.141	26.646	.0141
22	484	10,648	4.690	2.802	14.832	.0455	72	5,184	373,248	8.485	4.160	26.833	.0139
23	529	12,167	4.796	2.844	15.166	.0435	73	5,329	389,017	8.544	4.179	27.019	.0137
24	576	13,824	4.899	2.884	15.492	.0417	74	5,476	405,224	8.602	4.198	27.203	.0135
25	625	15,625	5.000	2.924	15.811	.0400	75	5,625	421,875	8.660	4.217	27.386	.0133
26	676	17,576	5.099	2.962	16.125	.0385	76	5,776	438,976	8.718	4.236	27.568	.0132
27	729	19,683	5.196	3.000	16.432	.0370	77	5,929	456,533	8.775	4.254	27.749	.0130
28	784	21,952	5.292	3.037	16.733	.0357	78	6,084	474,552	8.832	4.273	27.928	.0128
29	841	24,389	5.385	3.072	17.029	.0345	79	6,241	493,039	8.888	4.291	28.107	.0127
30	900	27,000	5.477	3.107	17.321	.0333	80	6,400	512,000	8.944	4.309	28.284	.0125
31	961	29,791	5.568	3.141	17.607	.0323	81	6,561	531,441	9.000	4.327	28.460	.0123
32	1,024	32,768	5.657	3.175	17.889	.0312	82	6,724	551,368	9.055	4.344	28.636	.0122
33	1,089	35,937	5.745	3.208	18.166	.0303	83	6,889	571,787	9.110	4.362	28.810	.0120
34	1,156	39,304	5.831	3.240	18.439	.0294	84	7.056	592,704	9.165	4.380	28.983	.0119
35	1,225	42,875	5.916	3.271	18.708	.0286	85	7,225	614,125	9.220	4.397	29.155	.0118
36	1,296	46,656	6.000	3.302	18.974	.0278	86	7,396	636,056	9.274	4.414	29.326	.0116
37	1,369	50,653	6.083	3.332	19.235	.0270	87	7,569	658,503	9.327	4.431	29.496	.0115
38	1,444	54,872	6.164	3.362	19.494	.0263	88	7,744	681,472	9.381	4.448	29.665	.0114
39	1,521	59,319	6.245	3.391	19.748	.0256	89	7,921	704,969	9.434	4.465	29.833	.0112
40	1,600	64,000	6.325	3.420	20.000	.0250	90	8,100	729,000	9.487	4.481	30.000	.0111
41	1,681	68,921	6.403	3.448	20.248	.0244	91	8,281	753,571	9.539	4.498	30.166	.0110
42	1,764	74,088	6.481	3.476	20.494	.0238	92	8,464	778,688	9.592	4.514	30.332	.0109
43	1,849	79,507	6.557	3.503	20.736	.0233	93	8,649	804,357	9.644	4.531	30.496	.0108
44	1,936	85,184	6.633	3.530	20.976	.0227	94	8,836	830,584	9.695	4.547	30.659	.0106
45	2,025	91,125	6.708	3.557	21.213	.0222	95	9,025	857,375	9.747	4.563	30.822	.0105
46	2,116	97,336	6.782	3.583	21.448	.0217	96	9,216	884,736	9.798	4.579	30.984	.0104
47	2,209	103,823	6.856	3.609	21.679	.0213	97	9,409	912,673	9.849	4.595	31.145	.0103
48	2,304	110,592	6.928	3.634	21.909	.0208	98	9,604	941,192	9.899	4.610	31.305	.0102
49	2,401	117,649	7.000	3.659	22.136	.0204	99	9,801	970,299	9.950	4.626	31.464	.0101
50	2,500	125,000	7.071	3.684	22.361	.0200	100	10,000	1,000,000	10.000	4.642	31.623	.0100

TABLE 2 Common Logarithms

x	0	1	2	3	4	5	6	7	8	9
1.0	.0000	.0043	.0086	.0128	.0170	.0212	.0253	.0294	.0334	.0374
1.1	.0414	.0453	.0492	.0531	.0569	.0607	.0645	.0682	.0719	.0755
1.2	.0792	.0828	.0864	.0899	.0934	.0969	.1004	.1038	.1072	.1106
1.3	.1139	.1173	.1206	.1239	.1271	.1303	.1335	.1367	.1399	.1430
1.4	.1461	.1492	.1523	.1553	.1584	.1614	.1644	.1673	.1703	.1732
1.5	.1761	.1790	.1818	.1847	.1875	.1903	.1931	.1959	.1987	.2014
1.6	.2041	.2068	.2095	.2122	.2148	.2175	.2201	.2227	.2253	.2279
1.7	.2304	.2330	.2355	.2380	.2405	.2430	.2455	.2480	.2504	.2529
1.8	.2553	.2577	.2601	.2625	.2648	.2672	.2695	.2718	.2742	.2765
1.9	.2788	.2810	.2833	.2856	.2878	.2900	.2923	.2945	.2967	.2989
2.0	.3010	.3032	.3054	.3075	.3096	.3118	.3139	.3160	.3181	.3201
2.1	.3222	.3243	.3263	.3284	.3304	.3324	.3345	.3365	.3385	.3404
2.2	.3424	.3444	.3464	.3483	.3502	.3522	.3541	.3560	.3579	.3598
2.3	.3617	.3636	.3655	.3674	.3692	.3711	.3729	.3747	.3766	.3784
2.4	.3802	.3820	.3838	.3856	.3874	.3892	.3909	.3927	.3945	.3962
2.5	.3979	.3997	.4014	.4031	.4048	.4065	.4082	.4099	.4116	.4133
2.6	.4150	.4166	.4183	.4200	.4216	.4232	.4249	.4265	.4281	.4298
2.7	.4314	.4330	.4346	.4362	.4378	.4393	.4409	.4425	.4440	.4456
2.8	.4472	.4487	.4502	.4518	.4533	.4548	.4564	.4579	.4594	.4609
2.9	.4624	.4639	.4654	.4669	.4683	.4698	.4713	.4728	.4742	.4757
3.0	.4771	.4786	.4800	.4814	.4829	.4843	.4857	.4871	.4886	.4900
3.1	.4914	.4928	.4942	.4955	.4969	.4983	.4997	.5011	.5024	.5038
3.2	.5051	.5065	.5079	.5092	.5105	.5119	.5132	.5145	.5159	.5172
3.3	.5185	.5198	.5211	.5224	.5237	.5250	.5263	.5276	.5289	.5307
3.4	.5315	.5328	.5340	.5353	.5366	.5378	.5391	.5403	.5416	.5428
3.5	.5441	.5453	.5465	.5478	.5490	.5502	.5514	.5527	.5539	.5551
3.6	.5563	.5575	.5587	.5599	.5611	.5623	.5635	.5647	.5658	.5670
3.7	.5682	.5694	.5705	.5717	.5729	.5740	.5752	.5763	.5775	.5786
3.8	.5798	.5809	.5821	.5832	.5843	.5855	.5866	.5877	.5888	.5899
3.9	.5911	.5922	.5933	.5944	.5955	.5966	.5977	.5988	.5999	.6010
4.0	.6021	.6031	.6042	.6053	.6064	.6075	.6085	.6096	.6107	.6117
4.1	.6128	.6138	.6149	.6160	.6170	.6180	.6191	.6201	.6212	.6222
4.2	.6232	.6243	.6253	.6263	.6274	.6284	.6294	.6304	.6314	.6325
4.3	.6335	.6345	.6355	.6365	.6375	.6385	.6395	.6405	.6415	.6425
4.4	.6435	.6444	.6454	.6464	.6474	.6484	.6493	.6503	.6513	.6522
4.5	.6532	.6542	.6551	.6561	.6571	.6580	.6590	.6599	.6609	.6618
4.6	.6628	.6637	.6646	.6656	.6665	.6675	.6684	.6693	.6702	.6712
4.7	.6721	.6730	.6739	.6749	.6758	.6767	.6776	.6785	.6794	.6803
4.8	.6812	.6821	.6830	.6839	.6848	.6857	.6866	.6875	.6884	.6893
4.9	.6902	.6911	.6920	.6928	.6937	.6946	.6955	.6964	.6972	.6981
5.0	.6990	.6998	.7007	.7016	.7024	.7033	.7042	.7050	.7059	.7067
5.1	.7076	.7084	.7093	.7101	.7110	.7118	.7126	.7135	.7143	.7152
5.2	.7160	.7168	.7177	.7185	.7193	.7202	.7210	.7218	.7226	.7235
5.3	.7243	.7251	.7259	.7267	.7275	.7284	.7292	.7300	.7308	.7316
5.4	.7324	.7332	.7340	.7348	.7356	.7364	.7372	.7380	.7388	.7396
x	0	1	2	3	4	5	6	7	8	9

TABLE 2 *(continued)*

x	0	1	2	3	4	5	6	7	8	9
5.5	.7404	.7412	.7419	.7427	.7435	.7443	.7451	.7459	.7466	.7474
5.6	.7482	.7490	.7497	.7505	.7513	.7520	.7528	.7536	.7543	.7551
5.7	.7559	.7566	.7574	.7582	.7589	.7597	.7604	.7612	.7619	.7627
5.8	.7634	.7642	.7649	.7657	.7664	.7672	.7679	.7686	.7694	.7701
5.9	.7709	.7716	.7723	.7731	.7738	.7745	.7752	.7760	.7767	.7774
6.0	.7782	.7789	.7796	.7803	.7810	.7818	.7825	.7832	.7839	.7846
6.1	.7853	.7860	.7868	.7875	.7882	.7889	.7896	.7903	.7910	.7917
6.2	.7924	.7931	.7938	.7945	.7952	.7959	.7966	.7973	.7980	.7987
6.3	.7993	.8000	.8007	.8014	.8021	.8028	.8035	.8041	.8048	.8055
6.4	.8062	.8069	.8075	.8082	.8089	.8096	.8102	.8109	.8116	.8122
6.5	.8129	.8136	.8142	.8149	.8156	.8162	.8169	.8176	.8182	.8189
6.6	.8195	.8202	.8209	.8215	.8222	.8228	.8235	.8241	.8248	.8254
6.7	.8261	.8267	.8274	.8280	.8287	.8293	.8299	.8306	.8312	.8319
6.8	.8325	.8331	.8338	.8344	.8351	.8357	.8363	.8370	.8376	.8382
6.9	.8388	.8395	.8401	.8407	.8414	.8420	.8426	.8432	.8439	.8445
7.0	.8451	.8457	.8463	.8470	.8476	.8482	.8488	.8494	.8500	.8506
7.1	.8513	.8519	.8525	.8531	.8537	.8543	.8549	.8555	.8561	.8567
7.2	.8573	.8579	.8585	.8591	.8597	.8603	.8609	.8615	.8621	.8627
7.3	.8633	.8639	.8645	.8651	.8657	.8663	.8669	.8675	.8681	.8686
7.4	.8692	.8698	.8704	.8710	.8716	.8722	.8727	.8733	.8739	.8745
7.5	.8751	.8756	.8762	.8768	.8774	.8779	.8785	.8791	.8797	.8802
7.6	.8808	.8814	.8820	.8825	.8831	.8837	.8842	.8848	.8854	.8859
7.7	.8865	.8871	.8876	.8882	.8887	.8893	.8899	.8904	.8910	.8915
7.8	.8921	.8927	.8932	.8938	.8943	.8949	.8954	.8960	.8965	.8971
7.9	.8976	.8982	.8987	.8993	.8998	.9004	.9009	.9015	.9020	.9025
8.0	.9031	.9036	.9042	.9047	.9053	.9058	.9063	.9069	.9074	.9079
8.1	.9085	.9090	.9096	.9101	.9106	.9112	.9117	.9122	.9128	.9133
8.2	.9138	.9143	.9149	.9154	.9159	.9165	.9170	.9175	.9180	.9186
8.3	.9191	.9196	.9201	.9206	.9212	.9217	.9222	.9227	.9232	.9238
8.4	.9243	.9248	.9253	.9258	.9263	.9269	.9274	.9279	.9284	.9289
8.5	.9294	.9299	.9304	.9309	.9315	.9320	.9325	.9330	.9335	.9340
8.6	.9345	.9350	.9555	.9360	.9365	.9370	.9375	.9380	.9385	.9390
8.7	.9395	.9400	.9405	.9410	.9415	.9420	.9425	.9430	.9435	.9440
8.8	.9445	.9450	.9455	.9460	.9465	.9469	.9474	.9479	.9484	.9489
8.9	.9494	.9499	.9504	.9509	.9513	.9518	.9523	.9528	.9533	.9538
9.0	.9542	.9547	.9552	.9557	.9562	.9566	.9571	.9576	.9581	.9586
9.1	.9590	.9595	.9600	.9605	.9609	.9614	.9619	.9624	.9628	.9633
9.2	.9638	.9643	.9647	.9652	.9657	.9661	.9666	.9671	.9675	.9680
9.3	.9685	.9689	.9694	.9699	.9703	.9708	.9713	.9717	.9722	.9727
9.4	.9731	.9736	.9741	.9745	.9750	.9754	.9759	.9763	.9768	.9773
9.5	.9777	.9782	.9786	.9791	.9795	.9800	.9805	.9809	.9814	.9818
9.6	.9823	.9827	.9832	.9836	.9841	.9845	.9850	.9854	.9859	.9863
9.7	.9868	.9872	.9877	.9881	.9886	.9890	.9894	.9899	.9903	.9908
9.8	.9912	.9917	.9921	.9926	.9930	.9934	.9939	.9943	.9948	.9952
9.9	.9956	.9961	.9965	.9969	.9974	.9978	.9983	.9987	.9991	.9996
x	0	1	2	3	4	5	6	7	8	9

TABLE 3 Exponential Functions

x	e^x	e^{-x}	x	e^x	e^{-x}	x	e^x	e^{-x}
0.00	1.0000	1.0000	0.55	1.7333	0.5769	3.6	36.598	0.0273
0.01	1.0101	0.9900	0.60	1.8221	0.5488	3.7	40.447	0.0247
0.02	1.0202	0.9802	0.65	1.9155	0.5220	3.8	44.701	0.0224
0.03	1.0305	0.9704	0.70	2.0138	0.4966	3.9	49.402	0.0202
0.04	1.0408	0.9608	0.75	2.1170	0.4724	4.0	54.598	0.0183
0.05	1.0513	0.9512	0.80	2.2255	0.4493	4.1	60.340	0.0166
0.06	1.0618	0.9418	0.85	2.3396	0.4274	4.2	66.686	0.0150
0.07	1.0725	0.9324	0.90	2.4596	0.4066	4.3	73.700	0.0136
0.08	1.0833	0.9231	0.95	2.5857	0.3867	4.4	81.451	0.0123
0.09	1.0942	0.9139	1.0	2.7183	0.3679	4.5	90.017	0.0111
0.10	1.1052	0.9048	1.1	3.0042	0.3329	4.6	99.484	0.0101
0.11	1.1163	0.8958	1.2	3.3201	0.3012	4.7	109.95	0.0091
0.12	1.1275	0.8869	1.3	3.6693	0.2725	4.8	121.51	0.0082
0.13	1.1388	0.8781	1.4	4.0552	0.2466	4.9	134.29	0.0074
0.14	1.1503	0.8694	1.5	4.4817	0.2231	5	148.41	0.0067
0.15	1.1618	0.8607	1.6	4.9530	0.2019	6	403.43	0.0025
0.16	1.1735	0.8521	1.7	5.4739	0.1827	7	1096.6	0.0009
0.17	1.1853	0.8437	1.8	6.0496	0.1653	8	2981.0	0.0003
0.18	1.1972	0.8353	1.9	6.6859	0.1496	9	8103.1	0.0001
0.19	1.2092	0.8270	2.0	7.3891	0.1353	10	22026	0.00005
0.20	1.2214	0.8187	2.1	8.1662	0.1225	11	59874	0.00002
0.21	1.2337	0.8106	2.2	9.0250	0.1108	12	162,754	0.000006
0.22	1.2461	0.8025	2.3	9.9742	0.1003	13	442,413	0.000002
0.23	1.2586	0.7945	2.4	11.023	0.0907	14	1,202,604	0.0000008
0.24	1.2712	0.7866	2.5	12.182	0.0821	15	3,269,017	0.0000003
0.25	1.2840	0.7788	2.6	13.464	0.0743			
0.26	1.2969	0.7711	2.7	14.880	0.0672			
0.27	1.3100	0.7634	2.8	16.445	0.0608			
0.28	1.3231	0.7558	2.9	18.174	0.0550			
0.29	1.3364	0.7483	3.0	20.086	0.0498			
0.30	1.3499	0.7408	3.1	22.198	0.0450			
0.35	1.4191	0.7047	3.2	24.533	0.0408			
0.40	1.4918	0.6703	3.3	27.113	0.0369			
0.45	1.5683	0.6376	3.4	29.964	0.0334			
0.50	1.6487	0.6065	3.5	33.115	0.0302			

TABLE 4 Factorials and Large Powers of 2

n	n!	2^n
0	1	1
1	1	2
2	2	4
3	6	8
4	24	16
5	120	32
6	720	64
7	5040	128
8	40,320	256
9	362,880	512
10	3,628,800	1024
11	39,916,800	2048
12	479,001,600	4096
13	6,227,020,800	8192
14	87,178,291,200	16,384
15	1,307,674,368,000	32,768
16	20,922,789,888,000	65,536
17	355,687,428,096,000	131,072
18	6,402,373,705,728,000	262,144
19	121,645,100,408,832,000	524,288
20	2,432,902,008,176,640,000	1,048,576

TABLE 5 Tables of Measures

LENGTH

1 kilometer (km)	=	1000 meters (m)
1 hectometer (hm)	=	100 meters
1 dekameter (dam)	=	10 meters
1 decimeter (dm)	=	0.1 meter
1 centimeter (cm)	=	0.01 meter
1 millimeter (mm)	=	0.001 meter

MASS OR WEIGHT

1 kilogram (kg)	=	1000 grams (g)
1 hectogram (hg)	=	100 grams
1 dekagram (dag)	=	10 grams
1 decigram (dg)	=	0.1 gram
1 centigram (dg)	=	0.01 gram
1 metric ton (MT or t)	=	1000 kilograms

AREA

1 hectare (ha) = 100 are (a), or 10,000 sq m (m²)
1 are (a) = 100 sq m (m²)
1 centare (ca) = 0.01 are, or 1 m²

The word "are" is pronounced "AIR."

VOLUME

1000 cubic centimeters (cc or cm³) = 1 liter (L)
1 cubic centimeter (cc) = 1 milliliter (mL)
1 mL of water weighs 1 g
1 stere = 1 cubic meter

Metric-American Conversions (Approximate)

LENGTH

1 m = 39.37 in. = 3.3 ft
1 in. = 2.54 cm
1 km = 0.62 mi
1 mi = 1.6 km
1 cm = 3/8 in.

MASS OR WEIGHT

1 kg = 2.2 lb
1 MT = 1.1 tons
1 lb = 454 g
1 oz = 28 g

AREA

1 hectare = 2.47 acres
1 are = 120 sq yd

VOLUME

1 liter = 1.057 qt = 2.1 pt
1 cup = 240 mL
1 ounce (liquid) = 30 mL
1 gallon = 3.78 liters
1 tablespoon = 15 mL
1 teaspoon = 5 mL

CHAPTER 1

Margin Exercises, Section 1.1

1. 1, 19   **2.** 0, 1, 19   **3.** $-6, 0, 1, 19$ **4.** All of them
5. Rational   **6.** Rational   **7.** Rational   **8.** Rational
9. Irrational   **10.** Irrational   **11.** 2   **12.** $\sqrt{3}$ **13.** 11.3
14. $\frac{3}{4}$ **15.** -12 **16.** -4.7 **17.** $-\frac{4}{5}$ **18.** -0.2 **19.** 5
20. 0   **21.** $-6, -6$ **22.** 8, 8   **23.** 3.4, 3.4   **24.** $2, -4$
25. -24 **26.** $\frac{21}{25}$ **27.** 144   **28.** 1.3   **29.** 17   **30.** $-\frac{11}{5}$
31. -13 **32.** 4   **33.** -3 **34.** $-\frac{8}{3}$ **35.** 2

Exercise Set 1.1, pp. 10–11

1. 3, 14   **3.** $\sqrt{3}, -\sqrt{7}, \sqrt[3]{2}$ **5.** $-6, 0, 3, -2, 14$
7. Rational   **9.** Rational   **11.** Rational   **13.** Irrational
15. Irrational   **17.** Irrational   **19.** Rational
21. Irrational   **23.** 12   **25.** 47   **27.** 7, 7   **29.** $-57, -57$
31. -87 **33.** -16 **35.** -10.3 **37.** $\frac{39}{10}$ **39.** 28
41. -49.2 **43.** 210   **45.** $-\frac{833}{5}$ **47.** 5   **49.** $-\frac{1}{7}$
51. $-\frac{3}{49}$ **53.** 25   **55.** -4 **57.** 18   **59.** -11.6
61. $-\frac{7}{2}$ **63.** (a) 1.96, 1.9881, 1.999396, 1.999962, 1.999990;
(b) $\sqrt{2}$ **65.** Identity $(+)$ **67.** Distributive
69. Commutativity $(+)$ **71.** Associativity $(\times)$
73. Inverse $(\times)$ **75.** $7 - 5 \neq 5 - 7$; $7 - 5 = 2, 5 - 7 = -2$
77. $16 \div (4 \div 2) \neq (16 \div 4) \div 2$; $16 \div (4 \div 2) = 8$,
$(16 \div 4) \div 2 = 2$ **79.** $|x - 3| = x - 3$ if $x - 3 \geqslant 0$;
$|x - 3| = -(x - 3)$ if $x - 3 < 0$ **81.** $\frac{371}{99}$ **83.** $\frac{12,346,418}{999,900}$

Margin Exercises, Section 1.2

1. 8^4 **2.** x^3 **3.** $(4y)^4$ **4.** $3 \cdot 3 \cdot 3 \cdot 3$, or 81
5. $5x \cdot 5x \cdot 5x \cdot 5x$, or $625x^4$
6. $(-5)(-5)(-5)(-5)$, or 625
7. $-[5 \cdot 5 \cdot 5 \cdot 5]$, or -625 **8.** 1   **9.** $25y^2$ **10.** $-8x^3$
11. 4^{-3} **12.** $\dfrac{1}{10^4}$, or $\dfrac{1}{10 \cdot 10 \cdot 10 \cdot 10}$, or $\dfrac{1}{10,000}$
13. $\dfrac{1}{4^3}, \dfrac{1}{4 \cdot 4 \cdot 4}, \dfrac{1}{64}$ **14.** 8^4 **15.** y^5 **16.** $-18x^{11}$

17. $-75x^{-14}$ **18.** $-10x^{-12}y^2$ **19.** $60y$ **20.** 4^3 **21.** 5^6
22. 10^{-14} **23.** 9^{-6} **24.** y^{11} **25.** $5y^{-1}$ **26.** $-2x^{-3}y^9$
27. 3^{49} **28.** 8^{-14} **29.** y^{-28} **30.** $8x^3y^3$ **31.** $16x^{-4}y^{14}$
32. $\frac{1}{27}x^{-12}y^{-6}$ **33.** $1000x^{-12}y^{21}z^{-6}$ **34.** 4.65×10^5
35. 3.789×10^3 **36.** 1.45×10^{-4} **37.** 6.7×10^{-10}
38. 0.0000467   **39.** 7,894,000,000,000   **40.**

41. x^8 **42.** $10m^2n^2|n|$ **43.** $\dfrac{2x^2|x|}{y^2}$

Exercise Set 1.2, pp. 16–17

1. 2^{-1} **3.** 1   **5.** 4^3 **7.** $6x^5$ **9.** $15a^{-1}b^5$ **11.** $72x^5$
13. $-18x^7yz$ **15.** b^3 **17.** x^3y^{-3} **19.** 1   **21.** $3ab^2$
23. $\frac{4}{7}xyz^{-5}$ **25.** $8a^3b^6$ **27.** $16x^{12}$ **29.** $-16x^{12}$
31. $36a^4b^6c^2$ **33.** $\frac{1}{25}c^2d^4$ **35.** 1   **37.** 32
39. $\frac{27}{4}a^8b^{-10}c^{18}$ **41.** $\frac{3}{4}xy$ **43.** 5.8×10^7 **45.** 3.65×10^5
47. 2.7×10^{-6} **49.** 2.7×10^{-2} **51.** $\$9.1 \times 10^{11}$
53. 400,000   **55.** 0.0062   **57.** 7,690,000,000,000
59. 9,460,000,000,000   **61.** $-25, 25$ **63.** $-1.1664, 1.1664$
65. $9|x||y|$ **67.** $3a^2|b|$ **69.** x^{8t} **71.** t^{8x} **73.** $(xy)^{ac+bc}$
75. $9x^{2a}y^{2b}$ **77.** $\$1044.65$
79. In $(x^3)^2$ exponents were added instead of multiplied; x^{10}
81. In 2^3 the base 2, and the exponent 3, were multiplied. In
$(x^{-4})^3$ the exponents were added. In $(y^6)^3$ the exponents were
subtracted. In $(z^3)^3$ the exponents were added; $8x^{-12}y^{18}z^9$

Margin Exercises, Section 1.3

1. 8, 6, 4, 9, 0; 9   **2.** 4, 4, 5, 6, 0; 6   **3.** $9x^3y^2 - 2x^2y^3$
4. $7xy^2 - 2x^2y$ **5.** $3x^4 \cdot \sqrt{y} + 2$ **6.** $-4x^3 + 2x^2 - 4x - \frac{3}{2}$
7. $5p^2q^4 + p^2q^2 - 6pq^2 - 3q + 5$
8. $-(5x^2t^2 - 4xy^2t - 3xt + 6x - 5)$,
$-5x^2t^2 + 4xy^2t + 3xt - 6x + 5$
9. $-(-3x^2y + 5xy - 7x + 4y + 2)$, $3x^2y - 5xy + 7x - 4y - 2$
10. $8xy^4 - 9xy^2 + 4x^2 + 2y - 7$
11. $7x^2y - 9x^3y^2 + 5x^2y^3 - x^2y^2 + 9y$

Exercise Set 1.3, pp. 20–21

1. 4, 3, 2, 1, 0; 4   **3.** 3, 6, 6, 0; 6   **5.** 5, 6, 2, 1, 0; 6
7. $3x^2y - 5xy^2 + 7xy + 2$
9. $-10pq^2 - 5p^2q + 7pq - 4p + 2q + 3$
11. $3x + 2y - 2z - 3$ **13.** $5x\sqrt{y} - 4y\sqrt{x} - \frac{2}{5}$
15. $-(5x^3 - 7x^2 + 3x - 6)$, $-5x^3 + 7x^2 - 3x + 6$
17. $-2x^2 + 6x - 2$ **19.** $6a - 5b - 2c + 4d$
21. $x^4 - 3x^3 - 4x^2 + 9x - 3$ **23.** $9x\sqrt{y} - 3y\sqrt{x} + 9.1$
25. $-1.047p^2q - 2.479pq^2 + 8.879pq - 104.144$

Margin Exercises, Section 1.4

1. $3x^3y^2 + 4x^2y^2 - xy^2 + 6y^2$
2. $2p^4q^2 + 3p^3q^2 + 3p^2q^2 + 2q^2$
3. $2x^3y - 4xy + 3x^3 - 6x$ **4.** $15x^2 - xy - 6y^2$
5. $6xy - 2\sqrt{2}x + 3\sqrt{2}y - 2$ **6.** $16x^2 - 40xy + 25y^2$
7. $4y^4 + 24x^2y^3 + 36x^4y^2$ **8.** $16x^2 - 49$ **9.** $25x^4y^2 - 4y^2$
10. $16y^4 - 3$ **11.** $4x^2 + 12x + 9 - 25y^2$ **12.** $25t^2 - 4x^6y^4$
13. $x^3 + 3x^2 + 3x + 1$ **14.** $x^3 - 3x^2 + 3x - 1$
15. $t^6 - 9t^4b + 27t^2b^2 - 27b^3$
16. $8a^9 - 60a^6b^2 + 150a^3b^4 - 125b^6$

Exercise Set 1.4, pp. 23–24

1. $6x^3 + 4x^2 + 32x - 64$
3. $4a^3b^2 - 10a^2b^2 + 3ab^3 + 4ab^2 - 6b^3 + 4a^2b - 2ab + 3b^2$
5. $a^3 - b^3$ **7.** $4x^2 + 8xy + 3y^2$ **9.** $12x^3 + x^2y - \frac{3}{2}xy - \frac{1}{8}y^2$
11. $2x^3 - 2\sqrt{2}x^2y - \sqrt{2}xy^2 + 2y^3$ **13.** $4x^2 + 12xy + 9y^2$
15. $4x^4 - 12x^2y + 9y^2$ **17.** $4x^6 + 12x^3y^2 + 9y^4$
19. $\frac{1}{4}x^4 - \frac{3}{5}x^2y + \frac{9}{25}y^2$ **21.** $0.25x^2 + 0.70xy^2 + 0.49y^4$
23. $9x^2 - 4y^2$ **25.** $x^4 - y^2z^2$ **27.** $9x^4 - 2$
29. $4x^2 + 12xy + 9y^2 - 16$ **31.** $x^4 + 6x^2y + 9y^2 - y^4$
33. $x^4 - 1$ **35.** $16x^4 - y^4$
37. $0.002601x^2 + 0.00408xy + 0.0016y^2$
39. $2462.0358x^2 - 945.0214x - 38.908$
41. $y^3 + 15y^2 + 75y + 125$ **43.** $m^6 - 6m^4n + 12m^2n^2 - 8n^3$
45. $a^{2n} - b^{2n}$ **47.** $x^{3m} - 3x^{2m}t^n + 3x^mt^{2n} - t^{3n}$ **49.** $x^6 - 1$
51. $16x^4 - 32x^3 + 16x^2$ **53.** $x^{a^2-b^2}$
55. $a^2 + b^2 + c^2 + 2ab + 2ac + 2bc$
57. $a^4 + 4a^3b + 6a^2b^2 + 4ab^3 + b^4$ **59.** $m^5 + t^5$
61. A term is missing—found by calculating twice the product of the terms; the first term is $9a^2$, not $3a^2$, where the 3 was not squared; $9a^2 + 6ab + b^2$
63. In step (1) $3x$ should be $6x$; the 2 and 3 were not multiplied. Similarly, -3 should be -12; the 4 and -3 were not multiplied. In step (2) $3x - 3$ is not x since they are not like terms; $6x^2 + 6x - 12$.

Margin Exercises, Section 1.5

1. $4x^2y(5x + 3)$ **2.** $(p + q)(2x + y + 2)$
3. $(4x - 3)(x + 5)$ **4.** $(x - 4)(x + 4)$
5. $(5y^2 + 4x)(5y^2 - 4x)$ **6.** $2(y^2 + 4x^2)(y - 2x)(y + 2x)$
7. $(x - \sqrt{3})(x + \sqrt{3})$ **8.** $(3y - 5)^2$ **9.** $(4x + 9y)^2$
10. $-3y^2(2x^2 - 5y^3)^2$ **11.** $(x + 7)(x - 2)$
12. $(3x + 2)(x + 1)$ **13.** $3(2x^2y^3 + 5)(x^2y^3 - 4)$
14. $(x - 2)(x^2 + 2x + 4)$ **15.** $(4 - t)(16 + 4t + t^2)$
16. $(3x + y)(9x^2 - 3xy + y^2)$
17. $(2m + 5t)(4m^2 - 10mt + 25t^2)$
18. $2y(4y^2 - 5x^2)(16y^4 + 20x^2y^2 + 25x^4)$

Exercise Set 1.5, pp. 27–28

1. $3ab(6a - 5b)$ **3.** $(a + c)(b - 2)$ **5.** $(x + 6)(x + 3)$
7. $(3x - 5)(3x + 5)$ **9.** $4x(y^2 - z)(y^2 + z)$ **11.** $(y - 3)^2$
13. $(1 - 4x)^2$ **15.** $(2x - \sqrt{5})(2x + \sqrt{5})$ **17.** $(xy - 7)^2$
19. $4a(x + 7)(x - 2)$ **21.** $(a + b + c)(a + b - c)$
23. $(x + y - a - b)(x + y + a + b)$
25. $5(y^2 + 4x^2)(y - 2x)(y + 2x)$ **27.** $(x + 2)(x^2 - 2x + 4)$
29. $3(x - \frac{1}{2})(x^2 + \frac{1}{2}x + \frac{1}{4})$ **31.** $(x + 0.1)(x^2 - 0.1x + 0.01)$
33. $3(z - 2)(z^2 + 2z + 4)$
35. $(a - t)(a + t)(a^2 - at + t^2)(a^2 + at + t^2)$
37. $2ab(2a^2 + 3b^2)(4a^4 - 6a^2b^2 + 9b^4)$
39. $(x + 4.19524)(x - 4.19524)$
41. $37(x + 0.626y)(x - 0.626y)$ **43.** $h(3x^2 + 3xh + h^2)$
45. $(y^2 + 12)(y^2 - 7)$ **47.** $(y + \frac{4}{7})(y - \frac{2}{7})$
49. $(t + 0.9)(t - 0.3)$ **51.** $(x^n + 8)(x^n - 3)$
53. $(x + a)(x + b)$ **55.** $(\frac{1}{2}t - \frac{2}{5})^2$
57. $(5y^m - x^n + 1)(5y^m + x^n - 1)$
59. $3(x^n - 2y^m)(x^{2n} + 2x^ny^m + 4y^{2m})$ **61.** $y(y - 1)^2(y - 2)$

Margin Exercises, Section 1.6

1. All real numbers except 3
2. All real numbers except -3 and -4
3. $\dfrac{x^2 + 2xy + y^2}{14x^3 - 7x}$ **4.** $\dfrac{x^2 + 2x - 8}{x^2 + 4x + 4}$
5. $\dfrac{x^2 + 5x + 6}{x^2 - 2x - 15}$; all real numbers except 5; all real numbers except 5 and -3
6. $\dfrac{3x + 2}{x + 2}$; all real numbers except 0 and -2; all real numbers except -2
7. $\dfrac{y + 2}{y - 1}$; all real numbers except 1 and -1; all real numbers except 1
8. $\dfrac{3x - 3y}{x + y}$ **9.** $\dfrac{2a^2b + 2ab^2}{a - b}$ **10.** $\dfrac{3x^2 + 4x + 2}{x - 5}$
11. $\dfrac{2x^2 + 11}{x - 5}$ **12.** $\dfrac{4x^2 - xy + 4y^2}{2(2x - y)(x - y)}$ **13.** $\dfrac{x - 6}{(x + 4)(x + 6)}$
14. $\dfrac{1}{a - x}$ **15.** $\dfrac{b + a}{b - a}$

Exercise Set 1.6, pp. 33–35

1. All numbers except 0, 1   **3.** All numbers except 0, 3, -2
5. $\dfrac{x - 2}{x + 3}$; all numbers except -3 **7.** $\dfrac{1}{x - y}$
9. $\dfrac{(x + 5)(2x + 3)}{7x}$ **11.** $\dfrac{a + 2}{a - 5}$ **13.** $m + n$ **15.** $\dfrac{3(x - 4)}{2(x + 4)}$
17. $\dfrac{1}{x + y}$ **19.** $\dfrac{x - y - z}{x + y + z}$ **21.** 1   **23.** $\dfrac{y - 2}{y - 1}$ **25.** $\dfrac{x + y}{2x - 3y}$
27. $\dfrac{3x - 4}{x^2 - 4}$ **29.** $\dfrac{3y - 10}{(y - 5)(y + 4)}$ **31.** $\dfrac{4x - 8y}{x^2 - y^2}$
33. $\dfrac{3x - 4}{(x - 2)(x - 1)}$ **35.** $\dfrac{5a^2 + 10ab - 4b^2}{(a - b)(a + b)}$
37. $\dfrac{11x^2 - 18x + 8}{(x + 2)(x - 2)^2}$ **39.** 0   **41.** $\dfrac{x + y}{x}$ **43.** $\dfrac{a^2 - 1}{a^2 + 1}$

45. $\dfrac{c^2 - 2c + 4}{c}$ **47.** $\dfrac{xy}{x - y}$ **49.** $x - y$ **51.** $\dfrac{x^2 - y^2}{xy}$

53. $\dfrac{1 + a}{1 - a}$ **55.** $\dfrac{b + a}{b - a}$ **57.** $2x + h$ **59.** $3x^2 + 3xh + h^2$

61. x^5

63. Step (1) uses the wrong reciprocal. The reciprocal of a sum is not the sum of the reciprocals. Step (2) would be correct if step (1) had been. Step (3) would be correct if steps (1) and (2) had been. Step (4) has an improper simplification of the b in the denominator; $\dfrac{12a}{b(4a + 3b)}$.

Margin Exercises, Section 1.7

1. Yes, no   **2.** No, yes   **3.** Yes, yes   **4.** Yes, yes
5. $|x + 2|$ **6.** $|x| \cdot |y - 2|$ or $|x(y - 2)|$ **7.** $|x + 2|$
8. $|x + 4|$ **9.** $-4xy$ **10.** $\sqrt{133}$ **11.** $\sqrt{x^2 - 4y^2}$
12. $\sqrt[4]{81}$, or 3   **13.** $10\sqrt{3}$ **14.** $6|y|$ **15.** $|x + 1|\sqrt{2}$
16. $2 \cdot \sqrt[3]{2}$ **17.** $(a + b) \cdot \sqrt[3]{a + b}$ **18.** $\dfrac{7}{8}$ **19.** $\dfrac{5}{|y|}$

20. $\dfrac{\sqrt{35}}{5}$ **21.** $\dfrac{\sqrt[3]{7}}{5}$ **22.** 5   **23.** $\dfrac{|x|}{5}$ **24.** $\dfrac{2x}{y}$

25. 3^{10} **26.** 3^4 **27.** 37.42 mph

Exercise Set 1.7, pp. 39-41

1. No, yes,   **3.** Yes, no   **5.** Yes, no   **7.** Yes, yes   **9.** 11
11. $4|x|$ **13.** $|b + 1|$ **15.** $-3x$ **17.** $|x - 2|$ **19.** 2

21. $6\sqrt{5}$ **23.** $3\sqrt[3]{2}$ **25.** $\dfrac{8|c|\sqrt{2}}{d^2}$ **27.** $3\sqrt{2}$ **29.** $2x^2y\sqrt{6}$

31. $3x\sqrt[3]{4y}$ **33.** $2(x + 4)\sqrt[3]{(x + 4)^2}$ **35.** $\sqrt{7b}$ **37.** 2

39. $\dfrac{1}{2x}$ **41.** $\sqrt{a + b}$ **43.** $\dfrac{3a\sqrt{2b}}{4b}$ **45.** $\dfrac{y \cdot \sqrt[3]{20x^2z^2}}{5z^2}$

47. $8x^2\sqrt[3]{2}$ **49.** $ab^2x^2y\sqrt{a}$
51. 1.57 sec, 3.14 sec, 8.88126 sec, 11.1016 sec
53. $10.124x^2y$ **55.** $\dfrac{0.5933a\sqrt{b}}{b}$ **57.** $h = \dfrac{a}{2}\sqrt{3}$ **59.** $\sqrt{2}\,s$

61. 8

Margin Exercises, Section 1.8

1. $-6\sqrt{5}$ **2.** $(10y + 7)\sqrt[3]{2y}$ **3.** $-4 - 9\sqrt{6}$
4. $\dfrac{\sqrt{3} + \sqrt{5}}{-2}$ **5.** $\dfrac{x - 7\sqrt{x} + 10}{x - 4}$ **6.** $\dfrac{1}{\sqrt{a + 2} + \sqrt{a}}$

7. $\dfrac{x - 5}{x + 2\sqrt{5x} + 5}$

Exercise Set 1.8, pp. 43-44

1. $-12\sqrt{5} - 2\sqrt{2}$ **3.** $19\sqrt[3]{x^2} - 3x$ **5.** $4y\sqrt{3} - 2y\sqrt{6}$
7. 1   **9.** $t - 2x\sqrt{t} + x^2$ **11.** $10\sqrt{7}$ **13.** x
15. $\dfrac{3(3 - \sqrt{5})}{2}$ **17.** $\dfrac{2\sqrt[3]{6}}{3}$

19. $\dfrac{8x - 20\sqrt{xy} - 6x\sqrt{y} + 15y\sqrt{x}}{4x - 25y}$ **21.** $\dfrac{2 - 5a}{6(\sqrt{2} - \sqrt{5a})}$

23. $\dfrac{x}{x + 2 - 2\sqrt{x + 1}}$ **25.** $\dfrac{a}{3(\sqrt{a + 3} + \sqrt{3})}$

27. $\dfrac{(2 + x^2)\sqrt{1 + x^2}}{1 + x^2}$

29. Let $a = 16$ and $b = 9$. Then $\sqrt{a + b} = 5$ and $\sqrt{a} + \sqrt{b} = 7$.

Margin Exercises, Section 1.9

1. $\sqrt{n^3}$, or $n\sqrt{n}$ **2.** $\dfrac{1}{\sqrt[7]{y^6}}$ **3.** 16   **4.** $\dfrac{1}{16}$

5. $(5ab)^{4/3}$ or $(5ab)\sqrt[3]{5ab}$ **6.** 8   **7.** $a^{2/3}$
8. $4^{1/6}$, $(2^2)^{1/6}$ or $2^{1/3}$ **9.** $5^{11/6}$ or $5\sqrt[6]{5^5}$ **10.** $\sqrt[4]{a^5}$ or $a\sqrt[4]{a}$
11. $\dfrac{1}{\sqrt[5]{x^6}}$ or $\dfrac{1}{x\sqrt[5]{x}}$ **12.** $\sqrt[4]{2^3} + \dfrac{1}{\sqrt[4]{2}}$ or $3\dfrac{\sqrt[4]{2^3}}{2}$ **13.** $\sqrt[6]{200}$
14. $\sqrt[6]{x^4y^3z^5}$ **15.** $\sqrt[4]{(x + y)}$

Exercise Set 1.9, pp. 46-47

1. $\sqrt[4]{x^3}$ **3.** $(\sqrt[4]{16})^3$ or $\sqrt[4]{(2^4)^3}$ or 8   **5.** $\dfrac{1}{5}$ **7.** $\dfrac{a}{b}\sqrt[4]{ab}$
9. $20^{2/3}$ **11.** $13^{5/4}$ **13.** $11^{1/6}$ **15.** $5^{5/6}$ **17.** 4
19. $2y^2$ **21.** $(a^2 + b^2)^{1/3}$ **23.** $3ab^3$ **25.** $\dfrac{m^2n^4}{2}$

27. $8a^{4/2}$ or $8a^2$ **29.** $\dfrac{x^{-3}}{3^{-1}b^2}$ or $\dfrac{3}{x^3b^2}$ **31.** $xy^{1/3}$ or $x\sqrt[3]{y}$

33. $\sqrt[6]{288}$ **35.** $\sqrt[12]{x^{11}y^7}$ **37.** $a\sqrt[6]{a^5}$
39. $(a + x)\sqrt[12]{(a + x)^{11}}$ **41.** 24.685   **43.** 43.138
45. 32.942   **47.** 5.56 ft   **49.** 7.07 ft   **51.** $a^{a/2}$

Summary and Review: Chapter 1, pp. 47-48

1. [1.1] 12, -3, -1, -19, 31, 0   **2.** [1.1] 12, 31
3. [1.1] All except $\sqrt{7}$, $\sqrt[3]{10}$ **4.** [1.1] All   **5.** [1.1] $\sqrt{7}$, $\sqrt[3]{10}$
6. [1.1] 0, 12, 31   **7.** [1.1] -4 **8.** [1.1] -8 **9.** [1.1] -5
10. [1.1] 30   **11.** [1.1] -6 **12.** [1.1] 153
13. [1.1] -3000 **14.** [1.1] $-\frac{3}{16}$ **15.** [1.1] $\frac{31}{24}$
16. [1.2] 3,261,000   **17.** [1.2] 0.00041
18. [1.2] 1.432×10^{-2} **19.** [1.2] 4.321×10^4
20. [1.2] $-14a^{-2}b^7$ **21.** [1.2] $6x^9y^{-6}z^6$ **22.** [1.7] 3

23. [1.7] -2 **24.** [1.6] $\dfrac{b}{a}$ **25.** [1.6] $\dfrac{x + y}{xy}$ **26.** [1.8] -4

27. [1.8] $25x^4 - 10x^2\sqrt{2} + 2$ **28.** [1.8] $13\sqrt{5}$
29. [1.4] $x^3 + t^3$ **30.** [1.4] $125a^3 + 300a^2b + 240ab^2 + 64b^3$
31. [1.3] $8xy^4 - 9xy^2 + 4x^2 + 2y - 7$
32. [1.5] $(x^2 - 3)(x + 2)$ **33.** [1.5] $3a(2a - 3b^2)(2a + 3b^2)$
34. [1.5] $(x + 12)^2$ **35.** [1.5] $x(9x - 1)(x + 4)$
36. [1.5] $(2x - 1)(4x^2 + 2x + 1)$
37. [1.5] $(3x^2 + 5y^2)(9x^4 - 15x^2y^2 + 25y^4)$
38. [1.9] $y^3 \cdot \sqrt[6]{y}$ **39.** [1.9] $\sqrt[3]{(a + b)^2}$ **40.** [1.9] $\sqrt[5]{b^7}$

41. [1.9] $\dfrac{m^4n^2}{3}$ **42.** [1.6] 3   **43.** [1.6] $\dfrac{x - 5}{(x + 3)(x + 5)}$

44. [1.8] $\dfrac{x - 2\sqrt{xy} + y}{x - y}$ **45.** [1.1] Inverses (+)
46. [1.1] Distributive   **47.** [1.1] Associative (×)
48. [1.1] Commutative (×)   **49.** [1.4] $x^{2n} + 6x^n - 40$
50. [1.4] $t^{2a} + 2 + t^{-2a}$ **51.** [1.4] $y^{2b} - z^{2c}$
52. [1.4] $a^{3n} - 3a^{2n}b^m + 3a^nb^{2m} - b^{3m}$ **53.** [1.5] $(y^n + 8)^2$
54. [1.5] $(x^t - 7)(x^t + 4)$
55. [1.5] $m^{3n}(m^n - 1)(m^{2n} + m^n + 1)$

CHAPTER 2

Margin Exercises, Section 2.1

1. $\{9\}$ 2. $\{0, -1\}$ 3. $\frac{4}{3}$ 4. $\frac{17}{2}$ 5. $-\frac{19}{8}$ 6. $\emptyset$ 7. $\emptyset$
8. $\{7, -\frac{3}{2}\}$ 9. $\{5, -4\}$ 10. $\{0, 5\}$ 11. $\{-\frac{1}{5}\}$
12. $\{0, -\frac{1}{3}, 4\}$ 13. $\{1, -1, -\frac{1}{5}\}$ 14. $\{x | x > \frac{3}{2}\}$
15. $\{y | \frac{22}{13} \le y\}$ 16. $\{x | x < 5\}$ 17. $\{x | x > \frac{5}{2}\}$
18. $\{y | y \ge -7\}$ 19. $\{x | x^2 = 5\}$ 20. $\{x | x \ge 2\}$
21. $\{x | x \ge -3\}$ 22. $\{x | \frac{11}{2} \ge x\}$

Exercise Set 2.1, pp. 55–56

1. 12 3. -6 5. 8 7. $\frac{4}{5}$ 9. 2 11. $-\frac{3}{2}$ 13. -2
15. $\frac{3}{2}, \frac{2}{3}$ 17. $0, 1, -2$ 19. $\frac{2}{3}, -1$ 21. $4, 1$
23. $-1, -2$ 25. $-\frac{5}{3}, 4, \frac{5}{2}$ 27. $0, \frac{1}{4}, -\frac{1}{4}$ 29. $\{0, 3\}$
31. $\{0, -\frac{1}{3}, 2\}$ 33. $\{\frac{3}{2}, -\frac{2}{3}, 1\}$ 35. $\{\frac{1}{2}, 0, -3\}$
37. $x > 3$ 39. $x \ge -\frac{5}{12}$ 41. $y \ge \frac{22}{13}$ 43. $x \le \frac{15}{34}$
45. $x < 1$ 47. $\{x | x > 2.5\}$ 49. $\{t | t^2 = 5\}$
51. $\{x | x \ge 3\}$ 53. $\{x | \frac{3}{4} \ge x\}$ 55. 0.7892
57. $\{0, 2.1522\}$ 59. $\{x | x < -0.7848\}$ 61. $\{-1, 1, -\frac{1}{7}\}$
63. $\{-2, 1, -1\}$ 65. $\{5\}$

Margin Exercises, Section 2.2

1. Yes   2. Yes   3. No   4. Yes   5. No   6. No
7. Add $5x^2$.   8. Add $-5x^2$.   9. $\emptyset$ 10. $\emptyset$ 11. $\{6\}$
12. $\{4\}$ 13. $\{6, -6\}$ 14. $\{\frac{16}{5}\}$

Exercise Set 2.2, pp. 60–61

1. Yes   3. No   5. No   7. $\{\frac{20}{9}\}$ 9. No solution
11. $\{286\}$ 13. $\{3, 2\}$ 15. $\{-2\}$ 17. $\{6\}$ 19. $\emptyset$
21. $\emptyset$ 23. $\{8, -5\}$ 25. $\{\frac{5}{3}\}$ 27. $\{0.94656\}$
29. $\{x | x \ne 3\}$ 31. $\{x | x \ne -2\}$ 33. $\{\frac{3}{2}\}$ 35. Identity
37. Identity   39. Not an identity

Margin Exercises, Section 2.3

1. $F = \frac{9}{5}C + 32$ 2. $r_2 = \dfrac{Rr_1}{r_1 - R}$ 3. 18% 4. 57 5. 27
6. $\$2450$ 7. $\$2565.78$ 8. $\$2637.93$ 9. 20 ft
10. 36 km/h   11. 375 km   12. 50 km/h, 60 km/h
13. $2\frac{2}{9}$ hr   14. Helen: 12 hr; Fran: 6 hr

Exercise Set 2.3, pp. 71–72

1. $w = \dfrac{P - 2l}{2}$ 3. $I = \dfrac{E}{R}$ 5. $T_1 = \dfrac{T_2 P_1 V_1}{P_2 V_2}$
7. $v_1 = \dfrac{H}{Sm} + v_2$ 9. $p = \dfrac{Fm}{m - F}$ 11. $x = \dfrac{5 + ab}{a - b}$
13. $x = -\dfrac{a}{9}$ 15. 44% 17. 6% 19. $\$14.500; \$16.095$
21. $\$650$ 23. $26°, 130°, 24°$ 25. 68 m, 93 m   27. 91%
29. 2 cm   31. $810,000$ 33. 12 km/h
35. A: 46 mph; B: 58 mph   37. 98.3 mi   39. $1\frac{34}{71}$ hr
41. 6.21 hr   43. (a) $\$1137.50$; (b) $\$1142.23$; (c) $\$1144.75$;
(d) $\$1147.37$; (e) $\$1147.40$ 45. 32 mph   47. 53 mph
49. $51\frac{3}{7}$ mph   51. $10:38\frac{2}{11}$

Margin Exercises, Section 2.4

1. $\pm\sqrt{3}$ 2. 0 3. $\pm\sqrt{\dfrac{\pi}{3}}$ 4. $\pm\sqrt{\dfrac{n}{m}}$ 5. $-4 \pm \sqrt{7}$
6. $5 \pm \sqrt{3}$ 7. $-3, -7$ 8. $4, (x + 2)^2$ 9. $9, (x - 3)^2$
10. $\frac{25}{4}, (x + \frac{5}{2})^2$ 11. $\frac{49}{4}, (x - \frac{7}{2})^2$ 12. $\frac{9}{64}, (x + \frac{3}{8})^2$
13. $\frac{1}{4}, (x - \frac{1}{2})^2$ 14. $-2 \pm \sqrt{7}$ 15. $4, 2$ 16. $2, 3$
17. $\dfrac{-1 \pm \sqrt{7}}{2}$ 18. $-1, \frac{1}{4}$ 19. $\frac{1}{2}, -4$ 20. $\dfrac{4 \pm \sqrt{31}}{5}$
21. One real solution   22. No real solutions
23. Two real solutions
24. $x^2 + \frac{7}{3}x - \frac{20}{3} = 0$, or $3x^2 + 7x - 20 = 0$
25. $x^2 - x - 56 = 0$ 26. $x^2 - (m + n)x + mn = 0$

Exercise Set 2.4, pp. 78–79

1. $\pm\sqrt{7}$ 3. $\pm\dfrac{\sqrt{5}}{3}$ 5. $\pm\sqrt{\dfrac{b}{a}}$ 7. $7 \pm \sqrt{5}$ 9. $h \pm \sqrt{a}$
11. $-3 \pm \sqrt{5}$ 13. $\{3, -10\}$ 15. $\dfrac{2 \pm \sqrt{14}}{5}$ 17. $-5, \frac{3}{2}$
19. $1, -5$ 21. $2, -\frac{1}{2}$ 23. $-1, -\frac{5}{3}$ 25. $6 \pm \sqrt{33}$
27. No real solution   29. Two real solutions
31. $x^2 + 2x - 99 = 0$ 33. $x^2 - 14x + 49 = 0$
35. $x^2 - \frac{4}{5}x - \frac{12}{25} = 0$ 37. $x^2 - \left(\dfrac{c + d}{2}\right)x + \dfrac{cd}{4} = 0$
39. $x^2 - 4\sqrt{2}x + 6 = 0$ 41. $1.1754, -0.4254$ 43. $2, -\frac{3}{2}$
45. $\dfrac{-3 \pm \sqrt{41}}{2}$ 47. $\frac{3}{2}, \frac{2}{3}$ 49. $-0.1 \pm \sqrt{0.31}$
51. $\dfrac{-1 \pm \sqrt{1 + 4\sqrt{2}}}{2}$ 53. $\dfrac{-\sqrt{5} \pm \sqrt{5 + 4\sqrt{3}}}{2}$
55. $\dfrac{\sqrt{6} \pm \sqrt{6 + 8\sqrt{10}}}{4}$ 57. $-2, \frac{3}{4}$ 59. $\dfrac{1 \pm \sqrt{113}}{2}$
61. $3 \pm \sqrt{5}$

Margin Exercises, Section 2.5

1. $r = \sqrt{\dfrac{3V}{\pi h}}$ 2. $t = \dfrac{-v_0 + \sqrt{v_0^2 + 64S}}{32}$ 3. 18.75%
4. $12 - 2\sqrt{22} \approx 2.619$ ft
5. (a) 4.33 sec; (b) 1.87 sec; (c) 44.9 m

Exercise Set 2.5, pp. 82–84

1. $d = \sqrt{\dfrac{kM_1M_2}{F}}$ 3. $t = \sqrt{\dfrac{2S}{a}}$ 5. $t = \dfrac{-v_0 \pm \sqrt{v_0^2 - 64s}}{-32}$
7. $n = \dfrac{3 + \sqrt{9 + 8d}}{2}$ 9. $i = -1 + \sqrt{\dfrac{A}{P}}$ 11. 18.75%
13. 11% 15. 9 17. 2 ft   19. 4.685 cm
21. A: 15 mph; B: 20 mph   23. (a) 3.91 sec; (b) 1.906 sec;
(c) 79.6 m   25. 3.237 cm   27. 7 29. 12 31. $2, -\dfrac{3}{k}$
33. $\dfrac{1}{m + n}, \dfrac{-2}{m + n}$ 35. 11.7% 37. $a_3 = \sqrt{a_1^2 + a_2^2}$

Margin Exercises, Section 2.6

1. $\emptyset$ **2.** 4 **3.** $\frac{17}{3}$ **4.** 5 **5.** $b = \sqrt{\dfrac{a^2}{A^2 - 1}}$, or $\dfrac{a}{\sqrt{A^2 - 1}}$

Exercise Set 2.6, p. 87

1. $\frac{5}{3}$ **3.** $\pm\sqrt{2}$ **5.** $\emptyset$ **7.** 4 **9.** $\emptyset$ **11.** -6 **13.** 3, -1
15. $\frac{80}{9}$ **17.** 62.4459 **19.** -8 **21.** 81 **23.** $\frac{1}{64}$
25. -125 **27.** $L = \dfrac{gT^2}{4\pi^2}$, $g = \dfrac{4L\pi^2}{T^2}$ **29.** 208 mi
31. 14,400 ft **33.** $5 \pm 2\sqrt{2}$ **35.** $-\frac{8}{9}$ **37.** 2
39. $\dfrac{-5 + \sqrt{61}}{18}$

Margin Exercises, Section 2.7

1. (a) 9; (b) $\sqrt{x} = 12 - x$; $(12 - x)^2 = x$; $x = 144 - 24x + x^2$;
$0 = x^2 - 25x + 144$; $0 = (x - 9)(x - 16)$. The procedure in
(a) was probably easier, since the factoring was easier.
2. $\pm\sqrt{\dfrac{5 + \sqrt{3}}{2}}, \pm\sqrt{\dfrac{5 - \sqrt{3}}{2}}$ **3.** $\pm\sqrt{3}, 0$ **4.** 125, -8
5. 350.44 ft

Exercise Set 2.7, pp. 91–92

1. 1, 81 **3.** $\pm\sqrt{5}$ **5.** -27, 8 **7.** 16 **9.** 7, 5, -1, 1
11. $1, 4, \dfrac{5 \pm \sqrt{37}}{2}$ **13.** $\pm\sqrt{2 + \sqrt{6}}$ **15.** $-\frac{1}{2}, \frac{1}{3}$
17. -1, 2 **19.** $-1 \pm \sqrt{3}, \dfrac{9 \pm \sqrt{89}}{2}$ **21.** $\frac{100}{99}$ **23.** $-\frac{6}{7}$
25. $1 \pm \sqrt{2}, \dfrac{-1 \pm \sqrt{5}}{2}$ **27.** 132.71 ft **29.** 2.0486
31. 1, 4 **33.** 19

Margin Exercises, Section 2.8

1. $y = 160x$ **2.** 4.5 kg **3.** 50 volts **4.** 176,250 tons
5. $y = \dfrac{6.4}{x}$ **6.** 7.5 hr **7.** $\dfrac{A_2}{A_1} = \dfrac{r_2^2}{r_1^2}$
9. $\dfrac{W_2}{W_1} = \dfrac{d_1^2}{d_2^2}$ **10.** $y = \dfrac{9}{x^2}$ **11.** $\dfrac{A_2}{A_1} = \dfrac{b_2 h_2}{b_1 h_1}$ **12.** $y = 7xz$
13. $y = 7\dfrac{xz}{w^2}$ **14.** 2 sec **15.** (a) 128 lb; (b) 4000 mi

Exercise Set 2.8, pp. 97–98

1. $y = \frac{3}{2}x$ **3.** $y = \dfrac{0.0015}{x^2}$ **5.** $y = \dfrac{xz}{w}$ **7.** $y = \dfrac{5}{4}\dfrac{xz}{w^2}$
9. y is doubled **11.** y is multiplied by $\dfrac{1}{n^2}$ **13.** 532,500 tons
15. L is multiplied by 16 **17.** 68.56 m **19.** 624.24 m²
21. 97 **23.** If p varies directly as q, then $p = kq$. Thus,
$q = \dfrac{1}{k}p$, so q varies directly as p. **25.** $\dfrac{\pi}{4}$

Margin Exercises, Section 2.9

1. $18{,}600 \dfrac{\text{m}}{\text{sec}}$ **2.** $0.5 \dfrac{\text{m}}{\text{sec}}$ **3.** 62 ft **4.** $\frac{23}{20}$ kg **5.** $105 \dfrac{\text{cm}}{\text{sec}}$
6. 12 yd **7.** 80 oz **8.** $\frac{7}{10}$ **9.** $11.25 \dfrac{\text{in.-lb}}{\text{hr}^2}$ **10.** $4 \dfrac{\text{lb}^2}{\text{m}^2}$
11. 1224 in. **12.** 58,080 ft **13.** 18,000 sec **14.** 20 yd
15. 36.96 km **16.** 100 hr **17.** $176 \dfrac{\text{ft}}{\text{sec}}$ **18.** 0.36 m²
19. $50 \dfrac{\text{g}}{\text{cm}^3}$ **20.** $300 \dfrac{\text{¢}}{\text{hr}}$

Exercise Set 2.9, pp. 101–102

1. 12 yd **3.** 48 hr **5.** 3 g **7.** 8 m **9.** 12 ft³
11. $\dfrac{7 \text{ kg}^2}{10 \text{ m}^2}$ **13.** $720 \dfrac{\text{lb-mi}^2}{\text{hr}^2\text{-ft}}$ **15.** $\dfrac{15 \text{ cm}^5\text{-kg}}{2 \text{ sec}^3}$ **17.** 6 ft
19. 172,800 sec **21.** 600 g/cm **23.** 2,160,000 cm²
25. 150¢/hr **27.** 6.228 L/hr **29.** 5,865,696,000,000 mi/yr
31. 1621.8 m/min **33.** 1638.4 km² **35.** 7.5 g, 1250 g
37. 1600 g **39.** 15 moles

Summary and Review: Chapter 2, pp. 101–103

1. [2.2] -1 **2.** [2.1] 3, $-\frac{2}{3}$, -2 **3.** [2.1] $\frac{4}{3}$, -2
4. [2.4] $\dfrac{-3 \pm \sqrt{13}}{2}$ **5.** [2.4] $\dfrac{3 \pm \sqrt{57}}{6}$ **6.** [2.2] $\frac{27}{7}$
7. [2.7] $\pm\sqrt{\dfrac{3 \pm \sqrt{5}}{2}}$ **8.** [2.7] 1 **9.** [2.7] $\pm\sqrt{3}$, 0
10. [2.7] -8, 125 **11.** [2.6] 5 **12.** [2.6] 0, 3
13. [2.4] 8, -2 **14.** [2.1] 6, -3 **15.** [2.1] -20
16. [2.1] -5, 3 **17.** [2.1] -2, 1 **18.** [2.1] $\{y \mid y > -2\}$
19. [2.1] $\{x \mid x \geqslant 5\}$ **20.** [2.1] $\{x \mid x \leqslant 4\}$
21. [2.4] No real solutions **22.** [2.4] Two real solutions
23. [2.4] $x^2 + \frac{5}{2}x - \frac{3}{2} = 0$ **24.** [2.6] $h = \dfrac{v^2}{2g}$
25. [2.3] $t = \dfrac{ab}{a + b}$ **26.** [2.3] 94% **27.** [2.3] $1\frac{1}{3}$ hr
28. [2.3] $1\frac{1}{2}$ hr **29.** [2.3] 60 **30.** [2.3] 4.5
31. [2.5] 80 km/h **32.** [2.5] 8, 15, 17
33. [2.5] $2 + 2\sqrt{2} \approx 4.8$ km/h **34.** [2.8] $y = \dfrac{0.5}{x^2}$
35. [2.8] $T = \dfrac{1}{180} \cdot \dfrac{x^2}{p}$ **36.** [2.8] $2.27 per share
37. [2.8] $s = 16t^2$; $7\frac{1}{2}$ sec **38.** [2.9] $166\frac{2}{3}\dfrac{\text{m}}{\text{min}}$ **39.** [2.2] No
40. [2.1] No **41.** [2.4] $-(a + c)$ **42.** [2.2] Yes
43. [2.8] $A = \dfrac{1}{4\pi} \cdot C^2$; $\dfrac{1}{4\pi}$ **44.** [2.6] 256 **45.** [2.2] No

CHAPTER 3

Margin Exercises, Section 3.1

1. (a) (Chuck, Deron), (Chuck, Vanessa), (Chuck, Elaine),
(Deron, Chuck), (Deron, Vanessa), (Deron, Elaine);
(b) No, the statement "Father is a brother of Chuck" is not true.
(c) Yes, the statement "Deron is a brother of Vanessa" is true.

2. (a) Yes, but one might have to do some research to verify it; (b) Yes, the cost of first-class postage was 22¢ in 1986.
3. (a) {(*d*, 1), (*d*, 2), (*e*, 1), (*e*, 2), (*f*, 1), (*f*, 2)}; (b) {(1, *d*), (1, *e*), (1, *f*), (2, *d*), (2, *e*), (2, *f*)}
4. {(1, 1), (1, 2), (1, 3), (1, 4), (2, 1), (2, 2), (2, 3), (2, 4), (3,1), (3, 2), (3, 3), (3, 4), (4, 1), (4, 2), (4, 3), (4, 4)}
5. {(Father, Father), (Father, Mother), (Father, Elaine), (Father, Vanessa), (Father, Chuck), (Father, Deron), (Mother, Mother), (Mother, Father), (Mother, Elaine), (Mother, Vanessa), (Mother, Chuck), (Mother, Deron), (Elaine, Elaine), (Elaine, Father), (Elaine, Mother), (Elaine, Vanessa), (Elaine, Chuck), (Elaine, Deron), (Vanessa, Vanessa), (Vanessa, Father), (Vanessa, Mother), (Vanessa, Elaine), (Vanessa, Chuck), (Vanessa, Deron), (Chuck, Chuck), (Chuck, Father), (Chuck, Mother), (Chuck, Elaine), (Chuck, Vanessa), (Chuck, Deron), (Deron, Deron), (Deron, Father), (Deron, Mother), (Deron, Vanessa), (Deron, Elaine), (Deron, Chuck)}
6. {(1, 1), (2, 2), (3, 3), (4, 4)}
7. {(2, 1), (3, 1), (3, 2), (4, 1), (4, 2), (4,3)}
8. Domain: {Chuck, Deron}; range: {Elaine, Vanessa, Chuck, Deron}
9. Domain: {1, 2, 3, 4}; range: {1, 2, 3, 4}
10. Domain: {2, 3, 4}; range: {1, 2, 3}
11. Domain: {1, 2}; range: {1, 2, 3}

Exercise Set 3.1, pp. 109–110

1. (a) {(Father, Elaine), (Father, Vanessa), (Father, Chuck), (Father, Deron)}; (b) yes; (c) no   **3.** (a) yes; (b) no
5. (a) {(0, a), (0, b), (0, c), (2, a), (2, b), (2, c)}; (b) {(a, 0), (a, 2), (b, 0), (b, 2), (c, 0), (c, 2)}; (c) {(0, 0), (0, 2), (2, 0), (2, 2)}; (d) {(a, a), (a, b), (a, c), (b, b), (b, a), (b, c), (c, c), (c, a), (c, b)}
7. {(−1, 0), (−1, 1), (−1, 2), (0, 1), (0,2), (1, 2)}
9. {(−1, −1), (−1, 0), (−1, 1), (−1, 2), (0, 0), (0, 1), (0, 2), (1, 1), (1, 2), (2, 2)}
11. {(−1, −1), (0, 0), (1, 1), (2, 2)}
13. Domain: {0, 5, 10, 15, 20}; range: {0, 490, 1270, 2000, 2790}
15. Domain: {−1, 0, 1}; range: {0, 1, 2}
17. Domain: {−1, 0, 1, 2}; range: {−1, 0, 1, 2}
19. (a) {(−1, −1), (−1, 0), (−1, 1), (−1, 2), (0, 0), (0, −1), (0, 1), (0, 2), (1, 1), (1, −1), (1, 0), (1, 2), (2, 2), (2, −1), (2, 0), (2, 1),}; (c) Domain: {0, 1}; range {0, 1, 2}

Margin Exercises, Section 3.2

1. (a)

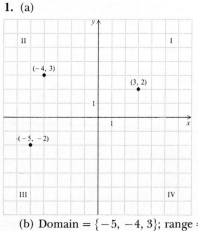

(b) Domain = {−5, −4, 3}; range = {−2, 2, 3}

2. Yes   **3.** No   **4.** No   **5.** Yes

6.

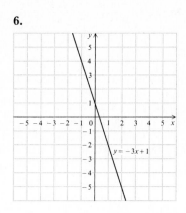

7.

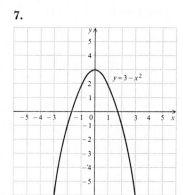

8. The shapes are the same, but this curve opens to the right instead of up.

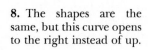

9.

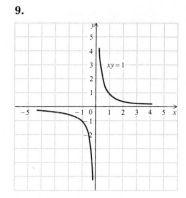

10. The shapes are the same, but this graph opens to the right instead of up.

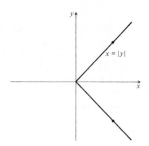

11.

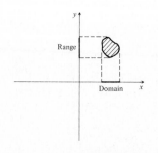

12.

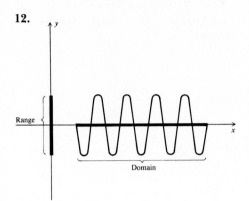

11.

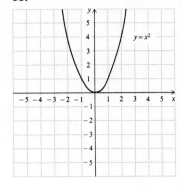

13.

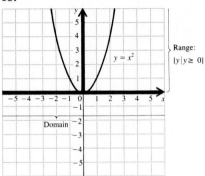

13.

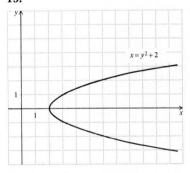

Exercise Set 3.2, pp. 116–117

1. Yes **3.** No **5.** No

7.

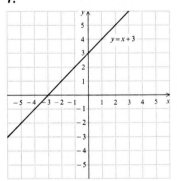

15.

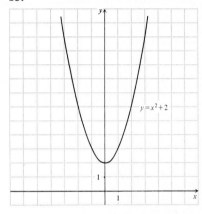

9.

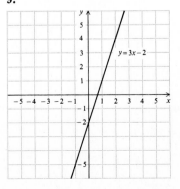

17.

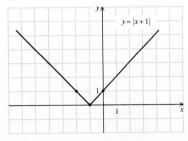

19.

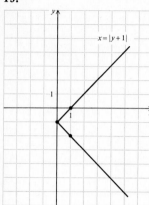

21.

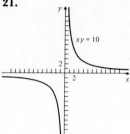

23.

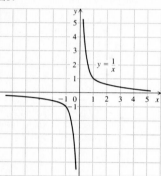

25.

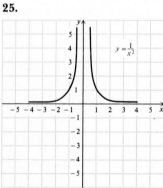

27.

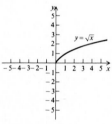

29.

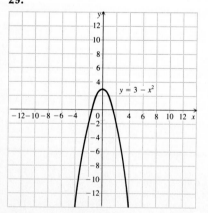

31. Same graphs

33. Domain $= \{x \mid 2 \leqslant x \leqslant 6\}$; range $= \{y \mid 1 \leqslant y \leqslant 5\}$

35. Domain = the set of real numbers; range $= \{y \mid y \leqslant 0\}$

37. Domain $= \{x \mid x \geqslant 0\}$; range = the set of real numbers

39. Horizontal line through $(0, 2)$

41. Line through $(0, 1)$ and $(-1, 0)$

43. Line through $(0, 0)$ and $(1, 2)$ **45.** See Exercise 11.

47.

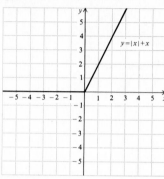

49.

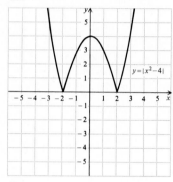

51.

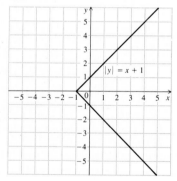

53.

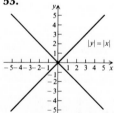

55.

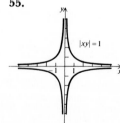

Margin Exercises, Section 3.3

1. (a), (b), (c)   **2.** 1, 4, 3, 4; domain $= \{1, 2, 3, 4\}$; range $= \{1, 3, 4\}$ **3.** 1, 3, 4.5

4. 4, $\sqrt{3}$, $\sqrt{-4}$ not possible   **5.** Each is 7; $\{7\}$.

6. (a) 1; (b) 4; (c) 4; (d) $12a^2 + 1$; (e) $3a^2 + 6a + 4$;

7. $6a + 3h$ **8.** $2a - 1 + h$ **9.** $\{x \mid x \neq -\frac{4}{3}$ and $x \neq -2\}$
10. $\{x \mid x \geqslant -2.5\}$ **11.** All real numbers
12. $f \circ g(x) = 2x^2 - 2;\ g \circ f(x) = 2x^2 - 8x + 8$
13. $u \circ v(x) = 18x^2 + 24x + 8;\ v \circ u(x) = 6x^2 + 2$

Exercise Set 3.3, pp. 124–126

1. (b), (c), (d)   **3.** (a) 0; (b) 1; (c) 57; (d) $5t^2 + 4t$;
(e) $5t^2 - 6t + 1$; (f) $10a + 5h + 4$ **5.** (a) 5; (b) -2; (c) -4;
(d) $4|y| + 6y$; (e) $2|a + h| + 3a + 3h$; (f) $\dfrac{2|a + h| + 3h - 2|a|}{h}$

7. (a) 3.14977; (b) 55.73147; (c) 3178.20675; (d) 1116.70323
9. (a) $\frac{2}{3}$; (b) $\frac{10}{9}$; (c) 0; (d) not possible

11. All real numbers   **13.** $\{x \mid x \neq 0\}$ **15.** $\{x \mid x \geqslant -\frac{4}{7}\}$
17. $\{x \mid x \neq 2, -2\}$ **19.** $\{x \mid x \neq -\frac{3}{4}, 2\}$
21. $f \circ g(x) = 12x^2 - 12x + 5;\ g \circ f(x) = 6x^2 + 3$

23. $f \circ g(x) = \dfrac{16}{x^2} - 1;\ g \circ f(x) = \dfrac{2}{4x^2 - 1}$

25. $f \circ g(x) = x^4 - 2x^2 + 2;\ g \circ f(x) = x^4 + 2x^2$

27. $0, -3, 3, 2$ **29.** $\dfrac{-1}{x(x + h)}$ **31.** $\dfrac{1}{\sqrt{x + h} + \sqrt{x}}$

33. $\{x \mid x \neq 2, -1$ and $x \geqslant -3\}$ **35.** All real numbers
37. Domain of $f \circ g$ is $\{x \mid x \neq 0\}$; domain of $g \circ f$ is
$\{x \mid x \neq \frac{1}{2}, -\frac{1}{2}\}$.

Margin Exercises, Section 3.4

1. (a) $(-3, 2)$; (b) $(4, -5)$

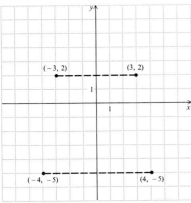

2. (a) $(4, -3)$; (b) $(3, 5)$

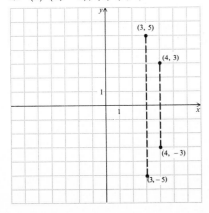

3. x-axis: no; y-axis: yes   **4.** x-axis: yes; y-axis: no
5. x-axis: yes; y-axis: yes   **6.** x-axis: yes; y-axis: yes
7. a-axis: no; b-axis: no   **8.** p-axis: no; q-axis; no

9. (a) $(-3, -2)$; (b) $(4, -3)$; (c) $(5, 7)$ **10.** Yes   **11.** Yes
12. Yes   **13** Yes
14. Yes   **15.** No

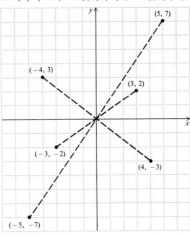

Exercise Set 3.4, pp. 132–133

1. x-axis, no; y-axis, yes; origin, no   **3.** x-axis, no; y-axis, yes;
origin, no   **5.** All yes   **7.** All yes   **9.** All no   **11.** All no
13. Yes   **15.** Yes   **17.** Yes   **19.** Yes   **21.** No   **23.** No

25.

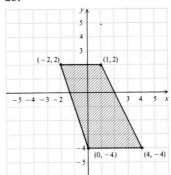

27.

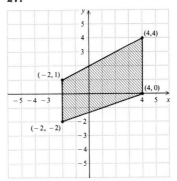

Margin Exercises, Section 3.5

1. (a) Same shape, but the second one is moved up 2 units;
(b) same shape as $y = x^2$, but moved down 2 units.
2. Same shape as $y = x^2$, but move down 3 units.

3.

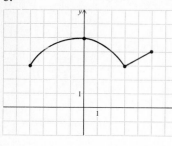

4.

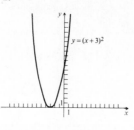

$y = (x + 3)^2$

13.

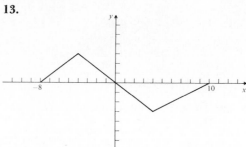

5.

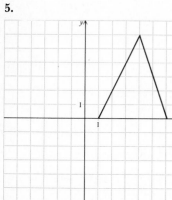

6.

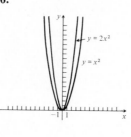

$y = 2x^2$

$y = x^2$

Exercise Set 3.5, pp. 139–140

1. and 3.

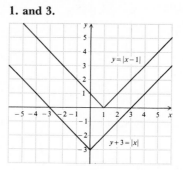

$y = |x - 1|$

$y + 3 = |x|$

7.

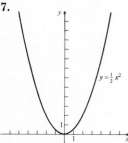

$y = \frac{1}{2}x^2$

8.

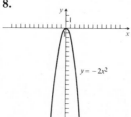

$y = -2x^2$

5. and 7.

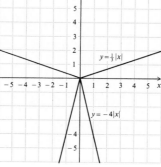

$y = \frac{1}{3}|x|$

$y = -4|x|$

9.

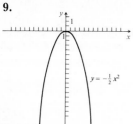

$y = -\frac{1}{2}x^2$

10.

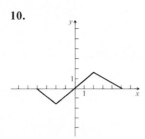

9. and 11.

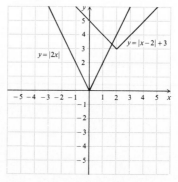

$y = |2x|$

$y = |x - 2| + 3$

11.

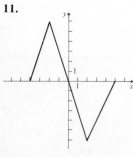

12.

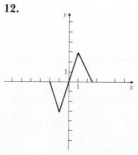

13.

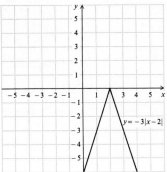

$y = -3|x-2|$

15. and 17.

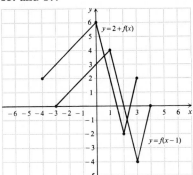

$y = 2 + f(x)$

$y = f(x-1)$

19.

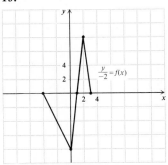

$\dfrac{y}{-2} = f(x)$

21.

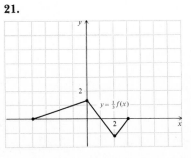

$y = \frac{1}{3}f(x)$

23.

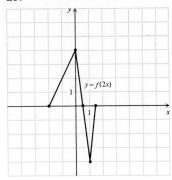

$y = f(2x)$

25.

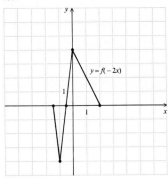

$y = f(-2x)$

27.

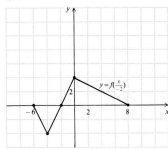

$y = f(\frac{x}{-2})$

29.

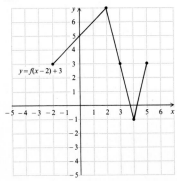

$y = f(x-2) + 3$

31.

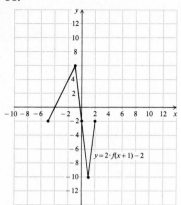

33.

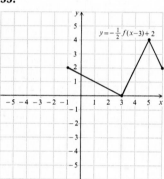

35.

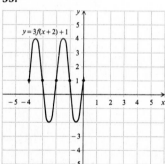

37. The graph is translated 1.8 units to the left, stretched vertically by a factor of $\sqrt{2}$, and reflected across the x-axis.

Margin Exercises, Section 3.6

1. (a) Yes; (b) yes; (c) no; (d) yes   **2.** (a) No; (b) yes;
(c) yes; (d) yes   **3.** (a) Odd; (b) odd; (c) even; (d) odd
4. (a) Neither; (b) even; (c) odd; (d) neither; (e) neither
5. $f(x) = f(x + 4), f(x) = f(x + 6)$ **6.** (a) Yes; (b) 3
7. $t(x) = t(x + 1)$; t is periodic. It does not have a period, because there is no *smallest* positive p for which $t(x + p) = t(x)$.
8. (a) (1, 5); (b) (−1, 4)   **9.** (a) (−2, 3); (b) (0, 1);
(c) $(-\frac{1}{4}, \sqrt{2})$ **10.** (a) [−1, 3]; (b) [−1, 3]; (c) [−1, 3);
(d) (−1, 3)   **11.** (a) $[4, 5\frac{1}{2}]$; (b) (−3, 0]; (c) $[-\frac{1}{2}, \frac{1}{2})$;
(d) (−π, π)   **12.** (a) No; (b) yes; (c) no; (d) yes; (e) no
13. Where $x = -3, 0, 2$ **14.** (a) Increasing;
(b) increasing; (c) decreasing; (d) neither
15. Increasing: [−3, 0]; decreasing: [0, 3]; there are many
answers.

16.

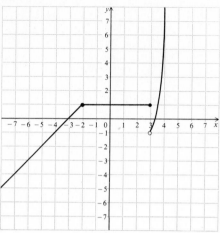

Exercise Set 3.6, pp. 146–149

1. (a) Even; (b) even; (c) odd; (d) neither   **3.** Neither
5. Even   **7.** Neither   **9.** Even   **11.** Odd   **13.** Neither
15. Odd   **17.** Even and odd   **19.** Even
21. (a) No; (b) yes; (c) yes; (d) no   **23.** 4
25. (a) (−2, 2); (b) (−5, −1); (c) [c, d]; (d) [−5, 1)
27. (a) (−2, 4); (b) $(-\frac{1}{4}, \frac{1}{4}]$; (c) [7, 10π); (d) [−9, −6]
29. (a) Yes; (b) yes; (c) no; (d) yes; (e) yes
31. Where x = −3 and x = 2   **33.** (a) Increasing;
(b) neither; (c) decreasing; (d) neither

35.

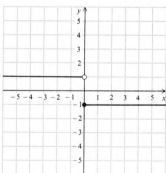

37.

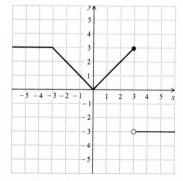

39.

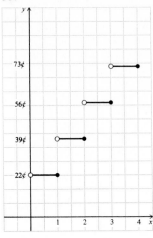

41. Increasing: $[0, 1]$; decreasing: $[-1, 0]$; many possible answers **43.** Increasing: (a), (e); decreasing: (b); neither: (c), (d), (f) **45.** (a) $[2, 3]$; (b) $(0, 9]$; (c) $(-6, 1)$

Summary and Review: Chapter 3, pp. 149–152

1. [3.1] $\{(1, 1), (1, 3), (1, 5), (1, 7), (3, 1), (3, 3), (3, 5), (3, 7), (5, 1), (5, 3), (5, 5), (5, 7), (7, 1), (7, 3), (7, 5), (7, 7)\}$
2. [3.1] Domain $= \{3, 5, 7\}$; range $= \{1, 3, 5, 7\}$

3. [3.2]

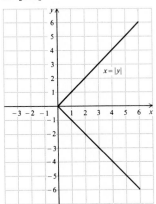

4. [3.2], [3.5]

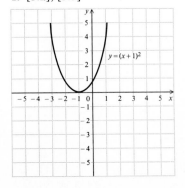

5. [3.2], [3.5]

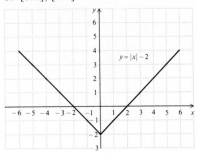

6. [3.2]

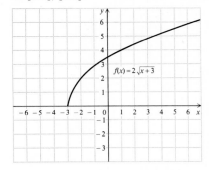

7. [3.2], [3.5]

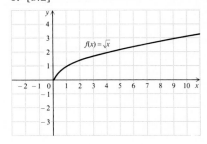

8. [3.2], [3.5]

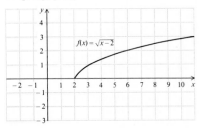

9. [3.2], [3.5]

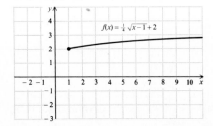

10. [3.2]

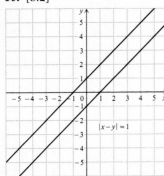

11. [3.4] (b), (d), (f)
12. [3.4] (a), (b), (d), (g)
13. [3.4] (b), (c), (d), (h)
14. [3.3] (b), (c)
15. [3.3] -3
16. [3.3] 9
17. [3.3] $2a + h - 1$
18. [3.3] 0
19. [3.3] 4
20. [3.3] $2\sqrt{a + 1}$
21. [3.3] $\{x \mid x \leqslant \frac{7}{3}\}$
22. [3.3] $\{x \mid x \neq 1, 5\}$
23. [3.3] $\{x \mid x \neq 0, 3, -3\}$
24. [3.3] $\{x \mid x < 0\}$

25. [3.3] $f \circ g(x) = \dfrac{4}{(3 - 2x)^2}$, $g \circ f(x) = 3 - \dfrac{8}{x^2}$

26. [3.3] $f \circ g(x) = 12x^2 - 4x - 1$, $g \circ f(x) = 6x^2 + 8x - 1$

27. [3.5] (a) $y = 1 + f(x)$ (b) $y = \frac{1}{2}f(x)$

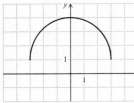
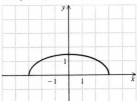

(c) $y = f(x + 1)$

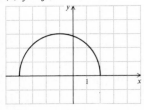

28. [3.6] (a), (b), (c)
29. [3.6] (e), (f)
30. [3.6] (d)
31. [3.6] (a), (c), (d)
32. [3.6] 2
33. [3.6] (a) Yes; (b) no

34. [3.6] (c), (d)   **35.** [3.6] (a)   **36.** [3.6] (b)
37. [3.6] $[-\pi, 2\pi]$ **38.** [3.6] $(0, 1]$

39. [3.6]

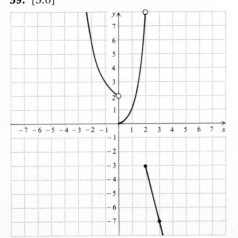

40. [3.5] Graph $y = f(x)$. Then reflect that portion that lies below the x-axis, across the x-axis.

CHAPTER 4

Margin Exercises, Section 4.1

1. (a), (b)

2.

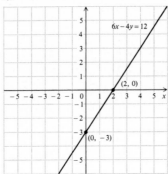

3.

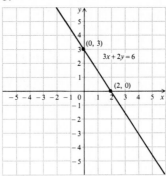

4.

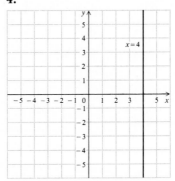

5.

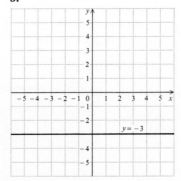

6. $m = 2$ **7.** $m = -2$ **8.** $m = 2$ **9.** $m = 6$ **10.** -1
11. $-\frac{1}{3}$ **12.** 0 **13.** m is undefined
14. 0 **15.** $y = -3x - \frac{23}{4}$ **16.** $y = \frac{x}{4} - 9$

17. $y = -\frac{x}{2} + \frac{5}{2}$ **18.** $y = -3x + 7$ **19.** $y = -\frac{10x}{3} + 4$

20. $m = -7;\ b = 11$ **21.** $m = 0;\ b = -4$
22. (a) $y = \frac{2}{3}x + 2$; (b) $m = \frac{2}{3};\ b = 2$

Exercise Set 4.1, pp. 161–162

1. (a), (b), (d), (h)   **3.** Line through $(0, -8)$ and $(3, 0)$
5. Line through $(0, 3)$ and $(-4, 0)$
7. Horizontal line through $(0, -2)$

9. Vertical line through $(3, 0)$ **11.** $\frac{1}{8}$ **13.** $\frac{1}{2}$ **15.** $\frac{1}{\pi}$

17. $y = 4x - 10$ **19.** $y = 2x - 5$ **21.** $y = \frac{x}{2} + \frac{7}{2}$

23. $m = 2;\ b = 3$ **25.** $m = -3;\ b = 5$ **27.** $m = \frac{3}{4};\ b = -3$
29. $m = 0;\ b = -\frac{10}{3}$ **31.** $y = 3.516x - 13.1602$
33. $y = 1.2222x + 1.0949$ **35.** $f(x) = mx$
37. $f(x) = x + b$ **39.** False   **41.** False   **43.** Yes
45. $\overline{AB}, \overline{DC}$: same slope; $\overline{BC}, \overline{AD}$: same slope
47. $F = \frac{9}{5}C + 32$ **49.** $P = mQ + b,\ m \neq 0.$
Then we can solve for Q: $Q = \dfrac{P}{m} - \dfrac{b}{m}.$

51. Grade $= 4\%$; $y = 4\%x$

Margin Exercises, Section 4.2

1. No   **2.** Yes   **3.** Perpendicular   **4.** Neither
5. Parallel   **6.** Parallel: $y - 4 = -2(x - 3)$,
or $y = -2x + 10$; perpendicular: $y - 4 = \frac{1}{2}(x - 3)$,
or $y = \frac{1}{2}x + \frac{5}{2}$ **7.** Parallel: $x = 5$; perpendicular: $y = -4$
8. $\sqrt{149}$ **9.** $6\sqrt{2}$ **10.** 16 **11.** 8 **12.** Yes   **13.** No
14. $(\frac{3}{2}, -\frac{5}{2})$ **15.** $(9, -5)$

Exercise Set 4.2, pp. 167–168

1. Neither   **3.** Perpendicular   **5.** $y = 3x + 3$ **7.** $x = 3$
9. $y = -3$ **11.** $y = -\frac{2}{5}x - \frac{31}{5}$ **13.** $y = 3$ **15.** $x = -3$
17. $y = 0.6114x + 3.4094$ **19.** 5 **21.** $3\sqrt{2}$ **23.** $\sqrt{a^2 + 64}$
25. $\sqrt{a^2 + b^2}$ **27.** $2\sqrt{a}$ **29.** 18.8061 **31.** Yes
33. $(-\frac{1}{2}, -1)$ **35.** $(a, 0)$ **37.** $(-0.4485, -0.2733)$
39. $y = -\frac{7}{3}x + \frac{22}{3}$ **41.** $(5, 0)$

Margin Exercises, Section 4.3

1. (a)

(b) upward; (c) y-axis, $x = 0$;
(d) 0; (e) $(0, 0)$

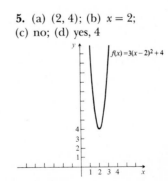

2. (a)

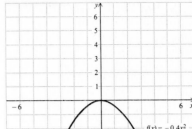

(b) downward;
(c) y-axis, $x = 0$;
(d) 0; (e) $(0, 0)$

3. (a) and (b)

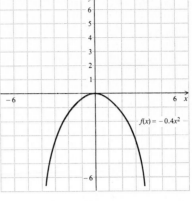

(c) $(2, 0)$ (d) $x = 2$; (e) 0;
(f) upward; (g) horizontal
translation to the right

4. (a) and (b)

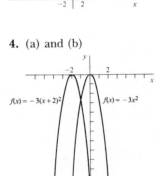

(c) $(-2, 0)$; (d) $x = -2$;
(e) 0; (f) downward;
(g) horizontal translation
to the left

5. (a) $(2, 4)$; (b) $x = 2$;
(c) no; (d) yes, 4

6. (a) $(-2, -1)$;
(b) $x = -2$; (c) yes, -1;
(d) no

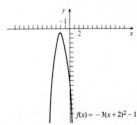

B

7. (a) $(5, \pi)$; (b) $x = 5$; (c) no; (d) yes, π
8. (a) $(5, 0)$; (b) $x = 5$; (c) yes, 0; (d) no
9. (a) $(-\frac{1}{4}, -6)$; (b) $x = -\frac{1}{4}$; (c) no; (d) yes, -6

4

10. (a) $(-9, 3)$; (b) $x = -9$; (c) yes, 3; (d) no
11. $f(x) = (x - 2)^2 + 3$ **12.** $f(x) = 3(x + 4)^2 - 38$
13. (a) $(\frac{3}{2}, -14)$, $x = \frac{3}{2}$; (b) -14 is a minimum
14. $(-\frac{5}{2}, \frac{43}{2})$; $\frac{43}{2}$ is a maximum

15.

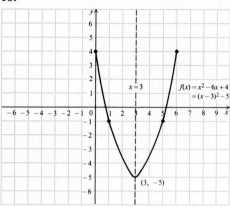

16.

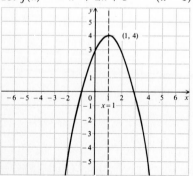

17. $(1 - \sqrt{6}, 0)$, $(1 + \sqrt{6}, 0)$
18. $(-1, 0)$, $(3, 0)$
19. $(-4, 0)$ **20.** None

Exercise Set 4.3, p. 175

1. (a) $(0, 0)$; (b) $x = 0$; (c) 0 is a minimum   **3.** (a) $(9, 0)$;
(b) $x = 9$; (c) 0 is a maximum   **5.** (a) $(1, -4)$; (b) $x = 1$;
(c) -4 is a minimum   **7.** (a) $f(x) = -(x - 1)^2 + 4$;
(b) $(1, 4)$; (c) 4 is a maximum   **9.** (a) $f(x) = (x + \frac{3}{2})^2 - \frac{9}{4}$;
(b) $(-\frac{3}{2}, -\frac{9}{4})$; (c) $-\frac{9}{4}$ is a minimum
11. (a) $f(x) = -\frac{3}{4}(x - 4)^2 + 12$; (b) $(4, 12)$;
(c) 12 is a maximum   **13.** (a) $f(x) = 3(x + \frac{1}{6})^2 - \frac{49}{12}$;
(b) $(-\frac{1}{6}, -\frac{49}{12})$; (c) $-\frac{49}{12}$ is a minimum

15. $f(x) = -x^2 + 2x + 3 = -(x - 1)^2 + 4$

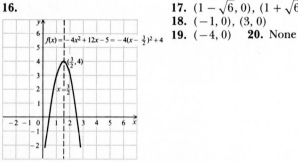

17. $f(x) = x^2 - 8x + 19 = (x - 4)^2 + 3$

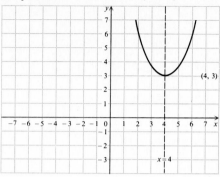

19. $f(x) = -\frac{1}{2}x^2 - 3x + \frac{1}{2} = -\frac{1}{2}(x + 3)^2 + 5$

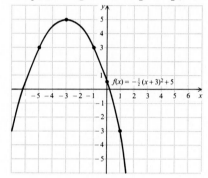

21. $f(x) = 3x^2 - 24x + 50 = 3(x - 4)^2 + 2$

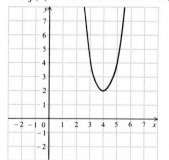

23. $(3, 0)$, $(-1, 0)$
25. $(4 \pm \sqrt{11}, 0)$
27. None

29. $f(x) = a\left[x - \left(-\dfrac{b}{2a}\right)\right]^2 + \dfrac{4ac - b^2}{4a}$

31.

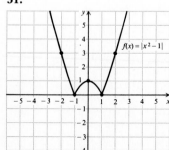

33. Minimum, -6.95
35. Minimum, -6.081;
± 2.466

Margin Exercises, Section 4.4

1. (a) $S = \frac{14}{15}d + 9\frac{2}{3}$; (b) 85   **2.** $P(x) = 60x - 15{,}000$; $x = 250$ is the break-even point.
3. Maximum height 12.4 m, at $t = \frac{2}{7}$.
Reaches ground in 1.88 sec.

Exercise Set 4.4, pp. 181–182

1. (a) $E = 0.15t + 72$; (b) 76.5, 77.85
3. (a) $R = -0.01t + 10.43$; (b) 9.75, 9.73; (c) 2063
5. (a) $C(x) = 40x + 22{,}500$; (b) $R(x) = 85x$;
(c) $P(x) = 45x - 22{,}500$; (d) profit = \$112,500; (e) $x = 500$;
(f) $x > 500$; (g) $x < 500$ **7.** (a) 1662.5 m after 15 sec;
(b) after 33.4 sec   **9.** 10×10 **11.** 30   **13.** \$3.67

Margin Exercises, Section 4.5

1. $\{-3, 4\}$ **2.** $\{2, d, e\}$ **3.**
4.   **5.**
6. $\emptyset$, nothing to graph   **7.** $\{1, 2, 3, 4, 5\}$
8. $\{-3, -4, 1, 2, 3, 4, 8, 9, 11\}$ **9.** $\{1, 2, a, b, c, d, e\}$
10.

11.   **12.**
13.   **14.** $4 < x < 8$ **15.** $-3 \leqslant x < 0$
16. $-5 \leqslant x \leqslant -2$ **17.** $-1 < x \leqslant -\frac{1}{4}$
18. $-\frac{1}{2} < x$ and $x < 1$ **19.** $-\frac{17}{3} \leqslant x$ and $x < -2$
20. $\frac{19}{4} \leqslant x$ and $x \leqslant \frac{37}{6}$
21. $\{x \mid -\frac{2}{3} < x \leqslant 4\} = \{x \mid -\frac{2}{3} < x\} \cap \{x \mid x \leqslant 4\}$
22. $\{x \mid -5 < x < 11\} = \{x \mid -5 < x\} \cap \{x \mid x < 11\}$
23. $\{x \mid -\frac{1}{3} \leqslant x \leqslant \frac{1}{6}\} = \{x \mid -\frac{1}{3} \leqslant x\} \cap \{x \mid x \leqslant \frac{1}{6}\}$
24. $\{x \mid x < -7 \text{ or } x > -1\} = \{x \mid x < -7\} \cup \{x \mid x > -1\}$
25. $\{x \mid x \leqslant -1 \text{ or } x > 4\} = \{x \mid x \leqslant -1\} \cup \{x \mid x > 4\}$
26. $\{x \mid x \geqslant \frac{5}{3} \text{ or } x \leqslant 1\} = \{x \mid x \geqslant \frac{5}{3}\} \cup \{x \mid x \leqslant 1\}$
27. $\{x \mid x < -\frac{11}{4} \text{ or } x \geqslant \frac{1}{4}\} = \{x \mid x < -\frac{11}{4}\} \cup \{x \mid x \geqslant \frac{1}{4}\}$

Exercise Set 4.5, pp. 186–187

1. $\{3, 4, 5\}$ **3.** $\{0, 2, 4, 6, 8, 9\}$ **5.** $\{c\}$
7. Entire set of real numbers
9. $-\frac{1}{2}$ and all numbers between $-\frac{1}{2}$ and $\frac{1}{2}$
11.
13.

15. $\emptyset$ **17.** $\{x \mid -3 \leqslant x < 3\}$ **19.** $\{x \mid 8 \leqslant x \leqslant 10\}$
21. $\{-7\}$ **23.** $\{x \mid -\frac{3}{2} < x < 2\}$ **25.** $\{x \mid 1 < x \leqslant 5\}$
27. $\{x \mid -\frac{11}{3} < x \leqslant \frac{13}{3}\}$ **29.** $\{x \mid x \leqslant -2 \text{ or } x > 1\}$
31. $\{x \mid x \leqslant -\frac{7}{2} \text{ or } x \geqslant \frac{1}{2}\}$ **33.** $\{x \mid x < 9.6 \text{ or } x > 10.4\}$
35. $\{x \mid x \leqslant -\frac{57}{4} \text{ or } x \geqslant -\frac{55}{4}\}$ **37.** $\{1\}$ **39.** $\{x \mid x > -\frac{1}{5}\}$
41. $\{x \mid x \geqslant -\frac{3}{2}\}$ **43.** $\{w \mid 4.885 \text{ cm} < w < 53.67 \text{ cm}\}$
45. $\{S \mid 97\% \leqslant S \leqslant 100\%\}$; yes   **47.** $\{x \mid x > 2\}$

Margin Exercises, Section 4.6

1. $\{-5, 5\}$ **2.** $\{\frac{1}{4}, -\frac{1}{4}\}$ **3.** $\{x \mid -5 < x < 5\}$
4. $\{x \mid -\frac{1}{4} \leqslant x \leqslant \frac{1}{4}\}$ **5.** $\{x \mid x \leqslant -5 \text{ or } x \geqslant 5\}$
6. $\{x \mid x < -\frac{1}{4} \text{ or } x > \frac{1}{4}\}$ **7.** $\{-5, 3\}$ **8.** $\{x \mid -9 < x < -5\}$
9. $\{x \mid x < 1 \text{ or } x > 5\}$ **10.** $\{-1, \frac{11}{3}\}$
11. $\{x \mid -\frac{3}{10} < x < \frac{1}{10}\}$ **12.** $\{x \mid x < -1 \text{ or } x > \frac{11}{3}\}$

Exercise Set 4.6, p. 190

1. $\{-7, 7\}$ **3.** $\{x \mid -7 < x < 7\}$ **5.** $\{x \mid x \leqslant -\pi \text{ or } x \geqslant \pi\}$
7. $\{-3, 5\}$ **9.** $\{x \mid -17 < x < 1\}$
11. $\{x \mid x \leqslant -17 \text{ or } x \geqslant 1\}$ **13.** $\{x \mid -\frac{1}{4} < x < \frac{3}{4}\}$
15. $\{-\frac{1}{3}, \frac{1}{3}\}$ **17.** $\{-1, -\frac{1}{3}\}$ **19.** $\{x \mid -\frac{1}{3} < x < \frac{1}{3}\}$
21. $\{x \mid -6 \leqslant x \leqslant 3\}$ **23.** $\{x \mid x < 4.9 \text{ or } x > 5.1\}$
25. $\{x \mid -\frac{7}{3} \leqslant x \leqslant 1\}$ **27.** $\{x \mid -\frac{1}{2} \leqslant x \leqslant \frac{7}{2}\}$
29. $\{x \mid x < -8 \text{ or } x > 7\}$ **31.** $\{x \mid x < -\frac{7}{4} \text{ or } x > -\frac{3}{2}\}$
33. $\{x \mid \frac{3}{8} \leqslant x \leqslant \frac{9}{8}\}$ **35.** $\emptyset$ **37.** $\emptyset$
39. $\{x \mid 1.9234 < x < 2.1256\}$ **41.** $\{x \mid 0.98414 \leqslant x \leqslant 4.9808\}$
43. $\{2, \frac{44}{5}\}$ **45.** $\{-4, 4\}$ **47.** $\{x \mid x \leqslant \frac{3}{2}\}$
49. $\{x \mid -\frac{9}{2} < x < \frac{11}{2}\}$ **51.** $\{x \mid x < -\frac{8}{3} \text{ or } x > -2\}$

Margin Exercises, Section 4.7

1. $\{x \mid -2 < x < 5\}$ **2.** $\{x \mid x > -1 + \sqrt{5} \text{ or } x < -1 - \sqrt{5}\}$
3. $\{x \mid x < -1 \text{ or } x > 4\}$ **4.** $\{x \mid -1 < x < 4\}$
5. $\{x \mid -3 < x < 0 \text{ or } x > 2\}$ **6.** $\{x \mid -11 \leqslant x < -4\}$
7. $\{x \mid x < 5 \text{ or } x \geqslant 10\}$

Exercise Set 4.7, pp. 194–195

1. $\{x \mid -1 < x < 2\}$ **3.** $\{x \mid x \leqslant -1 \text{ or } x \geqslant 1\}$
5. All real numbers   **7.** $\{x \mid 3 - \sqrt{5} < x < 3 + \sqrt{5}\}$
9. $\{x \mid -2 \leqslant x \leqslant 10\}$ **11.** $\{x \mid x < -2 \text{ or } x > 4\}$
13. $\{x \mid -3 < x < \frac{5}{4}\}$
15. $\left\{x \mid x < \dfrac{-1 - \sqrt{41}}{4} \text{ or } x > \dfrac{-1 + \sqrt{41}}{4}\right\}$
17. $\{x \mid -1 < x < 0 \text{ or } x > 1\}$
19. $\{x \mid x < -3 \text{ or } -2 < x < 1\}$ **21.** $\{x \mid x > 4\}$
23. $\{x \mid x < -\frac{2}{3} \text{ or } x > 3\}$ **25.** $\{x \mid \frac{3}{2} < x \leqslant 4\}$
27. $\{x \mid x \leqslant -\frac{5}{2} \text{ or } x > -2\}$ **29.** $\{x \mid x < -\frac{11}{7}\}$
31. $\{x \mid x > 1\}$ **33.** $\{x \mid x > 0 \text{ and } x \neq 2\}$
35. $\{x \mid x < 0 \text{ or } x \geqslant 1\}$
37. $\{x \mid x < -3 \text{ or } -2 < x < -1 \text{ or } x > 2\}$
39. $\{x \mid x < \frac{5}{3} \text{ or } x > 11\}$ **41.** $\emptyset$ **43.** $\{x \mid x < -\frac{1}{4} \text{ or } x > \frac{1}{2}\}$

45. $\{x \mid x \neq 0\}$ **47.** $\{x \mid -4 < x < -2 \text{ or } -1 < x < 1\}$
49. $\{h \mid h > -2 + 2\sqrt{6} \text{ cm}\}$ **51.** (a) 10, 35;
(b) $\{x \mid 10 < x < 35\}$; (c) $\{x \mid x < 10 \text{ or } x > 35\}$
53. (a) $\{k \mid k > 2 \text{ or } k < -2\}$; (b) $\{k \mid -2 < k < 2\}$
55. $\{x \mid -1 \leqslant x \leqslant 1\}$ **57.** $\{x \mid x \leqslant -3 \text{ or } x \geqslant 1\}$

Summary and Review: Chapter 4, pp. 195–196

1. [4.1]

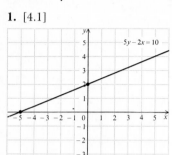

2. [4.1] $m = -2$;
y-intercept: $(0, -7)$
3. [4.1] 1
4. [4.1] $y = 3x + 5$
5. [4.1] $y = \frac{1}{3}x - \frac{1}{3}$
6. [4.2] $\sqrt{34}$
7. [4.2] $(\frac{1}{2}, \frac{11}{2})$
8. [4.2] $y = -\frac{2}{3}x - \frac{1}{3}$
9. [4.2] $y = \frac{3}{2}x - \frac{5}{2}$

10. [4.2] Parallel **11.** [4.2] Neither
12. [4.2] Perpendicular **13.** [4.3] (a) $f(x) = 3(x+1)^2 - 2$;
(b) $(-1, -2)$; (c) $x = -1$; (d) -2 is a minimum.
14. [4.3] (a) $f(x) = -2(x + \frac{3}{4})^2 + \frac{57}{8}$; (b) $(-\frac{3}{4}, \frac{57}{8})$;
(c) $x = -\frac{3}{4}$; (d) $\frac{57}{8}$ is a maximum.

15. [4.3]

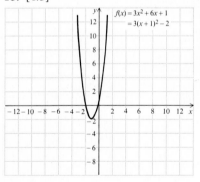

16. [4.3] None **17.** [4.5] $\{5\}$
18. [4.5] $\{3, 4, 5, 7, 8, 9, 11, 12, 13\}$
19. [4.5]
20. [4.5]
21. [4.5] $\{x \mid 2 \leqslant x \leqslant 4\}$ **22.** [4.6] $\{x \mid 1 < x < 11\}$
23. [4.6] $\{x \mid x > 0 \text{ or } x < -\frac{4}{3}\}$ **24.** [4.6] $\{x \mid -20 \leqslant x \leqslant 4\}$
25. [4.6] $\{2, -7\}$ **26.** [4.7] $\{x \mid -3 < x < 3\}$
27. [4.7] $\{x \mid x > 2 \text{ or } x < -\frac{1}{2}\}$
28. [4.7] $\{x \mid -4 < x < 1 \text{ or } x > 2\}$
29. [4.7] $\{x \mid x < -\frac{14}{3} \text{ or } x > -3\}$
30. [4.4] (a) $R = -0.4t + 260$; (b) 220 sec = 3:40
31. [4.4] 20×20 **32.** [4.4] (a) 20; (b) $\{x \mid x > 20\}$;
(c) $\{x \mid x < 20\}$ **33.** [4.5] $\{x \mid -2 \leqslant x < \frac{1}{3}\}$
34. [4.6] $\{x \mid x < -\frac{1}{2} \text{ or } x > \frac{1}{2}\}$ **35.** [4.7] $\{x \mid x < 2\}$
36. [4.1], [4.3] $m \neq 0$ **37.** [4.6] $\{x \mid \frac{1}{3} \leqslant x \leqslant 1\}$
38. [4.6] $\{x \mid -1 < x < \frac{3}{7}\}$

CHAPTER 5

Margin Exercises, Section 5.1

1. Yes **2.** No **3.** $(1, 2)$ **4.** $(4, -2)$ **5.** $(-2, 5)$
6. $(-3, 2)$ **7.** $(-\frac{1}{3}, \frac{1}{2})$ **8.** $(2, -1)$ **9.** $\frac{11}{2}, \frac{9}{2}$
10. $\frac{11}{2}, -\frac{9}{2}$ **11.** 10 km/h, 2 km/h **12.** 5 L, 3 L

Exercise Set 5.1, pp. 206–207

1. No **3.** $(-1, 3)$ **5.** $(\frac{39}{11}, -\frac{1}{11})$ **7.** $(-4, -2)$
9. $(-3, 0)$ **11.** $(10, 8)$ **13.** $(1, 1)$ **15.** $-\frac{11}{2}, -\frac{9}{2}$
17. 20 km/h, 3 km/h **19.** 12.5 L, 7.5 L **21.** 3 hr
23. 2 hr **25.** 12 white, 28 printed **27.** Paula is 32, Bob is 20
29. 76 m, 19 m **31.** 137 m, 55 m **33.** $(-12, 0)$
35. $(0.924, -0.833)$ **37.** 4 km **39.** 180 **41.** $\frac{7}{19}$
43. First train: 36 km/h; second train: 54 km/h **45.** $96
47. 4 boys, 3 girls **49.** $(-\frac{1}{4}, -\frac{1}{2})$
51. $\{(5, 3), (-5, 3), (5, -3), (-5, -3)\}$

Margin Exercises, Section 5.2

1. (a) No; (b) yes **2.** $(2, \frac{1}{2}, -2)$ **3.** A: 75, B: 84, C: 63
4. $f(x) = x^2 - 2x + 1$
5. $f(x) = \frac{5}{8}x^2 - 50x + 1150$; 510 accidents

Exercise Set 5.2, pp. 212–213

1. Yes **3.** $(3, -2, 1)$ **5.** $(-3, 2, 1)$ **7.** $(\frac{1}{2}, \frac{2}{3}, -\frac{5}{6})$
9. $(-2, 4, -1, 1)$; solution listed in the order in which the variables occur in the equations
11. $8, 21, -3$ **13.** $A = 30°, B = 90°, C = 60°$
15. 20 on Mon., 35 on Tues., 32 on Wed.
17. A: 2200, B: 2500, C: 2700 **19.** A: 10, B: 12, C: 15
21. $y = 2x^2 + 3x - 1$ **23.** (a) $E = -4t^2 + 40t + 2$; (b) $98
25. $(-1, \frac{1}{5}, -\frac{1}{2})$ **27.** A: 4 hr; B: 6 hr; C: 12 hr **29.** 180°

Margin Exercises, Section 5.3

1. $\emptyset$ **2.** Inconsistent **3.** Consistent **4.** No solution
(inconsistent system) **5.** Inconsistent **6.** Consistent

7. $\left(\dfrac{2y - 5}{3}, y\right)$ or $\left(x, \dfrac{3x + 5}{2}\right)$, $(1, 4)$, $(3, 7)$, $(0, \frac{5}{2})$, etc.

8. $\left(x, \dfrac{1 - 2x}{5}\right)$ or $\left(\dfrac{1 - 5y}{2}, y\right)$, $(-7, 3)$, $(3, -1)$, $(0, \frac{1}{5})$, etc.

9. $(-2z + 7, 3z - 5, z)$; $(7, -5, 0)$, $(5, -2, 1)$, $(3, 1, 2)$, etc.
10. $(-2z, 3z, z)$; $(-2, 3, 1)$, $(2, -3, -1)$, $(-4, 6, 2)$, etc.
11. $(0, 0, 0)$, only solution

Exercise Set 5.3, pp. 218–219

1. $\left(\dfrac{y + 5}{3}, y\right)$ or $(x, 3x - 5)$, $(0, -5)$, $(1, -2)$, $(-1, -8)$, etc.

3. $\emptyset$ **5.** $\left(\dfrac{5 - 2y}{3}, y\right)$ or $\left(x, \dfrac{5 - 3x}{2}\right)$

7. $\left(\dfrac{6y - 3}{4}, y\right)$ or $\left(x, \dfrac{4x + 3}{6}\right)$ **9.** $\emptyset$

11. $\left(\dfrac{10 + 11z}{9}, \dfrac{-11 + 5z}{9}, z\right)$; $(\frac{10}{9}, -\frac{11}{9}, 0)$, etc.

13. $(\frac{11}{9}z, \frac{5}{9}z, z)$; $(\frac{11}{9}, \frac{5}{9}, 1)$, etc.
15. $(4z - 5, -3z + 2, z)$; $(-1, -1, 1)$, etc. **17.** $(0, 0, 0)$

19. Consistent: 1, 5, 7, 11, 13, 15, 17, the others are inconsistent; dependent: 1, 5, 7, 11, 13, 15, the others are independent

21. $\left(\dfrac{724y + 9160}{2013}, y\right)$ or $\left(x, \dfrac{2013x - 9160}{724}\right)$

23. (a) $\emptyset$; (b) inconsistent; (c) dependent

Margin Exercises, Section 5.4

1. $(-\frac{63}{29}, -\frac{114}{29})$ **2.** $(-1, 2, 3)$

Exercise Set 5.4, pp. 221–222

1. $(\frac{3}{2}, \frac{5}{2})$ **3.** $(-1, 2, -2)$ **5.** $(\frac{1}{2}, \frac{3}{2})$ **7.** $(\frac{3}{2}, -4, 3)$
9. $(r - 2, 3 - 2r, r)$ **11.** $(-3, -2, -1, 1)$
13. 4 dimes, 30 nickels **15.** 10 nickels, 4 dimes, 8 quarters
17. 5 lb of $4.05; 10 lb of $2.70
19. $30,000 at $12\frac{1}{2}$%; $40,000 at 13%
21. $(1.0128, -4.8909)$ **23.** $(1.23, -2.11, 1.89)$

Margin Exercises, Section 5.5

1. -13 **2.** -2 **3.** $-2x + 12$ **4.** $(3, 1)$

5. $(-\frac{10}{41}, -\frac{13}{41})$ **6.** $\left(\dfrac{3\sqrt{2} + 4\pi}{2 + \pi^2}, \dfrac{4\sqrt{2} - 3\pi}{2 + \pi^2}\right)$ **7.** 93

8. 60 **9.** $x^3 - x^2$ **10.** $(1, 3, -2)$

Exercise Set 5.5, pp. 225–226

1. -11 **3.** $x^3 - 4x$ **5.** -109 **7.** $-x^4 + x^2 - 5x$

9. $(-\frac{25}{2}, -\frac{11}{2})$ **11.** $\left(\dfrac{4\pi - 5\sqrt{3}}{3 + \pi^2}, \dfrac{4\sqrt{3} + 5\pi}{-3 - \pi^2}\right)$

13. $(\frac{3}{2}, \frac{13}{14}, \frac{33}{14})$ **15.** $(\frac{1}{2}, \frac{2}{3}, -\frac{5}{6})$ **17.** $2, -2$
19. $\{x \mid x \leqslant -\sqrt{3} \text{ or } x \geqslant \sqrt{3}\}$ **21.** -34 **23.** 4

25. $\begin{vmatrix} L & -W \\ 2 & 2 \end{vmatrix}$ **27.** $\begin{vmatrix} a & b \\ -b & a \end{vmatrix}$ **29.** $\begin{vmatrix} 2\pi r & 2\pi r \\ -h & r \end{vmatrix}$

Margin Exercises, Section 5.6

1. No
2.
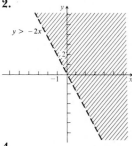
3.
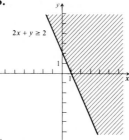
4.
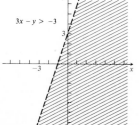
5.
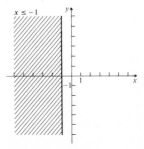

6.
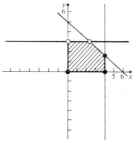
7.
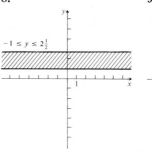

8.
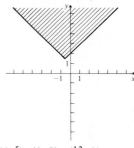
9. Vertex: $(-\frac{1}{2}, \frac{3}{2})$
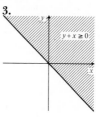

10. Vertices: $(0, 0)$, $(4, 0)$, $(4, \frac{5}{3})$, $(0, 3)$, $(\frac{12}{5}, 3)$
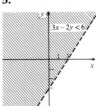

Exercise Set 5.6, p. 231

1.
3.
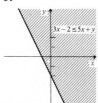

5.
7.
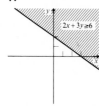

9.
11.
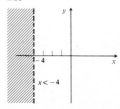

13.

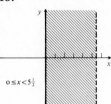

$0 \le x < 5\frac{1}{2}$

15.

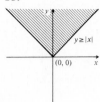

$y \ge |x|$

$(0, 0)$

31.

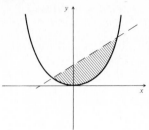

17.

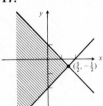

$(\frac{3}{2}, -\frac{1}{2})$

19.

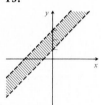

Margin Exercises, Section 5.7

1. Maximum 120, when $x = 3$, $y = 3$; minimum 34, when $x = 1$, $y = 0$ **2.** Maximum 45, when $x = 6$, $y = 0$; minimum 12, when $x = 0$, $y = 3$
3. Maximum \$23.70, by selling 50 hot dogs and 40 hamburgers

Exercise Set 5.7, pp. 234–235

1. Maximum 168, when $x = 0$, $y = 6$; minimum 0, when $x = 0$, $y = 0$ **3.** Maximum 152, when $x = 7$, $y = 0$; minimum 32, when $x = 0$, $y = 4$ **5.** Maximum 102, 8 of A, 10 of B
7. \$7000 bank X, \$15,000 bank Y, maximum income \$1395
9. Maximum \$192, 2 knits, 4 worsteds
11. Minimum \$460,000; 30 P1 airplanes, 10 P2 airplanes

21.

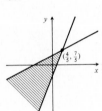

$(\frac{4}{5}, \frac{7}{5})$

23.

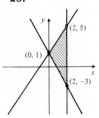

$(2, 5)$

$(0, 1)$

$(2, -3)$

Summary and Review: Chapter 5, pp. 235–236

1. [5.1] $(-2, -2)$ **2.** [5.1] $(-5, 4)$ **3.** [5.3] $\emptyset$
4. [5.3] $\emptyset$ **5.** [5.2] $(0, 0, 0)$
6. [5.2] $(13, 8, 2, -5)$, solution listed x, y, z, w
7. [5.3] Consistent: 1, 2, 5, 6; the others are inconsistent
8. [5.3] Dependent: none **9.** [5.1] 31 nickels, 44 dimes
10. [5.1] \$1600 at 10%, \$3400 at 10.5%
11. [5.2] A: 32°, B: 96°, C: 52°
12. [5.2] A: 74.5, B: 68.5, C: 82 **13.** [5.4] $(1, 2)$
14. [5.4] $(-3, 4, -2)$

15. [5.3], [5.4] $\left(\frac{z}{2}, -\frac{z}{2}, z\right)$; $(0, 0, 0)$, $(\frac{1}{2}, -\frac{1}{2}, 1)$,

$(1, -1, 2)$, etc. **16.** [5.4] $(1, -2, 3, -4)$
17. [5.2] $y = -x^2 - 2x + 3$ **18.** [5.5] 10 **19.** [5.5] -18
20. [5.5] $2x + 12$ **21.** [5.5] -6 **22.** [5.5] -16.588
23. [5.5] 0 **24.** [5.5] $(3, -2)$ **25.** [5.5] $(a, 0)$
26. [5.5] $(\frac{3}{2}, \frac{13}{14}, \frac{33}{14})$ **27.** [5.6] $(0, 9)$, $(2, 5)$, $(5, 1)$, $(8, 0)$
28. [5.7] Minimum $= 12$ at $(2, 0)$; maximum $= 60$ at $(10, 0)$
29. [5.7] Type A: 0; type B: 10; maximum score $= 120$ pts
30. [5.2] \$10,000 at 12%, \$12,000 at 13%, \$18,000 at $14\frac{1}{2}$%
31. [5.1] $(\frac{5}{18}, \frac{1}{7})$ **32.** [5.2] $(1, \frac{1}{2}, \frac{1}{3})$

33. [5.6]

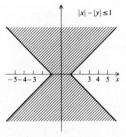

$|x| - |y| \le 1$

25.

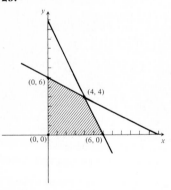

$(0, 6)$

$(4, 4)$

$(0, 0)$ $(6, 0)$

27.

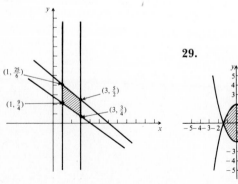

$(1, \frac{25}{6})$

$(3, \frac{5}{2})$

$(1, \frac{9}{4})$ $(3, \frac{3}{4})$

29.

34. [5.6]

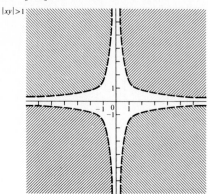

$|xy| > 1$

CHAPTER 6

Margin Exercises, Section 6.1

1. 3×2 **2.** 2×2 **3.** 3×3 **4.** 1×2 **5.** 2×1

6. 1×1 **7.** 2, 3, 6 **8.** $\mathbf{A} + \mathbf{B} = \begin{bmatrix} -2 & -6 \\ 13 & 0 \end{bmatrix} = \mathbf{B} + \mathbf{A}$

9. $\begin{bmatrix} 1 & 1 & -10 \\ 2 & 4 & -3 \end{bmatrix}$ **10.** $\mathbf{A} + \mathbf{0} = \begin{bmatrix} 4 & -3 \\ 5 & 8 \end{bmatrix} = \mathbf{0} + \mathbf{A} = \mathbf{A}$

11. $\begin{bmatrix} -1 & 4 & -7 \\ -2 & -4 & 8 \end{bmatrix}$ **12.** $\begin{bmatrix} -6 & 6 \\ 1 & -4 \\ -7 & 5 \end{bmatrix}$

13. $\begin{bmatrix} -2 & 1 & -5 \\ -6 & -4 & 3 \end{bmatrix}$ **14.** $\begin{bmatrix} 0 & 0 & 0 \\ 0 & 0 & 0 \end{bmatrix}$

15. $\begin{bmatrix} -1 & 4 & -7 \\ -2 & -4 & 8 \end{bmatrix}$ **16.** $\begin{bmatrix} 5 & -10 & 5x \\ 20 & 5y & 5 \\ 0 & -25 & 5x^2 \end{bmatrix}$

17. $\begin{bmatrix} t & -t & 4t & tx \\ ty & 3t & -2t & ty \\ t & 4t & -5t & ty \end{bmatrix}$ **18.** $[-13]$ **19.** $\begin{bmatrix} 12 \\ 13 \\ 5 \\ 16 \end{bmatrix}$

20. $\begin{bmatrix} 0 & 26 \\ -8 & 3 \\ -13 & 33 \\ -7 & 32 \end{bmatrix}$

21. $\mathbf{AB} = \begin{bmatrix} 2 & 8 & 6 \\ -29 & -34 & -7 \end{bmatrix}$; $\mathbf{BA}$ not possible

22. $[8 \ 5 \ 4]$ **23.** $\mathbf{AB} = \begin{bmatrix} -2 & 32 \\ 4 & 16 \end{bmatrix}$, $\mathbf{BA} = \begin{bmatrix} 8 & -13 \\ -16 & 6 \end{bmatrix}$

24. $\mathbf{AI} = \begin{bmatrix} 3 & 2 \\ -1 & 5 \end{bmatrix} = \mathbf{IA} = \mathbf{A}$

25. $\begin{bmatrix} 3 & 4 & -2 \\ 2 & -2 & 5 \\ 6 & 7 & -1 \end{bmatrix} \begin{bmatrix} x \\ y \\ z \end{bmatrix} = \begin{bmatrix} 5 \\ 3 \\ 0 \end{bmatrix}$

Exercise Set 6.1, pp. 246–247

1. $\begin{bmatrix} -2 & 7 \\ 6 & 2 \end{bmatrix}$ **3.** $\begin{bmatrix} 1 & 3 \\ 2 & 6 \end{bmatrix}$ **5.** $\begin{bmatrix} 9 & 9 \\ -3 & -3 \end{bmatrix}$ **7.** $\begin{bmatrix} 11 & 13 \\ 5 & 3 \end{bmatrix}$

9. $\begin{bmatrix} -4 & 3 \\ -2 & -4 \end{bmatrix}$ **11.** $\begin{bmatrix} 17 & 9 \\ -2 & 1 \end{bmatrix}$ **13.** $\begin{bmatrix} 0 & 0 \\ 0 & 0 \end{bmatrix}$

15. $\begin{bmatrix} 1 & 2 \\ 4 & 3 \end{bmatrix}$ or $\mathbf{A}$ **17.** $\begin{bmatrix} -5 & 4 & 3 \\ 5 & -9 & 4 \\ 7 & -18 & 17 \end{bmatrix}$

19. $\begin{bmatrix} -2 & 9 & 6 \\ -3 & 3 & 4 \\ 2 & -2 & 1 \end{bmatrix}$ or $\mathbf{C}$ **21.** $[-16]$ **23.** $[2 \ -19]$

25. $\begin{bmatrix} 3 & -2 & 4 \\ 2 & 1 & -5 \end{bmatrix} \begin{bmatrix} x \\ y \\ z \end{bmatrix} = \begin{bmatrix} 17 \\ 13 \end{bmatrix}$

27. $\begin{bmatrix} 1 & -1 & 2 & -4 \\ 2 & -1 & -1 & 1 \\ 1 & 4 & -3 & -1 \\ 3 & 5 & -7 & 2 \end{bmatrix} \begin{bmatrix} x \\ y \\ z \\ w \end{bmatrix} = \begin{bmatrix} 12 \\ 0 \\ 1 \\ 9 \end{bmatrix}$

29. $\begin{bmatrix} -40.19 & 37.94 & 142.24 \\ -36.78 & 16.63 & 119.62 \\ -1.66 & 14.97 & 12.65 \end{bmatrix}$

31. $(\mathbf{A} + \mathbf{B})(\mathbf{A} - \mathbf{B}) = \begin{bmatrix} -2 & 1 \\ 2 & -1 \end{bmatrix}$, $\mathbf{A}^2 - \mathbf{B}^2 = \begin{bmatrix} 0 & 3 \\ 0 & -3 \end{bmatrix}$

Margin Exercises, Section 6.2

1. $a_{11} = -8$, $a_{13} = 6$, $a_{22} = -6$, $a_{31} = -1$, $a_{32} = -3$

2. $M_{22} = \begin{vmatrix} -8 & 6 \\ -1 & 5 \end{vmatrix} = -34$, $M_{31} = \begin{vmatrix} 0 & 6 \\ -6 & 7 \end{vmatrix} = 36$,

$M_{13} = \begin{vmatrix} 4 & -6 \\ -1 & -3 \end{vmatrix} = -18$

3. $A_{22} = -34$, $A_{32} = 80$, $A_{13} = -18$

4. $|\mathbf{A}| = 0 \cdot A_{12} + (-6)A_{22} + (-3)A_{32} = -36$

5. $|\mathbf{A}| = 6 \cdot A_{13} + 7A_{23} + 5A_{33} = -36$

Exercise Set 6.2, pp. 250–251

1. $a_{11} = 7$, $a_{32} = 2$, $a_{22} = 0$
3. $M_{11} = 6$, $M_{32} = -9$, $M_{22} = -29$
5. $A_{11} = 6$, $A_{32} = 9$, $A_{22} = -29$ **7.** $|\mathbf{A}| = -10$
9. $|\mathbf{A}| = -10$ **11.** $M_{41} = -14$, $M_{33} = 20$
13. $A_{24} = 15$, $A_{43} = 30$ **15.** $|\mathbf{A}| = 110$ **17.** -195
19. Evaluate the determinant and compare with the two-point equation of a line.

Margin Exercises, Section 6.3

1. 0 **2.** 0 **3.** (a) $|\mathbf{A}| = -18$, $|\mathbf{B}| = 18$;
(b) The rows are interchanged. **4.** (a) $|\mathbf{C}| = -10$, $|\mathbf{D}| = 10$;
(b) The first and third columns are interchanged.

5. 0 **6.** $x = -4$ **7.** $x = 6$ **8.** 0 **9.** $\begin{vmatrix} -2 & 3 & 4 \\ -3 & 10 & 5 \\ 0 & 9 & 7 \end{vmatrix}$

10. 68 **11.** -195 **12.** $(b - a)(c - a)(b - c)$

Exercise Set 6.3, pp. 254–255

1. -70 **3.** -4 **5.** 9072 **7.** -153 **9.** 0 **11.** 0
13. $(x-y)(y-z)(x-z)$ **15.** $xyz(x-y)(y-z)(z-x)$

Margin Exercises, Section 6.4

1. (a) **A**; (b) **A**; (c) both equal **A**

2. (a) $\begin{bmatrix} 1 & 0 & 0 \\ 0 & 1 & 0 \\ 0 & 0 & 1 \end{bmatrix} = \mathbf{I}$; (b) **I**; (c) both equal **I**

3. $\mathbf{A}^t = \begin{bmatrix} -8 & -4 & 6 \\ 1 & 0 & 7 \\ -2 & -1 & 8 \end{bmatrix}$; $\mathbf{B}^t = \begin{bmatrix} -7 \\ 9 \\ 10 \\ 4 \end{bmatrix}$; $\mathbf{C}^t = \begin{bmatrix} -20 & 11 \end{bmatrix}$;

$\mathbf{D}^t = \begin{bmatrix} -4 & 1 & 0 \\ 5 & 0 & 1 \end{bmatrix}$ **4.** $\mathbf{A}^{-1} = \begin{bmatrix} 2 & -\frac{5}{2} \\ -1 & \frac{3}{2} \end{bmatrix}$ **5.** $\mathbf{A}^{-1} = \mathbf{A}$

6. $\mathbf{A}^{-1} = \begin{bmatrix} \frac{3}{8} & \frac{1}{8} & -\frac{1}{4} \\ -\frac{1}{8} & -\frac{3}{8} & \frac{3}{4} \\ -\frac{1}{4} & \frac{1}{4} & \frac{1}{2} \end{bmatrix}$ **7.** $\mathbf{A}^{-1} = \begin{bmatrix} -\frac{1}{2} & \frac{1}{2} & \frac{1}{2} \\ 1 & 0 & -1 \\ \frac{3}{2} & -\frac{1}{2} & -\frac{1}{2} \end{bmatrix}$

8. $\mathbf{A}^{-1} = \begin{bmatrix} 2 & -\frac{5}{2} \\ -1 & \frac{3}{2} \end{bmatrix}$ **9.** (a) $\begin{bmatrix} 4 & -2 \\ 1 & 5 \end{bmatrix}\begin{bmatrix} x_1 \\ x_2 \end{bmatrix} = \begin{bmatrix} -1 \\ 1 \end{bmatrix}$;

(b) $\mathbf{A} = \begin{bmatrix} 4 & -2 \\ 1 & 5 \end{bmatrix}$; (c) $\mathbf{A}^{-1} = \frac{1}{22}\begin{bmatrix} 5 & 2 \\ -1 & 4 \end{bmatrix}$;

(d) $x_1 = -\frac{3}{22},\ x_2 = \frac{5}{22}$

Exercise Set 6.4, pp. 261–262

1. $\mathbf{A}^{-1} = \begin{bmatrix} -3 & 2 \\ 5 & -3 \end{bmatrix}$ **3.** $\mathbf{A}^{-1} = \begin{bmatrix} 2 & -3 \\ -7 & 11 \end{bmatrix}$

5. $\mathbf{A}^{-1} = \begin{bmatrix} \frac{2}{11} & \frac{3}{11} \\ -\frac{1}{11} & \frac{4}{11} \end{bmatrix}$ **7.** $\mathbf{A}^{-1} = \begin{bmatrix} \frac{3}{8} & -\frac{1}{4} & \frac{1}{8} \\ -\frac{1}{8} & \frac{3}{4} & -\frac{3}{8} \\ -\frac{1}{4} & \frac{1}{2} & \frac{1}{4} \end{bmatrix}$

9. $\mathbf{A}^{-1} = \begin{bmatrix} \frac{1}{3} & 0 & \frac{1}{3} \\ -\frac{2}{5} & \frac{2}{5} & \frac{1}{5} \\ \frac{2}{15} & \frac{1}{5} & -\frac{1}{15} \end{bmatrix}$ **11.** $\mathbf{A}^{-1}$ does not exist.

13. $\mathbf{A}^{-1} = \begin{bmatrix} 1 & -2 & 3 & 8 \\ 0 & 1 & -3 & 1 \\ 0 & 0 & 1 & -2 \\ 0 & 0 & 0 & -1 \end{bmatrix}$ **15.–27.** See Exercises 1–13.

29. $(-\frac{1}{39}, \frac{55}{39})$ **31.** $(3, -3, -2)$

33. Find **AI** and **IA** and compare with **A**.

35. $\mathbf{A}^{-1}$ exists if and only if $xy \neq 0$. $\mathbf{A}^{-1} = \begin{bmatrix} x^{-1} & 0 \\ 0 & y^{-1} \end{bmatrix}$.

37. $\mathbf{A}^{-1}$ exists if and only if $xyzw \neq 0$.

$\mathbf{A}^{-1} = \begin{bmatrix} \frac{1}{x} & -\frac{1}{xy} & -\frac{1}{xz} & \frac{1}{xw} \\ 0 & \frac{1}{y} & 0 & 0 \\ 0 & 0 & \frac{1}{z} & 0 \\ 0 & 0 & 0 & \frac{1}{w} \end{bmatrix}$

Summary and Review: Chapter 6, pp. 262–263

1. [6.1] $\begin{bmatrix} 0 & -1 & 6 \\ 3 & 1 & -2 \\ -2 & 1 & -2 \end{bmatrix}$ **2.** [6.1] $\begin{bmatrix} -3 & 3 & 0 \\ -6 & -9 & 6 \\ 6 & 0 & -3 \end{bmatrix}$

3. [6.1] $\begin{bmatrix} -1 & 1 & 0 \\ -2 & -3 & 2 \\ 2 & 0 & -1 \end{bmatrix}$ **4.** [6.1] $\begin{bmatrix} -2 & 2 & 6 \\ 1 & -8 & 18 \\ 2 & 1 & -15 \end{bmatrix}$

5. [6.1] Not possible **6.** [6.1] $\begin{bmatrix} 2 & -1 & -6 \\ 1 & 5 & -2 \\ -2 & -1 & 4 \end{bmatrix}$

7. [6.4] $\begin{bmatrix} 1 & 2 & -2 \\ -1 & 3 & 0 \\ 0 & -2 & 1 \end{bmatrix}$ **8.** [6.4] $\begin{bmatrix} -1 & 1 & 0 \\ 0 & -2 & 1 \\ 6 & 0 & -3 \end{bmatrix}$

9. [6.4] $\begin{bmatrix} -\frac{1}{2} & 0 \\ \frac{1}{6} & \frac{1}{3} \end{bmatrix}$ **10.** [6.4] $\begin{bmatrix} 0 & 0 & \frac{1}{4} \\ 0 & -\frac{1}{2} & 0 \\ \frac{1}{3} & 0 & 0 \end{bmatrix}$

11. [6.4] $\begin{bmatrix} 1 & 0 & 0 & 0 \\ 0 & \frac{1}{9} & \frac{5}{18} & 0 \\ 0 & -\frac{1}{9} & \frac{2}{9} & 0 \\ 0 & 0 & 0 & 1 \end{bmatrix}$

12. [6.1] $\begin{bmatrix} 3 & -2 & 4 \\ 1 & 5 & -3 \\ 2 & -3 & 7 \end{bmatrix}\begin{bmatrix} x \\ y \\ z \end{bmatrix} = \begin{bmatrix} 13 \\ 7 \\ -8 \end{bmatrix}$ **13.** [6.2] -31

14. [6.2] -1 **15.** [6.2] 0 **16.** [6.2] 120

17. [6.3] $\begin{vmatrix} 5a & 5b & 5c \\ 3a & 3b & 3c \\ d & e & f \end{vmatrix} = 5(3)\begin{vmatrix} a & b & c \\ a & b & c \\ d & e & f \end{vmatrix} = 0$, since the first

two rows are the same. **18.** [6.3] $(b-a)(c-b)(c-a)$
19. [6.3] $(y-x)(z-x)(z-y)(yz+xz+yx)$
20. [6.3] $(b-a)(c-a)(d-a)(c-b)(d-b)(d-c)$
21. [6.4] $(-5, 4)$
22. [6.2] If a matrix has all 0's below the main diagonal, then its determinant is the product of the elements on the main diagonal. *Proof:* Expand about the first column.

CHAPTER 7

Margin Exercises, Section 7.1

1. (a) $x = 3y + 2$; (b) $x = y$; (c) $y^2 + 3x^2 = 4$;
(d) $x = 5y^2 + 2$; (e) $x^2 = 4y - 5$; (f) $yx = 5$
2. (a) $\boxed{(1, 4)}$ $\boxed{(2, 4)}$ $(3, 4)$ $(4, 4)$
$(1, 3)$ $\boxed{(2, 3)}$ $(3, 3)$ $(4, 3)$
$(1, 2)$ $(2, 2)$ $\boxed{(3, 2)}$ $\boxed{(4, 2)}$
$(1, 1)$ $(2, 1)$ $(3, 1)$ $\boxed{(4, 1)}$
(b) $\{(4, 1), (4, 2), (3, 2), (2, 3), (2, 4), (1, 4)\}$
(c) $\boxed{(1, 4)}$ $\boxed{(2, 4)}$ $(3, 4)$ $(4, 4)$
$(1, 3)$ $\boxed{(2, 3)}$ $(3, 3)$ $(4, 3)$
$(1, 2)$ $(2, 2)$ $\boxed{(3, 2)}$ $\boxed{(4, 2)}$
$(1, 1)$ $(2, 1)$ $(3, 1)$ $\boxed{(4, 1)}$
3. (d) They are reflections across the line $y = x$.

4. (a) 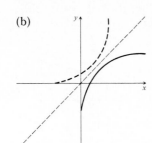 (b)

37. x-axis: no; y-axis: yes; origin: (no); $y = x$: (no)

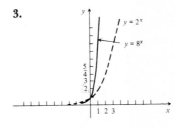

(c)

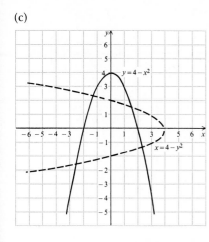

5. Yes   **6.** Yes
7. Yes   **8.** No
9. Yes   **10.** Yes
11. No   **12.** No
13. (a), (d)
14. $y = x^2 - 1$ is a
function; $x = y^2 - 1$
is not a function.

15. $g^{-1}(x) = \dfrac{x+2}{3}$

16. $f^{-1}(x) = \sqrt{x+1}$
17. $5, 5$ **18.** $a; a$

Margin Exercises, Section 7.2

1. (a) increasing; (b) set of all real numbers;
(c) set of all positive numbers; (d) 1; (e) 3.32
2. (a)-(d) all the same as Exercise 1; (e) 7.10; (f) 4^x

3.

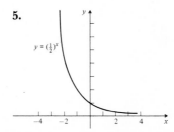

Exercise Set 7.1, pp. 271-273

1. $x = 4y - 5$ **3.** $y^2 - 3x^2 = 3$ **5.** $x = 3y^2 + 2$
7. $yx = 7$

9.

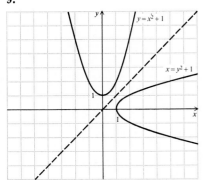

4.

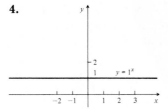

11.

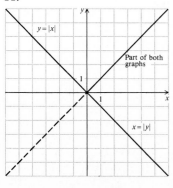

13. No   **15.** Yes
17. Yes   **19.** Yes
21. No   **23.** Yes
25. (a), (c)

27. $f^{-1}(x) = \dfrac{x-5}{2}$

29. $f^{-1}(x) = x^2 - 1$
31. $3; -125$
33. $12{,}053; -17{,}243$
35. 1.8

5.

6.

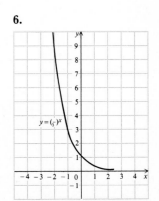

7. Domain: positive real numbers; range: all real numbers

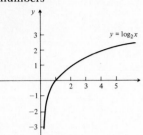

$y = \log_2 x$

8. Domain: positive real numbers; range: all real numbers

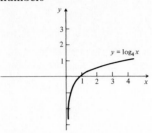

$y = \log_4 x$

5.

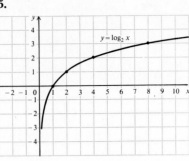

$y = \log_2 x$

7. $2^5 = 32$
9. $10^{-2} = 0.01$
11. $6^1 = 6$
13. $\log_6 1 = 0$
15. $\log_{6/5} \frac{25}{36} = -2$
17. $\log_5 \frac{1}{25} = -2$

9. $\log_6 1 = 0$ **10.** $\log_{10} 0.001 = -3$ **11.** $\log_{16} 2 = \frac{1}{4}$
12. $\log_{6/5} \frac{25}{36} = -2$ **13.** $2^5 = 32$ **14.** $10^3 = 1000$
15. $10^{-2} = 0.01$ **16.** $(\sqrt{5})^2 = 5$ **17.** $10,000$ **18.** 3
19. 4 **20.** 1 **21.** -2 **22.** 3 **23.** π **24.** 42 **25.** 37
26. M **26.** 3.2

Exercise Set 7.2, pp. 279–280

1. (a) and (b)

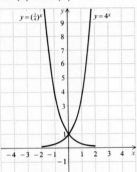

$y = (\frac{1}{4})^x$ $y = 4^x$

(c)

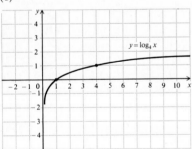

$y = \log_4 x$

3.

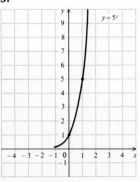

$y = 5^x$

19. $\log_e 1.0833 = 0.08$ **21.** $10,000$ **23.** $\frac{1}{2}$ **25.** 4
27. $\frac{1}{2}$ **29.** 2 **31.** 1 **33.** -3 **35.** $4x$ **37.** $\sqrt{5}$

39.

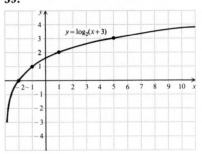

$y = \log_2(x + 3)$

41.

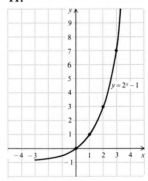

$y = 2^x - 1$

43.

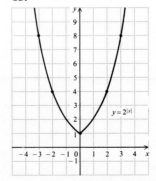

$y = 2^{|x|}$

45.

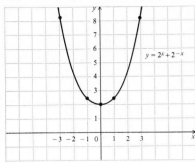

47. All real numbers
49. $\{x \mid x \neq 0\}$
51. $\{x \mid x > \frac{4}{3}\}$

53. $\{x \mid x < -3 \text{ or } x > 3\}$ **55.** $\{x \mid x \leqslant 0\}$ **57.** $\{x \mid x \geqslant 16\}$
59. $\{x \mid x > -3\}$ **61.** π^5

63.

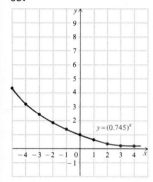

Margin Exercises, Section 7.3

1. $\log_a M + \log_a N$ **2.** $\log_5 25 + \log_5 5$ **3.** $\log_3 35$
4. $\log_a CABIN$ **5.** $5 \log_7 4$ **6.** $\frac{1}{2} \log_a 5$
7. (a) $\log_a M - \log_a N$; (b) $\log_c 1 - \log_c 4$
8. $\log_{10} 4 + \log_{10} \pi - \frac{1}{2} \log_{10} 23$
9. $\frac{1}{2}[3 \log_a z - \log_a x - \log_a y]$ **10.** $\log_a \dfrac{x^5 \sqrt[4]{z}}{y}$

11. (a) 0.954; (b) 0.1505; (c) 0.1003; (d) 0.176; (e) 1.585
12. 1 **13.** 0 **14.** 0 **15.** 1

Exercise Set 7.3, pp. 283–284

1. $2 \log_a x + 3 \log_a y + \log_a z$

3. $\log_b x + 2 \log_b y - 3 \log_b z$ **5.** $\log_a 4$ **7.** $\log_a \dfrac{2x^4}{y^3}$

9. $\log_a \dfrac{\sqrt{a}}{x}$ or $\frac{1}{2} - \log_a x$ **11.** $\log_a (x^2 - xy + y^2)$

13. $\frac{1}{2}[\log_a (1 - x) + \log_a (1 + x)]$ **15.** 0.602 **17.** 1.699
19. 1.778 **21.** −0.088 **23.** 1.954 **25.** −0.046
27. False **29.** True **31.** False **33.** False **35.** $\frac{1}{2}$
37. $\sqrt{7}$ **39.** −2, 0 **41.** $\{x \mid x > 0\}$

Margin Exercises, Section 7.4

1. 6.5480 **2.** −3.3038 **3.** 10,387.24 **4.** 0.003854
5. 11.1186 **6.** −6.0585 **7.** 36.5032 **8.** 0.05096
9. 0.7308 **10.** 5.12 **11.** 3.8048 **12.** −4.5467
13. 3730 **14.** 0.00144 **15.** 10.5611 **16.** 591

Exercise Set 7.4, pp. 289–290

1. 0.5934 **3.** 4.5751 **5.** 3.7117 **7.** −0.8496
9. −3.2905 **11.** −4.9087 **13.** 5.7498 **15.** 38.9054
17. −11.3349 **19.** Does not exist **21.** 44,915.89
23. 837,722.15 **25.** 1.0074 **27.** 0.0009676
29. 0.000001194 **31.** 1.0019 **33.** 1.0088
35. 1.000000065 **37.** 1.000000922 **39.** 8.0971
41. −2.0810 **43.** −7.5758 **45.** Does not exist
47. 66.56 **49.** −30.62 **51.** 32.9009 **53.** 2.8044
55. 7.0808×10^{13} **57.** 5.4516×10^7 **59.** 0.000000036
61. 0.000002674 **63.** 0.7987 **65.** 0.0128 **67.** 3.1271
69. 4.9159 **71.** 5.9661 **73.** −1.6180 **75.** −2.0894
77. Does not exist **79.** 7.57 **81.** 3.98 **83.** 862
85. 60,700 **87.** 0.00717 **89.** 0.0954 **91.** 0.00717
93. 0.0954 **95.** 8.2481 **97.** 8.8873 **99.** −0.8747
101. −4.8097 **103.** Does not exist **105.** 2.41
107. 1.82 **109.** 19.2 **111.** 119

113.   **115.**

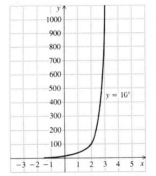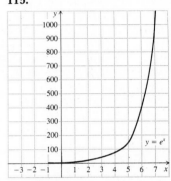

117. Let $x = \log_b M$. Then $b^x = M$, so that $\log_a M = \log_a b^x$, substituting b^x for M. By Theorem 5, $\log_a M = x \log_a b$. Thus, solving for x we get $x = \dfrac{\log_a M}{\log_a b}$. Recall that $x = \log_b M$ and substitute to obtain $\log_b M = \dfrac{\log_a M}{\log_a b}$.

Margin Exercises, Section 7.5

1. 2.8076 **2.** 2.9999 **3.** $x = \dfrac{\log (t + \sqrt{t^2 + 1})}{\log e}$ or
$\log_e (t + \sqrt{t^2 + 1})$ **4.** 125 **5.** 8.75 **6.** 2 **7.** 10 yr
8. $k \approx 0.02$; 1,239,000 **9.** 3.5 g **10.** 689 millibars
11. 34 db **12.** 60 db **13.** 7.8 **14.** (a) 68; (b) 54; (c) 40
15. $\log_2 7$ **16.** 3

Exercise Set 7.5, pp. 298–300

1. 5 **3.** $\frac{12}{5}$ **5.** $\frac{1}{2}$, −3 **7.** 2.7093 **9.** 10 **11.** 1
13. 5 **15.** 1; 100 **17.** 4 **19.** 4.6052 **21.** 4.0943

23. 2.3026 **25.** 140.67 **27.** $x = \ln \left(\dfrac{5}{2} t \pm \sqrt{\dfrac{25t^2}{4} + 1} \right)$

29. $x = \dfrac{1}{2} \ln \dfrac{t + 1}{t - 1}$ **31.** 11.9 yr **33.** $k = 0.02$;

$P = 1,239,000$ **35.** 0.4 gram **37.** 1.2 days
39. (a) 2973 yr; (b) 5135 yr **41.** 587 mb
43. (a) 82; (b) 68 **45.** 9 months **47.** 65 db **49.** 140 db
51. 6.7 **53.** $10^5 \cdot I_0$ **55.** (a) $k = 0.061$, $P = \$100e^{0.061t}$;
(b) $338.72; (c) 1978 **57.** 4.2 **59.** 1; 10,000 **61.** $\emptyset$

63. $-9; 9$ **65.** $\frac{7}{4}$ **67.** $\log_x y - \log_x a$
69. $t = \dfrac{\ln P - \ln P_0}{k}$ **71.** $t = -\dfrac{1}{k} \log \left[\dfrac{T - T_0}{T_1 - T_0} \right]$
73. $Q = a^b \cdot \sqrt[3]{y}$ **75.** $10; 100$
77. $2; 2.25; 2.48832; 2.593742; 2.704814; 2.716924$
79. 5 **81.** $x < 0.32193$ **83.** $x = 2.97435, y = -0.113$
85. $1, 4$

Summary and Review: Chapter 7, pp. 300–301

1. [7.1] $x = 3y^2 + 2y - 1$ **2.** [7.1] $x = \sqrt{y + 2}$
3. [7.1] (b), (e)   **4.** [7.1] (d)   **5.** [7.1] $f^{-1}(x) = (2x - 4)^2$
6. [7.1] $f^{-1}(x) = \sqrt{x - 2}$ **7.** [7.1] a **8.** [7.1] t

9. [7.2]

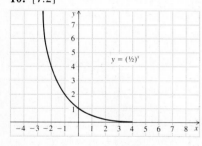

10. [7.2]

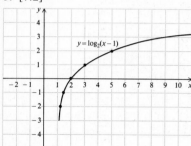

11. [7.2] $8^{-2/3} = \frac{1}{4}$
12. [7.2] $\log_7 x = 2.3$
13. [7.3] $\log_b \dfrac{a^{1/2} c^{3/2}}{d^4}$

14. [7.3] $\frac{2}{3} \log M - \frac{1}{3} \log N$ **15.** [7.3] 1.255
16. [7.3] 0.544 **17.** [7.3] -0.602 **18.** [7.3] 0.2385
19. [7.2] $x^2 + 1$ **20.** [7.2] $\sqrt{9}$ **21.** [7.2] 4 **22.** [7.2] $\frac{1}{2}$
23. [7.2] 3 **24.** [7.5] $\frac{1}{5}$ **25.** [7.5] 4.3820 **26.** [7.5] 1
27. [7.5] 9 **28.** [7.5] 3 **29.** [7.5] 3 **30.** [7.5] 5.7 yr
31. [7.5] 30 decibels   **32.** [7.5] 6.2 g   **33.** [7.4] -2.6655
34. [7.4] 6.1278 **35.** [7.4] -4.6910 **36.** [7.4] 11.3780
37. [7.4] -10.4919 **38.** [7.4] 24.5004 **39.** [7.4] 1068.32
40. [7.4] 0.000003441 **41.** [7.4] 1.0019
42. [7.4] 712.799 **43.** [7.4] $113,210.02$ **44.** [7.4] 0.0168
45. [7.5] $64, \frac{1}{64}$ **46.** [7.5] 1
47. [7.5]

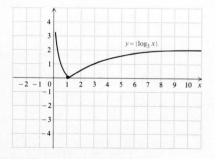

48. [7.5]

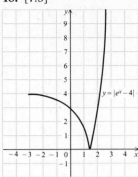

49. [7.5] $\{x \mid x > e^{6/5}\}$
50. [7.5] $\{x \mid x \neq \frac{1}{4} \ln 10\}$

CHAPTER 8

Margin Exercises, Section 8.1

1. $i\sqrt{6}$ **2.** $-i\sqrt{10}$ **3.** $2i$ **4.** $-5i$ **5.** $-\sqrt{10}$ **6.** $\sqrt{11}$
7. $i\sqrt{7}$ **8.** $7i$ **9.** $3i$ **10.** $(\sqrt{17} + 3)i$ **11.** i **12.** -1
13. $-i$ **14.** $12 + i$ **15.** $5 - i$ **16.** $2 + 14i$ **17.** 8
18. $-6 + 8i$ **19.** $3i$ **20.** $(x + 2i)(x - 2i)$
21. $(3 + yi)(3 - yi)$ **22.** Yes   **23.** $x = -1, y = 2$

Exercise Set 8.1, pp. 307–308

1. $i\sqrt{15}$ **3.** $4i$ **5.** $2i\sqrt{3}$ **7.** $9i$ **9.** $i(\sqrt{7} - \sqrt{10})$
11. $-\sqrt{55}$ **13.** $2\sqrt{5}$ **15.** $\sqrt{\frac{5}{2}}i$ **17.** $-\frac{3}{2}i$ **19.** -2
21. $6 + 5i$ **23.** 8 **25.** $2 + 4i$ **27.** $-4 - i$
29. $-5 + 5i$ **31.** $7 - i$ **33.** $-6 + 12i$ **35.** $-5 + 12i$
37. i **39.** $(2x + 5yi)(2x - 5yi)$ **41.** Yes
43. $x = -\frac{3}{2}, y = 7$ **45.** $-4 + 3i$
47. For example, $\sqrt{-1}\sqrt{-1} = i^2 = -1$, but
$\sqrt{(-1)(-1)} = \sqrt{1} = 1$.   **49.** $\begin{bmatrix} -1 & -2 \\ 3 + 3i & 3 + 12i \end{bmatrix}$

Margin Exercises, Section 8.2

1. $7 - 2i$ **2.** $6 + 4i$ **3.** $5i$ **4.** $-3i$ **5.** -3 **6.** 8
7. $\frac{9}{13} + \frac{7}{13}i$ **8.** $\frac{4}{13} + \frac{7}{13}i$ **9.** $\dfrac{1}{3 + 4i}, \dfrac{3}{25} - \dfrac{4}{25}i$
10. Both are $7 + 3i$ **11.** Both are $-13 - 11i$
12. $\overline{z^3} = \overline{z \cdot z \cdot z} = \bar{z} \cdot \bar{z} \cdot \bar{z} = \bar{z}^3$ **13.** $5\bar{z}^3 + 4\bar{z}^2 - 2\bar{z} + 1$
14. $7\bar{z}^5 - 3\bar{z}^3 + 8\bar{z}^2 + \bar{z}$

Exercise Set 8.2, p. 312

1. $\frac{1}{2} + \frac{7}{2}i$ **3.** $\frac{1}{3} + \frac{2}{3}i\sqrt{2}$ **5.** $2 - 3i$ **7.** $\frac{1}{5} + \frac{2}{5}i$ **9.** $-\dfrac{1}{2} - \dfrac{i}{2}$
11. $\frac{28}{65} - \frac{29}{65}i$ **13.** $-\frac{1}{2} + \frac{3}{2}i$ **15.** $\frac{5}{2} + \frac{13}{2}i$ **17.** $\frac{4}{25} - \frac{3}{25}i$
19. $\frac{5}{29} + \frac{2}{29}i$ **21.** $-i$ **23.** $\dfrac{i}{4}$ **25.** $3\bar{z}^5 - 4\bar{z}^2 + 3\bar{z} - 5$
27. $4\bar{z}^7 - 3\bar{z}^5 + 4\bar{z}$ **29.** $z = 1$ **31.** a **33.** $\dfrac{3 - i}{2 + i}$, or $1 - i$

Margin Exercises, Section 8.3

1. $2 + 5i$ **2.** $\dfrac{-1 + i \pm \sqrt{-18i}}{4}$ **3.** $\dfrac{-3 \pm 4i}{5}$

4. $x^2 - (1 + 2i)x + i - 1 = 0$ **5.** $x^3 - 2x^2 + x - 2 = 0$
6. $2 - i,\ -2 + i$

Exercise Set 8.3, pp. 315–316

1. $\frac{2}{5} + \frac{6}{5}i$ **3.** $\frac{8}{5} - \frac{9}{5}i$ **5.** $2 - i$ **7.** $\frac{11}{25} + \frac{2}{25}i$

9. $\dfrac{-1 + i \pm \sqrt{-6i}}{2}$ **11.** $-i,\ \dfrac{i}{2}$

13. $\dfrac{-1 - 2i \pm \sqrt{-15 + 16i}}{6}$ **15.** $1 \pm 2i$ **17.** $2 \pm 3i$

19. $-\dfrac{3}{2} \pm \dfrac{\sqrt{7}}{2}i$ **21.** $x^2 + 4 = 0$ **23.** $x^2 - 2x + 2 = 0$
25. $x^2 - 4x + 13 = 0$ **27.** $x^2 - 3x - ix + 3i = 0$
29. $x^3 - x^2 + 9x - 9 = 0$
31. $x^3 - 2x^2 i - 3x^2 + 5ix + x - 2i + 2 = 0$
33. $x^3 + x - 2x^2 i - 2i = 0$ **35.** $\sqrt{2} + \sqrt{2}i,\ -\sqrt{2} - \sqrt{2}i$
37. $2 + i,\ -2 - i$ **39.** $2,\ -1 \pm \sqrt{3}i$
41. $x = 2 + i,\ y = 1 - 3i$

Margin Exercises, Section 8.4

1.

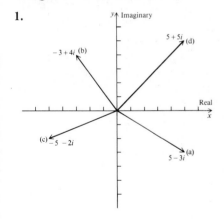

2.

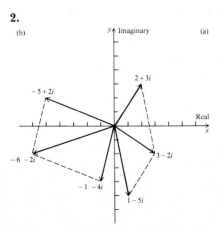

3. (a) 5; (b) 13
4. $1 - i$
5. $\sqrt{3} - i$
6. $\sqrt{2}$ cis $315°$
7. 6 cis $225°$
8. 20 cis $55°$
9. 4 cis $\dfrac{5\pi}{4}$
10. 4 cis $90°$
11. 2 cis $\dfrac{\pi}{4}$
12. $\sqrt{2}$ cis $285°$

Exercise Set 8.4, p. 321

1.   **3.**

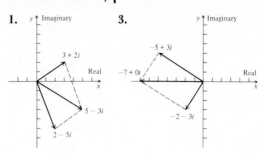

5.   **7.**

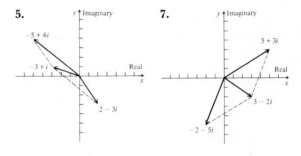

9. $\dfrac{3\sqrt{3}}{2} + \dfrac{3}{2}i$ **11.** $-10i$ **13.** $2 + 2i$
15. $-2 - 2i$ **17.** $\sqrt{2}$ cis $315°$ **19.** 20 cis $330°$
21. 5 cis $180°$ **23.** 4 cis $0°$, or 4   **25.** 8 cis $120°$
27. cis $270°$, or $-i$ **29.** 2 cis $270°$, or $-2i$
31. $z = a + bi,\ |z| = \sqrt{a^2 + b^2};\ -z = -a - bi,$
$|-z| = \sqrt{(-a)^2 + (-b)^2} = \sqrt{a^2 + b^2},\ \therefore\ |z| = |-z|$
33. $|(a + bi)(a - bi)| = |a^2 + b^2| = a^2 + b^2;$
$|(a + bi)^2| = |a^2 + 2abi - b^2| = |a^2 - b^2 + 2abi| =$
$\sqrt{(a^2 - b^2)^2 + (2ab)^2} = \sqrt{a^4 + 2a^2 b^2 + b^4} = a^2 + b^2$
35. $z \cdot w = (r_1 \text{ cis } \theta_1)(r_2 \text{ cis } \theta_2) = r_1 r_2 \text{ cis } (\theta_1 + \theta_2),\ |z \cdot w| =$
$\sqrt{[r_1 r_2 \cos (\theta_1 + \theta_2)]^2 + [r_1 r_2 \sin (\theta_1 + \theta_2)]^2} = \sqrt{(r_1 r_2)^2} =$
$|r_1 r_2|,\ |z| = \sqrt{(r_1 \cos \theta_1)^2 + (r_1 \sin \theta_1)^2} = \sqrt{r_1^2} = |r_1|,\ |w| =$
$\sqrt{(r_2 \cos \theta_2)^2 + (r_2 \sin \theta_2)^2} = \sqrt{r_2^2} = |r_2|.$ Then $|z| \cdot |w| =$
$|r_1| \cdot |r_2| = |r_1 r_2| = |z \cdot w|$

37.

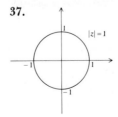

Margin Exercises, Section 8.5

1. 32 cis $270°$, or $-32i$ **2.** 16 cis $120°$, or $-8 + 8i\sqrt{3}$
3. (a) $1 + i,\ -1 - i$; (b) $\sqrt{5} + i\sqrt{5},\ -\sqrt{5} - i\sqrt{5}$

8

4. 1 cis 60°, 1 cis 180°, 1 cis 300°; or $\frac{1}{2} + \frac{\sqrt{3}}{2} i, -1, \frac{1}{2} - \frac{\sqrt{3}}{2} i$

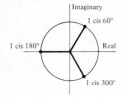

Exercise Set 8.5, p. 324

1. 8 cis π **3.** 64 cis π **5.** 8 cis 270°   **7.** $-8 - 8\sqrt{3}i$
9. $-8 - 8\sqrt{3}i$ **11.** i **13.** 1

15. $\sqrt{2}$ cis 60°, $\sqrt{2}$ cis 240°; or $\frac{\sqrt{2}}{2} + \frac{\sqrt{6}}{2} i, \frac{-\sqrt{2}}{2} - \frac{\sqrt{6}}{2} i$

17. cis 30°, cis 150°, cis 270°; or $\frac{\sqrt{3}}{2} + \frac{1}{2} i, \frac{-\sqrt{3}}{2} + \frac{1}{2} i, -i$

19. 2 cis 0°, 2 cis 90°, 2 cis 180°, 2 cis 270°; or 2, 2i, -2, $-2i$

21. $-1.366 + 1.366i, 0.366 - 0.366i$

Summary and Review: Chapter 8, pp. 324–325

1. [11.1] $-2\sqrt{10}i$ **2.** [11.1] $-4\sqrt{15}$ **3.** [11.1] $14 + 2i$
4. [11.1] $1 - 4i$ **5.** [11.1] $2 - i$ **6.** [11.2] $\frac{11}{10} + \frac{3}{10}i$
7. [11.1] No   **8.** [11.2] $\frac{6}{85} + \frac{7}{85}i$ **9.** [11.1] $x = 2, y = -4$
10. [11.2] $3\bar{z}^3 + \bar{z} - 7$ **11.** [11.3] $-\frac{7}{15} + \frac{3}{5}i$
12. [11.3] $\frac{2}{5} \pm \frac{1}{5}i$ **13.** [11.3] $\frac{(-3 \pm \sqrt{5})i}{2}$
14. [11.3] $x^2 - 2x + 5 = 0$ **15.** [11.3] $\sqrt{2} + \sqrt{2}i,$
$-\sqrt{2} - \sqrt{2}i$

16. [11.4]

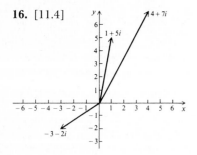

17. [11.4] $-\sqrt{2} + \sqrt{2}i$ **18.** [11.4] $\sqrt{2}$ cis 45°
19. [11.4] 70 cis 50°   **20.** [11.4] $\frac{\sqrt{2}}{2}$ cis 15°
21. [11.5] $4\sqrt{2}$ cis 135°   **22.** [11.5] $-\frac{81}{2} + \frac{81\sqrt{3}}{2} i$
23. [11.5] $\sqrt[6]{2}$ cis 15°; $\sqrt[6]{2}$ cis 135°, $\sqrt[6]{2}$ cis 255°

24. [11.5] 3, $\frac{3}{2}(-1 \pm \sqrt{3}i)$ **25.** [11.2] $5 - 9i$
26. [11.2] $1 + 2i$ **27.** [11.3] $x = 2 - i, y = -1 - 3i$

28. [11.4]

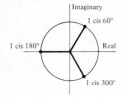

CHAPTER 9

Margin Exercises, Section 9.1

1. 5   **2.** 1   **3.** 2   **4.** 0   **5.** No degree   **6.** 7   **7.** 2, -2
8. $i, -i$ **9.** (a) Yes; (b) no   **10.** (a) Yes; (b) no; (c) no
11. $Q(x) = x^2 + 5x + 10, R(x) = 24,$
$P(x) = (x - 3)(x^2 + 5x + 10) + 24$

Exercise Set 9.1, p. 330

1. 4   **3.** 1   **5.** 2   **7.** 0   **9.** 2 yes; 3 no, -1 no
11. (a) Yes; (b) no; (c) no   **13.** $Q(x) = x^2 + 8x + 15,$
$R(x) = 0, P(x) = (x - 2)(x^2 + 8x + 15) + 0$
15. $Q(x) = x^2 + 9x + 26, R(x) = 48,$
$P(x) = (x - 3)(x^2 + 9x + 26) + 48$ **17.** $Q(x) = x^2 - 2x + 4,$
$R(x) = -16, P(x) = (x - 2)(x^2 - 2x + 4) - 16$
19. $Q(x) = x^2 + 5, R(x) = 0, P(x) = (x^2 + 4)(x^2 + 5) + 0$
21. $P(x) = (2x^2 - x + 1)(\frac{5}{2}x^5 + \frac{5}{4}x^4 - \frac{5}{8}x^3 - \frac{39}{16}x^2 - \frac{29}{32}x + \frac{113}{64}) +$
$\frac{171x - 305}{64}$ **23.** (a) -32; (b) -32; (c) -65; (d) -65

Margin Exercises, Section 9.2

1. $Q(x) = x^2 + 8x + 15, R(x) = 0$ **2.** $Q(x) = x^2 - 4x + 13,$
$R(x) = -30$ **3.** $Q(x) = x^2 - x + 1, R(x) = 0$
4. (a) $P(10) = 73120$; (b) $P(-8) = -37292$
5. (a) Yes; (b) no; (c) yes   **6.** No   **7.** No
8. (a) Yes; (b) $x^2 + 8x + 15$; (c) $(x - 2)(x + 5)(x + 3)$

Exercise Set 9.2, pp. 333–334

1. $Q(x) = 2x^3 + x^2 - 3x + 10, R(x) = -42$
3. $Q(x) = x^2 - 4x + 8, R(x) = -24$ **5.** $Q(x) = x^3 + x^2 +$
$x + 1, R(x) = 0$ **7.** $Q(x) = 2x^3 + x^2 + \frac{7}{2}x + \frac{7}{4}, R(x) = -\frac{1}{8}$
9. $Q(x) = x^3 + x^2y + xy^2 + y^3, R(x) = 0$
11. $P(1) = 0, P(-2) = -60, P(3) = 0$
13. $P(20) = 5,935,988, P(-3) = -772$ **15.** -3 yes, 2 no
17. -3 no, $\frac{1}{2}$ no   **19.** $P(x) = (x - 1)(x + 2)(x + 3);$ 1,
$-2, -3$ **21.** $P(x) = (x - 2)(x - 5)(x + 1);$ 2, 5, -1
23. $P(x) = (x - 2)(x - 3)(x + 4);$ 2, 3, -4
25. $P(x) = (x - 1)(x - 2)(x - 3)(x + 5);$ 1, 2, 3, -5
27. $-5 < x < 1$ or $x > 2$ **29.** $\frac{14}{3}$ **31.** $k = 0$

Margin Exercises, Section 9.3

1. 5, mult. 2; -6, mult. 1   **2.** -7, mult. 2; 3, multi. 1
3. -2, mult. 3; 3, mult. 1; -3, mult. 1

4. 4, mult. 2; 3, mult. 2   **5.** 1, -1, mult. 1
6. $x^3 - 6x^2 + 3x + 10$ **7.** $x^3 + (-1 + 5i)x^2 +$
$(-2 - 5i)x - 10i$ **8.** $x^5 + 6x^4 + 12x^3 + 8x^2$
9. $x^4 + 2x^3 - 12x^2 + 14x - 5$ **10.** $7 + 2i, 3 - \sqrt{5}$
11. $x^4 - 6x^3 + 11x^2 - 10x + 2$ **12.** $x^3 - 2x^2 + 4x - 8$
13. $-i, -2, 1$

Exercise Set 9.3, pp. 337–338

1. $-3(\text{m } 2), 1(\text{m } 1)$ **3.** $0(\text{m } 3), 1(\text{m } 2), -4(\text{m } 1)$
5. $x^3 - 6x^2 - x + 30$ **7.** $x^3 + 3x^2 + 4x + 12$
9. $x^3 - \sqrt{3}x^2 - 2x + 2\sqrt{3}$; no   **11.** $-3 - 4i, 4 + \sqrt{5}$
13. $x^3 - 4x^2 + 6x - 4$ **15.** $x^3 - 5x^2 + 16x - 80$
17. $x^4 + 4x^2 - 45$ **19.** $i, 2, 3$ **21.** $1 + 2i, 1 - 2i$
23. $4, i, -i$ **25.** $i, -i, 1 + \sqrt{2}, 1 - \sqrt{2}$
27. There is at least one complex value for sin x satisfying the equation (not necessarily a value of x).

Margin Exercises, Section 9.4

1. (a) $3, -3, 1, -1$; (b) $2, -2, 1, -1$; (c) $\frac{3}{2}, -\frac{3}{2}, 3, -3,$
$\frac{1}{2}, -\frac{1}{2}, 1, -1$; (d) $\frac{1}{2}, -3$ (e) $3 + \sqrt{10}, 3 - \sqrt{10}$
2. (a) $2, -2, 1, -1, 4, -4, 7, -7, 14, -14, 28, -28$;
(b) $1, -1$; (c) same as (a); (d) all coefficients positive;
(e) -7; (f) $2i, -2i$ **3.** (a) All coefficients positive; (b) none
4. (a) None; (b) yes, $\dfrac{-3 \pm i\sqrt{3}}{2}$, quadratic formula
5. (a) $-\frac{1}{6}, -\frac{4}{3}, \frac{1}{6}, \frac{1}{3}$; (b) 6; (c) $6P(x) = 6x^4 - x^3 - 8x^2 + x + 2$;
(d) $1, -1, -\frac{1}{2}, \frac{2}{3}$; (e) yes, $P(x) = 0$ and $6P(x) = 0$ are equivalent.

Exercise Set 9.4, pp. 341–342

1. $1, -1$ **3.** $\pm(1, \frac{1}{3}, \frac{1}{5}, \frac{1}{15}, 2, \frac{2}{3}, \frac{2}{5}, \frac{2}{15})$ **5.** $-3, \sqrt{2}, -\sqrt{2}$
7. $-\frac{1}{5}, 1, 2i, -2i$ **9.** $-1, -2, 3 + \sqrt{13}, 3 - \sqrt{13}$
11. $1, -1, -3$ **13.** $-2, 1 \pm i\sqrt{3}$ **15.** $\dfrac{1}{2}, \dfrac{1 \pm \sqrt{5}}{2}$
17. None   **19.** None   **21.** None   **23.** $-2, 1, 2$
25. 4 cm   **27.** 3 cm, $\dfrac{7 - \sqrt{33}}{2}$ cm

Margin Exercises, Section 9.5

1. 3   **2.** 2   **3.** Just 1   **4.** 5, 3, or 1   **5.** 2 or 0
6. 2 or 0   **7.** 1   **8.** 0   **9.** Answers may vary, but 2 or 3
will do.   **10.** Answers may vary but 75 will do.
11. Answers may vary, but -4 will do.
12. Answers may vary, but -2 will do.
13. Answers may vary, but -4 will do.
14. Positive roots, 3 or 1; negative roots, 1; upper bound 6
(answer may vary); lower bound -1 (answer may vary).
15. Positive roots, 1; negative roots, 1. Thus 2 nonreal roots;
upper bound 1 (answer may vary); lower bound -1 (in fact,
-1 is a root—answer may vary).

Exercise Set 9.5, p. 349

1. 3 or 1   **3.** 0   **5.** 2 or 0   **7.** 0   **9.** 3 or 1   **11.** 2 or 0
13. 3 (other answers possible)   **15.** 4   **17.** -1 **19.** -3

21. 3 or 1 positive; 1 negative; upper bound, 2; lower bound,
-3 **23.** 1 positive; 1 negative; 2 nonreal; upper bound, 2;
lower bound, -2 **25.** 2 or 0 positive; 2 or 0 negative;
upper bound, 4; lower bound, -3 **27.** 0 positive, 0 negative
29. Let $P(x) = x^n - 1$. There is one variation of sign, so there
is just one positive root. Since n is even, $P(-x) = P(x)$. Hence
$P(-x)$ has just one variation of sign, and there is just one
negative root. Zero is not a root, so the total number of real
roots is two.

Margin Exercises, Section 9.6

1.   **2.** (a)

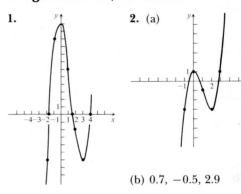

(b) $0.7, -0.5, 2.9$

Exercise Set 9.6, p. 355

1.   **3.**

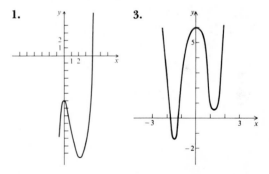

5.

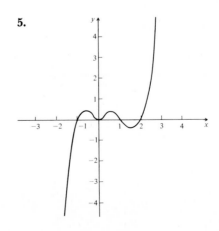

7.

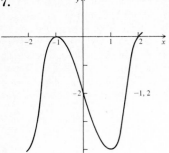

9.

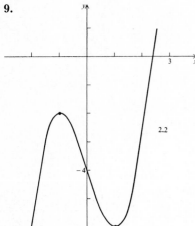

2.2

11. 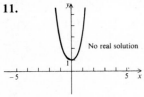 No real solution

13.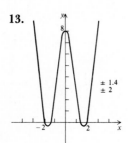
± 1.4
± 2

15.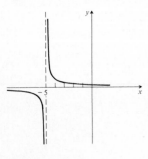
−1
± 1.4

17. 0.76   **19.** −1.27   **21.** 2.1

Margin Exercises, Section 9.7

1.

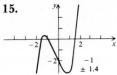

2.

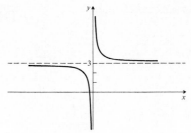

3.

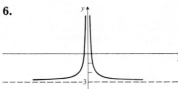

4.

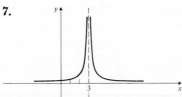

5.

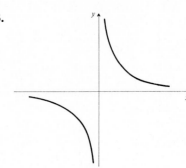

6.

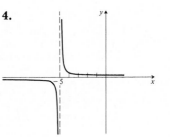

7.

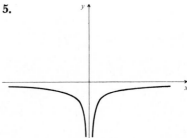

8. $x = 0$, $x = -2$, $x = 3$
9. (b) and (c)   **10.** $y = \frac{1}{2}$ **11.** $y = 3$ **12.** $y = 3x - 1$
13. $y = 5x$ **14.** $0, 3, -5$ **15.** $0, 1, -3$

16.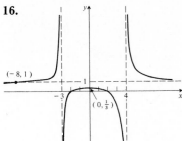

11.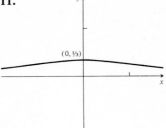

Exercise Set 9.7, pp. 363—364

1.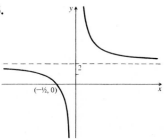

3.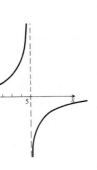

13.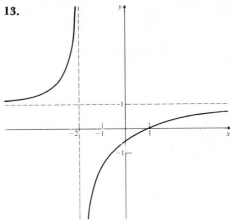

5.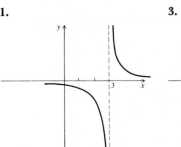

15.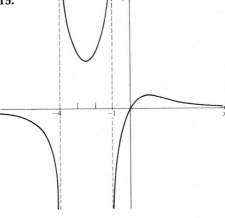

7.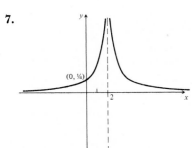

9.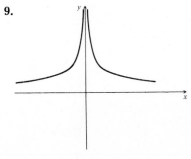

17.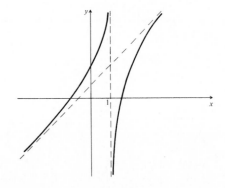

19.

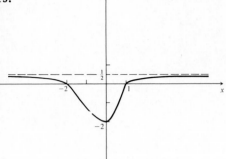

21.

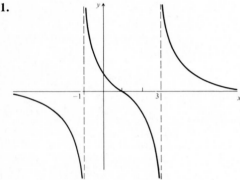

23.

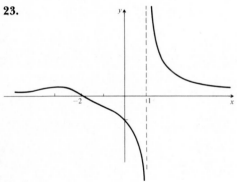

25.

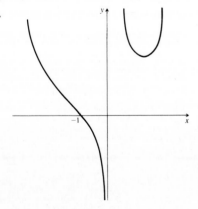

27.

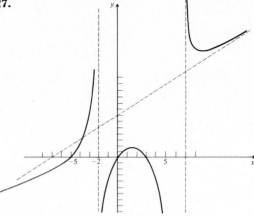

29.

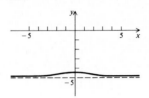

31.

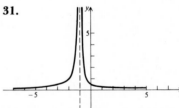

33.

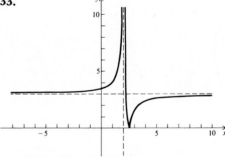

Margin Exercises, Section 9.8

1. $\dfrac{1}{3x+2} + \dfrac{4}{x-3}$ **2.** $\dfrac{1}{x+1} + \dfrac{2}{x-1} - \dfrac{1}{(x-1)^2}$

3. $\dfrac{x+2}{x^2+1} + \dfrac{1}{x+2}$ **4.** $3x+1 + \dfrac{4}{2x-1} - \dfrac{3}{x+5}$

Exercise Set 9.8, p. 367

1. $\dfrac{2}{x-3} + \dfrac{-1}{x+2}$ **3.** $\dfrac{-4}{3x-1} + \dfrac{5}{2x-1}$

5. $\dfrac{-3}{x-2} + \dfrac{2}{x+2} + \dfrac{4}{x+1}$ **7.** $\dfrac{-3}{(x+2)^2} + \dfrac{-1}{x+2} + \dfrac{1}{x-1}$

9. $\dfrac{3}{x-1} + \dfrac{-4}{2x-1}$ **11.** $x - 2 + \dfrac{-\frac{11}{4}}{(x+1)^2} + \dfrac{\frac{17}{16}}{x+1} + \dfrac{-\frac{17}{16}}{x-3}$

13. $\dfrac{-1}{x-3} + \dfrac{3x}{x^2+2x-5}$ **15.** $\dfrac{-2}{x+2} + \dfrac{10}{(x+2)^2} + \dfrac{3}{2x-1}$

17. $\dfrac{-\frac{1}{2a^2}\,x}{x^2+a^2} + \dfrac{\frac{1}{4a^2}}{x-a} + \dfrac{\frac{1}{4a^2}}{x+a}$

19. $\dfrac{-1}{x+1} + \dfrac{4}{x+4}$

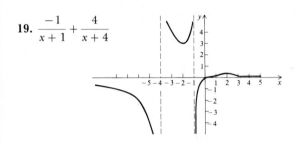

21. $\dfrac{-\frac{3}{25}}{\ln x + 2} + \dfrac{\frac{3}{25}}{\ln x - 3} + \dfrac{\frac{7}{5}}{(\ln x - 3)^2}$

Summary and Review: Chapter 9, pp. 368–369

1. [12.2] 0   **2.** [12.2] $Q = 2x^3 - 10x^2 + 27x - 59$, R = 119
3. [12.2] 88   **4.** [12.2] $(x-1)(x+3)(x+5)$; 1, −3, −5
5. [12.2] No   **6.** [12.3] 0, mult. 2; 3, mult. 2; −4, mult. 3; 4, mult. 1   **7.** [12.3] $x^3 - 3x^2 + 2x$
8. [12.3] $(x^2-1)(x-2)^2(x+3)^3$ **9.** [12.3] ± 3, $-3i$
10. [12.3] $-8 + 7i$, $10 - \sqrt{5}$
11. [12.4] $\pm(1, 2, 3, 4, 6, 12, \frac{1}{2}, \frac{3}{2})$ **12.** [12.4] −3, 4, $\pm 3i$
13. [12.5] 2 or none   **14.** [12.5] 4, 2, or none   **15.** [12.5] 2
16. [12.5] −2   **17.** [12.6] 1.41   **18.** [12.6] −0.9, 1.3, 2.5

19. [12.7]

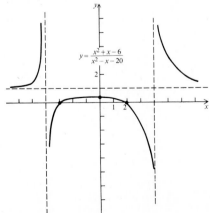

$y = \dfrac{x^2 + x - 6}{x^2 - x - 20}$

20. [12.8] $\dfrac{5}{x+1} - \dfrac{5}{x+2} - \dfrac{5}{(x+2)^2}$

21. [12.3] $(x-1)\left(x + \dfrac{1}{2} + i\dfrac{\sqrt{3}}{2}\right)\left(x + \dfrac{1}{2} - i\dfrac{\sqrt{3}}{2}\right)$

22. [12.2] 7   **23.** [12.3] 4   **24.** [12.2] −4

25. [12.8]

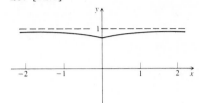

CHAPTER 10

Margin Exercises, Section 10.1

1. The union of the graphs of $y = -\frac{1}{2}x$ and $y = \frac{1}{2}x$
2. The union of the graphs of $y = 2$ and $y = -2$
3. Graph consists of a single line, $y = 3x$
4. Graph consists just of $(0, 0)$.
5. There is no real number solution, hence no graph.
6. $(x+3)^2 + (y-7)^2 = 25$ **7.** $(-1, 3), 2$ **8.** $(7, -2), 8$
9. $(x+1)^2 + (y-4)^2 = 41$

Exercise Set 10.1, pp. 378–379

1. The union of the graph of $y = x$ and $y = -x$
3. The union of the graphs of $y = -x$ and $y = \frac{3}{2}x$
5. The point $(0, 0)$ **7.** $x^2 + y^2 = 49$ **9.** $(-1, -3), 2$
11. $(8, -3), 2\sqrt{10}$ **13.** $(-4, 3), 2\sqrt{10}$ **15.** $(2, 0), 2$
17. $(-4.123, 3.174), 10.071$ **19.** $(0, 0), \frac{1}{3}$
21. $x^2 + y^2 = 25$ **23.** $(x-2)^2 + (y-4)^2 = 16$
25. 2.68 ft, 37.32 ft
27. (a) No; (b) $y = \pm\sqrt{4 - x^2}$;
(c) yes, domain $\{x \mid -2 \leqslant x \leqslant 2\}$, range $\{y \mid 0 \leqslant y \leqslant 2\}$;
(d) yes, domain $\{x \mid -2 \leqslant x \leqslant 2\}$, range $\{y \mid -2 \leqslant y \leqslant 0\}$
29. Yes   **31.** No   **33.** Yes   **35.** No

Margin Exercises, Section 10.2

1. (a) Stretched; (b) shrunk

2. $V(-3, 0)(3, 0)(0, -1)(0, 1)$;
$F(-2\sqrt{2}, 0)(2\sqrt{2}, 0)$

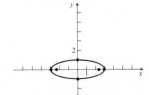

3. $V(-5, 0)(5, 0)(0, -3)(0, 3)$;
$F(-4, 0)(4, 0)$

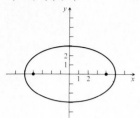

4. $V(-2, 0)(2, 0)(0, \sqrt{2})(0, -\sqrt{2})$;
$F(-\sqrt{2}, 0)(\sqrt{2}, 0)$

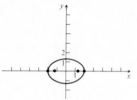

5. $C(0, 0)$; $V(-1, 0)(1, 0)(0, 3)(0, -3)$;
$F(0, 2\sqrt{2})(0, -2\sqrt{2})$

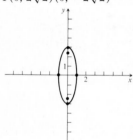

6. $C(0, 0)$; $V(3, 0)(-3, 0)(0, 5)(0, -5)$;
$F(0, 4)(0, -4)$

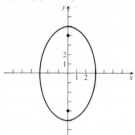

7. $C(0, 0)$; $V(\sqrt{2}, 0)(-\sqrt{2}, 0)(0, 2)(0, -2)$
$F(0, \sqrt{2})(0, -\sqrt{2})$

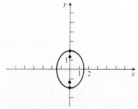

8. $C(-3, 2)$; $V(-2\frac{4}{5}, 2)(-3\frac{1}{5}, 2)(-3, 2\frac{1}{3})(-3, 1\frac{2}{3})$;
$F(-3, 2\frac{4}{15})(-3, 1\frac{11}{15})$

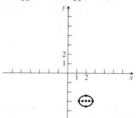

9. $C(2, -3)$; $V(2\frac{1}{3}, -3)(1\frac{2}{3}, -3)(2, -2\frac{4}{5})(2, -3\frac{1}{5})$;
$F(2\frac{4}{15}, -3)(1\frac{11}{15}, -3)$

Exercise Set 10.2, pp. 384–385

1. V: $(2, 0)$, $(-2, 0)$, $(0, 1)$, $(0, -1)$; F: $(\sqrt{3}, 0)$, $(-\sqrt{3}, 0)$

3. V: $(-3, 0)$, $(3, 0)$, $(0, 4)$, $(0, -4)$; F: $(0, \sqrt{7})$, $(0, -\sqrt{7})$

5. V: $(-\sqrt{3}, 0)$, $(\sqrt{3}, 0)$, $(0, \sqrt{2})$, $(0, -\sqrt{2})$; F: $(-1, 0)$, $(1, 0)$

7. V: $(\pm\frac{1}{2}, 0)$, $(0, \pm\frac{1}{3})$; F: $\left(\dfrac{\pm\sqrt{5}}{6}, 0\right)$

9. $C(1, 2)$; V: $(3, 2)$, $(-1, 2)$, $(1, 3)$, $(1, 1)$; F: $(1 \pm \sqrt{3}, 2)$

11. $C(-3, 2)$; V: $(2, 2)$, $(-8, 2)$, $(-3, 6)$, $(-3, -2)$;
F: $(0, 2)$, $(-6, 2)$

13. $C(-2, 1)$; V: $(-10, 1)$, $(6, 1)$, $(-2, 1 \pm 4\sqrt{3})$;
F: $(-6, 1)$, $(2, 1)$

15. $C(2, -1)$; V: $(-1, -1)$, $(5, -1)$, $(2, 1)$, $(2, -3)$;
F: $(2 - \sqrt{5}, -1)$, $(2 + \sqrt{5}, -1)$

17. $C(1, 1)$; V: $(0, 1)$ $(2, 1)$, $(1, 3)$, $(1, -1)$;
F: $(1, 1 + \sqrt{3})$, $(1, 1 - \sqrt{3})$

19. $C(2.003125, -1.00513)$; V: $(5.0234302, -1.00515)$,
$(-1.0171802, -1.00515)$, $(2.003125, -3.0186868)$,
$(2.003125, 1.0083868)$

21. $\dfrac{x^2}{4} + \dfrac{y^2}{9} = 1$ **23.** $\dfrac{(x-3)^2}{4} + \dfrac{(y-1)^2}{25} = 1$

25. $\dfrac{(x+2)^2}{\frac{1}{4}} + \dfrac{(y-3)^2}{4} = 1$

27. (a) No; (b) $y = \pm 3\sqrt{1 - x^2}$;

(c) yes, domain $\{x \mid -1 \leqslant x \leqslant 1\}$, range $\{y \mid 0 \leqslant y \leqslant 3\}$;

(d) yes, domain $\{x \mid -1 \leqslant x \leqslant 1\}$, range $\{y \mid -3 \leqslant y \leqslant 0\}$

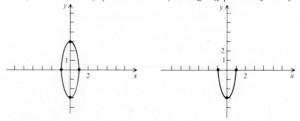

31. $x = a \cos t$, so $\frac{x^2}{a^2} = \cos^2 t$; similarly, $\frac{y^2}{b^2} = \sin^2 t$. Adding gives us $\frac{x^2}{a^2} + \frac{y^2}{b^2} = 1$.

Margin Exercises, Section 10.3

1. $V(3, 0)(-3, 0)$; $F(\sqrt{13}, 0)(-\sqrt{13}, 0)$;
A: $y = \frac{2}{3}x$, $y = -\frac{2}{3}x$

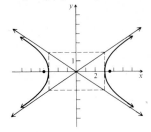

2. $V(4, 0)(-4, 0)$; $F(4\sqrt{2}, 0)(-4\sqrt{2}, 0)$;
A: $y = x$, $y = -x$

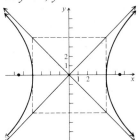

3. $V(0, 5)(0, -5)$; $F(0, \sqrt{34})(0, -\sqrt{34})$;
A: $y = \frac{5}{3}x$, $y = -\frac{5}{3}x$

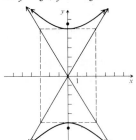

4. $V(0, 5)(0, -5)$; $F(0, 5\sqrt{2})(0, -5\sqrt{2})$;
A: $y = x$, $y = -x$

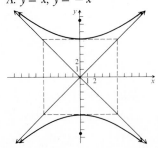

5. $C(1, -2)$; $V(6, -2)(-4, -2)$;
$F(1 + \sqrt{29}, -2)(1 - \sqrt{29}, -2)$;
A: $y = \frac{2}{5}x - \frac{12}{5}$, $y = -\frac{2}{5}x - \frac{8}{5}$

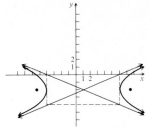

6. $C(-1, 2)$; $V(-1, 5)(-1, -1)$; $F(-1, 7)(-1, -3)$;
A: $y = \frac{3}{4}x + \frac{11}{4}$, $y = -\frac{3}{4}x + \frac{5}{4}$

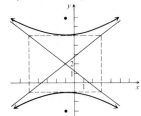

7.   **8.**

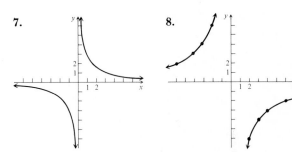

Exercise Set 10.3, p. 392

1. $C(0, 0)$; V: $(-3, 0)$, $(3, 0)$; F: $(-\sqrt{10}, 0)$, $(\sqrt{10}, 0)$;
A: $y = \frac{1}{3}x$, $y = -\frac{1}{3}x$
3. $C(2, -5)$; V: $(-1, -5)$, $(5, -5)$;
F: $(2 - \sqrt{10}, -5)$, $(2 + \sqrt{10}, -5)$;
A: $y = -\frac{x}{3} - \frac{13}{3}$, $y = \frac{x}{3} - \frac{17}{3}$
5. $C(-1, -3)$; V: $(-1, -1)$, $(-1, -5)$;
F: $(-1, -3 + 2\sqrt{5})$, $(-1, -3 - 2\sqrt{5})$;
A: $y = \frac{1}{2}x - \frac{5}{2}$, $y = -\frac{1}{2}x - \frac{7}{2}$
7. $C(0, 0)$; V: $(-2, 0)$, $(2, 0)$; F: $(-\sqrt{5}, 0)$, $(\sqrt{5}, 0)$;
A: $y = -\frac{1}{2}x$, $y = \frac{1}{2}x$
9. $C(0, 0)$; V: $(0, 1)$, $(0, -1)$; F: $(0, \sqrt{5})$, $(0, -\sqrt{5})$;
A: $y = -\frac{1}{2}x$, $y = \frac{1}{2}x$
11. $C(0, 0)$; V: $(-\sqrt{2}, 0)$, $(\sqrt{2}, 0)$; F: $(-2, 0)$, $(2, 0)$; A: $y = \pm x$
13. $C(0, 0)$; V: $(-\frac{1}{2}, 0)$, $(\frac{1}{2}, 0)$; F: $\left(-\frac{\sqrt{2}}{2}, 0\right)$, $\left(\frac{\sqrt{2}}{2}, 0\right)$;
A: $y = \pm x$ **15.** $C(1, -2)$; V: $(0, -2)$, $(2, -2)$;
F: $(1 - \sqrt{2}, -2)$, $(1 + \sqrt{2}, -2)$; A: $y = -x - 1$, $y = x - 3$

17. $C(\frac{1}{3}, 3)$; V: $(-\frac{2}{3}, 3)$; $(\frac{4}{3}, 3)$; F: $(\frac{1}{3} - \sqrt{37}, 3)$, $(\frac{1}{3} + \sqrt{37}, 3)$;
A: $y = 6x + 1$, $y = -6x + 5$ **19.** See text.   **21.** See text.
23. $C(1.023, -2.044)$; V: $(2.07, -2.044)$, $(-0.024, -2.044)$;

A: $y = x - 3.067$, $y = -x - 1.021$ **25.** $\dfrac{x^2}{4} - \dfrac{y^2}{9} = 1$

27. $\left(\dfrac{x - h}{a}\right)\left(\dfrac{x - h}{a}\right) - \left(\dfrac{y - k}{b}\right)\left(\dfrac{y - k}{b}\right) = 1$

$\therefore \dfrac{(x - h)^2}{a^2} - \dfrac{(y - k)^2}{b^2} = 1$

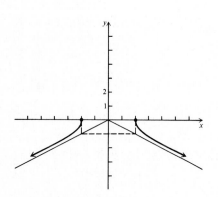

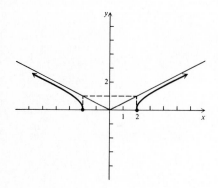

Margin Exercises, Section 10.4

1. $V(0, 0)$, $F(0, 2)$, D: $y = -2$

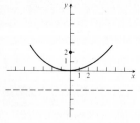

2. $V(0, 0)$, $F(0, \frac{1}{8})$, D: $y = -\frac{1}{8}$

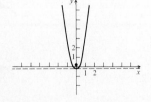

3. $V(0, 0)$, $F(-\frac{3}{2}, 0)$, D: $x = \frac{3}{2}$ **4.** $y^2 = 12x$ **5.** $x^2 = 2y$

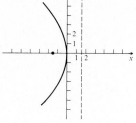

6. $y^2 = -24x$

7. $x^2 = -4y$

8. $V(-1, -\frac{1}{2})$, $F(-1, \frac{3}{2})$, D: $y = -\frac{5}{2}$

9. $V(2, -1)$, $F(1, -1)$, D: $x = 3$

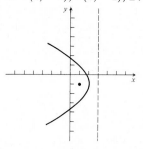

Exercise Set 10.4, p. 397

1. $V(0, 0)$; $F(0, 2)$: D: $y = -2$ **3.** $V(0, 0)$; $F(-\frac{3}{2}, 0)$; D: $x = \frac{3}{2}$
5. $V(0, 0)$; $F(0, 1)$; D: $y = -1$ **7.** $V(0, 0)$; $F(0, \frac{1}{8})$; D: $y = -\frac{1}{8}$
9. $y^2 = 16x$ **11.** $y^2 = -4\sqrt{2}x$ **13.** $(y - 2)^2 = 14(x + \frac{1}{2})$
15. V: $(-2, 1)$; F: $(-2, -\frac{1}{2})$; D: $y = \frac{5}{2}$
17. V: $(-1, -3)$; F: $(-1, -\frac{7}{2})$; D: $x = -\frac{5}{2}$
19. V: $(0, -2)$; F: $(0, -1\frac{3}{4})$; D: $y = -2\frac{1}{4}$
21. V: $(-2, -1)$; F: $(-2, -\frac{3}{4})$; D: $y = -1\frac{1}{4}$
23. V: $(5\frac{3}{4}, \frac{1}{2})$; F: $(6, \frac{1}{2})$; D: $x = 5\frac{1}{2}$
25. V: $(0, 0)$: F: $(0, 2014.0625)$; D: $y = -2014.0625$
27. The graph of $x^2 - y^2 = 0$ is two lines; the others are, respectively, a hyperbola, a circle, and a parabola.
29. $(x + 1)^2 = -4(y - 2)$
31. $(y - k)^2 - (x - h)(4p) = 0$; $(y - k)^2 = 4p(x - h)$

Margin Exercises, Section 10.5

1. $(4, 3)$, $(-3, -4)$ **2.** $(4, 7)$, $(-1, 2)$ **3.** $(-4, 4)$, $(2, 1)$
4. $(4, 3)$, $(-3, -4)$ **5.** $(4, 7)$, $(-1, 2)$ **6.** $(-4, 4)$, $(2, 1)$
7. $(-\frac{5}{7}, \frac{22}{7})$, $(1, -2)$ **8.** 7, 11   **9.** 5 ft, 12 ft

Exercise Set 10.5, pp. 400–401

1. $(-4, -3), (3, 4)$ **3.** $(0, -3), (4, 5)$ **5.** $(3, 0), (0, 2)$
7. $(-2, 1)$ **9.** $(3, 2), (4, \frac{3}{2})$

11. $\left(\dfrac{5 + \sqrt{70}}{3}, \dfrac{-1 + \sqrt{70}}{3}\right), \left(\dfrac{5 - \sqrt{70}}{3}, \dfrac{-1 - \sqrt{70}}{3}\right)$

13. $\left(\dfrac{15 + \sqrt{561}}{8}, \dfrac{11 - 3\sqrt{561}}{8}\right), \left(\dfrac{15 - \sqrt{561}}{8}, \dfrac{11 + 3\sqrt{561}}{8}\right)$

15. $9, 5$ **17.** 6 cm, 8 cm   **19.** 4 in., 5 in.
21. $(0.965, 4402.33), (-0.965, -4402.33)$

23. $2(L + W) = P, L + W = \dfrac{P}{2}, LW = A, L = \dfrac{P}{2} - W,$

$$W\left(\frac{P}{2} - W\right) = A, \ W^2 - \frac{WP}{2} + A = 0,$$

$$W = \frac{\dfrac{P}{2} \pm \sqrt{\left(\dfrac{P}{2}\right)^2 - 4A}}{2} = \frac{P}{4} \pm \frac{\sqrt{P^2 - 16A}}{4}$$

$$= \frac{1}{4}\left(P \pm \sqrt{P^2 - 16A}\right)$$

25. $(x - 2)^2 + (y - 3)^2 = 1$

Margin Exercises, Section 10.6

1. $(2, 0), (-2, 0)$ **2.** $(4, 0), (-4, 0)$ **3.** $(0, 2), (0, -2)$
4. $(2, 3), (2, -3), (-2, 3), (-2, -3)$
5. $(3, 2), (-3, -2), (2, 3), (-2, -3)$ **6.** 1 ft, 2 ft

Exercise Set 10.6, pp. 404–405

1. $(-5, 0), (4, 3), (4, -3)$ **3.** $(3, 0), (-3, 0)$
5. $(4, 3), (-4, -3), (3, 4), (-3, -4)$ **7.** No solution
9. $(\sqrt{2}, \sqrt{14}), (-\sqrt{2}, \sqrt{14}), (\sqrt{2}, -\sqrt{14}), (-\sqrt{2}, -\sqrt{14})$
11. $(1, 2), (-1, -2), (2, 1), (-2, -1)$
13. $(3, 2), (-3, -2), (2, 3), (-2, -3)$

15. $\left(\dfrac{5 - 9\sqrt{15}}{20}, \dfrac{-45 + 3\sqrt{15}}{20}\right), \left(\dfrac{5 + 9\sqrt{15}}{20}, \dfrac{-45 - 3\sqrt{15}}{20}\right)$

17. $(8.53, 2.53), (8.53, -2.53), (-8.53, 2.53),$
$(-8.53, -2.53)$ **19.** $13, 12$ and $-13, -12$
21. 1 m, $\sqrt{3}$ m   **23.** 16 ft, 24 ft
25. $(x + \frac{5}{13})^2 + (y - \frac{32}{13})^2 = \frac{5365}{169}$

Summary and Review: Chapter 10, pp. 405–406

1. [13.1]

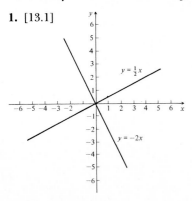

2. [13.1] $(x + 2)^2 + (y - 6)^2 = 13$
3. [13.1] Center: $(-1, 3)$; Radius: $\frac{3}{2}$

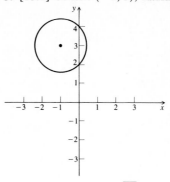

4. [13.1] $C(\frac{3}{4}, \frac{5}{4}); r = \dfrac{\sqrt{10}}{4}$

5. [13.1] $(x - 3)^2 + (y - 4)^2 = 25$
6. [13.1] $(x - 2)^2 + (y - 4)^2 = 26$

7. [13.2]

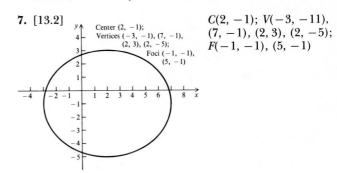

$C(2, -1); V(-3, -11),$
$(7, -1), (2, 3), (2, -5);$
$F(-1, -1), (5, -1)$

8. [13.2] $\dfrac{x^2}{9} + \dfrac{y^2}{16} = 1$ **9.** [13.3] $C(-2, \frac{1}{4}); V: (0, \frac{1}{4}), (-4, \frac{1}{4});$

$F: (-2 + \sqrt{6}, \frac{1}{4}), (-2 - \sqrt{6}, \frac{1}{4}); A: y - \frac{1}{4} = \pm\dfrac{\sqrt{2}}{2}(x + 2)$

10. [13.3]

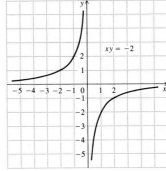

11. [13.3]

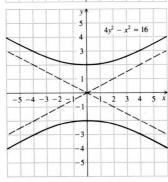

12. [13.4] $x^2 = -6y$
13. [13.4] $F: (-3, 0)$; $V: (0, 0)$; $D: x =$
14. [13.4] $V: (-5, 8)$; $F: (-5, \frac{15}{2})$; $D: y = \frac{17}{2}$
15. [13.6] $(-8\sqrt{2}, 8)$, $(8\sqrt{2}, 8)$

16. $[13.6] \left(3, \frac{\sqrt{29}}{2}\right), \left(-3, \frac{\sqrt{29}}{2}\right), \left(3, -\frac{\sqrt{29}}{2}\right), \left(-3, -\frac{\sqrt{29}}{2}\right)$

17. [13.6] $\frac{16\sqrt{6}}{7} \approx 5.6$ **18.** [13.5] 7, 4

19. [13.2] $x^2 + \frac{y^2}{9} = 1$

20. (a) Circle; (b) parabola; (c) ellipse; (d) parabola;
(e) hyperbola; (f) ellipse **21.** [13.6] $\frac{8}{7}$, $\frac{7}{2}$
22. [13.1] $(x - 2)^2 + (y - 1)^2 = 100$

CHAPTER 11

Margin Exercises, Section 11.1

1. (a) 1, 3, 5; (b) 67 **2.** (a) 1, $-\frac{1}{2}$, $\frac{1}{3}$, $-\frac{1}{4}$; (b) $-\frac{1}{48}$

3. $2n$ **4.** n **5.** n^3 **6.** $\frac{x^n}{n}$ **7.** 2^{n-1}

8. (a) 2, 2, 2, 2, 2; (b) -3, -9, -81, -6561, -43046721
9. $\frac{31}{32}$ **10.** $3 + 2\frac{1}{2} + 2\frac{1}{3}$ **11.** $5^0 + 5^1 + 5^2 + 5^3 + 5^4$

12. $8^3 + 9^3 + 10^3 + 11^3$ **13.** $\sum_{k=1}^{5} 2k$ **14.** $\sum_{k=1}^{4} k^3$

15. $\sum_{k=1}^{6} \frac{x^k}{k}$ **16.** $\sum_{k=1}^{4} (k + 1)$

Exercise Set 11.1, pp. 410–411

1. 4, 7, 10, 13; 31; 46 **3.** 1, $\frac{1}{2}$, $\frac{1}{3}$, $\frac{1}{4}$; $\frac{1}{10}$; $\frac{1}{15}$

5. $\frac{1}{2}$, $\frac{2}{3}$, $\frac{3}{4}$, $\frac{4}{5}$; $\frac{10}{11}$; $\frac{15}{16}$ **7.** $2n - 1$ **9.** $\frac{n+1}{n+2}$ **11.** $(\sqrt{3})^n$

13. $n - 1$ **15.** 4, $1\frac{1}{4}$, $1\frac{4}{5}$, $1\frac{5}{9}$ **17.** 64, 8, $2\sqrt{2}$, $\sqrt{2\sqrt{2}}$
19. 28 **21.** 30 **23.** $\frac{1}{2} + \frac{1}{4} + \frac{1}{6} + \frac{1}{8} + \frac{1}{10}$
25. $2^0 + 2^1 + 2^2 + 2^3 + 2^4 + 2^5$

27. $\log 7 + \log 8 + \log 9 + \log 10$ **29.** $\sum_{k=1}^{4} \frac{1}{k^2}$, $\frac{205}{144}$

31. $\sum_{k=1}^{6} (-1)^k 2^k$, 42 **33.** $\frac{3}{2}$, $\frac{3}{2}$, $\frac{3}{2}$, $\frac{3}{2}$, $\frac{3}{2}$ **35.** 1, 0, -1, 0, 1

37. π, 0, π, 0, π **39.** 2, 2.25, 2.37037, 2.441406, 2.488320,
2.521626; 14.071722
41. 2, 1.553774, 1.498834, 1.491398, 1.490379, 1.490238;
9.524623

Margin Exercises, Section 11.2

1. 50 **2.** 75 **3.** $a_1 = 7$, $d = 12$; 7, 19, 31, 43, ...
4. 20,100 **5.** 225 **6.** 455 **7.** 1665 **8.** 120
9. 7, 10, 13

Exercise Set 11.2, pp. 415–416

1. $a_{12} = 46$ **3.** 27th **5.** $a_{17} = 101$ **7.** $a_1 = 5$
9. $n = 28$ **11.** $a_1 = 8$; $d = -3$; 8, 5, 2, -1, -4 **13.** 670
15. 2500 **17.** 990 **19.** -264 **21.** 465
23. $5\frac{4}{5}$, $7\frac{3}{5}$, $9\frac{2}{5}$, $11\frac{1}{5}$ **25.** n^2 **27.** -10, -4, 2, 8

29. Sides are a, $a + d$, $a + 2d$, and $a^2 + (a + d)^2 = (a + 2d)^2$.

Solving, we get $d = \frac{a}{3}$. Thus the sides are a, $\frac{4a}{3}$, and $\frac{5a}{3}$ in the

ratio 3:4:5. **31.** 5200, 4687.5, 4175, 3662.5, 3150, 2637.5,
2125, 1612.5, 1100 **33.** \$51,679.65

Margin Exercises, Section 11.3

1. 2^8, or 256 **2.** -9375 **3.** \$476.41 **4.** 11,718 **5.** $\frac{341}{256}$
6. 363 **7.** \$10,485.75, or an approximation of this

Exercise Set 11.3, pp. 420–421

1. $a_{10} = \frac{81}{64}$ **3.** 1275 **5.** $\frac{547}{18}$ **7.** $\frac{63}{32}$ **9.** \$933.12

11. $\frac{1}{256}$ ft **13.** \$5866.60 **15.** $\frac{a_n}{a_{n+1}} = r$, so $\frac{a_n^2}{a_{n+1}^2} = r^2$;

hence a_1^2, a_2^2, ..., is geometric, with ratio r^2.

17. $\frac{a_n}{a_{n+1}} = r$; $\ln(a_n) - \ln(a_{n+1}) = \ln r$; hence $\ln a_1$,

$\ln a_2$, ..., is arithmetic.
19. $G = pr$, $q = pr^2$; $\sqrt{pq} = \sqrt{p^2 r^2} = pr = G$.
21. 10,485.76 inches

Margin Exercises, Section 11.4

1. No **2.** No **3.** Yes **4.** $\frac{3}{2}$ **5.** 2 **6.** $3\frac{1}{5}$ **7.** $\frac{5}{11}$
8. $\frac{59}{11}$ **9.** \$3333.33

Exercise Set 11.4, p. 425

1. No **3.** Yes **5.** Yes **7.** Yes **9.** 8 **11.** 125 **13.** 2
15. $\frac{160}{9}$ **17.** \$12,500 **19.** $\frac{7}{9}$ **21.** $\frac{7}{33}$ **23.** $\frac{170}{33}$
25. \$5.33 $\times 10^{10}$ **27.** 3.33 $\times 10^6$, $66\frac{2}{3}$%

Margin Exercises, Section 11.5

1. $1^2 + 1 > 1 + 1$ or $2 > 2$; $2^2 + 1 > 2 + 1$ or $5 > 3$;
$3^2 + 1 > 3 + 1$ or $10 > 4$; $4^2 + 1 > 4 + 1$ or $17 > 5$;
$5^2 + 1 > 5 + 1$ or $26 > 6$ **2.** $1 = \frac{1(1 + 1)}{2}$ or $1 = 1$;

$1 + 2 = \frac{2(2 + 1)}{2}$, or $1 + 2 = 3$; $1 + 2 + 3 = 6$;

$1 + 2 + 3 + 4 = 10$; $1 + 2 + 3 + 4 + 5 = 15$
3. $S_1: x \leq x$, $S_2: x \leq x^2$. Both obviously true if $x > 1$. Thus basis
step is complete. $S_k: x \leq x^k$, $S_{k+1}: x \leq x^{k+1}$. Assume $S_k: x \leq x^k$.
We know by hypothesis that $1 < x$. Multiply the inequalities to
get $x \cdot 1 \leq x^k \cdot x$, or $x < x^{k+1}$. We have arrived at S_{k+1}, hence
have shown that $S_k \rightarrow S_{k+1}$ for all natural numbers k. We can
now conclude that $x \leq x^k$ for all natural numbers n.
4. (a) $2 \cdot 1 = 1(1 + 1)$, $2 + 4 = 2(2 + 1)$;
(b) $2 + 4 + \cdots + 2k = k(k + 1)$;
(c) $2 + 4 + \cdots + 2(k + 1) = (k + 1)[(k + 1) + 1]$;
(d) $2 \cdot 1 \overset{?}{=} 1(1 + 1)$, $2 = 1 \cdot 2 = 2$;
(e) Assume S_k as hypothesis:

$$2 + 4 + \cdots + 2k = k(k + 1).$$

Add $2(k + 1)$ on both sides:

$$2 + 4 + \cdots + 2k + 2(k + 1) = k(k + 1) + 2(k + 1)$$
$$= (k + 1)(k + 2) \quad \text{Simplifying}$$

We now have

$$2 + 4 + \cdots + 2(k + 1) = (k + 1)(k + 2)$$

or

$$(k + 1)[(k + 1) + 1].$$

This is S_{k+1}. Hence $S_k \to S_{k+1}$ for all k. Finally, we conclude that $2 + 4 + \cdots + 2n = n(n + 1)$ for all natural numbers n.

5. S_1: $2 = \dfrac{1(3 + 1)}{2}$. True. S_k: $\displaystyle\sum_{p=1}^{k} (3p - 1) = \dfrac{k(3k + 1)}{2}$, or

$$2 + 5 + 8 + \cdots + 3k - 1 = \dfrac{k(3k + 1)}{2}.$$

S_{k+1}: $\displaystyle\sum_{p=1}^{k+1} (3p - 1) = \dfrac{(k + 1)[3(k + 1) + 1]}{2}$, or

$2 + 5 + \cdots + 3k - 1 + 3(k + 1) - 1 =$

$\dfrac{(k + 1)[3(k + 1) + 1]}{2}$. Assume S_k. Then add $[3(k + 1) - 1]$

on both sides:

$$2 + 5 + \cdots + [3(k + 1) - 1]$$
$$= \dfrac{k(3k + 1)}{2} + [3(k + 1) - 1]$$
$$= \dfrac{k(3k + 1) + 2[3(k + 1) - 1]}{2}$$
$$= \dfrac{3k^2 + 7k + 4}{2} = \dfrac{(k + 1)(3k + 4)}{2}$$
$$= \dfrac{(k + 1)[3(k + 1) + 1]}{2}$$

We have arrived at

$$\sum_{p=1}^{k+1} (3p - 1) = \dfrac{(k + 1)[3(k + 1) - 1]}{2}.$$

This is S_{k+1}. So $S_k \to S_{k+1}$ for all k. We conclude that

$$\sum_{p=1}^{n} (3p - 1) = \dfrac{n(3n + 1)}{2}$$

for *all* natural numbers n.

Exercise Set 11.5, pp. 429–430

1. $1^2 < 1^3$, $2^2 < 2^3$, $3^2 < 3^3$, etc.

3. A polygon of 3 sides has $\dfrac{3(3 - 3)}{2}$ diagonals. A polygon of 4 sides has $\dfrac{4(4 - 3)}{2}$ diagonals, etc.

5. S_n: $1 + 2 + 3 + \cdots + n = \dfrac{n(n + 1)}{2}$

S_1: $1 = \dfrac{1(1 + 1)}{2}$

S_k: $1 + 2 + 3 + \cdots + k = \dfrac{k(k + 1)}{2}$

S_{k+1}: $1 + 2 + 3 + \cdots + k + (k + 1) = \dfrac{(k + 1)(k + 2)}{2}$

1. Basis step: S_1 true by substitution.

2. Induction step: Assume S_k. Deduce S_{k+1}.
Starting with the left side of S_{k+1}, we have

$$\underbrace{1 + 2 + 3 + \cdots + k} + (k + 1)$$
$$= \dfrac{k(k + 1)}{2} + (k + 1) \qquad \text{(by } S_k\text{)}$$
$$= \dfrac{k(k + 1) + 2(k + 1)}{2} \qquad \text{(adding)}$$
$$= \dfrac{(k + 1)(k + 2)}{2} \qquad \text{(distributive law)}$$

7. S_n: $1 + 5 + 9 + \cdots + (4n - 3) = n(2n - 1)$
S_1: $1 = 1(2 \cdot 1 - 1)$
S_k: $1 + 5 + 9 + \cdots + (4k - 3) = k(2k - 1)$
S_{k+1}: $1 + 5 + 9 + \cdots + (4k - 3) + [4(k + 1) - 3]$
$\qquad\qquad\qquad\qquad = (k + 1)[2(k + 1) - 1]$
$\qquad\qquad\qquad\qquad = (k + 1)(2k + 1)$

1. Basic step: S_1 true by substitution.

2. Induction step: Assume S_k. Deduce S_{k+1}.
Starting with the left side of S_{k+1}, we have

$$\underbrace{1 + 5 + 9 + \cdots + (4k - 3)} + [4(k + 1) - 3]$$
$$= k(2k - 1) + [4(k + 1) - 3] \qquad \text{(by } S_k\text{)}$$
$$= 2k^2 - k + 4k + 4 - 3$$
$$= (k + 1)(2k + 1)$$

9. S_n: $\dfrac{1}{1 \cdot 2} + \dfrac{1}{2 \cdot 3} + \cdots + \dfrac{1}{n(n + 1)} = \dfrac{n}{n + 1}$

S_1: $\dfrac{1}{1 \cdot 2} = \dfrac{1}{1 + 1}$

S_k: $\dfrac{1}{1 \cdot 2} + \dfrac{1}{2 \cdot 3} + \cdots + \dfrac{1}{k(k + 1)} = \dfrac{k}{k + 1}$

S_{k+1}: $\dfrac{1}{1 \cdot 2} + \dfrac{1}{2 \cdot 3} + \cdots + \dfrac{1}{k(k + 1)} + \dfrac{1}{(k + 1)(k + 2)}$

$\qquad = \dfrac{k + 1}{k + 2}$

2. Induction step: Assume S_k. Deduce S_{k+1}. Add

$$\dfrac{1}{(k + 1)(k + 2)}$$

to both sides and simplify the right side.

11. *2. Induction step:* Assume S_k. Deduce S_{k+1}. Now

$$k < k + 1 \qquad \text{(by } S_k\text{)}$$
$$k + 1 < k + 1 + 1 \qquad \text{(adding 1)}$$
$$\therefore k + 1 < k + 2$$

13. S_1: $3^1 < 3^{1+1}$
S_k: $3^k < 3^{k+1}$
$3^k \cdot 3 < 3^{k+1} \cdot 3$
$3^{k+1} < 3^{(k+1)+1}$

15. S_1: $1^3 = \dfrac{1^2(1+1)^2}{4} = 1$

S_k: $1^3 + 2^3 + \cdots + k^3 = \dfrac{k^2(k+1)^2}{4}$

$1^3 + 2^3 + \cdots + (k+1)^3 = \dfrac{k^2(k+1)^2}{4} + (k+1)^3$

$$= \dfrac{(k+1)^2}{4}(k^2 + 4(k+1))$$

$$= \dfrac{(k+1)^2(k+2)^2}{4}$$

17. S_1: $1 + \frac{1}{1} = 1 + 1$

S_k: $\left(1 + \dfrac{1}{1}\right) \cdots \left(1 + \dfrac{1}{k}\right) = k + 1$

Multiply by $\left(1 + \dfrac{1}{k+1}\right)$:

$$\left(1 + \dfrac{1}{1}\right) \cdots \left(1 + \dfrac{1}{k+1}\right) = (k+1)\left(1 + \dfrac{1}{k+1}\right)$$

$$= (k+1)\left(\dfrac{k+1+1}{k+1}\right)$$

$$= (k+1) + 1$$

19. S_1: $a_1 = \dfrac{a_1 - a_1 r}{1 - r} = \dfrac{a_1(1 - r)}{1 - r} = a_1$

S_k: $a_1 + \cdots + a_1 r^{k-1} = \dfrac{a_1 - a_1 r^k}{1 - r}$

Add $a_1 r^k$.

$$a_1 + \cdots + a_1 r^k = \dfrac{a_1 - a_1 r^k}{1 - r} + a_1 r^k \dfrac{1 - r}{1 - r}$$

$$= \dfrac{a_1 - a_1 r^k + a_1 r^k - a_1 r^{k+1}}{1 - r}$$

$$= \dfrac{a_1 - a_1 r^{k+1}}{1 - r}$$

21. 2. *Induction step:* Assume S_k. Deduce S_{k+1}.
Starting with the left side of S_{k+1} we have:

$[r(\cos\theta + i\sin\theta)]^{k+1}$
$= [r(\cos\theta + i\sin\theta)]^k [r(\cos\theta + i\sin\theta)]$
$= r^k[\cos k\theta + i\sin k\theta][r(\cos\theta + i\sin\theta)]$
$= r^{k+1}(\cos k\theta + i\sin k\theta)(\cos\theta + i\sin\theta)$
$= r^{k+1}[\cos k\theta \cos\theta + i\sin k\theta \cos\theta + i\cos k\theta \sin\theta$
$\qquad\qquad\qquad\qquad\qquad - \sin k\theta \sin\theta]$
$= r^{k+1}[(\cos k\theta \cos\theta - \sin k\theta \sin\theta) + i(\sin k\theta \cos\theta$
$\qquad\qquad\qquad\qquad\qquad\qquad + \cos k\theta \sin\theta)]$
$= r^{k+1}[\cos(k\theta + \theta) + i\sin(k\theta + \theta)]$
$= r^{k+1}[\cos(k+1)\theta + i\sin(k+1)\theta]$

23. 2. *Induction step:* Assume S_k. Deduce S_{k+1}.
We start with the left side of S_{k+1}.

$\underbrace{\cos x \cdot \cos 2x \cdots \cos 2^{k-1}x} \cdot \cos 2^k x$

$= \dfrac{\sin 2^k x}{2^k \sin x} \cdot \cos 2^k x$

$= \dfrac{\sin 2^k x \cos 2^k x}{2^k \sin x}$

$= \dfrac{\frac{1}{2}\sin 2(2^k x)}{2^k \sin x}$ $\qquad$ (since $\sin 2\theta = 2\sin\theta\cos\theta$)

$= \dfrac{\sin 2^{k+1} x}{2^{k+1} \sin x}$

25. S_n: $\left(1 - \dfrac{1}{2^2}\right)\left(1 - \dfrac{1}{3^2}\right) \cdots \left(1 - \dfrac{1}{n^2}\right) = \dfrac{n+1}{2n}$

S_2: $1 - \dfrac{1}{2^2} = \dfrac{2+1}{2 \cdot 2}$

S_k: $\left(1 - \dfrac{1}{2^2}\right)\left(1 - \dfrac{1}{3^2}\right) \cdots \left(1 - \dfrac{1}{k^2}\right) = \dfrac{k+1}{2k}$

S_{k+1}: $\left(1 - \dfrac{1}{2^2}\right)\left(1 - \dfrac{1}{3^2}\right) \cdots \left(1 - \dfrac{1}{k^2}\right)\left(1 - \dfrac{1}{(k+1)^2}\right)$

$$= \dfrac{k+2}{2(k+1)}$$

1. *Basis step:* S_2 is true by substitution.
2. *Induction step:* Assume S_k. Deduce S_{k+1}.
Starting with the left side of S_{k+1} we have

$\underbrace{\left(1 - \dfrac{1}{2^2}\right)\left(1 - \dfrac{1}{3^2}\right) \cdots \left(1 - \dfrac{1}{k^2}\right)}\left(1 - \dfrac{1}{(k+1)^2}\right)$

$= \dfrac{k+1}{2k}\left(1 - \dfrac{1}{(k+1)^2}\right)$

$= \dfrac{k+1}{2k} - \dfrac{1}{2k(k+1)}$

$= \dfrac{(k+1)(k+1) - 1}{2k(k+1)}$

$= \dfrac{k^2 + 2k + 1 - 1}{2k(k+1)}$

$= \dfrac{k^2 + 2k}{2k(k+1)}$

$= \dfrac{k(k+2)}{2k(k+1)}$

$= \dfrac{k+2}{2(k+1)}$

27. S_2: $\overline{z_1 + z_2} = \bar{z}_1 + \bar{z}_2$:

$\overline{(a + bi) + (c + di)} = \overline{(a + c) + (b + d)i}$
$\qquad\qquad\qquad\qquad = (a + c) - (b + d)i$
$\overline{a + bi} + \overline{(c + di)} = a - bi + c - di$
$\qquad\qquad\qquad\qquad = (a + c) - (b + d)i.$
S_k: $\overline{z_1 + z_2 + \cdots + z_k} = \bar{z}_1 + \bar{z}_2 + \cdots + \bar{z}_k.$
$\overline{(z_1 + z_2 + \cdots + z_k) + z_{k+1}} = \overline{(z_1 + z_2 + \cdots + z_k)}$
$\qquad\qquad\qquad\qquad + \overline{z_{k+1}}$ $\qquad$ (by S_2).
$\qquad\qquad = \bar{z}_1 + \bar{z}_2 + \cdots + \bar{z}_k$
$\qquad\qquad\qquad\qquad + \overline{z_{k+1}}$ $\qquad$ (by S_k).

29. S_1: i is either i or -1 or $-i$ or 1.
S_k: i^k is either i or -1 or $-i$ or 1.
$i^{k+1} = i^k \cdot i$ is then $i \cdot i = -1$ or $-1 \cdot i = -i$ or
$-i \cdot i = 1$ or $1 \cdot i = i.$

31. S_1: 2 is a factor of $1^2 + 1$.
$\quad S_k$: 2 is a factor of $k^2 + k$.
$\quad (k+1)^2 + (k+1) = k^2 + 2k + 1 + k + 1$
$\qquad\qquad\qquad\qquad\quad = k^2 + k + 2(k+1)$.
By S_k, 2 is a factor of $k^2 + k$, hence 2 is a factor of the
the right-hand side, so is a factor of
$(k+1)^2 + (k+1)$.

Summary and Review: Chapter 11, pp. 431–432

1. [14.2] $3\frac{3}{4}$ **2.** [14.2] $a + 4b$ **3.** [14.2] 531
4. [14.2] 465   **5.** [14.2] 11   **6.** [14.2] -4
7. [14.3] $n = 6$, $S_n = -126$ **8.** [14.3] $a_1 = 8$, $a_5 = \frac{1}{2}$
9. [14.4] No   **10.** [14.4] Yes   **11.** [14.4] $\frac{3}{8}$ **12.** [14.4] $\frac{211}{99}$
13. [14.2] $5\frac{4}{5}, 6\frac{3}{5}, 7\frac{2}{5}, 8\frac{1}{5}$ **14.** [14.3] ≈ 50 ft
15. [14.2] \$7.38, \$1365.10   **16.** [14.3] 7,680,000

17. [14.4] $\dfrac{64\sqrt{2}}{\sqrt{2}-1}$ in., or $128 + 64\sqrt{2}$ in.   **18.** [14.4] $23\frac{1}{3}$ cm

19. [14.5] S_n: $1 + 4 + 7 + \cdots + (3n - 2) = \dfrac{n(3n-1)}{2}$

$\qquad S_1$: $1 = \dfrac{1(3-1)}{2}$

$\qquad S_k$: $1 + 4 + 7 + \cdots + (3k - 2) = \dfrac{k(3k-1)}{2}$

$\qquad S_{k+1}$: $1 + 4 + 7 + \cdots + [3(k+1) - 2]$
$\qquad\qquad = 1 + 4 + 7 + \cdots + (3k - 2) + (3k + 1)$
$\qquad\qquad = \dfrac{(k+1)(3k+2)}{2}$

1. *Basis step:* $1 = \dfrac{2}{2} = \dfrac{1(3-1)}{2}$ is true.

2. *Induction step:* Assume S_k. Add $(3k + 1)$ to both sides.

$1 + 4 + 7 + \cdots + (3k - 2) + (3k + 1)$
$\qquad\qquad\qquad = \dfrac{k(3k-1)}{2} + (3k + 1)$
$\qquad\qquad\qquad = \dfrac{k(3k-1)}{2} + \dfrac{2(3k+1)}{2}$
$\qquad\qquad\qquad = \dfrac{3k^2 - k + 6k + 2}{2}$
$\qquad\qquad\qquad = \dfrac{3k^2 + 5k + 2}{2}$
$\qquad\qquad\qquad = \dfrac{(k+1)(3k+2)}{2}$

20. [14.5] S_1: $1 = \dfrac{3^1 - 1}{2}$; S_2: $1 + 3 = \dfrac{3^2 - 1}{2}$

$\qquad S_k$: $1 + 3 + 3^2 + \cdots + 3^{k-1} = \dfrac{3^k - 1}{2}$

2. *Induction step:* Assume S_k. Add 3^k on both sides.

$\qquad 1 + 3 + \cdots + 3^{k-1} + 3^k$
$\qquad\qquad = \dfrac{3^k - 1}{2} + 3^k = \dfrac{3^k - 1}{2} + 3^k \cdot \dfrac{2}{2}$
$\qquad\qquad = \dfrac{3 \cdot 3^k - 1}{2} = \dfrac{3^{k+1} - 1}{2}$

21. [14.5] S_n: $\left(1 - \dfrac{1}{2}\right)\left(1 - \dfrac{1}{3}\right) \cdots \left(1 - \dfrac{1}{n}\right) = \dfrac{1}{n}$

$\qquad S_2$: $\left(1 - \dfrac{1}{2}\right) = \dfrac{1}{2}$

$\qquad S_k$: $\left(1 - \dfrac{1}{2}\right)\left(1 - \dfrac{1}{3}\right) \cdots \left(1 - \dfrac{1}{k}\right) = \dfrac{1}{k}$

$\qquad S_{k+1}$: $\left(1 - \dfrac{1}{2}\right)\left(1 - \dfrac{1}{3}\right) \cdots \left(1 - \dfrac{1}{k}\right)\left(1 - \dfrac{1}{k+1}\right)$

$\qquad\qquad\qquad = \dfrac{1}{k+1}$

1. *Basis step:* S_2 is true by substitution.
2. *Induction step:* Assume S_k. Deduce S_{k+1}.
   Starting with the left side of S_{k+1} we have:

$\underbrace{\left(1 - \dfrac{1}{2}\right)\left(1 - \dfrac{1}{3}\right) \cdots \left(1 - \dfrac{1}{k}\right)}\left(1 - \dfrac{1}{k+1}\right)$

$= \dfrac{1}{k} \cdot \left(1 - \dfrac{1}{k+1}\right) \qquad \text{(by } S_k\text{)}$

$= \dfrac{1}{k} \cdot \left(\dfrac{k+1-1}{k+1}\right)$

$= \dfrac{1}{k} \cdot \dfrac{k}{k+1}$

$= \dfrac{1}{k+1} \qquad \text{(simplifying)}$

22. [14.1] 5; 51; 5203; 54,142,419
23. [14.1] $\sum\limits_{n=1}^{7} (n^2 - 1)$ or $\sum\limits_{n=0}^{6} n(n+2)$ **24.** [14.5] S_1 fails.
25. [14.3] $\dfrac{a_{k+1}}{a_k} = r_1$, $\dfrac{b_{k+1}}{b_k} = r_2$, so $\dfrac{a_{k+1}b_{k+1}}{a_k b_k} = r_1 r_2$ (constant)

26. [14.2] $a_{k+1} - a_k = d$, so $\dfrac{c_{k+1}}{c_k} = \dfrac{b^{a_{k+1}}}{b^{a_k}} = b^{a_{k+1} - a_k} = b^d$

(constant)   **27.** [14.2] (a) a_n is all positive or all negative;
(b) always; (c) always; (d) $a_n = k$, k a constant;
(e) $a_n = k$, k a constant; (f) $a_n = k$, k a constant
28. [14.2] $-2, 0, 2, 4$ **29.** [14.4] $0.27, 0.0027, 0.000027$
30. [14.4] $\frac{1}{2}, -\frac{1}{6}, \frac{1}{18}$

CHAPTER 12

Margin Exercises, Section 12.1

1. $3 \cdot 2 \cdot 1$, or 6; $3 \cdot 3 \cdot 3$, or 27   **2.** $4 \cdot 2 \cdot 5$, or 40
3. $5 \cdot 4 \cdot 3 \cdot 2 \cdot 1$, or 120   **4.** $5 \cdot 4 \cdot 3 \cdot 2 \cdot 1$, or 120
5. $3 \cdot 2 \cdot 1$, or 6   **6.** $5 \cdot 4 \cdot 3 \cdot 2 \cdot 1$, or 120
7. $6 \cdot 5 \cdot 4 \cdot 3 \cdot 2 \cdot 1$, or 720   **8.** $4 \cdot 3 \cdot 2 \cdot 1$, or 24
9. $6 \cdot 5 \cdot 4 \cdot 3 \cdot 2 \cdot 1$, or 720
10. $8 \cdot 7 \cdot 6 \cdot 5 \cdot 4 \cdot 3 \cdot 2 \cdot 1$, or 40,320   **11.** 362,880
12. 40,320   **13.** 18!   **14.** (a) $10! = 10 \cdot 9!$; (b) $20! = 20 \cdot 19!$

15. $_7P_3 = 7 \cdot 6 \cdot 5 = 210$; $_7P_3 = \dfrac{7!}{4!} = \dfrac{7 \cdot 6 \cdot 5 \cdot 4 \cdot 3 \cdot 2 \cdot 1}{4 \cdot 3 \cdot 2 \cdot 1} =$

$7 \cdot 6 \cdot 5 = 210$ **16.** (a) $_{10}P_4 = 10 \cdot 9 \cdot 8 \cdot 7 = 5040$;
(b) $_8P_2 = 8 \cdot 7 = 56$; (c) $_{11}P_5 = 11 \cdot 10 \cdot 9 \cdot 8 \cdot 7 = 55,440$;
(d) $_nP_1 = n$; (e) $_nP_2 = n(n-1) = n^2 - n$
17. $_{12}P_5 = 12 \cdot 11 \cdot 10 \cdot 9 \cdot 8 = 95,040$
18. $_{10}P_6 = 10 \cdot 9 \cdot 8 \cdot 7 \cdot 6 \cdot 5 = 151,200$

19. $_4P_4 \cdot {}_3P_3 = 4! \cdot 3! = 144$ **20.** $\dfrac{11!}{1!4!4!2!} = 34,650$

12

21. $\dfrac{6!}{3!2!1!} = 60$ **22.** $\dfrac{8!}{3!3!2!} = 560$ **23.** $6! = 720$

24. $11! = 39,916,800$ **25.** $26^5 = 11,881,376$

Exercise Set 12.1, p. 440

1. $4 \cdot 3 \cdot 2$, or 24 **3.** $_{10}P_7 = 10 \cdot 9 \cdot 8 \cdot 7 \cdot 6 \cdot 5 \cdot 4$, or $604,800$

5. 120; 3125 **7.** 120; 24 **9.** $\dfrac{5!}{2!1!1!1!} = 5 \cdot 4 \cdot 3 = 60$

11. $9 \cdot 9 \cdot 8 \cdot 7 \cdot 6 \cdot 5 \cdot 4$, or $544,320$ **13.** $\dfrac{9!}{2!3!4!} = 1260$

15. $12!$, or $479,001,600$ **17.** (a) 120; (b) 3840

19. $52 \cdot 51 \cdot 50 \cdot 49 = 6,497,400$

21. $\dfrac{24!}{3!5!9!4!3!} = 16,491,024,950,400$

23. $4! = 24$, $8! \div 3! = 6720$, $\dfrac{13!}{2!2!2!2!2!} = 194,594,400$

25. $80 \cdot 26 \cdot 9999 = 20,797,920$ **27.** 11 **29.** 9

Margin Exercises, Section 12.2

1. (a) 6; (b) 1; (c) 4; (d) 1 **2.** (a) $5 \cdot 4 \cdot 3$, or 60;
(b) *ABC, BCA, CAB, BAC, ACB, CBA, ABD, BDA, DAB, DBA,
BAD, ADB, ABE, BEA, EAB, EBA, BAE, AEB, ACD, CDA, DAC,
DCA, CAD, ADC, ACE, CEA, EAC, ECA, CAE, AEC, ADE,
DEA, EAD, EDA, EDB, DAE, AED, BCD, CDB, DBC, DCB,
CBD, BDC, BCE, CEB, EBC, ECB, CBE, BEC, BDE, DEB,
EBD, DBE, BED, CDE, DEC, ECD, EDC, DCE, CED*; (c) 10;
(d) $\{A, B, C\}$, $\{A, C, D\}$, $\{B, C, D\}$, $\{D, C, E\}$, $\{A, C, E\}$,
$\{A, B, D\}$, $\{A, B, E\}$, $\{B, C, E\}$, $\{B, E, D\}$, $\{A, D, E\}$
3. (a) 45; (b) 45; (c) 21; (d) 1 **4.** 36, 36 **5.** $_{12}C_5 = 792$
6. $_{12}C_3 \cdot {}_8C_2 = 6160$

Exercise Set 12.2, pp. 442–443

1. 126 **3.** 1225 **5.** $\dfrac{n(n-1)(n-2)}{3!}$ **7.** 8855 **9.** 210

11. $\dbinom{8}{2} = 28$, $\dbinom{8}{3} = 56$ **13.** $\dbinom{10}{7} \cdot \dbinom{5}{3} = 1200$

15. $\dbinom{58}{6} \cdot \dbinom{42}{4}$ **17.** $\dbinom{4}{3} \cdot \dbinom{48}{2} = 4512$ **19.** $2,598,960$

21. $\dbinom{8}{3} = 56$ **23.** $\dbinom{5}{2}\dbinom{8}{2} = 280$ **25.** 5 **27.** 7

Margin Exercises, Section 12.3

1. $x^{10} - 5x^8 + 10x^6 - 10x^4 + 5x^2 - 1$

2. $16x^4 + 32x^3\dfrac{1}{y} + 24x^2\dfrac{1}{y^2} + 8x\dfrac{1}{y^3} + \dfrac{1}{y^4}$

3. $x^6 - 6\sqrt{2}x^5 + 30x^4 - 40\sqrt{2}x^3 + 60x^2 - 24\sqrt{2}x + 8$

4. $-1512x^5$ **5.** $5670x^4$ **6.** $8064y^5$ **7.** $243x^5$

8. $2^6 = 64$ **9.** 2^{50}

Exercise Set 12.3, pp. 446–447

1. $15a^4b^2$ **3.** $-745,472a^3$ **5.** $1120x^{12}y^2$

7. $-1,959,552u^5v^{10}$

9. $m^5 + 5m^4n + 10m^3n^2 + 10m^2n^3 + 5mn^4 + n^5$

11. $x^{10} - 15x^8y + 90x^6y^2 - 270x^4y^3 + 405x^2y^4 - 243y^5$

13. $x^{-8} + 4x^{-4} + 6 + 4x^4 + x^8$

15. $\dbinom{n}{0} - \dbinom{n}{1} + \dbinom{n}{2} - \dbinom{n}{3} + \cdots + (-1)^n\dbinom{n}{n}$

17. $140\sqrt{2}$ **19.** $9 - 12\sqrt{3}\,t + 18t^2 - 4\sqrt{3}\,t^3 + t^4$ **21.** 128

23. 2^{26}, or $67,108,864$ **25.** $-7 - 4\sqrt{2}\,i$

27. $\sin^7 t - 7\sin^5 t + 21\sin^3 t - 35\sin t + 35\csc t -$

$21\csc^3 t + 7\csc^5 t - \csc^7 t$ **29.** $\displaystyle\sum_{r=0}^{n}\dbinom{n}{r}(-1)^r a^{n-r}b^r$

31. -3 **33.** 5 **35.** $2^7 - 1$, or 127

Margin Exercises, Section 12.4

1. $\frac{82}{100}$ **2.** 73.4%; no **3.** 2074 **4.** 14.7%, 50.4% **5.** $\frac{1}{2}$

6. (a) $\frac{1}{13}$; (b) $\frac{1}{4}$; (c) $\frac{1}{2}$; (d) $\frac{2}{13}$ **7.** $\frac{6}{11}$ **8.** 0 **9.** 1

10. $\frac{11}{850}$ **11.** $\frac{15}{91}$ **12.** $\frac{6}{13}$ **13.** $\frac{1}{6}$

Exercise Set 12.4, pp. 454–456

1. 0.57, 0.43 **3.** 0.075, 0.134, 0.057, 0.071, 0.030

5. 0.633 **7.** $.054$ **9.** 52 **11.** $\frac{1}{4}$ **13.** $\frac{1}{13}$ **15.** $\frac{1}{2}$

17. $\frac{2}{13}$ **19.** $\frac{5}{7}$ **21.** 0 **23.** $\frac{11}{4165}$ **25.** $\frac{28}{65}$ **27.** $\frac{1}{18}$

29. $\frac{1}{36}$ **31.** $\frac{30}{323}$ **33.** $\frac{9}{19}$ **35.** $\frac{1}{38}$ **37.** $\frac{1}{19}$ **39.** $2,598,960$

41. (a) 36; (b) 1.39×10^{-5}

Summary and Review: Chapter 12, pp. 456–457

1. [15.1] $6! = 720$ **2.** [15.1] $9 \cdot 8 \cdot 7 \cdot 6 = 3024$

3. [15.2] $\dbinom{15}{8} = 6435$ **4.** [15.1] $24 \cdot 23 \cdot 22 = 12,144$

5. [15.1] $\dfrac{9!}{1!4!2!2!} = 3780$ **6.** [15.1] 36

7. [15.1] (a) $_6P_5 = 720$; (b) $6^5 = 7776$; (c) $_5P_4 = 120$;
(d) $_3P_2 = 6$ **8.** [15.1] $6! = 720$ **9.** [15.3] $220a^9x^3$

10. [15.3] $\dbinom{18}{11}a^7x^{11}$, or $\dfrac{18!}{11!7!}a^7x^{11}$

11. [15.3] $m^7 + 7m^6n + 21m^5n^2 + 35m^4n^3 + 35m^3n^4 +$
$21m^2n^5 + 7mn^6 + n^7$ **12.** [15.3] $x^8 + 12x^6y + 54x^4y^2 +$
$108x^2y^3 + 81y^4$ **13.** [15.3] $-6624 + 16,280i$

14. [15.3] $\cos^8 t + 8\cos^6 t + 28\cos^4 t + 56\cos^2 t + 70 +$
$56\sec^2 t + 28\sec^4 t + 8\sec^6 t + \sec^8 t$

15. [15.4] $\frac{86}{206} \approx 0.42$, $\frac{97}{206} \approx 0.47$, $\frac{23}{206} \approx 0.11$

16. [15.4] $\frac{1}{12}$, 0 **17.** [15.4] $\frac{1}{4}$ **18.** [15.4] $\frac{6}{5525}$

19. [15.3] $\left(\log\dfrac{x}{y}\right)^{10}$ **20.** [15.2] 36 **21.** [15.2] 14

INDEX